Wonders and the Order of Nature

1150–1750

Wonders and the

Order of Nature

1150–1750

Lorraine Daston

Katharine Park

ZONE BOOKS · NEW YORK

2001

© 1998 Lorraine Daston and Katharine Park
ZONE BOOKS
40 White Street, 5th Floor
New York, NY 10013

First Paperback Printing

Printed in Canada.

Distributed by The MIT Press,
Cambridge, Massachusetts, and London, England

Library of Congress Cataloging-in-Publication Data

Daston, Lorraine, 1951–
      Wonders and the order of nature, 1150–1750 / Lorraine
Daston, Katharine Park.
            p. cm.
      Includes bibliographical references and index.
      ISBN 0-942299-91-4 (pbk).
      1. Science, Medieval. 2. Science–History–16th–17th–18th
century. 3. Science–Philosophy–History. I. Park, Katharine,
1950–   . II. Title. Q124.97.D38   1997
306.4′2–dc21   96-49411
                                                           CIP

For Marty and Gerd and Thalia

Contents

Preface

Curiosity is a vice that has been stigmatized in turn by Christianity, by philosophy, and even by a certain conception of science. Curiosity, futility. The word, however, pleases me. To me it suggests something altogether different: it evokes "concern"; it evokes the care one takes for what exists and could exist; a readiness to find strange and singular what surrounds us; a certain relentlessness to break up our familiarities and to regard otherwise the same things; a fervor to grasp what is happening and what passes; a casualness in regard to the traditional hierarchies of the important and the essential.

I dream of a new age of curiosity. We have the technical means for it; the desire is there; the things to be known are infinite; the people who can employ themselves at this task exist. Why do we suffer? From too little: from channels that are too narrow, skimpy, quasi-monopolistic, insufficient. There is no point in adopting a protectionist attitude, to prevent "bad" information from invading and suffocating the "good." Rather, we must multiply the paths and the possibility of comings and goings.[1]

Michel Foucault, "The Masked Philosopher"

This book began some twenty years ago when we were enrolled in a graduate seminar on seventeenth-century metaphysics. Our seminar read the usual authors, but where our classmates saw arguments, we saw monsters — lots of them, everywhere. Bacon, Hobbes, Leibniz, Locke — all put monsters on the front lines of their campaigns to reform natural philosophy, explain religion, explore the relationships between art and nature, or challenge natural kinds. Our enthusiasm was not infectious. During the next five years, while writing sober dissertations on Renaissance medicine and Enlightenment probability theory, we spent many hours reading treatises and broadsides with titles like *The Hog-Faced Woman* and *A Thousand Notable Things*. When we published the fruits of our research in a 1981 article, it was very nearly the only historical treatment of monsters in English published in the previous fifty years.[2]

That situation has changed dramatically. Monsters are no longer a solitary pursuit. Books appear on the topic almost weekly, or so it seems as we try to keep abreast of the literature. Marvels, prodigies, and *Wunderkammern* are all the rage. What has changed in the last fifteen years? Has Foucault's "new age of curiosity" finally dawned? Certainly Foucault's own historico-critical work on deviance and normalcy has contributed to the fascination with the extraordinary and the marginal, as have the reflections of anthropologists like Mary Douglas on entities that straddle cultural categories. More generally, the last twenty years have seen a deep questioning of ideals of order, rationality, and good taste — "traditional hierarchies of the important and the essential" — that had seemed self-evident to intellectuals since the origins of the modern Republic of Letters in the late seventeenth century. Wonder and wonders have risen to prominence on a wave of suspicion and self-doubt concerning the standards and sensibilities that had long excluded them (and much else) from respectable intellectual endeavors like our seminar. Hence a history of the marvelous is also a history of the pursuits and, more pointedly, the nonpursuits of intellectuals: of why curiosity flows into some channels and not others.

We too have changed since our initial explorations of the topic in graduate school. Our subsequent researches and the evolution of the historiography of science have wrought transformations in our work. We soon realized that sixteenth- and seventeenth-century monsters were part of a coherent and long-lived cluster of wonders, persisting from late antiquity through at least the Enlightenment, which embraced a crowd of other strange objects and phenomena and from which they could only artificially be detached. Thus this book greatly expands our original chronology and subject matter in order to situate this cluster in its larger and longer intellectual and cultural context. Only more gradually did we come to query the narrative that had propelled the article, a linear story that took monsters from prodigies to wonders to naturalized objects. In the context of the history of science as practiced in the mid-1970s, the logic of this narrative seemed irresistible. But the work of many scholars within and beyond the history of science — in cultural history, in philosophy, in the history of art and literature, in sociology and anthropology — has since challenged the inevitability of that account of scientific change. Sentimentally, perhaps, we have kept monsters at the heart of this book, revisited and rethought in Chapter Five. But in that chapter and throughout the book, we have abandoned a plot of linear, inexorable naturalization for one of sensibilities that overlapped and recurred like waves.

Sensibility was invisible in the history of science of the 1970s (our article included). It pervades our book, where we have written as much about wonder as about wonders. Both of us recall overwhelming experiences of wonder: the sun apparently rises blazing in the west, a string of seven or eight huge meteorites glides impossibly slowly across the sky, an aurora borealis fills the heavens with curtains of iridescent light. These (rather than any pioneer spirit) surely inclined us toward such topics in the first place. But we learned through many years of reading and writing on the history of wonderful objects that the passion of wonder itself — visceral, immediate, vertiginous — also had its history. From the contrast between powerfully felt emotions and a history of their mutability we concluded that there is nothing "mere" about cultural constructs. They are as real as bricks.

From the very beginning, back in our seminar with Bacon and Leibniz, we have studied wonder and wonders in collaboration. Collaboration is rare among humanists, but we recommend it heartily. Our differences of field and approach (medievalist and *dix-huitièmiste*, social historian of medicine and intellectual historian of mathematics) have taught both of us volumes. Not only this project, but all of our subsequent work has benefited from our prolonged, intense discussion about what was interesting and how it was interesting. We still don't agree on some of the most basic issues. One of us believes that wonders appeal because they contradict and destabilize; the other, because they round out the order of the world. Perhaps more important, though, is our shared childhood love of E. Nesbit's *The Story of the Amulet*, for our choice of history as a vocation and marvels as an avocation.

Because this book was indeed an avocation, written in the interstices of our other projects and obligations, it has taken a long time to finish. Year after year, we visited this or that library to collect materials, hammered out one or another chapter outline during the few days we could get together, and endlessly discussed what we had read and written over enough tea to float an ocean liner. Our friends and colleagues mercifully stopped asking when the book would appear; our publisher must have despaired. In addition to our other scholarly commitments, the distance that separated us during most of the book's gestation slowed our progress. But we also confess to a certain inner resistance to speeding the book on its way. "For all knowledge and wonder (which is the seed of knowledge)," wrote Francis Bacon, "is an impression of pleasure in itself."[3] We lingered over the pleasure of wonders, a pleasure all the keener for being shared.

We are very grateful to all the people and institutions who supported this project. The National Science Foundation, the National Endowment for the Humanities, the Guggenheim Foundation, the Bunting Institute of Radcliffe College, the Center for Advanced Study in Behavioral Sciences, Wellesley College, and the University of Chicago Humanities Institute generously provided us with the time and means to conduct research and to write. The Max Planck Institute for the History of Science in Berlin made it possible for us to be together, anomalously, long enough to complete a full draft. Mario Biagioli, Marie-Noëlle Bourguet, Caroline Bynum, Joan Cadden, Mary Campbell, William Clark, Alix Cooper, Mordechai Feingold, Paula Findlen, Peter Galison, Michael Hagner, Anke te Heesen, Doris Kaufmann, Ursula Klein, Ramona Naddaff, Brian Ogilvie, Dorinda Outram, Krzysztof Pomian, Rosamond Purcell, Antoine Schnapper, Michael Shank, and Nancy Siraisi read the manuscript in part or in its entirety. They have saved us from many errors and various forms of *folie à deux*. We hope we have done justice to their thoughtful criticisms and suggestions. Francesca Bordogna, Marilyn Cooper, Elisabeth Mitchell, Barbara Naddeo, and Natascha Pflaumbaum were imaginative research assistants. Stephan Schmal tackled with care and patience the Herculean task of unifying the manuscript: many libraries in many places make for many editions of Aquinas and Bacon. We thank Meighan Gale and Amy Griffin at Zone Books for their help and patience, and our copy-editor Amy Johnson for her many improvements of our manuscript. There was, all things considered, remarkably little grumbling on the part of our nearest and dearest about how much of us the project took away from them. We affectionately dedicate the book to them, as well as to Nesbit's Psammead, emblem of all that is sharp-tongued but well-meaning, furry, and wondrous.

At the Limit

> Yesterday, when I was about to go to bed, an amanuensis of mine, accustomed
> to make observations, informed me, that one of the servants of the house,
> going upon some occasion to the larder, was frighted by something luminous,
> that she saw (not withstanding the darkness of the place) where the meat had
> been hung up before. Whereupon, suspending for a while my going to rest, I
> presently sent for the meat into my chamber, and caused it to be placed in a
> corner of a room capable of being made considerably dark, and then I plainly
> saw, both with wonder and delight, that the joint of meat did, in divers places,
> shine like rotten wood or stinking fish; which was so uncommon a sight, that
> I had presently thoughts of inviting you to be a sharer in the pleasure of it.[1]

On seeing the glowing veal shank discovered by his terrified servant,
Robert Boyle's first response was wonder. His second was immediately to
investigate the matter, despite the lateness of the hour and the head cold
he had caught trying out a new telescope. Even as he was undressing for
bed, he called for another leg of veal "ennobled with this shining faculty"
to be brought into his chamber. The pleasure of the "uncommon sight"
sustained him into the early morning hours.

Boyle and many of his contemporaries saw wonder as a goad to inquiry,
and wonders as prime objects of investigation. René Descartes called wonder the first of the passions, "a sudden surprise of the soul which makes it
tend to consider attentively those objects which seem to it rare and extraordinary."[2] Francis Bacon included a "history of marvels" in his program
for reforming natural philosophy.[3] Their focus on wonder and wonders in
the study of nature marked a unique moment in the history of European
natural philosophy, unprecedented and unrepeated. But before and after
this moment, wonder and wonders hovered at the edges of scientific inquiry. Indeed, they defined those edges, both objectively and subjectively.
Wonders as objects marked the outermost limits of the natural. Wonder
as a passion registered the line between the known and the unknown.
This book is about setting the limits of the natural and the limits of

the known, wonders and wonder, from the High Middle Ages through the Enlightenment. A history of wonders as objects of natural inquiry is therefore also a history of the orders of nature.[4] A history of wonder as a passion of natural inquiry is also a history of the evolving collective sensibility of naturalists. Pursued in tandem, these interwoven histories show how the two sides of knowledge, objective order and subjective sensibility, were obverse and reverse of the same coin rather than opposed to one another.

To study how naturalists over some six centuries have used wonders to chart nature's farthest reaches reveals how variously they construed its heartland. Medieval and early modern naturalists invoked an order of nature's customs rather than natural laws, defined by marvels as well as by miracles. Although highly ordered, this nature was neither unexceptionably uniform nor homogeneous over space and time. Wonders tended to cluster at the margins rather than at the center of the known world, and they constituted a distinct ontological category, the preternatural, suspended between the mundane and the miraculous. In contrast, the natural order moderns inherited from the late seventeenth and eighteenth centuries is one of uniform, inviolable laws. On this view, nature is everywhere and always the same, and its regularities are ironclad. Wonders may occasionally happen, but they occupy no special geographical region, nor can they lay claim to any special ontological status outside the strictly natural. Only a miracle — a divine suspension of natural laws — can in principle break this order. To tell the history of the study of nature from the standpoint of wonders is to historicize the order of nature and thereby to pose new questions about how and why one order succeeds another.

As theorized by medieval and early modern intellectuals, wonder was a cognitive passion, as much about knowing as about feeling. To register wonder was to register a breached boundary, a classification subverted. The making and breaking of categories — sacred and profane; natural and artificial; animal, vegetable and mineral; sublunar and celestial — is the Ur-act of cognition, underpinning all pursuit of regularities and discovery of causes. The passion of wonder had a mixed reception among late medieval and Renaissance natural inquirers, scorned by some as a token of ignorance and praised by others, following Aristotle, as "the beginning of philosophy."[5] All, however, agreed that wonder was not simply a private emotional experience but rather, depending on context, a prelude to divine contemplation, a shaming admission of ignorance, a cowardly flight into fear of the unknown, or a plunge into energetic investigation. Such states were charged with meaning for the image and conduct of naturalists as a group. Since the Enlightenment, however, wonder has become a disreputable passion in workaday science, redolent of the popular, the amateurish, and the childish. Scientists now reserve expressions of wonder for their personal memoirs, not their professional publications. They may acknowledge wonder as a motivation, but they no longer consider it part of doing science.

The history of wonder, however, extends beyond the history of its role in the study of nature and its positive or negative valuation therein. Wonder has its own history, one tightly bound up with the history of other cognitive passions such as horror and curiosity — passions that also traditionally shaped and guided inquiry into the natural world. Not only the valuation of these emotions, but also their proximity and distance from one another, and even their texture as felt experience, have changed with context and over time. The domain of wonder was broad, and its contexts were as various as the annual fair, the nave of a cathedral, the princely banquet hall, the philosopher's study, or the contemplative's cell. Context colored emotion. Wonder fused with fear (for example, at a monstrous birth taken as a portent of divine wrath) was akin but not identical to wonder fused with pleasure (at the same monstrous birth displayed in a *Wunderkammer*). In the High Middle Ages wonder existed apart from curiosity; in the sixteenth and seventeenth centuries, wonder and curiosity interlocked. Estrangement and alliance shaped the distinctive objects and the subjective coloring of both passions. Thus in writing a history of wonder as a passion, we have attempted to historicize the passions themselves.

To this end, we have adopted one fundamental principle: to attend as precisely as possible to what our sources meant by the passion of wonder and by wonders as objects. We here diverge from most recent students of the pre-modern marvelous, who have tended to define their subject in terms of "what we *now* call marvels," in the words of Jacques Le Goff.[6] This corresponds to a loose category coextensive with what might in English be called the fictional or fantastic and is defined mainly in privative terms as that which is excluded by modern views of the rational, the credible, and the tasteful: the products of imagination, the inventions of folklore and fairy tales, fabulous beasts of legend, freaks of sideshows and the popular press, and, more recently, the uncanny in all its forms. Because this view of wonders was a creation of Enlightenment thinkers, it is hardly surprising that, as Le Goff himself notes, medieval writers "did not possess a psychological, literary, or intellectual category" corresponding to the modern *merveilleux*.[7] Accounts of the subject based on this anachronistic definition are evocative for modern readers, but they lack historical coherence and precision.

What words did medieval and early modern Europeans use for the modern English "wonder" and "wonders"? In Latin, the emotion itself was called *admiratio* and the objects, *mirabilia*, *miracula*, or occasionally *ammiranda*. These terms, like the verb *miror* and the adjective *mirus*, seem to have their roots in an Indo-European word for "smile."[8] (The Greek *thauma*, on the other hand, found its origin in a verb "to see.")[9] The etymological ties between wonder and smiling persisted in the romance languages (*merveille* in French, *meraviglia* in Italian, *marvel* in English from c. 1300), though not in the German *Wunder* — a word of mysterious origin that may have to do with intricacy or complexity — or the English *wonder*.[10] We have followed late medieval and early modern English writers in employing the Germanic *wonder* and the romance *marvel* interchangeably in our translations and in our own prose.

Except for this difference between the Germanic and the romance roots, however, the vocabulary of wonder had a unified profile from at least the twelfth or thirteenth century in all the linguistic traditions we have studied. This argues for a strong common understanding. First, the words for passion and objects were, if not identical, then closely related, signaling the tight links between subjective experience and objective referents. Second, these languages all blurred the sacred and the secular objects of wonder — the miraculous and the marvelous. This suggests the impossibility of wholly divorcing these two kinds of wonders in the dominant Christian culture, although theologians and philosophers upheld an analytical distinction between them; the realms of the supernatural and preternatural can be differentiated in order to focus on the latter, as we have done in this study, but only with considerable care. Despite this difficulty, we have restricted this study to natural wonders, marvels rather than miracles. Finally, from at least the twelfth century the vernacular terms for wonder, like the Latin, admitted a spectrum of emotional tones or valences, including fear, reverence, pleasure, approbation, and bewilderment. Beginning in the late fifteenth or sixteenth centuries, these different flavors of wonder acquired different names: *admiration* and *astonishment* in English, for example, *Bewunderung* and *Staunen* in German, and *étonnement* and *admiration* in French. This multiplication and refinement of vocabulary signals the prominence of the passion and its nuances in the early modern period. Thus wonder was from at least the High Middle Ages a well-defined but also an extraordinarily rich and complex emotion, with associations that crystallized into separate terms over the course of time.

The tradition had a strong coherence, which rested in both the objects of wonder and the passion that they inspired. The canon of natural wonders had a stable core throughout the period we have studied (and indeed back into the Hellenistic period), with a penumbra that expanded and contracted as ideas, experiences, and sensibilities changed. At the center lay the most enduring marvels, like African pygmies, the mysterious lodestone, the glowing carbuncle, or the properties of petrifying springs. Over the course of time, some objects dropped out of this canon for various reasons. The basilisk was debunked, comets were explained, and unicorn horns became too common, even before they were reclassified as narwhal tusks: wonders had to be rare, mysterious, and real. At the same time, such new objects joined the canon of wonders as monstrous births, recuperated from the canon of horrors, and the louse, a marvel only under the microscope. Reassessing the meaning (and thus the emotional import) of an object or revealing a previously hidden characteristic could make it grounds for wonder. The passion and the objects mutually defined each other, a process in which neither remained static.

In placing wonder and wonders at the center of our narrative, we have had to challenge the traditional historiography of science and philosophy in fundamental ways. Most obviously, we have let go of not only the usual periodization, which divorces the medieval from the early modern study of nature, but also the much more basic ideas of distinct stages, watersheds, new beginnings, and punctual or decisive change. These narrative conventions, imported into intellectual history from eighteenth- and nineteenth-century political historiography, only distort the nonlinear and nonprogressive cultural phenomena we describe. For the most part our story is not punctuated by clearly distinguished epistemes or turning points, but is instead undulatory, continuous, sometimes cyclical.

It is not that the six hundred years we discuss saw no changes or that we are talking of the stasis of the *longue durée*. In our story, individuals change their minds, have remarkable experiences, and make extraordinary discoveries, which dramatically alter the known world. Social and intellectual communities and institutions appear and disappear or develop new allegiances and agendas over a decade or a generation. Since our study encompasses much of western Europe and spans a range of cultural environments, change was always happening somewhere: from the beginning of our period to the end, the canon of wonders was constantly shifting its contents and its meaning in innumerable ways. But change happened smoothly and continuously in its general outlines. The multiplicity of approaches in the interpretation of nature, the layering of cultural levels, the differences between national or linguistic traditions, the gap between the rear guard (usually) at the periphery and the avant-garde (usually) at the center — all acted to smooth out the watersheds and blur the borders between epistemes that are often projected onto this more complicated historical reality. As a result, our readers must be willing to abandon conventional periodization and a strictly linear narrative. In order to follow the substantive and chronological contours of the history of wonder and wonders we have integrated both periods and topics usually kept asunder — collecting and romances, travel and court spectacle, medical practice and popular prophecy, natural philosophy and aesthetic theory.

Despite these departures from historical convention, our story interweaves and intersects with many important and familiar narratives of high medieval and early modern European historiography: the rise of universities, the age of European exploration, the course of the Scientific Revolution, secularization, the rise of absolutism, and the like. Rather than rejecting or supplanting such narratives, we have used our sometimes unfamiliar material as seventeenth-century philosophers used marvels: to "break up our familiarities," as Foucault put it in our epigraph, "and to regard otherwise the same things." We do not propose wonders as the newest key to early modern science and philosophy, nor do we offer our own story as an alternative grand narrative for the Scientific Revolution, as Frances Yates did for magical Hermeticism.[11] But the history of science does look different when organized around ontology and affects rather than around disciplines and institutions.

Our study is in some ways unusually broad — contextually, chronologically, and geographically — but we have set limits as to who and when. Our book focuses on wonder and wonders as an elite tradition, engaging the attention of princes, clerical administrators, preachers, teachers, court artists and storytellers, naturalists, and theologians. We have begun with the mid-twelfth century for two related reasons. First, the dramatic increase in the number of ancient sources available provided the base for a rich ramification and elaboration of the ancient tradition of writing on wonders. Second, the coeval rise of cities and of royal and imperial bureaucracies, the creation of courts as centers of literary, artistic, and philosophical culture, the emergence of schools and, later, universities as centers of formal learning — all combined to create literate, wealthy, and powerful audiences for wonder and wonders. At the other terminus, we have taken our study well into the eighteenth century in order to trace and analyze the process by which wonder and wonders faded from prominence in elite circles as favored objects of contemplation and appreciation. How marvels fell from grace in European high culture has less to do with some triumph of rationality — whether celebrated as enlightenment or decried as disenchantment — than with a profound mutation in the self-definition of intellectuals. For them wonder and wonders became simply

18

vulgar, the very antithesis of what it meant to be an *homme de lumières*, or for that matter a member of any elite.

This marked the end of the long history of wonder and wonders as cherished elements of European elite culture, and therefore also the one sharp rupture in our narrative. During the period from the twelfth through the late seventeenth century, wonder and wonders — far from being primarily an element of "popular" culture, much less a site of popular resistance to elite culture[12] — were partly constitutive of what it meant to be a cultural elite in Europe. In the hands of medieval abbots and princes, natural wonders such as ostrich eggs, magnets, and carbuncles represented the wealth of their possessors and their power over the natural and the human world. In the hands of philosophers, theologians, and physicians, they were recondite objects of specialized knowledge that transcended prosaic experience. In the hands of sixteenth- and seventeenth-century virtuosi and collectors, they became occasions for elaborate exercises in taste and connoisseurship. All of these groups separated themselves from the vulgar in their physical access to marvels, in their knowledge of the nature and properties of these marvels, and in their ability to distinguish things that were truly wonderful from things that were not. When marvels themselves became vulgar, an epoch had closed.

In laying out this long, sinuous history of wonders, we have organized our book along only roughly chronological lines. Key themes such as the shaping role of court culture, the lure of the exotic, the practices of collecting, the forms of scientific experience, the unstable boundary between marvels and miracles, recur throughout. Chapter One discusses writing on extraordinary natural phenomena in the literature of travel and topography, chronicles, and encyclopedias, which, we argue, constituted the core tradition of medieval reflection on wonders. Chapter Two treats wonders as objects, both textual and material, and describes the way in which they were used for purposes ranging from religious meditation to court ritual, while Chapter Three turns to the culture of thirteenth- and fourteenth-century natural philosophy and its rejection of both wonder and wonders as an integral part of the study of the natural order. Chapter Four shows how various groups of intellectuals, especially court physicians, professors of medicine and natural history, apothecaries, and authors of texts in popular philosophy, rehabilitated wonders for both natural philosophical contemplation and empirical investigation.

Chapter Five, on monstrous births, is our only extended case study. The pivot of the book's argument, it spans the period from the late Middle Ages through the Enlightenment and rehearses the multiple meanings of wonders as religious portents, popular entertainment, philosophical challenge, and aesthetic affront. Monsters elicited wonder at its most iridescent, linked sometimes to horror, sometimes to pleasure, and sometimes to repugnance.

Chapter Six describes how the preternatural became a central element in the reform of natural history and natural philosophy in seventeenth-century scientific societies, while Chapter Seven examines how the early modern *Wunderkammern*, in blurring the ancient opposition between art and nature, served as an inspiration for the union of these ontological categories in the natural philosophy of Bacon and Descartes. Chapter Eight charts the shifting relationships between the two cognitive passions of wonder and curiosity, showing how they briefly meshed into a psychology of scientific inquiry in the seventeenth century — given triumphant expression in the passage from Boyle with which we began this Introduction — only to drift rapidly apart thereafter. Chapter Nine, finally, recounts how wonder and wonders became vulgar, at once metaphysically implausible, politically suspect, and aesthetically distasteful.

All of the chapters are the products of joint research, discussions, and writing, but Park had primary responsibility for Chapters One through Four, and Daston for Chapters Six through Nine. We wrote Chapter Five together.

The enduring fascination exerted by wonders cries out for explanation. How did a miscellany of objects become and remain so emotionally charged? Wonders and wonder limned cognitive boundaries between the natural and the unnatural and between the known and the unknown. They also set cultural boundaries between the domestic and the exotic and between the cultivated and the vulgar. All of these boundaries were electric, thrilling those who approached them with strong passions; to run up against any of these limits was necessarily to challenge the assumptions that ruled ordinary life. No one was ever indifferent to wonders and wonder. Neither the medieval and Renaissance princes who coveted them, nor the readers of romances and travelogues who dreamed with them, nor the Enlightenment philosophes who despised them could be neutral about wonders: markers of the outermost limits of what they knew, who they were, or what they might become.

CHAPTER ONE

The Topography of Wonder

When high medieval European writers invoked wonders, what exactly did they have in mind? Gervase of Tilbury, an early thirteenth-century English noble and imperial counselor resident at Arles, was eager to explain. He devoted the third and longest section of his *Otia imperialia*, written around 1210 and dedicated to Emperor Otto IV, to what he called "the marvels of every province — not all of them, but something from each one" (fig. 1.1).[1] After some introductory remarks, he set out a catalogue of a hundred and twenty-nine such marvels, beginning with the magnet, an Indian stone with the mysterious property of attracting iron, and ending with a spring near Narbonne that changed place whenever something dirty was put into it. In between, he wrote of a garden planted by Vergil in Naples that contained an herb that restored sight to blind sheep; Veronica's napkin, still imprinted with Christ's likeness, in St. Peter's; the portents at the death of Caesar; the sagacity of dolphins; a race of Egyptian people twelve feet high with white arms and red feet, who metamorphosed into storks; the phoenix; *dracs*, who lived in the Rhone and lured women and children by taking the form of gold rings; and werewolves, whose sighting Gervase described as "a daily event in these parts." Gervase protested the truth of all these phenomena, noting he had tested or witnessed many of them himself.

At first glance, this list appears incoherent. It included plants, animals, and minerals; specific events and exotic places; miracles and natural phenomena; the distant and the local; the threatening and the benign. Furthermore, Gervase had compiled his wonders from a wide range of sources. Many (the dolphins, the phoenix, the portents) came from classical texts, while others were obviously biblical or belonged to the capacious Christian corpus of wonder-working sites, images, and relics. Still others, like the werewolves and *dracs*, had their roots in Germanic, Celtic, or other local oral traditions. Yet for all their diversity, Gervase stressed the coherence of this catalogue of wonders, locating it in the emotion

Figure 1.1. The marvels of Naples
Jouvenel des Ursins group, *Livre des merveilles du monde*, trans. and comp. Harent of Antioch, Acc. no. MS 461, fol. 15v, Pierpont Morgan Library, New York (c. 1460).[1]

This illustration to the French translation of Gervase's *Otia imperialia* shows the wonders of Naples, in the province of Campania. These included several magical inventions attributed to Vergil, notably a bronze fly that prevented any other flies from entering the city and a bronze statue of a man with a trumpet that repulsed the south wind, so that the ash and cinders from Vesuvius (shown in the background with a flaming top) were blown away from the fields surrounding the city. Notable among the natural marvels of the region, in addition to Vesuvius itself, were the thermal and therapeutic baths of Pozzuoli, shown as two square basins on the flank of the volcano (see also fig. 4.1), and a bean plant (before the city gate) that caused anyone who ate its fruit to experience the feelings of the person who had picked the beans.

evoked by all of them. "And since the human mind always burns to hear and take in novelties," he wrote in his preface,

> old things must be exchanged for new, natural things for marvels [*mirabilia*], and (among most people) familiar things for the unheard of.... Those things that are newly created please naturally; those things that have happened recently are more marvelous if they are rare, less so if they are frequent. We embrace things we consider unheard of, first on account of the variation in the course of nature, at which we marvel [*quem admiramur*]; then on account of our ignorance of the cause, which is inscrutable to us; and finally on account of our customary experience, which we know differs from others'.... From these conditions proceed both miracles [*miracula*] and marvels [*mirabilia*], since both culminate in wonder [*admiratio*].[2]

In this passage, Gervase summarized the principal commonplaces of the high medieval understanding of wonder. First, he traced the emotion to two roots: experience of the novel or unexpected, and ignorance of cause. Marvels were either rare phenomena, astounding by their unfamiliarity (for example, the phoenix of the Atlas Mountains, which immolated itself periodically only to rise again), or more common but puzzling, counterintuitive, or unexplained phenomena (for example, the attractive properties of the magnet or ghostly appearances of the dead). As a result, Gervase emphasized in his reference to "customary experience," wonder was always relative to the beholder; what was novel to one person might be familiar to another, and what was mysterious to one might be causally transparent to someone better informed.[3] For this reason, Gervase continued in a passage cribbed from Augustine, we do not find it marvelous that lime catches fire in cold water, because it forms part of our everyday experience, but if we were told that some stone from India behaved in exactly the same way, we would either dismiss the story as incredible or be "stupefied with wonder."[4] Throughout this passage, finally, Gervase emphasized the tight links between wonder, pleasure, and the insatiable human appetite for the rare, the novel, and the strange. It is certainly for this reason that he placed wonders at the apex of a work designed ostensibly for the emperor's entertainment and relaxation.

Like his analysis of wonder, Gervase's list of marvels was broadly typical of contemporary learned literature. It represented, if not a fixed canon of individual phenomena, then certainly a canon of the *types* of things that thirteenth-century readers would expect to find in such a list. This canon was not a medieval invention. Gervase's wonders were for the most part the classic wonders of Greek and Roman paradoxography, a literary genre that had grown out of the Aristotelian project of compiling descriptive histories of natural phenomena and had coalesced in the third century B.C.E. in the form of catalogues of things that were surprising, inexplicable, or bizarre.[5] The purpose of the original Greek texts is unclear, but they may have served as commonplace books for rhetoricians. In any case, paradoxographical material later made its way into Roman encyclopedic writing, including the works of the Pliny the Elder and Solinus, who were well known in the medieval Latin West.

Gervase of Tilbury's account of the world's wonders took the form of a catalogue made up of extremely brief and descriptive entries, with no attempt to relate them either spatially or chronologically or to analyze or explain them in any way. The wonders themselves were overwhelmingly topographical in nature; that is to say, they were linked to particular places (the "provinces" of Gervase's subtitle) and often to particular topographical features, such as caves and springs, rocks and lakes. The magnet was indigenous to India, for example, and the phoenix to the Atlas Mountains, while there were mountains in Wales so wet that the land moved under travelers' feet.[6] Such wonders were, in other words, particular, localized, and concrete. Yet despite these similarities to ancient paradoxography, Gervase's work differed from it in important ways. He introduced his discussion by analyzing the emotion of wonder: its association with novelty, its pleasurable nature, its causes, and its universal appeal. Even more striking, he did not simply repeat the canonical marvels he found in earlier writers—a signal feature of ancient paradoxography—but sought to supplement them with wonders of his own. Many of these came from personal experience, which explains the strong showing of the region around Arles.[7] These differences mark an important feature of the later medieval tradition of what we will call topographical wonder: its emphasis on verification through personal experience and oral report. Less a purely erudite tradition than its ancient forebear, it had more room for development and growth.

Wonders of this sort were not confined to catalogues of *mirabilia* like Gervase's but appeared in recognizable clusters in medieval works of many different sorts: encyclopedias (together with the related genres of bestiary, lapidary, and herbal), chronicles, topographical treatises, travel narratives, and the literature of romance. The variety of this literature reflects the growing medieval audience for wonders. In the early thirteenth century, when Gervase was writing, marvels were largely confined to Latin culture, with the important exception of vernacular romance. The authors of this material were for the most part clerics, often in the employ of princely patrons, secular and ecclesiastical. Gervase himself claimed to be writing for the emperor, having previously served as courtier and counselor to England's Henry II, another European monarch with a taste for natural wonders. Henry was the dedicatee of another of the earliest Latin works to take up the topic, Gerald of Wales's topography of Ireland, which its author presented to Archbishop Baldwin in 1188.[8]

From the twelfth century on, marvels also figured prominently in vernacular romances. This signals a growing audience for wonders that included not only clerics and princes but also the knightly and eventually the bourgeois readers of that genre. By the middle of the fourteenth century, various earlier Latin books of marvels, including Gervase's and Gerald's, had been translated into the vernacular, and other writers had begun to produce original vernacular topographical books of wonders, culminating in the spectacularly popular *Mandeville's Travels*.[9] Some of these, like *Mandeville's Travels* and the book of Marco Polo, were in turn translated into Latin. Indeed, as a sign of the growing appetite for wonders, earlier books with no reference to wonders in their titles were renamed to underscore the marvels they contained. One example was Polo's originally rather prosaically titled *Devisament dou Monde* (*Description of the World*), repackaged in Latin as *Liber Milionis de magnis mirabilibus mundi* (*"Million's" Book of the Great Wonders of the World*); the author's nickname reflected his reputation for exaggeration.[10] This chapter and the one that follows focus on the diverse environments in which wonders, in all their myriad incarnations, were enthusiastically compiled, collated, analyzed, and multiplied in monasteries and convents, the households of urban lay readers, and the high and late medieval courts.

Marvels on the Margins

Like the ancient paradoxographers, medieval writers on topographical wonders depicted the margins of the world as a privileged place of novelty, variety, and exuberant natural transgression. In the circular mental map of medieval geography, the central territories — the Holy Land, Europe, and the Mediterranean — had their marvels, but they were far outstripped in this respect by the periphery: the territories and islands bathed by the great ocean thought to cover most of the globe. As the fourteenth-century English monk Ranulph Higden put it in his world history, "At the farthest reaches of the world often occur new marvels and wonders, as though Nature plays with greater freedom secretly at the edges of the world than she does openly and nearer us in the middle of it."[11] Gerald of Wales was of the same opinion: "Just as the countries of the East are remarkable and distinguished for certain prodigies [*ostentis*] peculiar and native to themselves, so the boundaries of the West also are made remarkable by their own wonders of Nature [*naturae miraculis*]. For sometimes tired, as it were, of the true and the serious, [Nature] draws aside and goes away, and in these remote parts indulges herself in these shy and hidden excesses."[12]

Gerald and Ranulph both had Ireland in mind, but most other medieval writers agreed that the most wonderful wonders lay in the far South and East, in Africa and India. Richard of Holdingham's great Hereford map, produced in England in the 1280s, illustrated the extraordinary fauna of these regions in vivid detail (fig. 1.2). Although these "marvels of the East," as they are usually called by modern scholars, had a long and rich history reaching to Hellenistic times and beyond, medieval authors and mapmakers knew the tradition largely through a variety of later Roman writers, notably Pliny, Solinus, Augustine of Hippo, Isidore of Seville, and the authors of a large body of literature associated with the figure of Alexander the Great.[13] The Islamic world had its own well-developed tradition of paradoxography, also shaped by Greek sources, but its central texts were never translated into Latin, and it had relatively little influence in the West.[14]

Up to this point we have emphasized general characteristics of the high and late medieval literature of topography and travel, but the genre also shifted and changed over the course of the Middle Ages. The exotic Eastern races had not always elicited the enthusiastic and appreciative response found among Gervase and his contemporaries. Early medieval writers tended to follow the Alexander tradition, which reflected the imperialist aims of its hero by portraying the East as adversary and prey.[15] The two most widely copied early medieval treatises on the eastern races, the *Liber monstrorum* and *Tractatus monstrorum*, both probably written in the eighth century, stressed the threatening nature of their material. The anonymous author of the first described his subject as the "three types of things on earth that provoke the greatest terror in the human race, monstrous human births, the horrible [*horribilibus*] and innumerable types of wild beasts, and the most terrible [*dirissimis*] kinds of serpents and vipers," and he compared his task in writing to diving, terror-struck, into a dark sea full of monsters.[16] The creatures in his catalogue included cannibals, harpies, crocodiles, boa constrictors, and enormous ants. One entry will suffice to give the flavor of this work: "There is a certain people of mixed nature who live on an island in the Red Sea. They are said to be able to speak in the tongues of all nations; in this way they astonish men who come from far away, by naming their acquaintances, so that they may surprise them and eat them raw."[17]

This tone of suspicion, if not outright paranoia, also marked various versions of the roughly contemporary *Tractatus monstrorum*, and the images in an eleventh-century manuscript of the treatise (fig. 1.3.1) showed beings that were at least barbaric and disturbing, if not outright dangerous.[18] In addition to its didactic uses — such beings were easily moralized by poets and sermonizers — this literature may have functioned as recreational reading in the monastic environment in which it circulated, as the number of illustrated manuscripts suggests. But its pleasures seem to have had less to do with wonder as Gervase would later come to understand it, an emotion rooted in the appreciation of novelty and difference, than with reading of foreign dangers in the safety and familiarity of home.

Traces of this earlier attitude also shaped Gerald of Wales's *Topographia Hibernie*, composed for Henry II (whom Gerald called "our western Alexander")[19] shortly after the king sent him to Ireland in 1185 as part of the conquest of that island. Gerald emphasized Ireland's natural marvels: he used the language of wonder appreciatively and specifically to refer to its lack of poisonous reptiles, for example, its petrifying wells, and its dramatic tides. But he was far more severe about its inhabitants, whom he described as barbarians, "adulterous, incestuous, unlawfully conceived and born, outside the law, and shamefully abusing nature herself in spiteful and horrible practices," a fact that explained the unusual number of animal-human hybrids and defective births.[20] When nature acted "against her own laws," as Gerald put it, she produced wonders in a land worthy of conquest, whereas when the Irish transgressed the moral order, they produced horrors, signs of the depravity that justified their dispossession.[21]

Gerald's juxtaposition of wonder and horror marks his book as transitional in the genre of medieval topographical writing, portraying the exotic as at times fascinating and at times opaque and dangerous, to be dominated or kept at arm's length. Gerald was also the earliest writer to inject the element of contemporary experience into this tradition: unlike the earlier treatises on monsters, which were purely literary compilations, his *Topographia* bristled with personal observations of flora and fauna, criticisms of the blunders of Bede and Solinus, and invocations of the testimony of reliable men.[22] The travel literature of the next two centuries shared this emphasis (sometimes more rhetorical than real) on eyewitness experience, but as time went on its authors increasingly characterized the margins of the globe as unambiguously "wonderful," using language even more extravagant than that of Gervase. From this point on, it is possible to talk without anachronism about the "marvels of the East."

This new attitude coincided with the great age of eastern travel and the opening of trade routes facilitated by the Mongol peace. From the

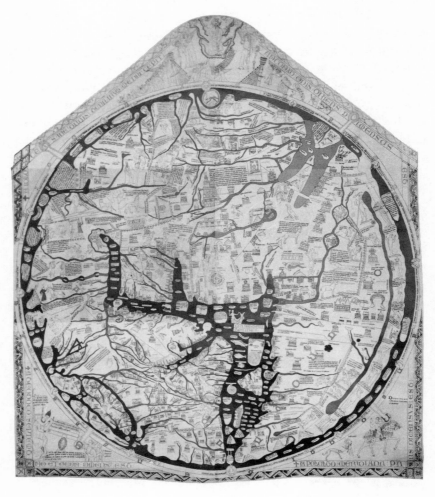

1.2.1 1.2.2

Figure 1.2. The marvels of Africa
1.2.1–2. Richard of Holdingham, Hereford Map, Hereford Cathedral, Hereford (1280s).

Four feet across, the Hereford map is the largest surviving example of the great medieval world maps, sometimes hung in churches and palaces as representations of the earth and symbols of divine creation. Asia appears at the top and Europe and Africa in the lower left- and right-hand sections, separated by the the Mediterranean. Asia and Africa teem with wonderful vegetable, animal, and human species, shown in captioned drawings. The detail of Africa (fig. 1.2.2) includes, for example, a winged salamander (a "venomous dragon) next to an anthropomorphized mandrake root (a "marvelously powerful plant"). The strip of land on the far right includes a number of the monstrous human races, including the one-legged Sciopodes; the hermaphroditic Androgynes; the Himantopodes, here shown on all fours; and the headless Blemmyes.

28

Similiter ibi nascuntur cenocephali quos
nos conopoenas appellamus habentes caballi
equorum aprorum dentes canina capita igne
& flamma flantes hec e. ciuitas uicina diues om
nib; bonis plena dexteriore parte ducit illa
tpmr abaegypto ⁊

Þac ⁊ þylce þan þeod cende healfhundingas þa
þindon hatene conopoenas hi habbad horses
manan ⁊ eoferes tuxas ⁊ hunda heafda ⁊
heopa oruð byð þylce fyres lig. þas land beoð
neah ðam buruȝum þe beoð eallum woruldwelum ȝefylled
þis rið healfe aegiptna landes ⁊

�7naliqua nascuntur homines statura pedum
vi. barbas habentes usq; adgenua comas
usq; adtalos quihomo dum appellantur
& pisces crudos manducat ⁊

Ðonsumon lande beoð men afcende ða beoð
on lenge fif fotmæla lange hi habbad hwitne
ꝇ ꝥ eow þæde ⁊ read oð hela⁊ homo dubii hi
þindon hatene þ e þ beoð ꝥylice ⁊ þ e hwi man
fixeon hi libbad ⁊ þa etað ⁊

Capisfluuius meodem loco apellat̃ gorgoneus
ibi nascunt formice statura canum habentes
pedes qrsi locuste rubro colore nigroq; fodientes
auru. & quod pnoctem fodiunt subtenu
pferat foras usq; dieihoram quinta hom̃ es
aute̅ quiaudaces sunt illud tollere.sic
tollunt aput camelos masculos & feminas illas
que habent foetas. foetas aut transflumen
gargulum alligatos relinquunt & camelisfoe
minis aurum imponunt Ille aut pietate ad
suospullosfestinantes.ibi masculi romaneñ ⁊

1.3.2

Figure 1.3. Cynocephali
1.3.1. *Tractatus de diversis monstris*, Cotton MS Tiberius B.V., fol. 80r, British Library, London (English, 11th century).
1.3.2. Boucicaut workshop, *Livre des merveilles du monde (Marco Polo)*, MS fr. 2810, fol. 76v, Bibliothèque Nationale, Paris (c. 1412–13).

The dog-headed Cynocephali were among the most widely discussed and variously described of the exotic human races. As these examples show, the images in books of wonders could be at odds with their texts. Thus the painter of this eleventh-century illumination (fig. 1.3.1) omitted the boarlike tusks, fiery breath, and rich city of the Cynocephali mentioned in his text.[2] Conversely, the illustrator of the relevant passage of Marco Polo's narrative, in the early fifteenth-century *Livre des merveilles du monde* (fig. 1.3.2), represented the dog-headed inhabitants of the Andaman Islands as apparently civilized and rational, where Polo had described them as "cruel" cannibals who liked their strangers raw and highly spiced.

1240s through the later fourteenth century, when the growing hostility of the Mongols in Asia and the Mamelukes in Egypt acted once again to isolate the West, Europeans set out for Asia in increasing numbers as missionaries, ambassadors, explorers, and entrepreneurs. Partly in response to a growing appetite for marvels on the part of their compatriots, many of these travelers produced descriptions of their experiences.[23] For example, Pope Innocent IV had sent the Franciscan John of Pian di Carpini as emissary to the Tartars in 1245. On his return, according to his fellow Franciscan and Italian chronicler Salimbene, he "had written a big book about the deeds of the Tartars and other wonders of the world, and when people tired him with questions on the subject, he had it read out loud, as I myself heard and saw on several occasions.... The friars read his book in his presence, and he interpreted and explained whatever seemed unclear."[24]

The relations produced by these travelers, unlike Gerald's *Topographia Hibernie*, did not reflect a straightforward imperialist program; the failure of successive waves of Christian Crusaders to expand and consolidate even their relatively modest territorial conquests in the Holy Land had scotched any serious European designs on the lands of India and beyond. Rather, these works expressed more limited aspirations: conversion of heathen peoples, formation of military and diplomatic alliances with the Mongols or Tartars, personal profit, or simple adventure. It is as if the failure of the early Crusading movement, together with Europeans' recognition of their own economic and technological inferiority, freed them to see the East not just as a reservoir of potential religious and military adversaries, but as a figure of aspiration and desire. Exotic, fertile, beautiful, and fabulously wealthy, it shaped their imaginations through the gems, precious metals, spices, and luxury textiles that made their way onto the western market in increasing volume thanks to the efforts of Mongol, Arab, and Italian traders. Thus the Venetian merchant Marco Polo, who had made two extended Asian voyages in the second half of the thirteenth century, stressed the natural endowments of the lands he visited. Writing about the Indian Kingdom of Quilon, for example, he effused:

> The country produces a diversity of beasts different [*devisees*] from those of all the rest of the world. There are black lions with no other visible colour or mark. There are parrots of many kinds. Some are entirely white — as white as snow — with feet and beaks of scarlet. Others are scarlet and blue — there is no lovelier sight than these in the world. And there are some very tiny ones, which are also objects of great beauty. Then there are peacocks of another sort than ours and much bigger and handsomer, and hens too that are unlike ours. What more need I say? Everything there is different from what it is with

us and excels both in size and beauty. They have no fruit the same as ours, no beast, no bird.[25]

What was the source of this extravagant and aestheticizing language that allowed Marco Polo and other topographical and travel writers to conceive and to articulate a new and more positive attitude toward the East? The answer is suggested by the circumstances under which Polo's narrative was composed. In fact, Polo was its author in only a dilute sense: while a prisoner of war in Genoa in 1298, he had told his story to a fellow detainee and writer of romances, Rustichello of Pisa, who wrote it up in French, embellishing it with many of the chivalric formulas developed in his earlier work.[26] That is, the medieval rhetoric of the marvelous was first elaborated in the twelfth- and thirteenth-century literature of romance—in its rhapsodic descriptions of Eastern luxuries, its emphasis on quest and adventure, its exploitation of the unexpected, its taste for exotic settings, its reliance on magical natural objects, its constant invocation of wonder and wonders, described in terms of diversity, and its association of those wonders with wealth and power.[27] Compare Polo's prologue, which touted the "very great marvels and great diversity [*grandismes mervoilles et les grant diversités*]" in his book with the opening of the romance *Cleomadès* by Rustichello's contemporary Adenet le Roi: "I have begun yet another book, one that is very marvelous and varied [*divers*].... The tale is of great worth and most pleasing to hear. It is so varied and marvelous that I believe that no one ever heard one so unusual [*diverse*] as this."[28] Gerald of Wales and Gervase of Tilbury had made considerable use of romance associations in their topographies, but it was only after Rustichello's collaboration with Marco Polo that the romantic rhetoric of wonder became an established feature of contemporary exotic travel narratives. Absent from the mid-thirteenth century accounts of the friars William of Rubruck and John of Pian di Carpini, this rhetoric informed the narratives attributed to Odoric of Pordenone (c. 1330), Jordan of Sévérac (c. 1330), and above all John Mandeville (c. 1357), where it was carried to unprecedented heights.[29]

As Marco Polo's ecstatic description of the natural wonders of Quilon indicates, late-thirteenth-century authors portrayed difference—in its manifold senses of diversity, novelty, unfamiliarity, and unlikeness—as a highly positive quality: the source of pleasure and delight. In the words of the author of *Mandeville's Travels*, "many men have a great liking to hear speak of strange things of diverse countries."[30] European observers did not accept everything Eastern as worthy of praise. They found much to blame, especially in the idolatry and, occasionally, the alleged cannibalism of the peoples they described.[31] Nonetheless, their great interest in the strange races and species of Asia and Africa was not primarily an expression of anxiety, as a number of scholars have recently argued,[32] and an obsessive attempt to objectify and exorcise pervasive fears of disorder, ignorance, dispossession, or sin. European authors certainly used the exotic races to test and explore fundamental boundaries in their own culture — between male and female, wild and civilized, human and animal — as is clear from the prominence in travel narratives of beings such as centaurs, satyrs, hermaphrodites, and cross-dressers. But these did not appear as scandalous or pathological, as seriously challenging European values or establishing a rival norm. They were too remote and strange to present a real alternative and thus a conceptual or political threat.[33] At their most transgressive, they served to satirize courtly and aristocratic culture or to figure a fantasy realm of freedom from the sexual restrictions and pervasive poverty of European culture. The wonders of the East had overwhelmingly positive associations; liberating precisely on account of their geographical marginality — unlike, say, the real and proximate difference represented by a resident Jewish population — they were viewed with a relatively benign and tolerant eye.[34]

This embracing of exoticism and variety in travel and topographical writing was strongly reinforced by the influence of the first-century Roman writer Pliny the Elder, whose massive *Natural History* in thirty-seven books served as the principal source from which medieval writers took, either directly or indirectly, much of their own material.[35] Like Marco Polo and his fellows, Pliny loved mirabilia: exotic species, strange topographical features, and unlikely and counterintuitive phenomena.[36] He underscored the wonderfulness of exotic eastern peoples, like the Psilli, whose bodies secreted a poison deadly to snakes, or the Indian Sciopodes, each with a single enormous foot. "These and similar varieties of the human race have been made by the ingenuity of Nature as toys for herself and marvels [*miracula*] for us," he wrote. "And indeed who could possibly recount the various things she does every day and almost every hour? Let it suffice for the disclosure of her power to have included whole races of mankind among her marvels."[37] Like the medieval romances and travel books, Pliny's work was permeated by an aesthetic and a sensibility that stressed the variety and diversity of nature.[38]

In thirteenth- and fourteenth-century topographical literature, this sensibility found expression in the topos of the relativity of the marvelous. James of Vitry, bishop of Acre in the thirteenth century, sketched this theme in his history of the East:

We know that all the works of God are marvelous [*mirabilia*], although those who are accustomed to look on them often are not moved by wonder [*admiratione*]. For perhaps the cyclops, who all have one eye, marvel as much at those who have two eyes as we marvel at them, or at others with three eyes. Just as we consider the pygmies dwarfs, so they would judge us giants, if they saw one of us among them.... We consider the black Ethiopians ugly, but among them, the blackest is judged the most beautiful. We do not marvel at many things in our lands that the peoples of the East, if they heard of them, either would not believe or would consider to be marvels.[39]

Among the many natural wonders to which westerners were accustomed, James cited the absence of snakes in Ireland, the midnight sun of Thule (or Iceland), fiery Mt. Etna in Sicily, men with tails in Britain, women with huge goiters in the Burgundian Alps, and broadly familiar phenomena like congenital lepers, the properties of quicksilver, and the loyalty of dogs.

Other authors made the same point, comparing Europe's elusive barnacle geese (thought to grow on trees) to the Scythian lambs of Mongolia. In the words of the Franciscan Odoric of Pordenone, who traveled to Asia in 1314,

Another very marvelous thing can be said, that I did not see but heard from persons worthy of credit. For it is said that in the great kingdom of Candelis there are Mountains called the Capei mountains. It is said that very large gourds grow there, which open when they are ripe, and inside is found a little animal, like a small lamb.... And although this may perhaps seem incredible, nonetheless, it can be true, just as it is true that in Ireland are trees that produce birds.[40]

The same story appeared in many versions of *Mandeville's Travels*, where it inspired a charming illustration — an emblem of the relativity of marvels — in the great early fifteenth-century *Livre des merveilles du monde* owned by the Duke of Berry (fig. 1.4.2).[41] This showed representatives of Europe and Asia (identified primarily by their headgear) exchanging a branch laden with barnacle geese for a Scythian lamb, in an act of apparent equality.

The empathy involved in this representation, which encouraged readers to place themselves in the minds of exotic people and to imagine their reactions, differentiated the later medieval literature of marvels from the earlier books of monsters, which presented the inhabitants of the East as barbarians or enemies to be conquered and feared. This change in attitude grew out of shifting circumstances, as western Europeans found them-

35

1.4.1

1.4.2

Figure 1.4. The marvels of Europe and Asia
1.4.1. *Bestiary*, MS Bodley 764, fol. 58v, Bodleian Library, Oxford (first half of 13th century).
1.4.2. Boucicaut workshop, *Livre des merveilles du monde (John Mandeville)*, MS fr. 2810, fol. 210v, Bibliothèque Nationale, Paris (c. 1412–13).

As described in an illustrated English bestiary (fig. 1.4.1), the barnacle goose, native to Ireland, was produced by nature "in a way which contradicts her own laws."[3] Straddling the animal-vegetable divide, it was considered analogous to the Scythian lamb, native to certain parts of Asia, which was said to be born from large gourds that grew on trees. Natural wonders were also commodities, and in this illustration to a famous French manuscript of exotic travel narratives produced in the early fifteenth-century workshop of the Boucicaut Master (fig. 1.4.2), European merchants on a journey to Asia exchange wonders with their Oriental counterparts.

selves in increasing political, commercial, and military contact with what had once been the distant East. As Fulk of Chartres wrote in his chronicle, the creation of the twelfth-century Crusading kingdoms of Outremer, however small and precarious, had the result that "God has transferred the Occident into the Orient, for we who were Occidentals have now become Orientals."[42] James of Vitry and other Europeans who resided for an extended period in the Holy Land practiced such acts of mental accommodation daily. Nor were churchmen and Crusaders the only ones whose experience argued for a broader view. Merchants such as Marco Polo understood the way in which highly localized conditions of supply and demand created value, just as they created wonder; pepper and indigo, exotic and expensive commodities in Europe, were common in Quilon. Like merchants, diplomats and missionaries were also forced to adopt an attitude of civility toward unfamiliar peoples and to accept strange customs, if only for instrumental reasons.

This set of attitudes shaped the illustrations in the splendid collection of travel narratives in French translation that John the Fearless, Duke of Burgundy, presented in 1413 to John, Duke of Berry. Among other works, this manuscript included *Mandeville's Travels* and the works of Marco Polo and Odoric, and its images, from the workshop of the famous Boucicaut Master, contrasted dramatically with those in the eleventh-century *Tractatus monstrorum*. For example, where the earlier manuscript showed the dogheaded Cynocephalus as an uncivilized and ravening monster, the later one represented him as an urbane and productive citizen in European dress, engaged in commercial activity and flanked by a castle of French architecture and design (fig. 1.3.2). The message of this last image, like many others in the same manuscript, was ambiguous. On the one hand, it vividly illustrated the commonality fundamental to the subjectivity of wonder: the society of the Cynocephali might be just as civilized and prosperous as that of their counterparts in the West. On the other hand, this does not mean that western Christian rulers had abandoned all military designs on eastern territories. The Duke of Berry's manuscript was produced in the context of the Crusading aims of the house of Burgundy, and the conquest of America in the next century produced representations of that continent's natural wonders and its human horrors as striking as anything in Gerald of Wales.

Despite their seductiveness, the marvels of the East did not ultimately represent an autonomous alternative — a standpoint from which effectively to challenge the centrality and normative character of the natural and moral order in the West. In the early fourteenth century, Jordan of Sévérac had spent several years as bishop in India, and he filled his *Mira-*

bilia descripta with enthusiastic accounts of that land's variety, beauty, and "many and boundless marvels." Yet his final, rather defensive reflections left no room for doubt concerning his ultimate allegiance: "One general remark I will make in conclusion," he wrote, "to wit, that there is no better land or fairer, no people so honest, no victuals so good and savoury, dress so handsome, or manners so noble, as here in our own Christendom; and, above all, we have the true faith, though ill it be kept."[43]

Wonders of Creation

Although some of the most famous medieval writers on topography and travel — Jordan, William of Rubruck, and Odoric of Pordenone — were Christian clerics, their works were secular in orientation and portrayed the marvels of creation with little reference to the Creator himself. A second strain of the paradoxographical tradition, however, used natural wonders for religious ends. The authors of encyclopedias, bestiaries, and similar collections emphasized the symbolic uses of wonders as keys to scripture, repositories of moral lessons, and testimonies to the benevolence and omnipotence of the Christian God. To understand this tradition, we must circle back again to an earlier period — to the time of Augustine, whose own discussions of wonders suggest a healthy interest in paradoxographical material, taken largely from the works of Pliny.

Augustine's most influential reflections on wonders appeared in his discussion of the pains of hell in Book XXI of *De civitate Dei*, where his immediate aim was to convince skeptics that God could make human bodies burn forever. He began by invoking two classic mirabilia, the salamander and Mt. Etna, both of which burned continually without being consumed. But these two examples did not exhaust Augustine's interest in wonders, for he proceeded to list a host of equally puzzling but unrelated phenomena: the incorruptibility of peacock meat, the hardness of diamond, and the marvelous properties of the magnet.[44] Many examples later, Augustine arrived at his more general point: the omnipotence of God. "For God is certainly called Almighty for one reason only," he wrote, "that he has the power to do whatever he wills, and he has the power to create so many things which would be reckoned obviously impossible, if they were not displayed to our senses or else reported by witnesses who have always proved reliable."[45] In this connection, Augustine stressed the relativity of wonder (in the passage cited by Gervase of Tilbury), noting that many of the things his contemporaries took for granted (such as the combustibility of lime in water) were at least as marvelous as the things mentioned above. His ultimate claim, however, was far more ambitious than Gervase's brief remarks on the psychology of wonder. Augustine's entire treatment of wonders in these chapters culminated in the argument that there was no inherent way to distinguish between apparently commonplace and apparently marvelous phenomena, since all depended directly on divine will. According to Augustine, such phenomena showed that everything created by God was wonderful, including what appeared to be his commonest and most pedestrian works. The order of the world was infinitely mutable, depending on God's purposes at the time, and explanations based on natural causes were unrevealing and entirely beside the point:

> So, just as it was not impossible for God to set in being natures according to his will, so it is afterwards not impossible for him to change those natures which he has set in being, in whatever way he chooses. Hence the enormous crop of marvels, which we call "monsters," "signs," "portents," or "prodigies"; if I chose to recall them and mention them all, would there ever be an end to this work?[46]

Augustine used the tropes of ancient paradoxography to recast and reshape the emotion of wonder. By placing his remarks in a vigorous discussion of the eternal torments that awaited the unbeliever, he made of wonder a serious and sobering emotion, dissolving its links with the more frivolous sorts of pleasure rooted in the experience of novelty and stressing instead its affinity with religious awe. But these reflections did not exhaust the Christian uses of wonders, at least not in Augustine's eyes. A basic appreciation of the marvelousness of creation sufficed for ordinary believers, but preachers, teachers, and exegetes, whose responsibility it was to interpret the Bible for the others, required more specialized knowledge of the properties of natural things. This was because the Bible was written in figurative language, with many metaphors and similes taken from the natural world. As Augustine put it in *On Christian Doctrine*,

> An ignorance of things makes figurative expressions obscure when we are ignorant of the natures of animals, or stones, or plants, or other things which are often used in the Scriptures for purposes of constructing similitudes. The well-known fact that a serpent exposes its whole body in order to protect its head from those attacking it illustrates the sense of the Lord's admonition that we be wise like serpents.... The same thing is true of stones, or of herbs or of other things that take root. For a knowledge of the carbuncle which shines in the darkness also illuminates many obscure places in books where it is used for similitudes, and an ignorance of beryl or of diamonds frequently closes the doors of understanding.[47]

As these examples suggest, Augustine laid considerable emphasis on familiarity with mirabilia, whose striking and memorable nature made them particularly suited to rhetorical uses of this sort.

These hermeneutic concerns underpinned a large body of medieval Christian writing on the wonders of the natural world. In some cases, medieval Christian authors took pre-existing catalogues of marvels and glossed them to bring out their moral sense. Thus the fourteenth- or fifteenth-century commentator on a twelfth-century poem enumerating the wonders of the world took the fearsome serpent Iaculus to stand for "wrath and mental furor," while the poisonous plant Sardonia, which caused its victims to die laughing, showed that "the joys of this world bring death." "Pliny once wrote these things," noted its author. "He told of wonders, and I speak of morals."[48] Similar material appeared in medieval bestiaries, herbals, and lapidaries (themselves the descendants of late antique originals), though to a varying degree.[49] Herbals, compilations of plants and their properties for medicinal purposes, tended to be practical in orientation and focused on relatively common plants. Bestiaries, by contrast, regularly juxtaposed wonderful animals with familiar ones, often giving them a moral or allegorical interpretation. For example, one of the most famous illustrated medieval bestiaries, produced in early thirteenth-century England, included such beings as the Ethiopian satyr and the Arabian phoenix, as well as a description of barnacle geese and the osprey taken from Gerald of Wales. In these works, many of the newer marvels imported from contemporary topographical writing into the traditional repertory derived from the *Physiologus* had not yet acquired allegorical meanings, unlike standbys such as the panther, elephant, or dog (fig. 1.5).[50] Richest of all in wonders were the lapidaries, treating as they did the marvelous powers of (mostly) exotic eastern stones and gems. The wide circulation, in the fifteenth century, of a lapidary attributed to Mandeville suggests the links between this literature and the literature of eastern travel.[51]

Herbals, bestiaries, and lapidaries were not only repositories of practical information, recreational reading, and religious lessons, but also preserved the kind of information that Augustine considered necessary for biblical exegesis. But Augustine's program was most clearly realized in the clerical encyclopedias of the thirteenth century. These works embraced all kinds of natural history, commonplace and marvelous, together with the general Augustinian message that everything in creation was wonderful — so that the world itself was, as Augustine put it, "beyond doubt a marvel greater and more wonderful than all the wonders with which it is filled."[52] The three most influential and widely circulated of these

1.5.1

1.5.2

Figure 1.5. Bestiary animals
1.5.1–2. *Bestiary*, MS Bodley 764, fols. 7v and 25r, Bodleian Library, Oxford (first half of 13th century).

According to this thirteenth-century bestiary, the breath of the panther is so sweet that "when the other animals hear his voice they gather from far and near, and follow him wherever he goes.... Thus our Lord Jesus Christ, the true panther, descended from heaven and saved us from the power of the devil" (fig. 1.5.1). The Indian manticore (fig. 1.5.2), in contrast, appeared as an unallegorized (and terrifying) wonder, devoid of moral meaning: "it has a triple row of teeth, the face of a man, and grey eyes; it is blood-red in colour and has a lion's body, a pointed tail with a sting like that of a scorpion, and a hissing voice. It delights in eating human flesh. Its feet are very powerful and it can jump so well that neither the largest of ditches nor the broadest of obstacles can keep it in."[4]

Latin encyclopedias were all composed in the 1240s by friars and teachers at schools of religious orders: Thomas of Cantimpré, lector in various Dominican convents; his fellow Dominican Vincent of Beauvais, who taught at the Cistercian abbey of Royaumont; and Bartholomaeus Anglicus, who composed his treatise "on the properties of things" for his students at the Dominican provincial *studium* of Magdeburg. These schools prepared their students for the practical and pastoral activities of spiritual outreach for which the orders of friars had recently been founded: converting and disciplining heretics, instructing the laity, reforming morals, and especially, preaching. The encyclopedias thus functioned as textbooks for those whose task was to communicate the biblical message in the wider world.[53]

The authors of these works were explicit about their aims, which they described with frequent references to Augustine. But they differed from one another in the aspects of Augustine's program they chose to stress. Bartholomaeus Anglicus specifically emphasized exegesis; he wrote that his work was to help its audience to understand the "enigmas of scripture, which have been transmitted and veiled by the holy spirit in the symbols and figures of the properties of natural and artificial things." For this reason, he included only those phenomena (many of which were relatively commonplace) mentioned in the Bible and its glosses.[54] Thomas of Cantimpré and Vincent of Beauvais cast their nets more broadly, emphasizing Augustine's more general point: the Christian's duty to wonder at creation and, by extension, at its Creator. As a result, they included many more phenomena than Bartholomaeus, describing these with frequent and explicit references to wonder. Of the "sea monsters" in his sixth book, for example, Thomas wrote: "they have been given by the omnipotent God to the wonder of the globe. For in this sense they appear very wonderful, since they are rarely offered to the sight of men. In truth it can be said, that God hardly acted so marvelously in any other things under heaven, except human nature, in which the imprint of the Trinity is to be seen. For what under heaven can appear more wonderful than the whale?"[55] Vincent explained why he had chosen the mirror as his governing metaphor: "mirror refers to almost everything worthy of contemplation [*speculationis*], that is of wonder [*admirationis*] or imitation."[56] His *Speculum naturale* was truly Plinian in scope, with close to four thousand chapters divided into thirty-two books, though still organized on biblical principles according to the order of the six days of creation.

Vincent invoked wonder at every opportunity, not only in his accounts of standard exotic mirabilia, but also in his rhapsodic descriptions of more commonplace things. He exclaimed over the extraordinary range in the size of living things, from the elephant to the gnat or the tiger to the turtle ("we wonder at the greatness of the one, the smallness of the other"); the remarkable generation of like by like; the extraordinary variety and beauty of the colors of flowers and gems.[57] Citing Augustine in Book XXI of *De civitate Dei*, he presented the created world as a spectacle of wonders, engineered for human pleasure and delight.[58] In emphasizing the wonderfulness of all creation, Vincent reflected Augustine's elaboration of wonder as a religious emotion. His rhapsodic tone, however, as well as his tendency to reserve his most ringing prose for exotica and other rarities, betrays the influence of the medieval topographical tradition. Augustine's wonder, introduced and framed in terms of God's power to subject the unfaithful to eternal torture, was appreciative but tinged with fear. Vincent's wonder, in contrast, was largely informed by pleasure, like that of Marco Polo, Odoric, or Mandeville.

Despite the non-Augustinian tone of some of their entries, the thirteenth-century encyclopedias were certainly used for Augustinian ends. Vincent's *Speculum* appeared in the 1307 catalogue of the library of the Dominican convent at Dijon, in the same section as biblical concordances, commentaries, and exegetical works, while the encyclopedias of Thomas and Bartholomaeus circulated in abridged manuscript versions for pulpit use.[59] Their contents, supplemented by material from Solinus, Pliny, Gervase of Tilbury, and the bestiary literature, appeared in compendia for preachers like the *Gesta Romanorum*. Compiled in the early fourteenth century for the use of English Franciscans, this included chapters on the basilisk, Armenian dragons, wonder-working springs in Sicily, and the exotic human races. Each of these received a moralized Christian meaning: the dog-headed Cynocephali signified ascetic preachers in hair shirts, for example, while the enormous ears of the Scythians stood for willingness to hear the word of God.[60]

Interest in this material was not confined to clerics. Vincent's elephantine encyclopedia remained untranslated, but over the course of the fourteenth and fifteenth centuries the works of Thomas and, especially, Bartholomaeus Anglicus found a broader, lay audience in new vernacular versions.[61] Bartholomaeus' work underwent a shift as it moved into the lay world. In his preface the author had emphasized only the religious and exegetical uses of his book. The prologues added by vernacular translators eliminated these passages entirely, substituting for them a far broader portrayal of the *De proprietatibus* as a general encyclopedia — in the words of its French translator, Jean Corbechon, "a general summa containing all things."[62] This shift also appeared in the illustrations that accompanied the most lavish French manuscripts, commissioned by princely patrons;

1.6.2

Figure 1.6. The audiences of Bartholomaeus Anglicus
1.6.1. Boucicaut workshop, Bartholomaeus Anglicus, *Le livre des propriétés des choses*, trans.
Jean Corbechon, MS 251, fol. 190v, Fitzwilliam Museum, Cambridge (c. 1415)
1.6.2. Boucicaut workshop, Bartholomaeus Anglicus, *Le livre des propriété des choses*, trans.
Jean Corbechon, MS fr. 9141, fol. 197r, Bibliothèque Nationale, Paris (c. 1410–14).

Although the illustrations in both of these early fifteenth-century manuscripts of Bartholomaeus Anglicus come from the same workshop, they show two different audiences for his work. In one (fig. 1.6.1), probably commissioned by Amadeus VIII, Count of Savoy, Bartholomaeus lectures on the properties of water to a group of four soberly dressed clerics and scholars. In the other, made for Béraud III, Count of Clermont, he demonstrates the properties of fish to a noble in aristocratic garb (fig. 1.6.2). Other images in the second manuscript include among Bartholomaeus' pupils King Charles VI and Béraud himself.

whereas earlier copies showed the philosopher lecturing on the properties of natural objects to attentive groups of students and clerics, later ones portrayed noble listeners, splendid in fur and jewels, including King Charles VI himself (fig. 1.6).[63]

Thus the Augustinian framework of the thirteenth-century encyclopedias fell away as their audience expanded and shifted to include lay and vernacular readers. In the process, the marvelous natural phenomena they contained shed their vestigial associations with the fear of divine retribution, to emerge as objects of unadulterated pleasure and fascination. In Book XXXII of his *Speculum naturale*, however, Vincent hinted at a more somber side of wonders when he wrote of monstrous births, in the sense of eunuchs or children with extra or missing fingers. Passing to the general topic of portents, he noted that God used such births, together with dreams and oracles, to reveal his future intentions and to warn of catastrophes to come.[64] Vincent devoted very little space to this topic, which fit poorly with his rapt vision of creation. But its presence signals another, darker strand in the wonder tradition that even he could not ignore.

Prodigious Individuals and Marvelous Kinds

Despite the variety of their interests, virtually all of the authors of high and late medieval topographical and encyclopedic treatises made a clear distinction between extraordinary individuals and marvelous species. Writing about snakes, for example, Thomas of Cantimpré distinguished the two-headed serpent he called the "ansibena" (an exotic species with a head at each end, as described by Solinus and Pliny) from the two-headed snakes mentioned by Aristotle, which occur "by a monstrosity of birth." "According to Aristotle," Thomas continued, "the cause of this monstrosity is as follows: if two seeds have been contained in one uterus without a wall between them, then a monstrous snake will be generated. However, this happens rarely among snakes, since their uteruses are extremely well disposed for multiple births."[65]

In stressing this difference, Thomas implicitly rejected the Augustinian interpretation of wonders. Augustine had omitted any reference to natural causes and denied any significance to the distinction between exotic species and monstrous individuals, in the service of a theology that deemphasized the autonomy of the natural order and treated each thing in the created universe as a direct and separate manifestation of divine will. Early medieval writers on marvels tended to follow Augustine in this matter. Thus the author of the eighth-century *Liber monstrorum* included among his races of Cyclops, Sirens, and seahorses a hermaphrodite he knew personally, as well as two examples of monstrous births described by Augustine: a man with two chests and two heads, and a man with un-usually shaped hands and feet.[66]

By the time of Thomas of Cantimpré, however, the penetration of ideas of natural order from Greek and Arabic natural philosophical sources, newly translated into Latin beginning in the twelfth century, had ren-dered the profound distinction between exotic species and monstrous individuals increasingly evident to philosophically trained European read-ers and writers. This distinction rested on a view of nature that treated it no longer as immediately reflecting divine commands — the Augustinian position — but as possessed of an independent internal order located in the chains of causes that produced particular phenomena. God had cre-ated the physical universe and the causal principles that moved it, notably the forms, the elements, the "prime qualities" of hot, cold, wet, and dry. He retained the prerogative to suspend this order at any moment, produc-ing miracles and other supernatural events, but under normal circumstances the physical world reflected the operation of autonomous sequences of causes and effects. In this view nature became an agent, predictable within certain limits — the powerful goddess personified in works like Alan of Lille's *De planctu Naturae* and Jean de Meun's *Roman de la rose*.[67] Despite their Augustinian allegiances, Thomas, Vincent, and Bartholomaeus Angli-cus all subscribed to this new view of the natural order, peppering their discussions with hardheaded explanations based on natural causation and derived from Galen, Aristotle, and the like.

This new set of ideas drove a wedge between the monstrous individual and the wonderful species. Both continued to qualify as wonders because of their rarity, but they otherwise differed in almost every way.[68] The lat-ter was a permanent and regular (if rare or exotic) feature of the physical world, generated by natural causes, while the former was a unique, super-natural, and usually ephemeral creation, directly dependent on the will of God. This difference appeared clearly in Gerald of Wales's *Topographia Hibernie*, one of the last medieval writers to intermingle both types of wonder in a single work. When Gerald described the unfamiliar species or permanent topographical oddities of Ireland — its barnacle geese and ospreys, its petrifying wells and excessive tides — he used the language of the marvelous. Describing such things benignly and appreciatively as "the wonderful works of nature at play," he noted that they derived their appeal from the natural human inclination to treasure the unfamiliar: "Only what is unusual and infrequent excites wonder or is regarded as of value."[69] He was much less sanguine, however, when he discussed mon-strous births or other singular events, such as the fact that ravens and owls bore their young out of season during the winter of 1185, which he saw as possibly foretelling "the occurrence of some new and premature evil."[70]

Although monstrous births occasionally appeared in topographical writing, they differed fundamentally from the staples of that genre both in meaning and in the emotions they evoked. The permanent topographical feature or exotic species was a natural marvel in the strict sense. A regular anomaly, possessed of a stable form and properties, it expressed rather than violated the created order of nature, enhancing the beauty and diversity of the world. Marvels of this sort might be unequally distributed geographically, but there was always some group of inhabitants for whom they represented nature's usual course. The appropriate Christian response to this kind of phenomenon was clearly appreciative wonder. In contrast, monstrous births and other *individual* anomalous occurrences (comets, meteorites, snow in summer, rains of blood) demanded another response entirely. Where Augustine, following Pliny and Cicero, had rejected the common interpretation of all such events as portents of immediately impending evil, his medieval successors reverted to the well-established religious tradition of prodigies they had inherited from both Greco-Roman paganism and the Hebrew Bible; these treated such occurrences as divine messages and signs of things (usually undesirable) to come.[71] As single and unique events, often of brief duration, their extraordinary nature was unmitigated and absolute, in contrast to that of the exotic species, which was relative to the experience of the observer; thus according to the common formula, the former were produced "against nature" (*contra naturam*).

This phrase, frequently invoked, also required clarification, and Isidore of Seville's somewhat tortured explanation of portents remained standard throughout the medieval period:

> Portents, according to Varro, are those things that appear to be produced against nature. But they are not against nature, since they happen by the will of God, since nature is the will of the Creator of every created thing. For this reason, pagans sometimes call God nature and sometimes, God. Therefore the portent does not happen against nature, but against that which is known as nature [*contra quam est nota natura*]. Portents and omens [*ostenta*], monsters and prodigies are so named because they appear to portend, foretell [*ostendere*], show [*monstrare*] and predict future things.... For God wishes to signify the future through faults in things that are born, as through dreams and oracles, by which he forewarns and signifies to peoples or individuals a misfortune to come.[72]

At the beginning of this passage, Isidore laid out the original Augustinian line on monsters: since God is responsible for the creation of every being in the universe, these only *appear* to be against nature and should consequently provoke no alarm. But toward the end, he came down squarely on the side of the monster as a special sign sent by God to warn of approaching evil. As later medieval encyclopedists, building on Isidore's account, hastened to point out, such individual monsters might well have natural causes — this is the thrust of Thomas of Cantimpré's reference to Aristotle on two-headed snakes — but in the case of portents, those natural causes combined at God's extraordinary will to produce a singular, and significant, effect.

We will treat the problem of causes in greater detail in Chapter Three. Here it is enough to say that, as Isidore's account suggests, the associations of the portent, unlike those of the marvelous race or species, were overwhelmingly negative and not in the least romantic or exotic. As a result, writers described them in entirely different terms. After a long and wonder-filled account of the East, replete with marvelous races of all sorts, James of Vitry began his complementary book on the West with the other sort of marvel:

> The head and mother of the faith is Jerusalem, as Rome is the head and mother of the faithful. For as the pain of the head echoes in the other members, so the Lord has indicated his wrath and indignation with various afflictions and scourges, because after the Holy Land came into the hands of the impious on account of our sins, God, the just avenger, the lord of punishment, has scourged the whole world, afflicting it with various troubles.... Monsters of vices and prodigies of abominations sprang up miserably and covered the entire globe.[73]

If the marvelous races were a phenomenon of the margins, an embellishment and completion of the natural order, individual monsters erupted in the Christian center, brought on by its corruption and sin. They were suspensions of that order, signs of God's wrath and warnings of further punishment; the appropriate reaction was not pleased and appreciative wonder, but horror, anxiety, and fear.

This reading of individual anomalous phenomena as prodigies was common in medieval Europe. Annals and chronicles were full of eclipses, conjoined twins, unseasonable thunderstorms, examples of peculiar animal behavior, and the like, often presented implicitly or explicitly as the precursors of dramatic, local, and usually catastrophic events: assassinations, epidemics, famines, fires, and wars.[74] Guibert of Nogent, a twelfth-century French abbot, reported numerous such occurrences as foreshadowing the 1116 revolt of Laon:

One man saw a moon-shaped ball fall over Laon, which meant that a sudden rebellion would arise in the city.... Moreover, I learned from the monks of Saint-Vincent that a tumult of evil spirits (as they thought) was heard, and flames appeared in the air at night in the city. Some days before, a baby was born who was double down to the buttocks; that is, he had two heads and two bodies right down to the loins, each with its own arms; double above, he was single below. After he was baptized, he lived three days. In short, many portents were seen to occur which left no doubt that they presaged the great disaster which followed.[75]

According to the English chronicler William of Malmesbury, among many others, the terrifying appearance of a brilliant comet heralded the Norman conquest of England and the downfall of King Harold in 1066.[76] The comet was immortalized in the Bayeux Tapestry (fig. 1.7), while Guibert's double infant — or one very like it — was depicted on a capital in the church of the Magdalene at Vézelay (fig. 1.8.1).

The principal difference between prodigies and marvelous species lay in their signification rather than their form. (Formally, for example, there is little to distinguish the individual hermaphrodite from the Androgynes of Africa, a whole race of beings of doubled sex.) The wonders of the East and other topographical marvels had no particular intrinsic meaning. God had made them at the beginning of time for his own reasons; they simply *were*, like foxes or Frenchmen or the Rock of Gibraltar, and they symbolized at most the power and wisdom of their Creator. They were "monsters" only by association — the term was only intermittently applied to such phenomena after the twelfth century — since they were not created to show (*monstrare*) anything in particular. Their only meaning was allegorical: they could be read as figures of some higher theological or moral truth, as when the author of the *Gesta Romanorum* moralized the exotic races or Gerald of Wales used the story of the spontaneously generated barnacle geese to argue (against the "unhappy Jew") the truth of the virgin birth of Christ.[77]

Monstrous individuals, portents, and prodigies, on the other hand, were rarely read allegorically and were treated not as symbols but as signs.[78] Temporary deviations from the natural order, they were deliberate messages, fashioned by God to communicate his pleasure or (much more frequently) his displeasure with particular actions or situations (such as the loss of the Holy Land to the infidels, in James of Vitry). Most monsters functioned solely as signifiers; for this reason, according to Isidore of Seville, they usually died immediately after birth.[79] They presaged divine punishment, which could be forestalled only by rapid repentance.

Figure 1.7. A prodigious comet
The Bayeux Tapestry, Centre Guillaume-le-Conquérant, Bayeux (Norman, late 11th century).

As part of the story of the Norman Conquest of England in 1066, the Bayeux Tapestry shows the comet — now known as Halley's Comet — that announced King Harold's defeat by William the Conqueror. A group of Englishmen "wonder at the star," according to the embroidered caption, while a messenger gives an attentive Harold the bad news. The ships in the border below the doomed king, presumably representing the Norman fleet, illustrate the impending political and military disaster.

53

1.8.1

1.8.2

Figure 1.8. Conjoined twins as prodigies
1.8.1. Capital, Basilica of the Sainte-Madeleine, Vézelay (c. 1120–32).
1.8.2. Relief from the façade of the Hospital of Santa Maria della Scala, Museo di San Marco, Florence (c. 1317).

Sculptures of this sort show the importance attached to publicly memorializing prodigious events so that their lessons might be taken to heart. Twins similar to those illustrated in the relief also appear in fig. 5.8, although the meanings attributed to them changed dramatically from the fourteenth to the eighteenth century.

The animal-human composite was one of the commonest types of monster to appear in medieval literature, and it shows with particular clarity the distinction between a monstrous individual and an exotic race. The significance of composite beings was encoded in their meaning rather than in their form. Hybrid races, such as the dog-headed Cynocephali or the horse-bodied Onocentaurs, were grounds for wonder; if they evoked fear, it was only because they were physically dangerous. The hybrid individual, on the other hand, evoked a horror arising clearly and explicitly from the violation of sexual norms. Mary Douglas's discussion of category-crossing as pollution, often invoked in this context, is not enough to account for this repulsion; if it were, medieval writers would have been as horrified by the Cynocephali as by individual hybrids of human and dog. As Arnold Davidson has argued, "horror is appropriate only if occasioned by a normative cause, the violation of some norm, as when the human will acts contrary to the divine will" — most notably through "unnatural" and bestial sex.[80]

Gerald of Wales's treatment of Irish hybrids makes this point clearly: while describing a creature born "from the intercourse of a stag with a cow" — animals not bound by the sexual prohibitions that governed humans — he merely noted that the nature of the cow was dominant in its progeny, which therefore "stayed with the herd" (fig. 1.9.1).[81] But when he came in the next chapter to a woman who had sex with a goat (fig. 1.9.2) — a wonder in and of itself — he erupted into invective:

> How unworthy and unspeakable! How reason succumbs so outrageously to sensuality! That the lord of the brutes, losing the privileges of his high estate, should descend to the level of the brutes, when the rational submits itself to such shameful commerce with a brute animal! ... Perhaps we might say that nature makes known her indignation and repudiation of the act in verse:
>> Only novelty pleases now: new pleasure is welcome;
>>> Natural love is outworn
>> Nature pleases less than art; reason, no longer reasoning
>>> Sinks in shame.[82]

Nature, here personified, was a highly normative concept for the writers of the High and later Middle Ages, but the norms she embodied were moral rather than physical. When, in Gerald's descriptions, the Irish embraced their own barbarous customs as "another nature," the author indicated their departure from the moral order.[83] The deformities of their offspring manifested that departure only incidentally, and the very same physical features, when manifested in a whole Indian or African people, did not count as deformities at all. The same assumptions informed the works of later topographical writers like Marco Polo and Odoric of Pordenone, who reserved their scorn or revulsion for the idolatry or cannibalism of Eastern races rather than their physical peculiarities, which they interpreted neutrally as stemming from climatic differences. When such peculiarities appeared in a European context and in isolated individuals, however, they had another meaning entirely. They corresponded to a rupture in the moral order, the product and the sign of sin.

For all these reasons, portents and prodigies, although occasionally classified with marvels as deviations from the familiar course of nature, evoked only horror. They also had an urgency that was completely lacking in the exotic races, as is clear from the case of a monstrous infant born near Florence in 1317. Giovanni Villani described the event in his chronicle:

> In the said year, in January, . . . there was born in Terraio di Valdarno di sopra a boy with two bodies; he was brought to Florence and lived more than twenty days. Then he died in the Florentine hospital of Santa Maria della Scala, first one body and then the other. And when it was proposed to bring him alive to the then priors, as a wonder [per maraviglia], they refused to allow him in the palace [of the city government], fearing and suspecting such a monster, which according to the ancients signifies future harm wherever it is born.[84]

Twenty-five years later, Petrarch vividly recalled the effect of this event in his own family. "Some friends in Florence sent a picture [of the child] to us in France, where we were staying," he wrote, "and a huge crowd of people came just to see it. I was seven years old when I saw the image in the hands of my father. When I asked what it was, he told me, showing it to me, and ordered me to remember it and tell the story to my (as he said then) sons. And I will indeed tell it to my nephews."[85] The birth of the twins was considered so significant that a commemorative bas-relief portrait was erected outside the hospital where they died (fig. 1.8.2).[86]

Local monstrous births, unlike distant peoples, required immediate decisions regarding attitude and behavior. Writers like Augustine and Thomas of Cantimpré could speculate at length on the human status of the monstrous races,[87] but the parents, the midwife, and the parish priest had to determine if a monstrous baby was human and should be baptized — and if so, whether as one person or two. The Florentine priors and population had to decide if they needed to engage in formal acts of public and private penitence to ward off whatever disaster was in store. For this reason, a monstrous birth required documentation, written and pictorial, and raised important questions concerning the credibility of the report.

1.9.1

1.9.2

Figure 1.9. Hybrids and unnatural sex
1.9.1–2. Gerald of Wales, *Topographia Hibernie*, MS 700, National Library of Ireland, Dublin
(13th century).

Regarding sex between humans and animals Gerald wrote, "Such crimes have been attempted
not only in modern times but also in antiquity, which is praised for its greater innocence and simplicity.... And so it is written in Leviticus: 'If a woman approaches any beast to have intercourse
with him, ye shall kill the woman, and let the beast die the death.' The beast is ordered to be
killed, not for the guilt, from which he is excused as being a beast, but to make the remem-
brance of the act a deterrent, calling to mind the terrible deed."[5]

Wonder and Belief

Credibility was less of an issue in the high medieval topographical literature, where mineral, plant, and animal species were concerned. For one
thing, the wonders of the far East and West lay very far away and had no
practical implications for most European readers and writers. This blunted
the question of their authenticity. In the absence of regular networks of
commerce and communication, the margins of the world occupied a
space almost wholly discontinuous from that of the European center. Like
romances, medieval books of topography and travel offered pleasure and
entertainment. They enlarged their readers' sense of possibility, allowing
them to fantasize about alternative worlds of barely imaginable wealth,
flexible gender roles, fabulous strangeness and beauty. Like novels or
movies today, they demanded emotional and intellectual consent rather
than a dogmatic commitment to belief.

Furthermore, medieval readers and writers shared an approach to truth
more complicated and multivalent than the post-seventeenth-century
obsession with the literal fact, the rise of which we discuss in Chapter
Six.[88] For them, truth could exist on various levels, both literal and figurative. Moral or spiritual meaning was at least as important as descriptive
accuracy, and wonder, as Caroline Bynum has argued, was a "significance
reaction."[89] James of Vitry reflected this more capacious approach to matters of truth and belief when he concluded an extended discussion of the
exotic Indian races with the following reflections:

> I have gone beyond sequential history [*praeter historiae seriem*] in including the
> preceding material in the present work. I have taken it partly from the his-
> tories of the Orientals and world maps, and partly from the works of the
> blessed Augustine and Isidore and also from the books of Pliny and Solinus. If
> by chance it appears incredible to some, I do not compel anyone to believe it;
> let everyone follow his own judgment. However, I do not consider that there
> is any danger in believing things that are not in opposition to the faith. For we
> know that all the works of God are wonderful, although those who observe
> them frequently, through familiarity and custom, are not moved to wonder.[90]

In this passage James drew a distinction between the limits of narrative
(or "sequential") history and the more ambitious enterprise in which he
was involved. Unlike contemporary chroniclers and annalists, or writers
of romances and travel narratives, he aimed to present a spatial and temporal map of the world, drawn from many textual sources and laden with
theological meaning. In invoking the authority of Augustine, Isidore,
Pliny, and Solinus, he was doing more than simply appealing to the power

60

of the written word in a society where literacy and textuality (especially in Latin) were associated with authority, although that was certainly also the case.[91] In many respects, he was expressing an attitude toward credibility and truth in narrative that resembled that of the ancient historians, including natural historians, and paradoxographers, as analyzed by Paul Veyne.[92] Like their ancient predecessors, most medieval encyclopedists and cosmographers saw themselves in the first instance as philologists, engaged in collecting and transmitting existing testimony, without constantly evaluating its truth or plausibility. As Veyne points out, this approach accommodated a much more heterogeneous audience than the modern historian envisages, embracing amateurs in search of entertainment, rhetoricians looking for striking images or examples, as well as natural philosophers and professional politicians or military men. In deference to this range of interests, ancient historians typically reserved their own opinions; "they do not express the truth itself," as Veyne put it, "it is up to their readers to form their own idea."[93] In this respect James of Vitry, like Vincent of Beauvais and the other encyclopedists, was only following the example set by Pliny and Solinus.

But as the passage from James's history intimates, Christian theology and piety added new presumptions in favor of belief. The world of medieval Christians, even more than that of ancient writers, was a world of wonders, heterogeneous over both time and space. Spatially, it had never been uniform: although it was governed by an overarching causal order, that order produced dramatic differences from place to place, most notably between the exotic margins and the Mediterranean center. But Christianity added a temporal dimension to this variability, emphasizing sudden irruptions of the marvelous into the course of everyday life in the form of miracles, prodigies, and other forms of divine communication. Thus myth and marvel were no longer confined to a distant land or a discontinuous past, as for Veyne's ancient writers. The dogmas of divine providence and omnipotence meant that from a strictly theological point of view, the most reliable antique criterion of plausibility — whether a particular report was consistent with the regularities of common experience — had largely lost its force.[94]

Quite the contrary, in fact: James of Vitry's world, like Augustine's, was such a perpetual spectacle of marvels that the principal danger for Christian observers, their sensibility dulled by familiarity, was to believe too little rather than too much. Belief in causally incomprehensible and naturally impossible events was the duty of the pious; such belief, with its concomitant emotion of wonder, bespoke a laudable stance of obedience, humility, and faith. In the famous formula of Tertullian, "It is certain because it is impossible."[95] Some Christian writers, especially those most influenced by Augustine, saw skepticism concerning wonders as the hallmark of the narrow-minded and suspicious peasant, trapped in the bubble of his limited experience, while belief characterized the pious, the learned, and the theologically informed.[96]

James of Vitry also followed Augustine in using the relativity of wonder to compel belief, a topos already present in Gervase of Tilbury's *Otia imperialia* and one that became standard in late medieval topographical literature. As Gervase had argued, Europeans were surrounded by wonderful phenomena, like the properties of lime or quicksilver, which only daily familiarity rendered unmarvelous and therefore credible; for the inhabitants of India or the Atlas, the lodestone and the phoenix were equally banal. Odoric used the same reasoning when he argued for the plausibility of the Eastern Scythian lamb by comparing it to the Western barnacle goose.[97]

Yet despite these cultural presumptions in favor of belief, medieval readers and writers clearly knew that reports of wonders could be falsified or mistaken. To counter such charges, for example, some manuscripts of Odoric of Pordenone's narrative appended an oath of accuracy that Odoric had supposedly taken before his father superior: "I, Friar Odoric of Friuli and of the Franciscan order, testify and bear witness to the reverend father Guidotto, minister of the province of Saint Anthony (being required by him on my obedience to do so) that I either saw with my own eyes all the things I have written above, or heard them from men worthy of credit."[98] Similarly, the prologue of Marco Polo's *Travels* protested its veracity, while the Latin translation of Mandeville circulated together with what purported to be a papal certificate declaring it to be true.[99]

Why were travel writers so much more concerned with credibility than the ancient and medieval compilers of encyclopedias, cosmographies, bestiaries, and the like? In part, they seem to have imported these issues from romance, albeit without any of the subtlety and irony with which they were treated by an author like Chrétien de Troyes.[100] Romance and travel narratives, like chronicles, belonged to what James of Vitry called "sequential history." They told specific stories, set in a particular time and place, rather than laying out a general cosmographical structure freighted with moral and theological meaning. Furthermore, because the appeal of such stories lay largely in the novelty and implausibility of their material, truth to fact was of greater concern. In part, too, their authors had to compensate for the fact that, as they were ostensibly recording their own experiences, they could invoke the authority of Pliny, Solinus, and Augustine only as confirmation. For travel writers, unlike encyclopedists and cosmographers, the margins of the world were topologically continuous with the European center; their own experience and credibility were at stake, and they needed to present their narratives as both literally and morally true.

For all these reasons, the travel narratives laid great emphasis on eyewitness experience. Gerald of Wales introduced his section on Irish mirabilia by noting, "I am aware that I shall describe some things that will seem to the reader to be either impossible or ridiculous. But I protest solemnly that I have put down nothing in this book the truth of which I have not found out either by the testimony of my own eyes, or that of reliable men found worthy of credence and coming from the districts in which the events took place."[101] Using these criteria, Gerald rejected observations by both Solinus and Bede concerning Ireland's bees and vineyards: "Neither would it be strange if these authors sometimes strayed from the path of truth," he wrote, "since they knew nothing by the evidence of their eyes, and what knowledge they possessed came to them through one who was reporting and was far away. For it is only when he who reports a thing is also one that witnessed it that anything is established on the sound basis of truth."[102]

This and many similar passages in the works of Gerald and his contemporaries belie facile generalizations about the medieval reliance on textual authority. Like Augustine,[103] medieval travel writers believed that the most credible evidence for marvels, in the absence of divine revelation, was personal, sensory experience (preferably of sight or touch); next best was the testimony of eyewitnesses certifiable as what Odoric called "persons worthy of credit" and Gerald, "reliable men." This set the travel writers apart from encyclopedists and compilers of bestiaries and lapidaries, whose enterprise fell into the very different "program of truth" (to use Veyne's phrase) of James of Vitry. The travel writers all stressed the claims of direct experience, continually lamenting the inadequacy of language to communicate the wonders they had personally observed. "'Tis marvellous," wrote Jordan of Sévérac about the many trunks of the banyan tree, "and truly this which I have seen with mine eyes, 'tis hard to utter with my tongue."[104] Likewise, Odoric concluded the account of this travels with the following words: "I have omitted many other things that I have not had set down, because they would seem almost incredible to others, unless they had seen them with their own eyes."[105]

Some of the earliest European travelers to Asia were indeed shaken by the lack of evidence for the exotic human races. On his return from visiting the Tartars in the 1250s, William of Rubruck wrote, "I asked about the monsters, or monstrous men, about which Pliny and Solinus wrote. They told me they had never seen such creatures, which led me to wonder greatly if it were true." Thirty years later, John of Montecorvino noted, "I asked and searched at length, but I found nothing."[106]

Emperor Frederick II and Gervase of Tilbury went one step further, setting out systematically to test the authenticity of certain natural wonders. Gervase tried in vain to tempt flies to land in a refectory renowned for repelling the insects, while, according to Albertus Magnus, Frederick sent messengers with orders to bring back one of his own gloves, after having immersed it in what was supposed to be a petrifying lake in Gothia.[107] He dispatched yet others to collect driftwood from the shores of northern Europe, in search of the famous barnacle geese (fig. 1.4.1). He reported the results of this last experiment in his great treatise, *De arte venandi cum avibus* (c. 1245–50): "On them we saw a kind of shellfish clinging to the wood. In none of their parts did these shellfish exhibit any form of a bird and, because of this, we do not believe this opinion unless we have a more convincing demonstration of it. It seems to us that this opinion arose because barnacle geese are born in such remote places that men are ignorant of where they nest."[108] In both these instances, Frederick II clearly privileged the sensory evidence of things even above the testimony of eyewitnesses. In this he, like the authors of travel narratives, echoed developments in contemporary legal practice as it moved from a probative regime based on proof (in the form of ordeals, single combats, or oaths) to one based on evidence, in which physical evidence trumped both oral testimony and written report.[109]

Relatively few people made concerted attempts to verify natural wonders, however. Most travelers hesitated to deny flatly the reality of wonders described by authorities like Pliny and Augustine and deeply embedded in the pictorial and intellectual tradition. Instead, they were content either to identify the wonders they knew from books with those they *had* observed — Marco Polo's description of what he calls the "unicorn" of Sumatra resembles nothing so much as a rhinoceros[110] — or to assume that if they had not personally seen the Scythian lambs or islands of Cynocephali, they had simply not gone quite far enough east. The fact that they had witnessed many actual marvels described in the wonder literature — elephants and rhinoceroses, orangutans and black people — only made the others all the more plausible.

In contrast to wonders as exotic species, wonders as prodigious individuals engaged yet another program of truth. Marvelous species belonged to the margins of the world, occupying for most European readers (and for all but a handful of travel writers) a twilight world of consent rather than belief. Prodigies, on the other hand, were phenomena of the center, engaging immediate human interests: they happened in a space and time continuous with daily reality, and they *mattered* on a scale and in a way the Cynocephali did not. At stake was more than a single merchant's or churchman's reputation for veracity: God sent prodigies as signs of disapproval and of imminent and widespread disaster, which could be averted only by swift and decisive action. Petrarch's relatives, like the rectors of the foundling hospital in Florence, had a strong investment in disseminating the news of monstrous births in Tuscany and ensuring that they were believed and commemorated in accurate detail.

The literal truth of prodigies was vitally important, as were the precise date and place of their occurrence, which determined what they meant and for whom. Contemporary accounts of individual portents showed even greater concern for questions of evidence than appears in travel writing, let alone in encyclopedias, bestiaries, and other such compilations. Like Villani on the Tuscan monster of 1317, their authors were often careful to record precisely not only the date and place (much easier for local phenomena than Eastern marvels, which were in any case timeless), but also the names of participants and witnesses. In addition, images were produced not only after the descriptions of observers, but often also by the observers themselves. The chronicle compiled by an anonymous fifteenth-century Parisian contains such examples. Describing a bloody spring that appeared under the bridge at the gate of Saint-Honoré in 1421, he specified the exact duration of the phenomenon (from the Sunday of the feast of Saints Peter and Paul to the following Wednesday),[111] and he was even more precise concerning the local birth of a pair of conjoined twins:

> Item, on June 6, 1429, two children were born at Aubervilliers who were exactly as you see in this image, for I myself truly [*pour vray*] saw them and held them in my hands: they had as you see two heads, four arms, two necks, four legs, four feet, but only one belly and one navel; two heads, two backs. They were christened, and were kept above ground for three days so that the people of Paris could see this great wonder. And truly [*pour vray*] more than ten thousand people, men and women, went from Paris to see them.... They were born at around seven in the morning and christened in the parish of Saint-Cristophe, and the one on the right was named Agnès and the one on the left Jehanne. Their father was Jean Discret and their mother Gillette, and they lived about an hour after baptism.[112]

As further testimony to the truth of this extraordinary event, the author included a sketch of the twins in his journal — the only time in a forty-five year period that he supplemented his written account with an image.

Such details do not mean that these accounts were necessarily any more reliable than those of Marco Polo or Odoric on the wonders of the East; the Parisian chronicler probably exaggerated the number of people who went to see the twins, and there are significant discrepancies between the descriptions of eyewitnesses and implausibilities in documentation produced immediately after the event.[113] But it is nonetheless clear that there was a heightening of concern when wonders moved from the margins to the center — from the world of otherness to the here and now.

Despite their fears, however, Europeans craved direct contact with wonders in all their myriad forms. Wanting to become eyewitnesses, they flocked to see the twins of Aubervilliers. The Parisian authorities had to close the gate and bridge of Saint-Honoré for two days to control the throng who wished to see the bloody spring. Examples of exotic species were equally fascinating and, in many cases, equally rare. Brought back from the East by travelers and merchants, imitated (or even forged) by the ingenious, and collected by the wealthy and the powerful, they formed part of a social and material culture of the marvelous that complemented the textual tradition.

CHAPTER TWO

The Properties of Things

In addition to being textual objects — things to think about and think with — natural wonders were also things in and of themselves: gems with marvelous properties, exotic plants and animals, and even human beings of unusual or unfamiliar appearance. Steeped in the aura of romance, these wonders called up images of the East, with its opulence, its refinement, and its fabulous wealth. Like other luxury products — drugs and spices, jewelry and metalwork, silk and other fine textiles — that formed the basis of the growing commercial contacts between Europe and Asia over the course of the High Middle Ages, the value of such mirabilia sprang in part from their scarcity in the European market. As a result, wonders were also commodities: to be bartered, bought, sold, collected, and sometimes literally consumed.

The desire for wonderful things was closely related to the textual tradition; texts both shaped and reflected the taste of their readers, creating the background against which objects were coveted or appraised. When Suger of Paris admired the gems of the reliquaries preserved in his monastery's treasure, he evaluated them in terms of the lapidary literature, as we will describe shortly. Frederick II supplemented his own treatise on birds, based partly on his experience and partly on his reading, with an extensive menagerie stocked with hunting leopards, Arabian mules, an elephant, a giraffe, and a racing camel given to him by the Sultan El-Kamil.[1] Many of the most admired and collected naturalia, such as griffin claws and unicorn horns, came ostensibly from animals known only or primarily through images and texts.

The collecting and manipulation of marvelous objects was thus tied to the elite literary and intellectual culture that we described in Chapter One. Yet the material culture of the medieval marvelous was more complicated and more resonant than the world of its texts. Texts could be read, individually or communally, and the associated images pondered and admired; actual objects from the exotic margins additionally carried with them the sense of unmediated contact with another world. Magical powers were often attributed to rare and exotic natural objects — for example, unicorn horns or Eastern gems such as the agate or the beryl were said to aid in childbirth, neutralize poison, or cure disease. In addition, wonders could be — and often were — put on display, used to inspire or impress, and manipulated as part of the pageantry of temporal and ecclesiastical power. The possession and control of wonders represented (and, in part, constituted) the wealth and power of those who owned them; on a more abstract level, their rarity or uniqueness reflected the rarity and uniqueness of their proprietors, conceived in terms of nobility and cultivation. In the eyes of high and late medieval beholders, there was a hierarchy of natural objects that corresponded to and, in some ways, naturalized the hierarchy of persons: the vegetable kingdom, for example, ascended from lowly tubers and root vegetables, associated with peasants, to rare and exotic fruits.[2] More than anything else, wonders were nature's noblest creations, and they enveloped those around them with an aura of nobility and might.

For these reasons marvels had strong courtly associations, though the cultural purposes served by those associations shifted with the gradual transformation of the European elites. Initially the monopoly of kings and high nobility, both lay and ecclesiastical, wonders increasingly fascinated the growing ranks of urban patricians and professionals as well. In this chapter we will describe the late medieval culture of marvelous objects, both natural (as we depict in our first section) and artificial (which drew on the properties of natural wonders and imitated their magical powers, as we analyze in our second). We will end with a case study in the culture of the marvelous that focuses on the ways in which the fifteenth-century dukes of Burgundy mobilized wondrous objects, texts, and rituals in a brilliant display of cultivation and power.

Collecting Wonders

Medieval collections bore little resemblance to early modern or modern museums. They functioned as repositories of wealth and of magical and symbolic power rather than as microcosms, sites of study, or places where the wonders of art and nature were displayed for the enjoyment of their proprietors and the edification of scholars and amateurs. The treasure of the French royal abbey of Saint-Denis, although virtually unparalleled in its richness, was broadly typical of other medieval collections both in the kinds of objects it contained and in the presence of objects and materials that belonged to the canon of natural wonders.[3] While some medieval collections were the work of individual princes and prelates, most others were, like that of Saint-Denis, the fruit of generations or even centuries of accumulation by royal families and ecclesiastical institutions. Like the other great royal and church treasures, the Saint-Denis collection included hundreds of objects. At its core lay the relics of the Christian saints and the jeweled reliquaries that housed them, as well as a wealth of liturgical objects (elaborately worked missals, for example, together with reliquaries, vessels, embroidered miters, pallia, and other vestments), coronation regalia and the insignia of the kings of France (golden spurs, jeweled swords and scabbards, scepters, and a number of crowns), and jewelry and other small decorative objects (brooches, cameos, a set of eleventh-century ivory chess pieces). But the collection also contained a number of mirabilia and exotica (fig. 2.1), such as the so-called horn of Roland, carved from an elephant tusk; a griffin claw (or bison horn) in an early thirteenth-century French mounting; and the abbey's famous unicorn horn (or narwhal tusk), originally mounted on a column of gilded copper inside the abbey church and measuring six-and-a-half feet long.[4]

Saint-Denis's priceless unicorn horn was a real rarity, but elephant tusks, griffin claws, and griffin eggs were a staple of collections of this sort, together with various magical Eastern stones. For example, the eleventh-century monastery of Limburg, attached to the Salian dynasty, owned two nautilus shells set in gold and silver and six ivory hunting horns,[5] while a 1383 inventory of the shrine of Saint Cuthbert at Durham listed, in addition to hundreds of relics: two griffin claws (one currently in the collection of the British Museum); an "eagle stone" (a kind of African geode described by Pliny); a "beryl, white and hollow, of wonderful structure"; and no less than eleven "griffin eggs," including one "ornamented and cut in two."[6] Griffin eggs (or ostrich eggs) were a perennial favorite in medieval collections — they seem to have been one of the more easily available and less expensive mirabilia — and this last item was doubtless one of the many ostrich-egg reliquaries that appeared in increasing numbers from the ninth century on (fig. 2.2.1).[7] Later goldsmiths incorporated even more exotic objects into their reliquaries: not only griffin eggs and claws, but also coconuts, coral, nautilus shells, and sharks' teeth.[8] Thus wonders were collected for their own sake, like Saint-Denis's unicorn horn, or used to ornament receptacles for the still more wonderful relics of Christian saints.

Relics belonged to a different realm from marvels, that of the divine or miraculous rather than that of the natural order. Nonetheless, the two had clear affinities. Both tended to have Eastern origins: important relics were often gifts from Oriental sovereigns or from European princes who had acquired them on pilgrimage or on Crusade. They could also be bought

2.1.1

Figure 2.1. Wonders in the treasure of Saint-Denis

2.1.1. Boucicaut workshop, *Livre des merveilles (Marco Polo)*, MS fr. 2810, fol. 88r, Bibliothèque Nationale, Paris (c. 1412–13).

2.1.2. Griffin claw, Bibliothèque Nationale, Cabinet des Médailles, Paris (mounting from first half of 13th century).

2.1.3. Horn of Roland, Bibliothèque Nationale, Cabinet des Médailles, Paris (11th century).

The mounted griffin claw and carved elephant tusk (figs. 2.1.2–3) owned by the abbey of Saint-Denis, together with its unicorn horn, not shown here, were typical of the objects in many ecclesiastical collections. Their physical presence was taken to confirm the existence of animals otherwise known only through the accounts of Eastern travelers like Marco Polo. This image from the *Livre de merveilles* (fig. 2.1.1) shows the island of Madagascar as described by Polo, with its ferocious griffins and the elephants on which they preyed — all represented in accordance not with Polo's descriptions, but with contemporary artistic convention.

2.1.2

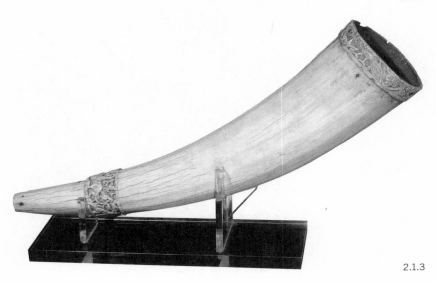

2.1.3

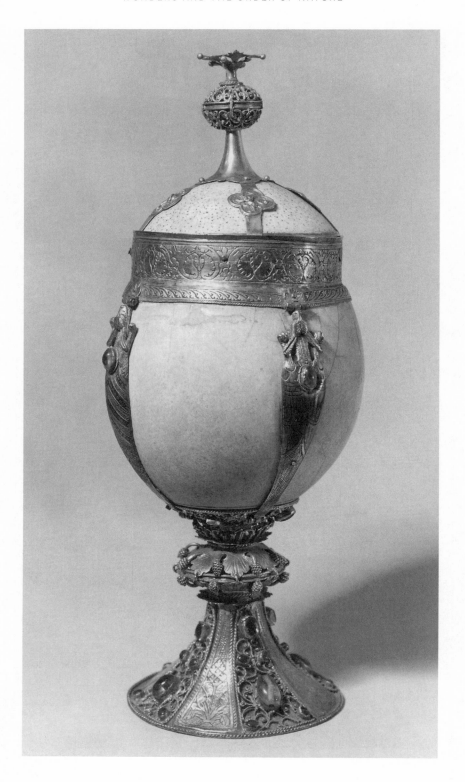

2.2.1

2.2.2

Figure 2.2. Ostriches and ostrich eggs
2.2.1. Ostrich-egg vessel, Halberstadt Cathedral Treasure, Halberstadt (first decades of 13th century).
2.2.2. *Bestiary*, MS Bodley 764, fol. 67r, Bodleian Library, Oxford (first half of 13th century).

Like griffins and elephants, ostriches produced exotic objects that could be made into reliquaries and ritual vessels (fig. 2.2.1). Their sacred aura was enhanced by the symbolic associations of the ostrich. According to this English bestiary (fig. 2.2.2), whose artist had evidently never seen an ostrich, the bird awaits the appearance of the Pleiades each year. "When, in about the month of June, it sees those stars, it digs in the earth, lays its eggs and covers them in sand. When it gets up from that place, it at once forgets them and never returns to its eggs.... If the ostrich thus knows its proper time, and forgets its offspring, laying aside earthly things to follow the course of heaven, how much more, o man, should you turn to the prize of the summons from on high, for which God was made man."[6]

on the open market, particularly after the sack of Constantinople in 1204, when objects from the Byzantine imperial collection fed an already thriving European trade in Eastern relics and other objects.[9] Both were rare, therefore expensive, and both had marvelous properties to heal or harm. Furthermore, both evoked wonder, which embraced both the pleasure afforded by the exotic, novel, and luxurious, and the awestruck veneration owed to the omnipotent Christian God.

There is no evidence that such substances were prized for cognitive or philosophical reasons. Medieval inventories and descriptions of these treasures showed little interest in classification or in relationships among the objects they contained. Instead, they tended to emphasize the preciousness of the materials of which those objects were made, their monetary value, and their provenance.[10] The medieval collection, in other words, was not a *musaeum* but a *thesaurus* ("treasure" — the term most commonly used to refer to it) in the sense of a repository of economic and spiritual capital.[11] Thus throughout the Middle Ages (and long thereafter) the abbots of Saint-Denis treated the monastery's treasure as a monetary reserve, selling off its contents as needed to weather famines, ransom vassals, buy off foreign occupiers, or defray their own expenses. Other churches mounted traveling exhibitions of their treasures to generate income by stimulating the generosity of the faithful, or used them as collateral in raising loans.[12] From this point of view, rare substances such as unicorn horns and ostrich eggs could be as valuable as gold and gems.[13]

More importantly, medieval collections were also reservoirs of power. This was true not only in a symbolic sense — many of the objects were gifts from patrons, vassals, and friends (often on the occasion of marriage contracts, treaties, or oaths of allegiance), which both cemented and represented political, social, and military alliances — but in a literal sense as well: they contained magical objects that controlled mighty natural and supernatural forces. Such objects included not only the religious relics that formed the core of most medieval treasures — relics were a principal channel through which divine power was made available to ordinary Christians, to heal, promote fertility, and so forth — but also many of the natural substances mentioned above. Indeed, most of the naturalia in medieval collections were renowned for their occult properties. Chief among them was the unicorn horn, which was thought to neutralize poison either by contact or when administered in powdered form. The Saint-Denis horn may well have been used in this way in the Middle Ages: as late as 1657, when skepticism about the existence of the unicorn itself was mounting, petitioners to the abbey were allowed to drink water in which the horn had soaked.[14] Similarly, one of the two great unicorn horns in the treasure

74

of the basilica of San Marco in Venice was given to the church in 1488 by the son of a wealthy jewel merchant and bore a silver gilt handle with the inscription, "John Paleologus, Emperor. Unicorn horn good against poison." The horn itself shows clear signs of having been scraped – a practice later forbidden except by unanimous vote of the Venetian Council of Ten.[15]

Private collectors also prized unicorn horns as antidotes to poison, a testimony to the climate of fear in many late medieval courts. Thus the 1413 inventory of John, Duke of Berry included among its many references to this marvelous substance "a great golden cup that breaks down into three parts, and has at the bottom a piece of unicorn horn and other things against poison that the king of England gave to the duke."[16] Other marvelous substances also acted against poison: the same inventory mentions the Duke's bezoar, "decorated with gold, hanging from three little golden chains, which the Maréchal of Boucicaut sent as a gift to Monseigneur,"[17] as well as a "large greenish stone, unset, which protects from poison." (Next to this last item a scribe noted, "this stone was afterward found to be extremely ordinary and of little value, and on account of that it was not appraised, but given to several servants of the said duke.")[18] Other rare natural objects were thought not to neutralize poison but to reveal it, usually by sweating in its presence. "Serpents' tongues" (or fossilized sharks' teeth) were particularly prized for this function, and they often adorned saltcellars or decorative poison detectors known as "espreuves".[19] The inventory drawn up for Charles V of France in 1380 reveals additional wonderful naturalia that were often collected for their magical properties: stones to aid in childbirth (usually eagle-stones), a belt made of lionskin ("for diseases of the kidneys"), and an engraved stone against gout.[20]

According to contemporary theory, the magical powers possessed by such things were, in contrast to those of relics, purely natural in origin.[21] The East was supposedly fertile in such objects: according to the thirteenth-century natural philosopher William of Auvergne, "In parts of India and other adjoining regions, there is a great quantity of things of this sort, and on account of this, natural magic particularly flourishes there."[22] Gerald of Wales was even more explicit, arguing that the marvels of the spartan but salubrious West – by which he meant Ireland – were superior to all the fabulous gems and spices of the East, which he called a "well of poisons." He attributed the concentration of magical natural wonders in Asia and Africa to the work of a provident creator: "Nature has indeed provided that where there were many evils, there should be many remedies against those evils."[23]

The power of precious stones, like that of other marvelous natural objects, was not only symbolic and magical; it had a moral and religious dimension as well.[24] The contents of medieval treasures served as objects of meditation. It was well known that God had proclaimed his might in the book of nature as well as in scripture, and encyclopedic and clerical writers followed Augustine in especially praising the wonders of nature as testimony to the wonderfulness of their omnipotent creator.[25] Collecting practices could reflect the same attitude. Thus in 1276, the wife of the Hapsburg ruler Rudolf I sent a porcupine to the Dominicans of Basel, according to a contemporary chronicler, so that they "might wish to recognize the marvelous nature of God."[26]

The great twelfth-century abbot of Saint-Denis, Suger of Paris, testified to his own use of the treasure of Saint-Denis for similar purposes. "We often contemplate these different ornaments, both new and old, out of sheer affection for the church our mother," he wrote in a treatise on his activities as abbot (c. 1148). He went on to describe how, whenever he beheld on the altar the "wonderful" (*ammirabilem*) jeweled cross of Saint Eloi and the shining screen reliquary known as the Crista (fig. 2.3), he was reminded of a famous passage in Ezekiel:

> Then I say, sighing deeply in my heart: *Every precious stone was thy covering, the sardius, the topaz, and the jasper, the chrysolite, and the onyx, and the beryl, the sapphire, and the carbuncle, and the emerald.* To those who know the properties of precious stones it becomes evident, to their greatest wonder [*ammiratione*], that none of these is missing except the carbuncle, but that they abound most copiously. Thus, when — out of my delight in the beauty of the house of God — the loveliness of the many-colored gems has called me away from external cares, and worthy meditation has induced me to reflect on the diversity of their sacred virtues, transferring that which is material to that which is immaterial: then it seems to me that I see myself dwelling, as it were, in some strange region of the universe which neither exists entirely in the slime of the earth nor entirely in the purity of Heaven; and that, by the grace of God, I can be transported from this inferior to that higher world in an anagogical manner.[27]

In addition to magical powers, the "properties" of rare and exotic Eastern gems included their scriptural associations and the fact that each corresponded to a particular Christian virtue in the lapidary tradition; this allowed them to serve as mediating objects that bridged the physical and the spiritual realms, elevating Suger, through the emotion of wonder, to a state of ecstatic devotion.[28]

Despite their powerful spiritual qualities, the wonders of the Crista were not generally available for popular contemplation. As abbot, Suger had access to them at his leisure, so he could meditate on them in what he called "silence and perpetual remoteness." As he noted, "the curious are not admitted to the sacred objects,"[29] though he gladly made exceptions for princes, prelates, nobles, and other benefactors of the abbey. But ordinary laymen had to wait for one of the special festivals when the treasure was exhibited to the avid multitude, resulting in intense and sometimes rowdy scenes (fig. 2.4). According to Suger, it often happened that "no one among the countless thousands of people because of their very density could move a foot [and] no one, because of their very congestion, could [do] anything but stand like a marble statue, stay benumbed or, as a last resort, scream."[30]

Such public occasions were extremely rare, however, and the very limited role of display reflected the special nature of the medieval collection. As spiritual and economic capital, it had to be hoarded and secured from theft, while as a collection of magical objects, it had to be protected from exhaustion and overuse. As Richard Trexler has argued, "indiscriminate and repeated unveiling of social and divine power by random individuals decreased the devotion of the people and therewith the social efficacy of the object. The less dignified the viewer . . . , the less devotion it could command."[31] Thus when the king arrived for the consecration of the rebuilt church of Saint-Denis the populace was kept outside, according to Suger, "with canes and sticks."[32] By restricting access to the treasure, the abbot enforced its rarity and novelty, guaranteeing that the wonder it elicited would remain intact.

If the bejeweled objects in Suger's treasure were too precious for unlimited popular viewing, there were nonetheless other natural wonders permanently on display. Some churches owned giant eggs, teeth, and bones, which were suspended from walls and ceilings.[33] Examples included Saint-Denis's unicorn horn on its column; three "giant's teeth" displayed in the fourteenth century in the church of the Annunziata in Trapani (according to Boccaccio, the rest of the giant had fallen into dust); two whale ribs given to the church of All Saints in Wittenberg by the duke of Pomerania in 1331; and the countless ostrich eggs that were a stock part of such displays (fig. 2.5).[34] For the most part—and with the signal exception of the unicorn horn—these things had neither magical properties nor great monetary value, and they were left relatively unadorned. But this does not mean that they lacked emotional effect. Writing on church treasures in the later thirteenth century, Bishop Guillaume Durand gave such practices a convincing rationale, observing that "in some churches they are accustomed to hang two ostrich eggs and other things of this sort, things that prompt wonder [admirationem] and are

77

2.3.1   2.3.2

Figure 2.3. The Crista of Saint-Denis
2.3.1. M. Félibien, *Histoire de l'abbaye de Saint-Denis* (Paris, 1706), Plate IV.
2.3.2. Roman intaglio, Bibliothèque Nationale, Cabinet des Médailles, Paris (late 1st century).
2.3.3. Etienne-Eloi de Labarre, watercolor of the Crista, Bibliothèque Nationale, Cabinet des Estampes, Paris (1794).

Part of the treasure of Saint-Denis since at least the beginning of the tenth century, the ninth-century Crista was melted down in 1794, after the revolutionary Commission for the Arts had ordered it documented in this watercolor (fig. 2.3.3). The engraving from Félibien's book (fig. 2.3.1) shows it as it as it appeared in the early eighteenth century, stored with other items of the treasure — some dating from Suger's abbacy — in one of the abbey's five armoires. Many early medieval works of this sort incorporated engraved gems and cameos from the Roman era; in this case the Crista's finial includes an intaglio showing Julia, daughter of the Emperor Titus (fig. 2.3.2).

2.3.3

78

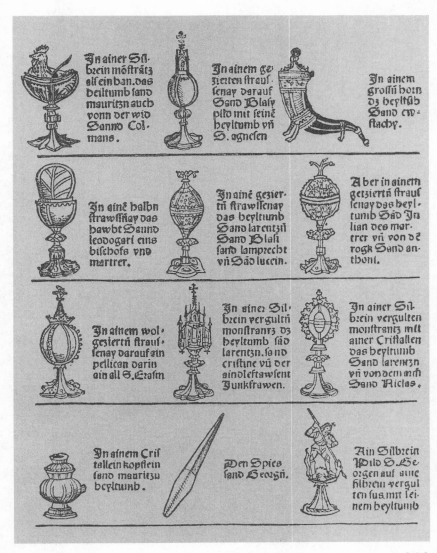

2.4.1   2.4.2

Figure 2.4. Church treasure on display
2.4.1–2. *Wiener Heiltumbuch* (Vienna, 1502).

In the late fifteenth and sixteenth centuries, inventories and advertisements of the contents of the treasures of a number of well-endowed German churches were issued as printed books. In 1485, the cathedral of St. Stephan in Vienna erected a separate building to house its collection and to serve as a stage for its periodic display (fig. 2.4.1). Among many other items, the cathedral owned five reliquaries made from ostrich eggs, here described in some detail (fig. 2.4.2), and one—made from the kind of horn often identified as a griffin claw (cf. fig. 2.1.2)—that contained the relics of Saint Eustachius.[7]

2.5.1   2.5.2

Figure 2.5. Hanging wonders
2.5.1. Detail of Vittore Carpaccio, *The Vision of the Prior of Sant'Antonino in Castello*, Galleria dell'Accademia, Venice (c. 1490–1500).
2.5.2. Mounted ostrich egg, Museo degli Argenti, Palazzo Pitti, Florence (late 15th and 16th century).

On the left of this church interior (fig. 2.5.1), Carpaccio shows numerous ostrich eggs hung among ex-votos, many in the form of parts of the body, donated in gratitude for cures and other miraculous deliverances. Several ostrich eggs mounted for hanging survive in church treasures from the late Middle Ages and Renaissance, including this Italian example (fig. 2.5.2). Such eggs might be imported from north Africa by Italian merchants or produced on the spot by the ostriches that formed part of the menageries of wealthy collectors, like the Strozzi in Florence.[8]

2.6.1

Figure 2.6. More hanging wonders
2.6.1. Wooden crocodile, Portal of the Lizard, Cathedral of Seville, Seville.
2.6.2. Stuffed crocodile, Oiron, Chapel of the Château of Oiron.

Some medieval and Renaissance churches displayed not only ostrich eggs, but other, more spectacular natural wonders. The crocodile that gave its name to the Portal of the Lizard (fig. 2.6.1),
which leads from the cloister into the Seville cathedral proper, was given to King Alfonso X in
1260 by the Sultan of Egypt. When it died, it was dried and hung before the portal, where it was
eventually joined by other items of interest such as an elephant tusk, visible on the wall behind
it, and the reported tomb of the Cid.[9] The crocodile eventually decayed and was replaced by a
wooden replica. The provenance of the crocodile in the sixteenth-century chapel of Oiron (fig.
2.6.2) is unknown.

84

2.6.2

rarely seen, so that by them the people are drawn into church and are more affected."[35] Like Suger, Durand emphasized the religious dimensions of wonder in its double aspect of pleasure inspired by rarity and awe inspired by the remarkable multiplicity and variety of God's creation. Like Suger with his gems, furthermore, he stressed the moral and exemplary aura of natural marvels, invoking the ostrich as an emblem of conversion in terms familiar from bestiaries and encyclopedias (fig. 2.2.2).[36]

Up to this point, we have emphasized the commonalities in the assumptions and practices that explain the presence of wonders in medieval collections. There were, however, significant shifts from the High to the later Middle Ages. For one thing, over the course of the fourteenth and fifteenth centuries the numbers and types of collectors grew and diversified. In the twelfth century, great collections were the monopoly of monarchs like Henry the Lion and wealthy religious foundations such as the monastery of Saint-Denis.[37] Krzysztof Pomian has called them "collections without collectors," the work of dynasties and institutions rather than individuals.[38] During the fourteenth and fifteenth centuries, other groups began to make their mark in this area. These included, in northern Europe, nobles below the rank of royalty — John, Duke of Berry, is the salient example — and, especially in Italy, members of the new urban elites, whose fortunes reflected a thriving commercial economy and whose tastes reflected the novel values and interests of an emergent urban and civic culture.[39] The members of this group ranged from urbanized nobles and merchant princes, such as Pope Paul II or Lorenzo di Piero de' Medici (both eager to model themselves on northern royalty and aristocracy) to more modest collectors like Petrarch and the Venetian notary Oliviero Forzetta, whose fourteenth-century inventory also survives.[40]

Although these new kinds of collections showed strong continuities with the earlier examples in their emphasis on small objects and precious, often exotic materials, there were nonetheless important changes. Alongside the relics and regalia — sometimes displacing them altogether — we find antiquities, works of contemporary artists, and (in some cases) an increasing number of purely natural, albeit still wonderful, objects, apparently valued for their own sake and not for their jeweled settings or their magical powers. The practice of suspending natural items in churches gathered momentum in the course of the fifteenth century, when ostrich eggs and whale ribs were joined by meteorites and, most notably, crocodiles (fig. 2.6).[41] Among private collectors, John, Duke of Berry, owned a scattering of unusual natural objects: an ostrich egg, a snail shell, seven boars' tusks, a porcupine quill, a giant's molar, a large serpent's jaw, a coconut shell, a number of pieces of red coral, a white bearskin, and at

Figure 2.7. The Duke of Berry as connoisseur
Boucicaut workshop, Bartholomaeus Anglicus, *Le livre des propriétés des choses*, trans. Jean Corbechon, MS fr. 9141, fol. 235v, Bibliothèque Nationale, Paris, (c. 1410–14).

John, Duke of Berry, was a famous collector and connoisseur of jewels and gems. Book XVI in the French translation of Bartholomaeus Anglicus prepared for Béraud III (see fig. 1.6.2), on the properties of precious stones and metals, is illustrated by a painting that depicts the Duke studying a number of gems offered to him for purchase.

least three whole unicorn horns.[42] Although some of these were magical objects (the horns, the coral) and were appraised accordingly, others were listed as being "of no or little value."[43] The presence of these last suggests that the duke had at least a passing interest in natural mirabilia and *curiosa* for their own sake. To these we can add the various animals in his famed menagerie, which included an ostrich, a dromedary, a monkey, and twelve peacocks.[44]

The ownership of rare and unusual objects served to reinforce social, political, and religious hierarchies. Why was this the case? In part, of course, the answer lay in their expense. Not all scarce objects were costly, but all costly objects were scarce, and they could therefore be used as a symbol and (to a select audience of the rich and powerful) a display of wealth. In part, too, it lay in the particular charisma of natural wonders, many of which, as we have argued, were thought to have magical and protective powers. The owner of such objects could control that power, which was normally thought to be dependent on physical contact or proximity, by restricting access to certain chosen individuals, unleashing it on the unwary, or monopolizing it for himself.[45] In addition to the ability to possess and deploy, there was the ability to appreciate and interpret — skills which, as unusual natural objects came to enjoy an established place in the realm of religious and political symbolism, acquired a meaning of their own. Men like Suger of Paris and the Duke of Berry flaunted their own connoisseurship, as distinguishing them from their peers. Not only did they have the intellectual formation that allowed them to identify the specific magical properties, moral meanings, and scriptural associations of gems and other wonderful natural objects, as Suger did with the Crista; they could also make fine judgments as to what was a proper object of wonder and what was not (fig. 2.7).[46] "Marvel not at the gold and the expense," wrote Suger (of the doors of his new church, though the same considerations could apply equally to smaller objects) "but at the craftsmanship of the work."[47]

Artificial Marvels

The objects in medieval collections testify to the close association between the wonders of nature and the wonders of human art. Sometimes the line between these was blurry — the thirteenth-century philosopher Albertus Magnus was doubtless not the only person to question whether an ancient cameo was a work of art, like a medal, or a work of nature, like a fossil.[48] Often wonders of art and nature were combined in the same piece. Like the wonders of nature, the wonders of art were highly textualized objects, associated with the marvels of the East. The far West possessed only one really famous artificial marvel: Stonehenge, whose stones were supposed to have been quarried by giants in Africa and first erected in Ireland before being transported to England by the enchanter Merlin.[49] In general, the marvels of art came from Africa and Asia, lands believed far to surpass Europe not only in natural variety and fertility, but also in fertility of human invention.

We have already described typical members of the textual canon of natural wonders as it appears in the works of medieval writers; there was a similar canon of wonders of art. This was well established by the middle of the twelfth century, as appears in two of the earliest great French romances: *Eneas* and the *Roman de Troie*. As described in *Eneas*, for example, the walls of the citadel of Carthage were made of marble and adamant painted a hundred colors and topped with three rows of magnets to incapacitate armed attackers; "the magnet is of such a nature," wrote the poet, "that no armed man could approach without being drawn to the stones, and as many halbardiers as came were always attracted to the wall."[50] The tomb of the Amazon Camilla incorporated "a hundred marvels [*cent mervoilles*]," including not only spectacular carved ornaments, but also its own defensive magnets, a magic mirror that revealed the approach of enemies, a sarcophagus hermetically sealed with cement made of ground gems moistened with serpents' blood, a cushion for Camilla's head stuffed with caladrius feathers, an ever-burning lamp made of asbestos, and a metal archer set to loose an arrow and extinguish the lamp should the tomb be disturbed.[51]

Benoît de Sainte-Maure's *Roman de Troie* featured a similar but even more astounding array of artificial wonders, including not only tombs "rich and strange and marvelous [*merveilos*]" of the sort described in *Eneas*, but also a seamless magic cloth decorated with jewels and fabricated in India "by necromancy and marvel" from the precious multicolored skin of the dindialos, a rare eastern animal hunted by the Cynocephali.[52] Hector's sickroom, the famous alabaster "Chamber of Beauties," was described at length by Benoît:

> [This] glistens with Arabian gold and the twelve twin stones which God decided were the loveliest of all when he gave them the name "precious stones" — sapphire and sard, topaz, chrysoprase, chrysolite, emerald, beryl, amethyst, jasper, ruby, precious sardonyx, bright carbuncle and chalcedony.... No other source of light was needed, for the Chamber on a dark night far outshines the very brightest summer day. The windows are made of green chrysoprase and sard and fine almandite and the frames are moulded in Arabian gold. I do not intend to recount or to speak of the many sculptures

and statues, the images and paintings, the marvels and the tricks there were in various places, for it would be tiresome to listen to.[53]

Nonetheless, Benoît went on to describe the four greatest "marvels and tricks" in Hector's chamber: four life-sized gold and silver automata, perched on top of columns made of precious materials. One held a mirror that allowed people to adjust their dress and behavior; one played every sort of instrument and periodically strewed the room with flowers (later swept up by a mechanical eagle); one had a censer made of topaz and filled with salubrious and sweet-smelling gums; and a female acrobat "performed and entertained and danced and capered and gambolled and leapt all day long on top of the pillar, so high up that it is a wonder it did not fall."[54]

What made a work of art marvelous, as characterized in these texts? In the first place, many artificial wonders exploited and depended for their operation on the magical properties of exotic substances that were already part of the canon of natural wonders: magnetic lodestones or luminescent carbuncles, the prophetic caladrius or the mysterious dindialos, precious spices or medicinal gums. To fabricate artificial wonders required access to such products, as well as deep knowledge of their natural powers. Thus the writers of romance invariably described their makers as scholars and magicians rather than as artisans; Benoît called them "wise and learned men, well versed in the magic arts."[55] Most wonders of art were what we might call wonders of engineering. In addition to being decorative, they harnessed powerful natural forces to produce astounding effects. Like natural wonders, these heterogeneous creations were united by the psychology of wonder, drawing their emotional effect from their rarity and the mysteriousness of the forces and mechanisms that made them work.

This wonderful technology had a particularly courtly character, expressed in the aesthetic that governed its creations. These were beautiful, intricate, precious, expensive — more akin to the work of the jeweler than that of the blacksmith. The marvels of art did not include quotidian, if undeniably important, inventions like the windmill and the plough; rather, they performed functions associated with the military and social elites.[56] Many served defensive or military purposes: *Eneas*' magnetic fortifications, burning mirrors (capable of setting fire to an entire army — a perennial favorite), and automata that warned of impending attack. Others intimidated or provided personal protection, like the cloak lined with dindialos skin or the marvels in Camilla's tomb. Still others entertained — consider the emir of Babylon's "marvelous toys" in the early thirteenth-century *Aymeri de Narbonne*, which described a gilded copper tree covered with

90

pneumatically powered singing birds,[57] or the automata in Benoît's Chamber of Beauties.

These last show very clearly a central characteristic of many artificial marvels: their explicitly civilizing intent. Romances served, among other things, to foster and implant aristocratic and courtly ideals and behavior.[58] Marvels, the aristocracy of phenomena, played a fundamental part in this project by refining sensibilities, promising mastery (including self-mastery), and providing a window onto a more opulent world. According to Benoît, the automaton with the mirror "was there for the common benefit of all who entered the Chamber. They would look at their reflections and be immediately aware of what was unbecoming in their dress.... People there were hardly ever accused of unseemly behavior or foolish laughter: the mirror showed everything—attitudes, manners, complexion."[59] Similarly, the automaton with the censer "would watch people in the Chamber and convey to them by means of signs what they ought to do and what was most important for them.... It kept those who came into the Chamber, who entered or left it, from being disagreeable, uncourtly [vilains], or importunate; no-one could be irresponsible or foolish or uncourtly or senseless there, for the statue very cleverly kept them all from any uncourtly action [vilanie]."[60]

The wonders of art, then, like the wonders of nature, embodied a form of symbolic power—over nature, over others, and over oneself. Men versed in the knowledge of natural properties could use them to work marvels, turning day into night, controlling the weather, eliminating disease and decay. Artificial marvels also allowed lords to defeat their enemies and to enforce stringent standards of conduct in their dependents. Automata functioned as ideal servants: beings useful for the discipline and surveillance of others, and over whom their owners could have in turn perfect control. Finally, they helped their noble (and not-so-noble) readers and hearers to internalize increasingly stringent standards of courtly conduct intended to elevate them above the rest of society. The wonderful alabaster walls of the Chamber of Beauties modeled this exclusionary quality to perfection: "Anyone inside can see outside clearly, but no-one on the outside can see in, however hard he looks."[61]

As in the case of the monstrous races, attitudes toward the wonders of art shifted over the course of the Middle Ages. The earliest discussions were marked by emotional and temporal distance and by a degree of moral disdain. Consider, for example, the Seven Wonders of the World (the walls of Babylon, the Colossus of Rhodes, the lighthouse at Alexandria, and so forth), enumerated in the Hellenistic period and transmitted to the Latin West by the late Roman writers Martial and Cassiodorus.[62]

One early medieval treatise on this topic, of uncertain date but preserved in a twelfth-century French manuscript, included the traditional
seven together with biblical examples such as the temple of Solomon and
Noah's ark. After enumerating the remarkable features of each, the anonymous author ended on an elegiac note, describing these human artifacts
as "close to ruin," in contrast to the "wonders" of divine creation – the
sun, moon, and stars, the phoenix, Mount Etna – which "do not age, do
not decay, and are never attacked by time, until the Lord decides to end
the world."[63]

The late twelfth-century encyclopedist Alexander Neckam also stressed
the vanity and transitory nature of even the greatest works of human art,
notably the marvelous inventions commonly ascribed to Vergil and also
described by Gervase of Tilbury (fig. 1.1).[64] In addition to a meat market
whose contents never rotted, these included a bronze fly that chased away
other insects, a marvelous garden, and a noble Roman palace containing
wooden figures, each corresponding to an imperial province. Whenever
there was a rebellion, according to Neckam, the figure of the unruly
province would ring a bell, causing a brass horseman mounted on the
pediment to turn until he faced the source of the disorder. When asked
how long his automata would endure, the poet answered, "Until a virgin
shall bear a child" – which of course turned out to be only a few years
later.[65] In the same vein, Neckam berated human builders who would
compete with nature and usurp the wonder due to God. "O curiosity! o
vanity!" he lamented, "o vain curiosity! o curious vanity! Man, suffering
from the illness of inconstancy, 'destroys, builds, and changes the square
to round.'"[66]

If early medieval writers tended to displace artificial marvels to the
ancient past, high medieval writers projected them outward, to the margins of the world. This happened not only in romances, but also in the
literature of Eastern travel, which looked to romance for many of its
themes. The prominence of mechanical wonders in this literature also
reflected the fact that the construction of automata had flourished in the
Byzantine and Islamic world. The inhabitants of Latin Christendom knew
of such wonders primarily through the reports of pilgrims and other travelers, although such items were occasionally sent home by merchants or
received by European rulers as diplomatic gifts.[67] The imperial automata
of Byzantium had long since corroded into immobility, as described in
Robert of Clari's account of the Fourth Crusade,[68] but William of Rubruck
found a bonafide functioning automaton shortly after 1250 at the court of
the Great Khan. This silver tree was ornamented by silver lions and gilded
serpents, which belched mare's milk and other beverages, together with

an angel with a moveable trumpet connected by a pneumatic tube to a man hidden in a vault below:

> When the drink is wanted, the head butler cries to the angel to blow his trumpet. Then he who is concealed in the vault, hearing this, blows with all his might in the pipe leading to the angel, and the angel places the trumpet to his mouth, and blows the trumpet right loudly. Then the servants who are in the cellar, hearing this, pour the different liquors into the proper conduits, and the conduits lead them down into the bowls prepared for that, and then the butlers draw it and carry it to the palace to the men and women.[69]

William described this contraption in purely mechanical terms, but subsequent visitors to the Khan's court embellished their descriptions of other imperial automata with magical language recalling Benoît de Sainte-Maure's *Roman de Troie*. According to Marco Polo, Kublai Khan's famous levitating cups were made by *Bakshi* or enchanters — "those who are skilled in necromancy will confirm that it is perfectly feasible," he assured his readers — while Odoric of Pordenone described the dancing golden peacocks of the imperial palace in Peking as products of "the diabolical art or ...a device under the ground."[70]

Mandeville's Travels went the furthest in this respect, furnishing the Khan's palace with many such wonderful fantasies: an imperial hall, upholstered in panther skins and equipped with a convenient network of conduits for exotic beverages that fell into vessels of gold; golden thrones for the Khan and his family, set with gems and pearls and ornamented with the finest feathers; and a life-size artificial grapevine made of gold and precious stones. One version's account of Odoric's dancing peacocks summed up this vision of the East:

> And whether [they work] by craft or by necromancy I wot never, but it is a good sight to behold and fair, and it is a great marvel how it may be. But I have the less marvel because that they be the most subtle men in all sciences and in all crafts that be in the world; for of subtlety and of malice and of farcasting they pass all men under heaven. And therefore they say themselves that they see with two eyes and the Christian men see but with one, because that they be more subtle than they.[71]

As the references to malice and farcasting suggest, suspicions persisted concerning the marvels of Eastern art. (Farcasting allowed magicians to obtain knowledge of distant events and places through intermediary demons.) The necromancers of twelfth- and thirteenth-century romance

93

derived their powers in part from knowing the wonderful properties of lodestones or carbuncles, but also, at least implicitly, from commerce with demons, which they used to animate their statues and to accomplish their feats of force.[72]

While many Latin Christians lost themselves in wonder before the artificial marvels of the East and of antiquity, others, beginning in the thirteenth century, took them as a model and built around them fantasies of their own mastery. The English Franciscan Roger Bacon (d. 1292) incorporated this material into his systematic program for the advancement of learning. Inspired by his study of the *Secretum secretorum*, an Arabic compendium of political, medical, and magical arts purportedly composed by Aristotle to help Alexander the Great conquer the world, Bacon developed the notion of what he called experimental (or experiential) science (*scientia experimentalis*).[73] This aimed to harness the hidden powers of nature in order to produce startling and useful effects. He sketched his vision in the *Epistola de secretis operibus artis et naturae* (c. 1260), where he painted a vivid picture of the wonders such a study could produce. These included, among "infinite other marvels," many staples of romance: perpetual lamps ("for we know many things that are not consumed in flame, such as salamander skin, talc, and things of this sort");[74] distorting mirrors to confuse a hostile army ("for so Aristotle is said to have taught Alexander");[75] and innumerable mechanical devices. Bacon imagined ships and carts that moved without the help of oarsmen or draught animals, as well as flying machines and "instruments for going to the bottom of the sea or rivers without physical danger, for Alexander the Great used these to see the secrets of the sea" (fig. 2.8).[76]

Bacon's list of artificial marvels was clearly inspired by the marvels of romance, conceived as instruments of princely power; he later tried to realize this program, proposing to play Aristotle to Pope Clement IV's Alexander.[77] In the *Epistola* he legitimized Western aspiration to the wonders of Eastern art in two distinct ways. Not only did he set the wonders of art at the center of a program of religious and intellectual reform with impeccable Christian credentials, but he also attempted to dispel the demonic haze that surrounded them. "Although nature is powerful and wonderful," he argued, "nonetheless art using nature as an instrument is even more powerful through the force of nature [*virtute naturali*]. But whatever goes beyond the operation of nature or art is either inhuman or is a fiction and a fraud."[78] In other words, for Bacon art could perform its wonders using only natural forces, without invoking demons — a practice he considered both impious and ineffective.[79]

The increasing currency of actual artificial wonders on the European scene was fostered by, and doubtless in turn fostered, this more confident attitude toward the wonders of art as legitimate and accessible to contemporary Europeans, rather than the monopoly of some distant time or place. Thus it formed part of the Western technological revival that began to gather momentum at least as early as the eleventh century, inspired in many cases by the technical achievements of the Islamic world.[80] Bacon was an armchair inventor — the Mandeville of technology — and there is no indication that he produced any of the marvelous works he described. But he and his contemporaries observed the actual appearance and diffusion of devices only recently confined to the verses of romance. These included inventions that exploited the mysterious properties of traditional wonders such as magnets and lenses (the compass and, shortly after Bacon's death, eyeglasses) as well as automata and other flashy pneumatic and mechanical contraptions prominent among the marvels of the East.

The hydraulic clocks and cups of thirteenth-century Europe were pale reflections of their splendid Muslim counterparts, but they impressed and inspired contemporaries, as is clear from the notebook compiled by the French Cistercian Villard of Honnecourt during the second quarter of the thirteenth century. Villard included among the devices he had seen, heard of, or imagined on his extensive travels: a Tantalus cup in the tradition of Hero of Alexandria; a mechanical lectern in the shape of an eagle, which turned to face the reader of the gospel; and a primitive mechanical sundial in the form of an angel that rotated to follow the sun (fig. 2.9).[81] Villard's devices were crude and his understanding of them often primitive. Nonetheless, they mark the early stages of a process of rapid development that was to produce in little over a century marvels such as the late fourteenth-century silver gilt and enamel table fountain given to Abu al-Hamid II by the Duke of Burgundy (fig. 2.10),[82] as well as the elaborate automata and marionettes in Robert of Artois's park of Hesdin.

Robert had commissioned these "engiens d'esbattement" (machines for fun) in the late thirteenth century, and they appeared repeatedly in the account books of his daughter Mahaut, Countess of Artois, to whom he had bequeathed these fragile creations. From this source we know that they included a group of mechanical monkeys (horns were attached in 1312); an elephant, a goat, and a hydraulic stag; and — from 1344 — a carved tree covered with birds spouting water.[83] Later described by Guillaume of Machaut as "wonders, sports, artifices, machinery, watercourses, entertainments, and strange things,"[84] they had fallen into disrepair by 1432, when Philip the Good of Burgundy renovated them at considerable expense. The description of them in the Duke's accounts remains the most detailed to date of actual medieval European automata. A short extract will give the general idea:

2.8.1

Figure 2.8. The wonderful conveyances of Alexander the Great
2.8.1–2. Ulrich von Etzenbach, *Alexander*, Cod. Guelf. 1.2.5, fols. 128r, 129v, Herzog August Bibliothek, Wolfenbüttel (late 14th century).

The first image shows Alexander in his submarine, a glass sphere that Queen Roxa lowers into a sea filled with maritime monsters (fig. 2.8.1). He has taken with him a cat and a cock, whose crow will inform him if it is day or night; in case of trouble he plans to kill the animals that accompany him, forcing the sea, which cannot abide a corpse, to cast the vessel ashore. In the second, he soars above the earth in a flying machine powered by griffins (themselves natural marvels) which continually rise toward the large piece of meat that he holds above them on a pole (fig. 2.8.2).[10]

2.8.2

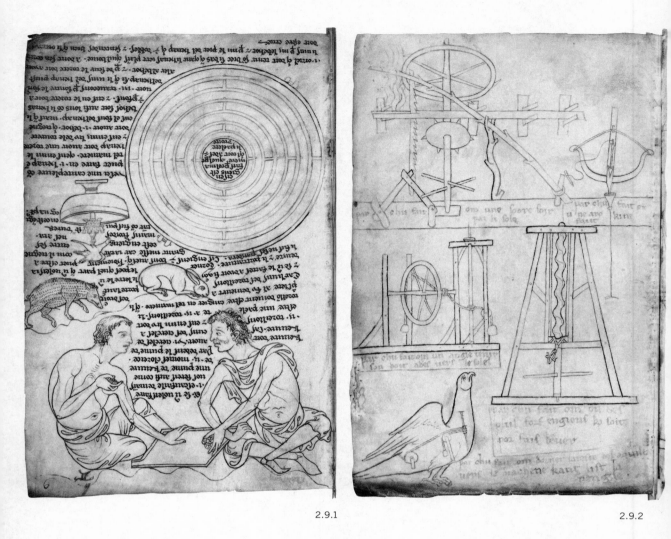

2.9.1   2.9.2

Figure 2.9. Villard's automata

2.9.1–2. Villard de Honnecourt, Sketchbook, MS fr. 19093, fols. 22r and 9r, Bibliothèque Nationale, Paris (second quarter of 13th century)

In his notebook, Villard de Honnecourt identified the automata in fig. 2.9.2 as follows: "How to make an angel keep pointing his finger toward the sun" (middle left, presumably a form of sundial); "How to make the eagle face the Deacon while the Gospel is being read" (lower left, equipment for a pulpit). The page shown in fig. 2.9.1 includes, among other things, a long description of a Tantalus cup, surmounted by a bird (here shown upside-down), which was designed to allow the wine to drain away into the foot of the cup as it was filled.[11]

98

Figure 2.10. Mechanical marvels and court spectacle
Table fountain of silver gilt and enamel, Cleveland Museum of Art, Cleveland (Burgundian, late
14th century).

Pneumatic and hydraulic devices were a common part of court spectacle, both real and imagined. This table fountain of Burgundian workmanship has thirty-two spouts, some in the shape of
gargoyles, lions, or dragons, through which poured wine or perfumed water that in turn rang bells
and turned metal wheels. In a reversal of earlier lines of influence (devices of this sort were associated with Muslim and Byzantine technology and had been given by Eastern rulers to more
backward Western Europeans), this device was delivered by a son of the duke of Burgundy to Abu
al-Hamid II.[12]

99

Item, there is a window where, when people wish to open it, a personage in front of it wets people and closes the window again in spite of them. Item, there is a lectern on which there is a book of ballades, and, when they try to read it, people are all covered with black, and, as soon as they look inside, they are all wet with water, when one so wishes. And there is another mirror where people are sent to look at themselves when they are besmirched, and, when they look into it, they are once more all covered with flour, and all whitened. Item, there is a personage of wood that appears above a bench in the middle of the gallery and fools [people] and speaks by a trick and cries out on behalf of *Monsieur le Duc* that everyone should go out of the gallery, and those who go because of that summons will be beaten by tall personages dressed like "sots" and "sottes," who will apply the rods aforesaid, or they will have to fall into the water at the entrance to the bridge, and those who do not want to leave will be so wetted that they will not know where to go to escape from the water.[85]

Clearly, Philip's "engiens" only roughly approximated the elegant "marvels and tricks" of Benoît's Chamber of Beauties. Nonetheless they still reflected (albeit in more bumptious form) both the automata of courtly literature and the aspirations to discipline and surveillance that those automata represented.

It was only in the late Middle Ages that European nobles and princes began explicitly to orchestrate the discourse of the marvelous as a coherent and comprehensive set of images and themes. To some extent this development was a general European phenomenon, corresponding to an overall growth in the extravagance and complexity of court life in the later fourteenth century, beginning with the Italian *principates* and the France of Charles V. But it reached its zenith in the mid-fifteenth century, in the smaller courts that hovered on the borders of the French kingdom, most notably those of Savoy, Anjou, and, especially, Burgundy under Dukes Philip the Good (1419–67) and Charles the Bold (1467–77). The dukes of Burgundy were the ultimate impresarios of the marvelous, both natural and artificial, and it is to them that we now turn.

Wonders at Court

The splendor of the fifteenth-century courts of Savoy, Anjou, and Burgundy stemmed from their wealth and from their political and cultural rivalry with the court of France.[86] One partisan Burgundian author — probably Georges Chastellain — contrasted the duke of Burgundy's court, where "each day are held festivals, jousts, tournaments, dances, and carols," with its French counterpart, devoted only to "sleeping, drinking,

and eating."[87] The princes of these relatively small but ambitious and ex-
pansionist principalities were virtuosi of the kind of courtly spectacle
required to reassure allies, impress subjects and rivals, and enhance their
prestige at home and abroad.[88] But their mobilization of the discourse of
wonders went well beyond a general commitment to magnificence and
liberality as part of the contemporary theater of power. Like René of
Anjou, Philip the Good of Burgundy and his son Charles made repeated
and specific use of the marvelous as an elaborate system of emblems and
signs to dramatize both their particular historical situation and their political aims.

This appears not only in these princes' self-conscious and programmatic revival of the culture of chivalric romance, with its jousts and tournaments, its perilous forests and magical springs, but also in their taste for
the exotic and their orientalizing style.[89] René, like Philip, owned a famous menagerie containing lions, leopards, monkeys, ostriches, drome-
daries (attended by Moorish keepers), civets, wild African goats, Turkish
or Indian hens, and an extraordinary "doe with horns."[90] His household
included dwarfs, giants, and other human prodigies — notably a famous
pinhead named Triboulet — as well as Moorish slaves, for whom he imported clothing from Africa and the Near East. He acquired Indian gems,
jewelry, and Eastern textiles from Italian merchants and through the
offices of his envoys and naval captains, while fashionable members of
his court adorned themselves in Turkish and Moorish cloaks, turbans,
and belts.[91]

The dukes of Burgundy had many of the same tastes. The late fourteenth-century markets of Bruges and Paris supplied Philip the Bold with
birds, spices, jewels, fruit, and other exotic luxury products. His grand-
son Philip the Good retained a group of dancers of the *morisque* — a dance
of Moorish origins highly fashionable in the courts of the mid-fifteenth
century — whom he dressed in costumes decorated with "Saracen" letters and collars "in the manner of animals."[92] Over the years, Philip the
Good's household included at least twenty-four dwarfs and fools; among
them were Madame d'Or, the blonde dwarf who belonged to Philip's
wife Isabella of Portugal, and his own giant Hans, whom one of his
chroniclers described as "the largest, without artifice, that I have ever
seen."[93] One of Philip's court painters, Hue of Boulogne, was charged not
only with outfitting his *morisque* dancers in their Eastern finery and looking after his renowned aviary, but also with renovating the automata of
Hesdin.[94]

The dukes' interest in wonders also took literary form, in the many
books they collected devoted to romance, the Alexander legend, and the

marvels of the East. These books included not only a significant number of works of travel and topography in French translation — *Mandeville's Travels*; Alexander's letter to Aristotle; the beautifully illustrated *Livre des merveilles du monde* that Duke John the Fearless gave to his uncle John, Duke of Berry (figs. 1.3.2, 1.4.2, 2.1.1); and works by Burgundian travelers to the Holy Land such as Bertrandon de la Broquière and Guillebert de Lannoy — but also many books connected with the Crusading movement, which obsessed Philip the Good. The 1467 inventory of Philip's library contained an entire rubric devoted to "medicine, astrology, and Outremer."[95]

The dukes' assiduous cultivation of all of the branches of the mirabilia tradition reflected a particular aesthetic, attracted not only to courtly refinement and Eastern luxury but also to the exotic and the strange. This aesthetic embodied specific economic and political realities and aspirations. It relied for expression on the wealth that flowed into the dukes' coffers from their rapidly expanding territories, most notably from the prosperous cities of Flanders and Brabant, and on the revived and expanding networks of trade and communication between western Europe and the eastern Mediterranean; by the mid-fifteenth century, oranges picked in Damascus could be sold in Bruges. Eastern commodities could represent princely magnificence: they were available, but only at great expense. The culture of the marvelous could also carry specific political messages. Philip the Good used it to remind his rivals and supporters of his illustrious marital alliance with the royal house of Portugal, with its ties to Africa and the Islamic world.

Philip, like other European princes, relied on his collections of exotica and other wonders to impress foreign visitors with his wealth. When Baron Leo of Rozmital, brother-in-law and emissary of the king of Bohemia, visited the duke's court in Brussels, his two chroniclers, Schaseck and Gabriel Tetzel, recorded their reactions in detail. For dinner at the ducal palace, according to Tetzel,

> a costly side-table had been set up overflowing with countless costly vessels and other things impossible to describe.... When my lord had eaten, the other lords led him again to the Duke. He dispatched him first with attendants to see his zoological garden, which is of vast proportions with many fountains and lakes, in which one found all manner of birds and animals which seemed strange to us. Afterward the Duke caused his treasure and jewels to be shown to my lord which are beyond measure precious, so much so that one might say that he far outdid the Venetians' treasure in precious stones and pearls. It is said that nowhere in the world were such costly trea-

sures, if only because of the hundred-thousand-pound weight of beaten gold and silver gilt vessels which we saw in many cabinets, and which were so abundant that we never thought to see the like.[96]

The treasures enumerated by Tetzel included a precious relic of the Holy Cross, encased in gold and jewels, and a jeweled replica of an ostrich feather for the duke's hat that was worth 50,000 crowns. Leo had come to Brussels under friendly auspices, but such a display of the duke's collection could also be used to impress potential enemies. In 1456, for example, Philip ordered a magnificent display of his treasure in the Hague, to show that he possessed the resources to conquer Utrecht.[97]

In concrete and physical form, therefore, mirabilia corresponded to liquidity and represented the wealth and power of the prince. Equally important to the duke's theater of power, however, was the actual display of wonders, which became increasingly central to the decorative schemes, dumb shows, and *tableaux vivants* that framed the great symbolic spectacles of his reign: weddings, funerals, state entries, and banquets for distinguished guests. Philip, his publicists, and his court officers used wonders to communicate in emblematic form a set of increasingly precise messages about the duke's aspirations to cultural and political leadership within the duchy and in western Christendom generally.

Philip's duchy — like the realm of René of Anjou — was an artificial construct, hastily assembled by three generations of a cadet branch of the house of Valois and cobbled together from territories originally subject to the French king and the Holy Roman Emperor. It did not correspond to any natural boundaries, whether topographical, historical, or linguistic; lacking even a formal name, it was usually referred to as the "grand duchy of the West." The elaboration of court life and spectacle, like the multiplication of court offices, bound nobles from the disparate territories of Burgundy to the duke and provided a seed around which might crystallize a nascent national aristocratic culture. At the same time, the emphasis on exoticism and artificiality reflected the duchy's liminal and "unnatural" position between the Empire and France.

An early example of Philip's mobilization of the symbolic discourse of wonders appears in the celebrations for his third marriage — to Isabella, daughter of King John I of Portugal, in January 1430. The match marked the entry of Philip's dynasty into the ranks of European royalty, at least by association, and the wedding banquet at Bruges was a royal show organized around the imagery of wonder. Jean Le Fèvre, lord of Saint-Rémy and later the first "king of arms" of Philip's Order of the Golden Fleece, left a detailed account.[98] The events included a splendid banquet where

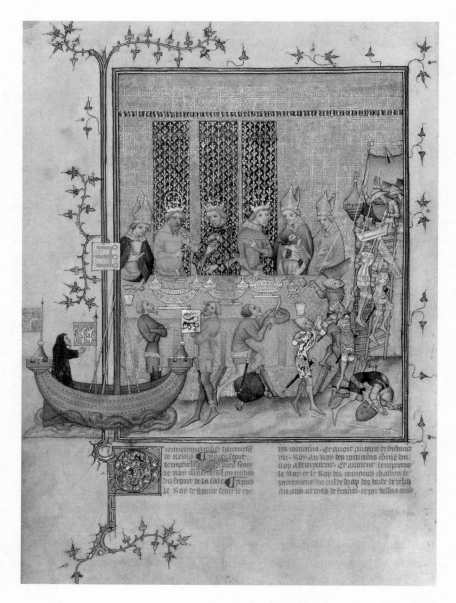

Figure 2.11. Staging the First Crusade
Chronique de Charles V, MS. fr. 2813, fol. 473v, Bibliothèque Nationale, Paris (c. 1378–79).

This painting illustrates a dramatic spectacle staged at the banquet given by Charles V of France for his uncle, Emperor Charles IV, on January 6, 1378. According to the text that accompanies it, the show reenacted the conquest of Jerusalem during the First Crusade by Godefroy de Bouillon: a beautifully decorated ship, containing twelve armed men and Peter the Hermit, was pulled across the hall to the great dais, followed by a model of the city of Jerusalem, complete with temple and occupied with exotically dressed Saracens. The two parties engaged in battle, resulting in a Christian victory. This kind of spectacle, or *entremets*, became a staple of Burgundian state banquets in the next century.[13]

each course (*mets*) was accompanied by what were called *entremets*, elaborate decorations or tableaux composed of both painted and living animals and people (fig. 2.11), such as a lady and unicorn or a group of wild men mounted on roast suckling pigs.[99] The climax was a great pie that opened to disgorge a live sheep, dyed blue and sporting gilded horns. The duke's giant, Hans, dressed in animal skins, then raced down the table and engaged in a mock battle with Isabella's dwarf, Madame d'Or. This spectacle emphasized the joining of the houses of Burgundy and Portugal, with a bow to Portugal's proximity to Africa, home of marvels. At the same time, it set up an image of exotic savagery as a carnivalesque foil to the self-conscious cultivation of the Burgundian court. The blue sheep anticipated the other climax of the wedding festivities: Philip's creation of the Order of the Golden Fleece, under the patronage of no less than Jason of Troy.[100]

Apparently modeled on the English Order of the Garter, the Order of the Golden Fleece was intended, according to its act of foundation, to encourage feats of chivalry.[101] It functioned to unite the disparate Burgundian nobility and bind them to the person of the duke. A central element in Philip's domestic program for the nobility, it relied on an elaboration of court life and the self-conscious revival of a romantic and nostalgic culture of chivalry to create a disciplined and service-oriented cadre, weaned from their previous rustic autonomy to a new status as a court elite.[102] At the same time the allusions to Jason, a romantic hero who sought the golden fleece in the Eastern reaches of Colchis, recalled one of Philip's other political and military priorities: the organization of a Crusade to reconquer the Holy Land and defeat the Turks.[103]

An obvious expression of Christian piety, this project also had dynastic roots. In 1396 Philip's father, John the Fearless, had been ignominiously captured by the Turks at the disastrous battle of Nicopolis in Hungary, and Philip saw the Crusade as an opportunity to restore the family honor. Cramped by older and better-established states in western Europe, he also viewed the eastern margins as a space of opportunity, where he could exercise leadership on a global scale. In the service of this ambition, he expanded his father's collection of topographical and Crusading literature, even going so far as to dispatch two of his subjects, Bertrandon de la Broquière and Guillebert de Lannoy, on a series of trips to the eastern Mediterranean for political and military recognizance.[104]

These themes of Eastern conquest received forceful expression in the most famous of Philip's banquets, the "Feast of the Pheasant" at Lille in 1454.[105] This event, organized to drum up noble support for his Crusade, showed a more pointed use of the discourse of wonders than the relatively early wedding feast of 1430. Not only was the display of wealth and splendor even more sumptuous and the entertainments even more elaborate, but the Eastern and exotic themes were much more clearly defined. The hall, hung with tapestries of the life of Hercules, was presided over by a large statue of a naked woman (presumably representing Constantinople, recently captured by the Turks). Spouting mead from her right breast, she was guarded by a live lion, presumably representing Duke Philip, with the motto, "Don't touch my lady."[106] The stationary *entremets* on the hall's three tables included (in the words of Olivier de la Marche, one of the participants) "a desert ... where there was a tiger, made marvelously life-like, which was fighting a great serpent"; a wild man on a camel, "who appeared to travel from land to land"; and "a marvelous forest, like a forest in India, within which were a number of strange animals of strange aspect, who moved by themselves, as if they were alive."[107] In addition to these automata, a number of live acts paraded among the tables between courses or flew overhead, notably a two-headed horse walking backward; a "deformed monster" called a *luytin*, which seems to have been a composite of human and griffin mounted on a boar; and a red dragon.[108] In addition, the banquet was interrupted by a series of three pantomimes showing episodes from Jason's sojourn in Colchis; after "looking around him, as if he had come to a strange land," he proceeded to vanquish two giant bulls and a serpent, using magical talismans given him by Medea.[109]

The climax of the banquet's entertainments was the appearance of a huge giant (presumably Hans again) dressed as an armed Saracen and leading an elephant, which de la Marche described as an "unusual beast" (*diverse bête*). This latter carried a castle containing the Holy Church herself (played by de la Marche, in female dress), who addressed a poetic appeal for salvation to "you knights who bear the Fleece."[110] At this point the court official known as Golden Fleece entered, carrying a live pheasant with a fabulously expensive jeweled collar, on which the Duke swore a vow to engage in a Crusade. He then invited all of those nobles present to follow suit, which many did, with more or less enthusiasm and élan.

The *entremets* associated with this banquet systematically mobilized the imagery of wonders in the service of empire-building and conquest. Indeed they signaled a shift in Latin views of the East. The hypercivilized Burgundians, surrounded by luxury and opulence, now presented themselves as having not only equaled but outstripped the Orientals, whose wonders paled beside their own. Like Gerald of Wales justifying the English pacification of Ireland, Philip and his festival organizers created a picture of the Turks that at once exoticized and dehumanized them, representing them as giants, monsters, or wild beasts, while tactfully reminding potential Crusaders that they stood to gain not only military and spiritual glory, but also some of the East's still-fabled wealth. Philip further demonstrated his financial ability to mount a Crusading expedition through a display of his own resources — the magnificent plate, the opulent entertainment, and his own lavish dress, which, according to observers, was decorated with hundreds of thousands of crowns' worth of gems.[111]

Most remarkable of all, however, was the way this spectacle explicitly manipulated the emotion of wonder to further one of the duke's particular aims. The ultimate point of the feast emerged only at the moment when Duke Philip responded to the Church's plea with his Crusader's vow: "After this," wrote Jehan de Molesme, "a number of people were very amazed and struck with wonder [*esbays et esmerveillez*], and believe me that a person would have had to have a remarkably hard heart not to have been at that moment softened and made tender."[112] In other words, the marvelous spectacle of the banquet seems to have had a double effect. On the one hand, it elicited emotions of wonder that could then be transferred to the person responsible for it, enhancing the charisma of the duke himself. If, as Lauro Martines has argued, court spectacle and ritual served ultimately to reflect a series of "self-images" of the prince,[113] then the strangeness and rarity of the banquet's wonders underscored the duke's unparalleled wonderfulness, as their expense and lavishness pointed to his liberality and wealth. Just as the audience was meant to wonder at the exploits of Hercules (in the tapestries that decorated the banqueting hall) and Jason (on its stage), so they were expected to identify both heroes with Philip himself.[114] In this way, the entertainment and decoration presented Philip as a new Alexander and future conqueror of the East.

On the other hand, the awesome spectacle also aimed to inspire (or weaken the resistance of) its noble guests — to "soften" their hard hearts, as Jehan de Molesme put it — and leave them prepared to give up their will to the duke and follow him on his Crusade. In this way, the spectacle functioned as a secular analogue to displays of relics or the hanging ostrich eggs that Guillaume Durand described as "things that prompt wonder and are rarely seen, so that by them the people are drawn into church and are more affected."[115] Its success in this respect was immediate: so many nobles rushed to commit themselves on the jeweled pheasant that not all could be accommodated that evening, and the rest were asked to submit their vows in writing the next day. (The elaborate conditions that many of them then set on their participation bespeak a number of second thoughts.)[116] As it turned out, their commitment was never tested. Philip was beset by financial difficulties and pressing local considerations — the death of Pope Nicholas V, a crisis in Franco-Burgundian relations, and the opportunity to conquer Utrecht — and his great Crusade, except for an abortive excursion led by his bastard son Anthony, remained an unrealized dream.[117]

In this case study, we have analyzed the discourse of the marvelous at the court of Philip the Good of Burgundy. We would argue, however, that this example can be generalized. Philip and his counselors were not alone in using the language and imagery of wonder and wonders to further political, military, and cultural goals. We find an important precursor at the court of Charles V of France in the later fourteenth century, and very similar practices at the fifteenth-century courts of Portugal, Savoy, and Anjou. Philip himself provided the model for his son Charles the Bold, whose wedding to Margaret of York in 1468 outdid even his father's in its display of exotica and other mirabilia, including six great unicorn horns mounted "like very long candles" on the dresser that held the plate.[118] The dukes' example was influential in many parts of Europe, especially the Hapsburg Empire, where it was taken over by Charles's son-in-law Maximilian and Maximilian's own successors Charles V and Philip II. These Hapsburg rulers adopted and elaborated the tradition of their Burgundian namesakes, increasingly keying it not to Eastern marvels but to the wonders of their new empire in the exotic West, identified no longer with Ireland but with America.

Europe's conquest of America enriched its vocabulary of wonders, though the syntax of the discourse retained its medieval structure for some time. Well into the sixteenth century, topographical marvels and the exotic products of both East and West kept their princely associations and their aura of romance; they continued to represent wealth, nobility, and colonialism — this last increasingly a matter not of aspiration but of fact. Collectors supplemented their rhinoceros teeth and unicorn horns with the carapaces of armadillos; the European luxury market expanded to include not just gems and spices, but new commodities such as tobacco and short-lived slaves of unfamiliar physiognomy and hue. These new wonders eventually wrought great changes in European culture, not only in its economy and its political order, but also in its interpretation of the natural world. But before turning to those latter changes, we must look once more to the Middle Ages and to the place of the old wonders in the tradition of natural philosophy.

CHAPTER THREE

Wonder Among the Philosophers

All of the intellectual traditions we have described to this point, from the mid-twelfth through the mid-fifteenth centuries, set the passion of wonder in a highly positive light. The proper response due to the Christian God and his marvelous creation, it partook at times of awe and at times of pleasure — often, indeed, of both. Always, however, it elevated those who experienced it. In front of the marvelous gems of the Crista, Suger of Paris found himself transported to a higher world, suspended between the earth and heaven, just as the nobles in the court of Philip the Good of Burgundy, dazzled by the pageantry in the Feast of the Pheasant, found themselves (temporarily) moved to commit body and goods to a Christian Crusade.

The contemporary natural philosophical tradition, however, presented wonder with more ambivalence. Consider, for example, the *Quaestiones naturales* of Adelard of Bath (fl. 1116–42), a generation before Suger. This work takes the form of a series of seventy-six questions on the natural order posed to Adelard by his nephew: Why do plants grow in places they have not been planted? Why don't human beings have horns? Why is the sea salty? Are the stars alive? Toward the end of the work, the talk turns to thunder; calling it "an object of wonder to all nations," the nephew asks Adelard about its cause. Adelard answers without hesitation — thunder is produced by the collision of frozen clouds — and then berates the nephew for his ignorance:

> Why is it that you so wonder at this thing? Why are you amazed, why are you confused?...I know that the darkness that holds you, shrouds and leads into error all who are unsure about the order of things. For the soul, imbued with wonder [*admiratione*] and unfamiliarity, when it considers from afar, with horror [*abhorrens*], the effects of things without [considering their] causes, has never shaken off its confusion. Look more closely, consider the circumstances, propose causes, and you will not wonder at the effects.[1]

In this passage, Adelard did not portray wonder as admirable — allied with pleasure or akin to veneration — but rather as next door to horror, the passion associated with monsters, prodigies, and other expressions of divine wrath. For Adelard, however, the horror did not lie in a justified sense of moral outrage or fear of divine retribution, but sprang from another source entirely: considering effects in ignorance of their natural causes. He associated wonder not with piety and reverence but with ignorance and superstition — what he called "confusion." For this reason, as Adelard repeatedly emphasized, he, as a philosopher, did not feel wonder, unlike his intellectually backward nephew, who had not yet learned to see the natural order as a complicated and semi-autonomous chain of secondary causes that depended distantly, but only distantly, on the First Cause, God. For Adelard, it was not enough to attribute the growth of plants to the "wonderful [*mirum*] effect of the wonderful divine will," as his nephew put it; one had to analyze the process further in terms of the interaction of the four elements.[2] When his nephew puzzled over the "wonderful nature of sight," and specifically the fact that it is easier to see something light when your eye is in darkness than vice versa, Adelard remarked dryly, "I do not wonder at your wonder, for the blind person speaks thus of light."[3]

In this chapter, we move from the court and the cloister to the classroom, from the literature of travel and romance to natural philosophy, where wonder and its objects acquired a set of meanings very different from those we have explored to date. We will argue that academic natural philosophers interpreted wonder as the usual response not only to the rare and unfamiliar, but also to the phenomenon of unknown cause. Imbued with a vision of their discipline that emphasized the certain causal knowledge of natural phenomena, they therefore rejected wonder as inappropriate to a philosopher, and they developed an armory of explanatory tactics to dispel it. At the same time they elaborated a view of the natural order as governed by "habits" or "rules," which it was the philosopher's task to explore. They did not reject wonders as illusory, still less as miraculous, but merely labeled them as *praeter naturam* (outside or beyond the course of nature) and therefore irrelevant to the natural philosopher's work. In this way, they marginalized both the passion of wonder and wonders as objects, in favor of a view that emphasized both the regularity of nature and the completeness of the philosopher's knowledge, marred by no unseemly gaps.

The Philosophers Against Wonder

The vehemence of Adelard's rejection of wonder reflects the situation of Latin philosophy in the twelfth century, when it was just emerging from a long eclipse. More than six hundred years after the collapse of the western Roman Empire had disrupted traditions of schooling and philosophical inquiry, small numbers of Latin intellectuals began to look east and south to Byzantium and the Islamic world, where philosophers had not only preserved but greatly expanded on the legacy of ancient Greek ideas. Adelard of Bath was one such pioneer. After studying and teaching at two of the most eminent French schools, Tours and Laon, he embarked on a long journey to Sicily and the Near East. Returning seven years later with a working knowledge of Arabic and a number of Arabic books, he devoted himself with the fervor of a Christian missionary to spreading the gospel of Greek and Arabic learning in the backward Christian West.[4] He wrote the *Quaestiones naturales* as part of this program, assigning to his nephew the part of the ignorant Latin intellectual, hidebound by religious authority and unfamiliar not only with the causes of natural phenomena, but also with the idea of natural inquiry itself. Adelard himself stood for the ideal of rational explanation, which he identified with Arabic learning and with the work of Aristotle and Plato. Thus Adelard's main target was the Augustinian tradition, which elevated wonder at the mighty works of God above the causal exploration of natural phenomena. For Adelard, that wonder prevented Christians from inquiring further, in the manner of Aristotle, Plato, and the Arabs, and consigned them to backwardness and intellectual sloth.

Although Adelard invoked the authority of Aristotle, most of whose works were still unavailable in Latin, his suspicion of wonder reveals, more than anything else, his ignorance of the Aristotelian corpus, for in the first book of the *Metaphysics* Aristotle had emphasized the centrality of wonder to the philosopher's task. "It is owing to their wonder that men both now begin and at first began to philosophize," he wrote; "they wondered originally at the obvious difficulties, then advanced little by little and stated difficulties about the phenomena of the moon and those of the sun and the stars, and about the genesis of the universe."[5] For Aristotle, wonder, which arose from ignorance about the causes of natural phenomena, led people to search for those causes and was therefore essential to the process of philosophical inquiry. Aristotle's Arabic followers echoed this opinion. According to the eleventh-century Arabic philosopher Avicenna, for example,

> the ignorance of the causes of the effects of the virtue in the magnet is not more marvelous than the ignorance of causes disposing a thing to redness or yellowness or body or soul. But wonder [*admiratio*] falls, indeed, from ordinary things [*consuetis*] and the soul neglects to inquire about them; but what

rarely exists does excite wonder and induces inquiry and speculation about its causes.[6]

In his treatise on the soul, too, Avicenna tied wonder unambiguously to philosophical inquiry, associating it with pleasure and laughter rather than anxiety, in direct opposition to Adelard's view.[7]

Adelard's patent unfamiliarity with even the rough outlines of Aristotelian philosophy might partially explain the discrepancy between his views and Aristotle's. But the jaundiced attitude toward wonder in Latin natural philosophy persisted throughout the thirteenth and fourteenth centuries. By that time, Aristotle's works had been not only translated but institutionalized as the basis of the undergraduate curriculum in the universities and schools of religious orders that had rapidly coalesced as centers of Latin philosophical speculation and teaching in Italy, southern England, and northern France.[8] The two principal concerns of these universities and schools (and of the "scholastic" learning they elaborated) were the assimilation of previously unknown Greek and Arabic works, newly available in Latin, and the training of teachers, lawyers, physicians, theologians, and the secular and ecclesiastical administrators who were rapidly developing into western Europe's most important non-military elite.

The disdain for wonder that characterized the scholastic environment was clearly expressed in a series of influential expositions of the *Metaphysics*, beginning with two composed by Roger Bacon, apparently in connection with his teaching at the university of Paris in the 1240s. In the first, Bacon (like Adelard) identified wonder (*admiratio*) with ignorance, although he acknowledged that it could raise people from brute stupidity. His attempts to distinguish between the more profound ignorance of animals or children, who can't know or don't care to know the causes of things, and the only slightly superior ignorance of the man who wonders, served to underscore the association of wonder with the uninformed mind.[9] He was even more explicit in his second commentary, explaining that Aristotle had not meant to say that wonder was in any strict sense the cause of philosophy but only its "occasion," moving the philosopher to flee from it as a frightened man flees from a battle.[10]

The link between wonder and fear also appears in the *Metaphysics* (c. 1260) of the German Dominican writer Albertus Magnus, who had briefly been a contemporary of Bacon at the University of Paris and whose massive corpus of expositions of Aristotle, more than any other work, stimulated and shaped European natural philosophy in the High and later Middle Ages. In his own commentary, Albertus described wonder as "shocked surprise [*agoniam*] and a suspension of the heart in amazement [*stupore*] before the sensible appearance of a great prodigy, so that the heart experiences systole. Thus wonder is somewhat similar to fear in the motion of the heart."[11] He indicated elsewhere the source of this rather compressed definition of wonder, placing it in the typology of fear he found in the eighth-century Christian writer John Damascene. (According to Damascene, the six varieties of fear included wonder, amazement, and "agony" or shocked surprise.) Albertus described both wonder and amazement as the result of an encounter with the unfamiliar; this produces a "flight of the heart in systole," since—the words echo Bacon's second *Metaphysics* commentary—"the heart flees the unfamiliar as it flees the bad and the harmful."[12]

Albertus transmitted his distaste for wonder to his most famous student, Thomas Aquinas. While not denying wonder's affinity with pleasure and inquiry, Aquinas nonetheless treated it in a minor key. "As sloth is to external behavior, so wonder and amazement are to the act of the intellect," he noted in his great summa of theology, though he differentiated between wonder and amazement just enough to save Aristotle's basic point.[13] Thomas's discussion of Jesus's wonder at the faith of the centurion (Matthew 8.10) was similarly ambivalent; whereas Augustine had celebrated this passage—"for that our Lord marveled means that we should marvel," he wrote in one of his commentaries on Genesis—Thomas shifted the verbal emphasis, classifying Christ's wonder as a "defect" (*defectum*), which he took on for human edification.[14] Penitentially salutary as an example of humility, it was nonetheless a sign of the fallen human state.

To what can we attribute this marked philosophical skepticism (or at best ambivalence) concerning the cognitive uses of the passion of wonder? In part, it may reflect the tradition of portents and prodigies, which focused on individual wonders as signs of divine wrath and therefore objects of fear.[15] But the dysphoric attitude toward wonder had another, more obvious source. If theologians associated wonder with fear, philosophers, as Adelard had already made clear, associated it with ignorance of causes. This was a grave failing in the thirteenth century, given the stringent Aristotelian definition of philosophy as certain causal knowledge, the social and cultural disdain for manual labor, and the institutional novelty of the university itself.

Albertus Magnus's discussion of wonder in his commentary on the *Metaphysics* illuminates these points. In this work he took care to distinguish the philosopher's goal from that of the *artifex*—the expert in one of the fields of applied knowledge, or "arts," such as practical medicine,

agriculture, dramaturgy, or building.[16] Philosophers aspired to *scientia* —
in Greek, *episteme* — defined by Aristotle as certain knowledge, in contrast to other forms of inquiry, which could only yield probable opinion.[17]
For Aristotle's thirteenth-century followers, *scientia* was the privileged
body of universal and necessary truths that could be known with absolute
certainty.[18]

Applying such a rigorous epistemological ideal to the shifting and irregular world of physical phenomena — as opposed to, for example, the
unchanging nature of God — was inherently problematic. In practice nature
appeared only to approximate regularity, and particular natural effects or
processes often fell short of this lofty ideal. As a result, Aristotle's medieval Latin followers (following the lead of their Arabic predecessors) had
to develop strategies to explain how a universal and certain "science" of
nature was in fact attainable, in view of the relative irregularity of particular phenomena. These strategies evolved over the course of the thirteenth and fourteenth centuries, but all rested on the notion that natural
philosophers did not, in fact, study particulars (where natural variability
resided) but focused instead on elaborating general statements about the
causes of certain *types* of phenomena — as opposed to particular phenomena — through a process of definition and deduction from those universal
principles.[19]

In many respects, medieval natural philosophers went well beyond
Aristotle in their commitment to the necessary and universal. In *De partibus animalium*, Aristotle had argued that the study of particular natural
phenomena, which the Greeks referred to as compiling "histories," was a
crucial preliminary to a genuinely philosophical account of the natural
world, since it supplied raw material for the abstraction of universals and
illustrations for causal arguments. His treatises on physics, meteorology,
zoology, psychology, and biology in fact overflow with careful observations of everything from sleeping to eclipses to the embryological development of chicks.[20] His medieval Latin commentators did not follow him
in this path. Even Albertus Magnus — in many ways closest to Aristotle
in his interest in particular phenomena — relegated all particulars to the
realm of the accidental and thence excluded them rigorously from philosophical consideration.[21] Thus, for example, while Aristotle saw his great
work of descriptive natural history, the *Historia animalium*, as the introduction to and indispensable foundation for his more philosophical works
on the topic, *De partibus animalium* and *De generatione animalium*, Albertus marginalized the study of particulars (even particular species) by placing them at the end of his works on animals, vegetable, and minerals. To
each of these he appended, almost as an afterthought, a section of short

entries devoted to particular animals, plants, and minerals and their properties. While sometimes impressive in their careful retailing of Albertus'
personal observations, these purely descriptive entries lacked any philosophical explanation and recall contemporary bestiaries, herbals, lapidaries, and encyclopedias, rather than works of causal analysis.[22]

Albertus underscored the marginal importance of particular species in
the study of philosophy in a somewhat apologetic introduction to Book VI
of his *De vegetabilibus* (before 1260), which was made up of alphabetically
ordered entries on particular plant species:

> I am satisfying the curiosity of students rather than philosophy, for there can
> be no philosophy of particulars, and in this sixth book I intend to set forth
> certain properties of particular plants.... Of those I will set forth, I have
> tested some myself, and I have taken others from the writings of those whom
> I judge not to write frivolously but only things tested by experience. For only
> experience provides certainty in such things, because it is impossible to con-
> struct a syllogism concerning particular natures.[23]

If the syllogism was the tool of the philosopher in pursuit of *scientia* or
certain knowledge, experience was the tool of the *artifex*, bound to the
world of probable opinion and to the use of the senses and the hands.
Albertus' attention to the healing properties of the plant species described
in Book VI — beginning with *abies* (fir) and ending with *zedoaria* (which
seems to be related to cedar) — and his account of the domestication of
wild plants in Book VII bespeak both practical and recreational ends. As
he put it, "it is pleasurable for the student to know the nature of things
and useful to the life and preservation of cities."[24] Pleasure, like wonder,
was for students, not philosophers.

Later medieval Latin philosophical writers chose to restrict themselves
even more strictly to their epistemological brief. In the two centuries after
Albertus, there were very few commentaries on works of Aristotelian
natural history — Aristotle's animal books, or the pseudo-Aristotelian
Problemata and *De plantis* — and these were never part of the regular reading for a university degree.[25] Alone of all the major works attributed (in
this case erroneously) to Aristotle, the one promisingly entitled *De mirabilibus auscultationibus* (*On Marvelous Things Heard*) received no medieval
commentaries. This work, a miscellany of paradoxographical lore, considered topics ranging from the Arabian camel's rejection of incest to the
revivifying qualities of the whirlpool of Cilicia; the lack of commentaries
implies that this work was either never officially taught or thought too
marginal for formal lectures on it to be recorded and preserved.[26]

Although more standard works like the *Meteorologica* or the *Parva Naturalia* contain a smattering of marvelous material, the orphan status of *De mirabilibus auscultationibus* stands as an emblem of the marginality not only of wonder but also of wonders in academic philosophy. Albertus Magnus and other Latin commentators on Aristotle had already excluded all particular species, however commonplace, as inadequately universal for philosophical analysis. This exclusion applied even more strongly to the large number of species, such as the lodestone or carbuncle, rendered wonderful by virtue of their occult qualities, since these could only be determined by experience rather than reasoning, as we will describe below. Still less amenable were the most wonderful wonders: rare or unique phenomena fascinating precisely because of their unknown causes and their violation of expectations about type. Conjoined twins, Mt. Etna, or an English cave that communicated with the Antipodes: such things could only be accounted for in natural terms as the products of chance, which Albertus Magnus described as coming into play "when something in the works of nature happens outside the intention of nature, such as a sixth finger, or two heads on one body, or the absence of a finger."[27] Although occurrences of this sort lay outside the ordinary course of nature, as Albertus went on to emphasize, this did not mean that they were not governed by the same kinds of causes that governed other natural phenomena. Rather, those causes had combined in unspecifiable and unforeseeable ways to produce something whose being was not only particular but utterly contingent — as far as it was possible to get from the philosophical realm of the necessary and the universal.

In fact, when Aristotle wrote that wonder was the beginning of philosophy, he had something very different in mind from medieval mirabilia, as a passage in the *De partibus animalium* makes clear: "Every realm of nature is wonderful. Absence of haphazard and conduciveness of everything to an end are to be found in nature's works in the highest degree, and the end for which those works are put together and produced is a form of the beautiful."[28] Thus Aristotle's ideal type for the objects of philosophical wonder was not the singularities of ancient paradoxography or the phenomena in the pseudo-Aristotelian *De mirabilibus auscultationibus* — in this he already distinguished himself from the Greek rhetorical tradition — but the celestial bodies, characterized by the beautiful and inexorable regularity of their motions and therefore a fine model of natural orderliness and a good foundation for universal and necessary knowledge. When medieval Latin writers thought of wonders, however, they did not imagine universal and stately celestial motions but precisely the kinds of phenomena we have described in the preceding chapters: the atypical, the marginal, the strange. The wonders described by Gervase of Tilbury, the shifting spring near Narbonne, the phoenix, Naples's volcano and its peculiar vegetation, would not have diverted the emperor if they had reflected common phenomena and repeated experience. Individual prodigies, like the monstrous births or rains of blood recorded by medieval chroniclers, drew their force and meaning precisely from their aura of singularity; each one arose from particular circumstances, never to be repeated exactly in the same configuration. Given this discrepancy between Aristotle's identification of natural wonderfulness with the regular and Latin scholastic philosophers' identification of it with the unusual, it is hardly surprising that the latter no longer embraced wonder in the way that Aristotle had.

In addition to minimizing the role of wonder in philosophy, Albertus Magnus expelled the rhetoric of wonder from his writing on natural history, which thus differs markedly from the bestiaries, lapidaries, and encyclopedias with which it shares much of its content. *De vegetabilibus* contains only a handful of phrases that echo the language of books of wonders, and even these are remarkably subdued: "One of the wonderful things about the balsam tree" is that it produces annually a virtually constant quantity of resin; the fig tree is "wonderful" in that fig rinds are bitter while the fruit is sweet.[29] Albertus followed the same practice in *De mineralibus*, where his dispassionate tone contrasts notably with the rhetoric of wonder in the lapidaries he used as his sources. In the few passages where he employed this rhetoric, his language was extraordinarily restrained or — even more telling — marked by outright skepticism. Thus of the ability of *liparea* (bitumen or sulfur) to protect wild animals from dogs and hunters, he commented, "This is very marvelous if it is true."[30] By emphasizing their implausibility, Albertus underscored the irrelevance of wonders in medieval natural philosophy; devoted to universals, regularities, and certain causal knowledge, natural philosophy excluded a priori anomalous and contingent phenomena of uncertain veracity and unknown cause.

All of this was intended to confer on the natural philosopher (together with the theologian and master of theoretical medicine) a virtual monopoly on absolute certainty — something that in the context of the medieval education acquired a social and professional as well as an epistemological cast. When Albertus Magnus, Thomas Aquinas, and Roger Bacon were writing, in the mid- to later thirteenth century, universities and the schools associated with the Dominican and Franciscan orders were comparatively new institutions, still developing the complicated structure of examinations and degrees that authorized certain types of learning and certified its possessors (at least in theory) to teach, to practice law and medicine, to serve as theological experts, and the like. The intellectual authority of graduates depended in good part on the special claim to certainty that defined the scholastic study of philosophy, which formed the core of the undergraduate curriculum and was considered preparatory for the studies of theology, medicine, and law. This view of authority also tended to discount knowledge gained in other ways (notably through experience) by associating it with subordinate groups such as old women and artisans;[31] the social ideology of the period identified their work with manual labor, which was strongly stigmatized in elite circles.[32]

The intellectual implications of this ideology were far-reaching. In addition to privileging a small group of knowers, it sanctioned a special type of knowledge (necessary and universal) and a special way of acquiring that knowledge, called *doctrina*. This referred to the transmission of knowledge through institutionally sanctioned texts and lecturers (*doctores* or masters).[33] Thus the natural philosopher did not concern himself with observing natural phenomena and providing new explanations for them. His job was rather to refine and distill the universal truths he found in books and received from his teachers, and in turn to transmit these truths to his students (fig. 3.1).

The ideology of *doctrina* may seem to be only tenuously related to wonder, but Albertus made the connection clear. Philosophy, he wrote in his commentary on the *Metaphysics*, "is doctrinal [*doctrinalis*], teaching by cause, and it must begin with causes."[34] He acknowledged that Aristotle identified wonder at effects as the force that drove the philosopher to seek for causes. But, Albertus went on to argue, this only applied to the philosophical primitive, the man "inventing" (*inveniens*) philosophy. The modern philosopher, who belonged to the perfected, "doctrinal" phase of philosophy — when the general outlines of causal explanation had been well established — had no need of wonder, which would only serve to associate him with the *artifex*, who trafficked not in causes but effects. Thus, Albertus explained, the master of philosophy felt no wonder as he taught, since he already knew the causes — which is what qualified him to teach. Similarly, the student felt no wonder as he was taught, since his teacher's exposition followed the order of certainty, from cause to effect.[35]

This intellectual program produced a remarkable body of sophisticated work, but one in which wonder and the wonderful had no place. In the medieval scholastic analysis, wonder became a taboo passion: the mark of the ignorant, the non-philosopher, the old woman, the empiric, all of whom were only one step up, as Bacon indicated, from animals and children. This also accounts to a large degree for its demeaning associations with laziness and fear. The sentiments expressed in the thirteenth-

Figure 3.1. Philosophy as *doctrina*
Nicole Oresme, trans., *Les éthiques d'Aristote*, MS 9505–06, fol. 2v, Bibliothèque Royale
Albert I, Brussels (Paris, after 1372).

Like Bartholomaeus Anglicus's *De proprietatibus rerum* (fig. 1.6) and various other works of natural and moral philosophy, King Charles V had Aristotle's *Ethics* translated into French in the
early 1370s; this illustrated manuscript of the work eventually found its way into the collection
of the dukes of Burgundy. The top register of the frontispiece shows the king receiving the translation from Oresme, next to a scene of the king with his family. The bottom register illustrates
part of Oresme's text, where he describes "which persons are appropriate to hear this science" —
the king and his counselors (on the left), rather than those "inexperienced in human affairs" and
hence "more tempted by bodily desires" (the youth being expelled on the right). As a whole, the
frontispiece indicates the bookish and doctrinal orientation of medieval philosophy.[14]

century *De mirabilibus mundi*, although attributed falsely to Albertus Magnus, were nonetheless not uncharacteristic of his view: the task of the wise man was "to make wonders cease."[36] The philosopher sees no marvels: to him, monstrous births or the rare meteorological and topographical particulars that so amaze the layperson appear only as the necessary (if unforeseeable) effects of familiar and universal causes. He takes pleasure in his work, as Roger Bacon emphasized, but that pleasure arises not from the process of inquiry into the unknown, but rather from the possession of knowledge already perfect and complete.[37]

The nervousness of thirteenth-century philosophers about wonder persisted, and their fourteenth-century followers dropped the topic at the first opportunity. In their commentaries on the *Metaphysics*, Albertus Magnus, Roger Bacon, and Thomas Aquinas were constrained, however uncomfortably, to accommodate Aristotle's discussion of wonder. Two generations later, when philosophical writers began to use a new translation of the *Metaphysics* that happened to omit the relevant passage altogether, John of Jandun, William of Ockham, and Jean Buridan could avoid the whole issue.[38]

Curiosity and the Preternatural

What was the vision of natural order that underpinned the epistemology of high medieval commentators on Aristotle, and what was the place of wonders in this order? Aristotle had provided a point of departure in Book VI of his *Metaphysics*, where he discussed the nature of the accidental, which he defined as that "which is capable of being otherwise than as it for the most part is."[39] As a result, he hastened to point out, "there is no science [in the sense of the Latin *scientia* or Greek *episteme*] of the accidental ... for all science is either of that which is always or of that which is for the most part."[40] As this last, rather imprecise, phrase suggests, neither Aristotle nor his medieval commentators thought of nature in the post-seventeenth-century sense, as governed by unbreakable laws.[41] Although scholastic philosophers sometimes used the word "law" (*lex*) in referring to the natural order, they almost always explicated it in the sense of "rule" (*regula*): medieval natural philosophers thought of nature as regular in her actions (rather than inexorable), governed by what were often referred to as habits (*habitus*), inclinations (*inclinationes*), or intentions (*intentiones*). Like any artisan, nature aimed at a certain uniform standard, but occasionally, for better or worse, she missed the mark, resulting in an "accidental" production, such as a baby with six fingers.[42]

Scholastic natural philosophers had to refine further Aristotle's notion of the natural in order to determine its relationship to miracles. This had not been an urgent issue for early medieval philosophers, in the light of Augustine's view that all of nature was immediately and miraculously dependent on God. Beginning in the twelfth century, however, Adelard of Bath and other Latin writers began to develop the idea of an autonomous natural order against which God's special, miraculous interventions were placed in sharp relief.[43] Thomas Aquinas' treatment of the subject, in Book III of his *Summa contra gentiles*, includes a particularly clear articulation of this distinction:

> The order imposed on things by God is based on what usually occurs, *in most cases*, in things, but not on what is always so. In fact, many natural causes produce their effects in the same way, frequently but not always. Sometimes, indeed, though rarely, an event occurs in a different way, either due to a defect in the power of an agent, or to the unsuitable condition of the matter, or to an agent with greater strength — as when nature gives rise to a sixth finger on a man.... So, if by means of a created power it can happen that the natural order is changed from what is usually so to what occurs rarely — without any change of divine providence — then it is more certain that divine power can sometimes produce an effect, without prejudice to its providence, apart from the order implanted in natural things [*praeter ordinem naturalibus inditum rebus*] by God. In fact, He does this at times to manifest His power.[44]

In this passage Aquinas distinguished between three types of physical occurrences. The first was *natural* in the sense used by Aristotle: "that which is always or that which is for the most part." But this natural order of things could be violated in either of two ways: 1) by chance, accidental, or otherwise unforeseeable events (the man with six fingers); or 2) by miracles, performed directly by God without mobilizing secondary causes. We will call this last category of phenomena *supernatural*; miraculous events were naturally impossible, "above nature." The intermediate category, however, which we will call the *preternatural* — from Aquinas' repeated phrase "*praeter naturae ordinem*" — was made up of unusual occurrences that nonetheless depended on secondary causes alone and required no suspension of God's ordinary providence. Mirabilia belonged to this last category, which was intimately associated with the passion of wonder.[45]

This set of distinctions housed a host of problems, most of which centered on the category of the preternatural. For one thing, the boundaries of the three realms, despite their metaphysical clarity, were extremely difficult to define in practical terms. The boundary between the natural and preternatural could be established only with reference to the frequency of phenomena, since both arose from the same kinds of secondary causes.

Preternatural phenomena were by definition rarer than natural ones ("that which is always or for the most part"). But as travel and topographical writers increasingly emphasized, rarity and the wonder it occasioned were often dependent on geography: a pygmy might appear to western Europeans as a preternatural marvel, but in pygmy land, the reverse would be true.[46] It was equally difficult to determine the boundary between preternatural and supernatural phenomena, both of which were rare and therefore inspired wonder, so that writers less precise than Aquinas often conflated the two. How then were the two to be distinguished? The centrality of wonder to this set of questions appears clearly in the discussion in Aquinas' *Summa contra gentiles*, which focused not on the objective criterion of rarity but on the subjective passion of wonder:

> Things that are at times divinely accomplished, apart from the generally established order in things, are customarily called *miracles*; for we observe the effect but do not know its cause. And since one and the same cause is at times known to some people and unknown to others, the result is that, of several who see an effect at the same time, some wonder, while others do not. For instance, the astronomer does not wonder when he sees an eclipse of the sun, for he knows its cause, but the person who is ignorant of this science must wonder, for he ignores the cause. And so, a certain event is wondrous [*mirum*] to one person, but not so to another. So, a thing that has a completely hidden cause is wondrous in an unqualified way, and this the name, *miracle*, suggests; namely, *what is of itself filled with admirable wonder*, not simply in relation to one person or another. Now, absolutely speaking, the cause hidden from every man is God.[47]

Thus the preternatural is wonderful only to the uninstructed, whereas the miraculous is wonderful to all. Aquinas underscored this distinction, noting that although "it may seem astonishing to ignorant people that a magnet attracts iron or that some little fish might hold back a ship" (by which he meant the marvelous remora), the truly miraculous is "what is done by divine power, which, being infinite, is incomprehensible in itself."[48]

In addition to finessing tricky questions as to how the preternatural, or marvelous, might in practice be distinguished from the supernatural, or miraculous, Aquinas' discussion of the difference between the reactions of the learned and the ignorant also pointed to associated ethical problems, at least in the context of the Christian tradition. As we have already indicated, Augustine had described wonder, which Aquinas associated with ignorance, as a highly salutary passion — the proper expression of humility before the omnipotence of God.[49] Augustine had contrasted wonder with curiosity, which he and other patristic writers viewed as a particularly heinous sort of sin. In his *Confessions*, Augustine described curiosity as a variety of lust: the "concupiscence of the eyes [*concupiscentia oculorum*]."[50] The eye in question was the eye of the mind, made to stand for all knowledge gleaned by the senses, and Augustine considered its temptations more dangerous than those of the eye of the body. Thus he castigated the "vain and curious desire [*cupiditas*], not to take delight in the flesh, but to have experiences through the flesh, which is masked under the title of knowledge [*scientiae*] and learning."[51] He enumerated as typical of the "disease" of curiosity a fascination with mangled corpses, magical effects and other marvelous spectacles (*in spectaculis . . . miracula*), and the obsession with religious prodigies and omens—all but the first signal examples of the preternatural, whose causes God had precisely and intentionally hidden from human eyes.[52] At best, such curiosity is perverted and futile; at worst it is a distraction from God and salvation.

By virtue of its proximity to lust and other appetites, Augustine portrayed curiosity as akin to bodily incontinence. Addressing God, he wrote, "You impose continence upon us, which binds us up and brings us into one [*colligimur et redigimur in unum*], whence we have been scattered into many."[53] Like other appetites, curiosity shattered self-mastery and with it the sense of self fostered by continence. Just as lust overwhelmed the body and scattered its energies, so curiosity waylaid the mind, dissipating concentration. Augustine admitted to being daily tempted by distractions that wrenched his attention from his prayers to the inanities of a lizard or a spider catching flies. Nor was it enough then to praise the "wonderworking creator [*creatorem mirificem*]," for "that was not why I paid attention to them."[54]

Augustine also associated curiosity with pride, identifying it in this context as a vice particular to the learned, since it prompts men to "investigate the works of nature, which do not concern us and which it is useless to know, but which people desire to know only for the sake of knowing."[55] Thus the power of astronomers to predict eclipses swells them with vainglory:

> The proud cannot find you [God], even though with curious skill [*curiosa peritia*] they number the stars and grains of sand, and measure the starry heavens, and track the courses of the planets. . . . People that do not know these things marvel [*mirantur*] and are amazed [*stupent*], and those that know them exult and are puffed up. And turning from your light through impious pride, they foresee an eclipse of the sun far in the future but even in the present do not see their own eclipse.[56]

For Augustine, the astronomers' presumption here led them into a two-fold trespass. On the one hand, it prevented them from feeling appropriate wonder when faced with the marvels of creation, such as a solar eclipse — wonder that the laudably ignorant felt in full measure. On the other hand it led them, and encouraged them to lead others, into error, usurping for themselves the wonder that ordinary Christians should direct not toward other humans, however learned, but rather should reserve for God.

These Augustinian readings of curiosity and wonder strongly influenced later Christian writers.[57] The twelfth-century monastic reformer Bernard of Clairvaux reiterated Augustine's identification of curiosity and pride, calling it (with reference to the falls of both Adam and Lucifer) "the beginning of all sin." Pope Innocent III, in the early thirteenth century, also railed against wise men who presumed to study the "height of the sky, the breadth of the earth, and the depth of the sea."[58] Such passages created a dilemma for Aquinas and other writers on natural philosophy, whose job it was to study and to teach the order of the physical world. It was one thing for Augustine to castigate the astronomer who inquired into the causes of eclipses as impious, curious, and lacking in seemly wonder. For Aquinas, however, as indicated in the passage of his *Summa contra gentiles* quoted above, the knowledge of the astronomer was clearly a positive trait, one that elevated him above the ignorance of the unlearned and defined him as wise. Thus the Latin natural philosopher had an ambivalent relationship to the preternatural and to the passion of wonder that defined its boundaries. As a Christian — often a cleric — he was committed to a tradition that saw humble and accepting wonder as the proper passion with which to regard natural phenomena, particularly marvelous natural phenomena. Because wonder was associated with the ignorance of causes, however, it was a peculiarly unsuitable passion for one whose entire discipline was organized around the causal knowledge of nature, if not the numbering of stars and grains of sand.

Aquinas and his contemporaries proved unable to resolve this dilemma in any satisfying way. Aquinas' solution, like that of Albertus Magnus (and a number of ancient writers), was to make a distinction between curiosity, which he retained as a vice, and acceptable forms of inquiry. For Albertus, curiosity was "the investigation of matters which have nothing to do with the thing being investigated or which have no significance for us," while he defined as prudence "those investigations that pertain to the thing or to us."[59] Aquinas' strategy was similar: for him curiosity, a vice, was aimless and half-hearted, while studiousness (*studiositas*), a virtue, was disciplined devotion to intellectual knowledge itself. On the basis of this distinction, he simply laid aside the heart of Augustine's argument, replacing it with another set of values, less sympathetic to wonder and more sympathetic to curiosity. For professional scholars and teachers, as Aquinas put it, "however much it abounds, knowledge of the truth is not bad, but good. The desire for a good is not wicked. Therefore no wrongful curiosity can attend intellectual knowledge."[60]

Albertus Magnus and Thomas Aquinas acknowledged the power of wonder but distanced themselves from it as academic philosophers. In the fourteenth century, wonder — whether romantic and paradoxographical wonder at exotic and unusual phenomena, Aristotelian wonder at ignorance of causes, or Augustinian wonder at the mysteries of creation — largely disappeared from the works of philosophical writers. The principal exception to this statement was the idiosyncratic Catalan philosopher, Ramon Lull, who attempted to combine all three types of wonder in his vernacular treatise, the *Llibre de meravelles* (c. 1310).

Lull's syncretism reflected his background: he was not only a man of noble birth with courtly connections, but a university lecturer and a Christian apologist as well. His prologue to this work outlined his peculiar mix of themes. Describing himself as "sad and melancholy," he lamented "how little the people of this world knew and loved God, who created this world and with great nobility and goodness gave it to men so that He would be much loved and known by them."[61] He went on to sketch a frame story drawn straight from the narrative and rhetorical conventions of topography and romance. Sent by his father to explore the world, Lull's hero, Felix, traveled "through woods, over hills and across plains, through deserts and towns, visiting princes and knights, in castles and cities; and he wondered at the wonders of this world, inquiring about whatever he did not understand and recounting what he knew."[62] Like a romance hero, "going throughout the world seeking wonders,"[63] Felix wandered in search of intellectual adventure, and each encounter became an opportunity for him to learn (and Lull to teach) the causes of natural phenomena, articulated for the most part in Aristotelian terms: how plants grow, the causes of the shapes of clouds, why the lodestone attracts iron, how alchemical transformations work. Yet the explanations never served to dispel Felix's wonder, which was ultimately rooted in an Augustinian reverence for divine creation rather than an Aristotelian ignorance of causes. In the end Felix died, still wondering, with his last breath, why God had not spared him to finish his work.

Lull's *Llibre de meravelles* was a valiant effort to bridge the gap between university natural philosophy, Augustinian and monastic values, and the literary tastes of courtly and urban elites. It was of very limited influence,

125

however, despite fifteenth-century translations into Spanish, Italian, and French. (The sole surviving French manuscript belonged, not surprisingly, to Louis de Bruges, lord of Gruthuyse, a close associate of Dukes Philip the Good and Charles the Bold.)[64] For other philosophers, particularly those aiming at a professional audience, curiosity retained its Augustinian stigma, while wonder receded as a cognitive and religious value. In this context, the preternatural shook off its mystery — emblem for Augustine of divine omnipotence — and became fair game in the philosopher's program to make wonders cease.

Making Wonders Cease

From the point of view of causal analysis, the preternatural posed special problems, largely because it was a negative category — and a negative category, furthermore, whose limits were defined in practical terms by a pair of unstable criteria both of which depended on the experience and knowledge of the viewer: that which was infrequently experienced or that at which the ignorant wondered. As a result, the preternatural consisted of a stratigraphy of heterogeneous phenomena, built up in layers from several different traditions with no internal coherence except their awkward relationship to *scientia* in the Aristotelian sense. These phenomena might include (depending on the author in question): conjuring tricks; natural substances (domestic and, especially, exotic) endowed with occult properties, as well as other staples of the ancient paradoxographical tradition; necromancy and other forms of demonic intervention; and chance or accidental phenomena as defined by Aristotle himself. Although thirteenth- and fourteenth-century philosophers might agree on the importance of supplying natural causes for such effects, they differed significantly as to the specific sorts of causes involved, particularly once the influx of Arabic philosophical writing in the twelfth century had vastly expanded the very limited repertory of causal mechanisms previously available to Latin writers.

Some kinds of causes proposed to explain wonders were relatively uncontroversial, at least in their general outlines. All academic writers on philosophy acknowledged the existence of tricksters or charlatans, who produced marvelous illusions by sleight of hand.[65] Similarly, all subscribed to the existence of chance or accidental phenomena — Aristotle's discussion of this point was unambiguous — defined as the result of a tangle of natural causes that had combined in an unpredictable way. As we have seen, this was the mechanism invoked by Thomas Aquinas to explain the birth of a six-fingered baby, while in his *Summa contra gentiles*, he used Aristotle's example of a treasure unearthed during the digging of a grave:

"now, the grave and the location of the treasure are one only accidentally," he wrote, "for they have no relation to each other."[66]

Generally accepted, too, by the middle of the thirteenth century was the idea, also traceable to Aristotle, that the heavens had impressed many natural substances with marvelous properties. Most Latin philosophers and medical theorists identified the principal intermediary in this process as the substance's "specific form," defined as the form possessed by particulars insofar as they belonged to a given species: that which confers sapphireness on a sapphire, or poppyness on a poppy. This doctrine had been elaborated by Arabic philosophers, notably Avicenna and Alkindi, and was used to account for the action of many mirabilia: the attractive powers of the lodestone, for example, the poisonous aura of the basilisk, the extraordinary strength of the remora. Such formal properties were often termed "occult," or hidden, to distinguish them from the "manifest" properties of substances, which were thought to arise from their complexion (their particular balance of hot, cold, wet, and dry) and therefore ultimately from the matter of which they were composed.[67] Whereas the manifest properties of things depended on sensible qualities, so that the manner of their action was perceptible (hence manifest), occult properties embraced more puzzling sorts of operation, involving action at a distance (the lodestone, the basilisk) or dramatic effects out of all proportion to the manifest cause (the remora). In addition to working through specific forms, some philosophers argued that the heavens might also impress other kinds of remarkable and unpredictable properties on matter. Albertus Magnus in particular frequently invoked this form of causal explanation: he used it in his *De mineralibus* to explain the birth of certain kinds of monsters (such as apparent animal-human hybrids), the appearance of figured stones (such as fossils), and local topographical phenomena (such as petrifying springs).[68] In the case of the last, he argued, each place on earth received a constellation of celestial rays unique in angle and direction, and this accounted for its unique properties — thereby accounting for the geographical variability in the natural order that was so fundamental to the topographical tradition.

In addition to chance and the occult properties impressed by the heavens, some writers on philosophy invoked another kind of causal process in their attempt to make wonders cease: the direct intervention of subcelestial intelligences, either human or demonic. Here Aristotle's Latin commentators differed sharply among themselves regarding the nature and admissibility of such action. Appealing to Avicenna, the author of the pseudo-Albertine *De mirabilibus mundi* held that certain extraordinary human souls, elevated by passion, could work immediately on other souls or material objects to alter them or force them to obey their will—a view that Albertus Magnus seems also to have shared.[69] Most other Christian writers considered this position theologically suspect, since it attributed the miraculous powers of saints and prophets to human abilities; they argued that the human soul could operate only through the intermediary of its body's humors and spirits—a position that was ultimately to prevail in the early modern period.[70]

For many of the same reasons, demonic action as an explanation for marvelous phenomena proved even more controversial (in addition to being wholly un-Aristotelian). Thomas Aquinas, for example, insisted that demons, like human souls, could work only through physical intermediaries such as vapors, the elements, or other applications of matter— a restriction that confined them to the domain of preternatural rather than supernatural action.[71] His contemporary at Paris, Siger Brabant, went considerably further, refusing categorically to admit demons into any strictly philosophical consideration and restricting them to the domain of entities whose existence was guaranteed by faith alone.[72]

Although they differed on individual points, most of Aristotle's Latin commentators agreed in rejecting what they considered to be popular superstition, which attributed inordinate power to magic and to the demons that were supposed to make it work. Although some philosophical writers (for example, William of Auvergne and Albertus Magnus) attempted to make a distinction between good magic, which used the occult properties of natural objects, and bad magic, which used demons, most followed Bacon and Aquinas in reserving the term "magic" for the latter, insisting on the very limited role of demonic action in the world, despite firm lay opinion to the contrary.[73] Thus Aquinas composed an erudite letter in response to a query from a "northern knight," in which he explained the strict limits placed by God on demonic action and the natural causes that could account for apparently demonic intervention, while Bacon denounced all so-called magic as the result either of natural forces or of sleight of hand.[74] Although neither would have gone as far as Siger in arguing that all natural phenomena could be explained without recourse to demons—or indeed divine intervention—they would certainly have concurred in his lament concerning vulgar credulity on this topic: "In matters where the truth is deeply hidden, the common folk [*vulgo*] are not to be believed.... And if you say that it is commonly believed, this is no proof, for many falsities are commonly believed."[75] It is for this reason that Albertus Magnus and Thomas Aquinas used examples such as the birth of a six-fingered baby or the discovery of a buried treasure as types of accidental occurrence: popular wisdom tended to attribute such things to the direct intervention of either God (in the form of a prodigy) or demons (in the form of magic), rather than recognizing them as coincidences and chance events.

Despite philosophers' commitment in principle to explain marvels by reference to natural, or secondary, causes and with minimal recourse to divine or demonic intervention, the actual study of particular wonders could never form part of natural philosophy, because these belonged to the realm of contingent effects, as we have indicated above. Some wonders (individual or unique phenomena) were the result of chance, while others (the properties of marvelous species) depended on imperceptible forms. In practical terms, this meant that the properties of any particular such place or substance could be determined only by experience, which was not part of the philosopher's brief. Even a philosopher fully versed in the doctrine of specific forms, for example, would be unable to figure out a priori that the sapphire was beneficial to eyesight, for example, or that the lodestone attracted iron, although he knew in theory that they worked through their specific forms; the knowledge of particular properties came only out of personal trial and error or the accumulated experience of generations, as recorded in books or enshrined in popular wisdom. As the author of the pseudo-Albertine *De mirabilibus mundi* put it:

> Certain things are to be believed only by experience, without reason, for they are concealed from people; others are to be believed only by reason, because we lack sensations of them. For although we do not understand why the lodestone attracts iron, nevertheless experience shows it, so that no one should deny it. And just as this is marvelous and established as certain only by experience, so likewise should one suppose [to be the case] in other things. One should not deny any marvelous thing because he lacks a reason for it, but should try it out [*experiri*]; for the causes of marvelous things are hidden, and follow from such diverse causes preceding them that human understanding, as Plato says, cannot apprehend them.[76]

Thus natural wonders often overlapped with "secrets" and "experiments" (*experimenta*), another group of phenomena accessible only to experience; these craft formulas, or proven recipes for medical and magical preparations, often drew on the occult properties of natural substances, and they were excluded from natural philosophy for the same reasons.[77]

Inevitably, this fundamental inaccessibility of marvelous phenomena to unaided reason placed serious limitations on the ability of the philosophers to make wonders cease — or at least to provide full and specific explanations for them. It meant that philosophers could only invoke general *types* of explanations (of the sort given above) to explain general *types* of marvelous phenomena. Albertus Magnus, for example, attributed petrifying springs and the powers of figured stones generally to the powers of the heavens, but he was unable in any specific case (for example, the spring in Gothia, or a remarkable cameo) to indicate which particular spheres or planets were at work and in what way. To our knowledge, neither he nor any of his philosophical colleagues — with the possible exception of the elusive Peter of Maricourt — ever even tried. (After a set of exhaustive experiments with lodestones, Peter speculated in 1269 that the fact that the lodestone had poles meant that its powers must derive from the north and south poles of the sphere of the fixed stars.)[78] More typical was the author of *De mirabilibus mundi*, who devoted the body of his work to listing a series of marvelous "secrets" that ranged from the power of the stone in the hoopoe's nest to confer invisibility to recipes for hallucinogenic preparations touted as capable of making people think that anyone they saw was an elephant, or that the house was full of snakes.[79] But despite the arsenal of general causal explanations elaborated in the first part of his treatise, the author made no effort to apply any of them to these particular examples.

Within these limits, however, when (and if) high and late medieval scholastic philosophical writers considered marvels, they did so in order to explain their marvelousness away. Despite their differences, Adelard of Bath, Roger Bacon, Albertus Magnus, Thomas Aquinas, and Siger of Brabant all shared the goal of removing mirabilia from the realm of the demonic and the supernatural, where they had been placed by the ignorant and the vulgar, and of ascribing them to natural causes alone.[80] This process served two linked agendas, one philosophical (focused on causal explanation) and one social and theological (focused on combating what were seen as erroneous, superstitious, and potentially heretical lay beliefs).

The most comprehensive and systematic such attempt was Nicole Oresme's *De causis mirabilium* (c. 1370). Trained in philosophy and theology at the University of Paris, Oresme had built a successful career as a master of theology before he was called to the service of the Dauphin who in 1364 became King Charles V of France. In his preface, Oresme was quite explicit about his aims in this treatise on wonders. "In order to set people's minds at rest to some extent," he wrote,

> I propose here, although it goes beyond what was intended, to show the causes of some effects which seem to be marvels and to show that the effects occur naturally, as do the others at which we commonly do not marvel. There is no reason to take recourse to the heavens, the last refuge of the weak, or

demons, or to our glorious God as if He would produce these effects directly, more so than those effects whose causes we believe are well known to us.[81]

As this passage indicates, Oresme had two goals. First, he wished to attack the entire edifice of astrology, including not only astrological divination, but also celestial influences in general. This was one of the central themes of his intellectual career, and it led him to reject even the doctrine of specific forms, a radical position in which he was followed by only a few writers, such as Henry of Hesse.[82] His second goal, by now familiar, was to combat the disturbing lay tendency to invoke God or demons, both as an explanation of preternatural phenomena and as a means of political manipulation. Oresme's concern with judicial astrology and sorcery were by no means merely theoretical: both practices seem to have flourished at the court of his royal patron, and as Charles V's advisor, Oresme was in direct competition with their practitioners.[83]

In a number of respects, Oresme's treatise continued the philosophical tradition we have been describing. In four chapters devoted, respectively, to marvels involving vision, hearing, touch and taste, and operations of the soul and body, Oresme constructed a complex edifice of causal explanations to account for anomalous phenomena of all sorts. Like his predecessors he continued to restrict himself to enumerating the general type of causal mechanism that he would expect to explain a general type of wonder, while implicitly acknowledging the impossibility of philosophically accounting for individual wonders such as the extent to which people differ in their need for food and drink. "Now who would render the causes and the particular differences in all these individual cases?," he asked. "Certainly God alone."[84]

But there are several elements of Oresme's work that set him apart from his thirteenth-century predecessors and looked forward to the changes we will trace in the next chapters of our book. In the first place, Oresme laid unusual emphasis on the variety and diversity of natural phenomena, perhaps because of his familiarity with court culture. The hallmark of Aristotelian writing on the natural order was its emphasis on the regularity — habitual if not inviolable — of nature in her "common course." Oresme, in contrast, dwelt more on contingency and diversity. Considering the variety of human tastes in food and sexual pleasure, he noted that

> if you knew which and how many conditions and circumstances are required for the desire of eating to be natural, and which and how many for the desire of sexual activity to be natural, then you would not marvel about diversity in these things.... This is to be noted: we must marvel more when nature pro-

ceeds and acts in so orderly a course than if it sometimes is altered or deviates from the usual, because for the first case many ordered causes are required etc. For you can hit the mark in only one way but miss in many.[85]

While Oresme's thirteenth-century Parisian predecessors attacked the wonderfulness of wonders by emphasizing the regularity and comprehensibility of nature, Oresme here took another tack, suggesting that diversity was so much the norm in nature that we should marvel not at the exception but the rule. He made the same point when he took up monstrous births, to which he denied any portentous or supernatural meaning, noting that "it is more marvelous how nature does not fail more often in this process, than how it often does so. And it is more marvelous how it completes all things in an ordered manner since error can happen from many causes but only in one way can it complete all things successfully — and for this one way many things are required."[86]

Second, we can find in Oresme's discussion of mirabilia a slightly more optimistic attitude than that of his philosophical predecessors toward the eventual ability of humans to determine the actual causes of particular marvels. Regarding the fact that certain houses repeatedly fall down, no matter how many times they are rebuilt, he noted that "it is not on this account necessary to have recourse to marvelous and unknown causes, since if you are willing to pay sufficient attention, the causes of the said effects are apparent": the builder may be incompetent or the materials unsuitable.[87] To whom then did the task of paying attention to such details fall? Oresme gave no indication in this passage on construction disasters, but he hinted at the answer in his prologue:

> As I have said, then, I shall only show in a general manner that [wonders] occur naturally, as do successful [*valentes*] physicians who compose general rules in medicine and leave specific cases to practicing physicians. For no physician would know how to say — if Sortes were ill — what kind of illness he has and how it will be cured, except by seeing him and considering the particulars. Similarly, successful moral philosophers like Aristotle and the rest wrote only general principles, and no law exists, as Aristotle said in the *Politics*, that does not need to be changed at some time.[88]

In identifying physicians — specifically practicing physicians — as those concerned with explaining particular natural phenomena, including wonders, Oresme was echoing a passage from Aristotle's *Ethics*.[89] But he may also have been aware that some of his contemporaries were beginning to take an interest in matters of this sort. We do not refer here to the fourteenth-century followers of William of Ockham and the *via moderna* — there is no evidence that any of these "nominalists," despite their empiricist epistemology, ever engaged in the empirical study of particulars[90] — but rather to a number of Italian medical writers (some of them with alchemical interests). Although these men had studied philosophy and had taught at universities, like Oresme, they served princely patrons whose needs and interests differed from those of academic philosophers and for whom wonders had a strong appeal. Immersed in elite medical practice, these physicians began to explore the therapeutic powers of particular marvels, giving them philosophical explanations when they could. In part as a result of their work, wonder and wonders began to be integrated into natural philosophy. By the middle of the sixteenth century, wonders lay at the heart of much philosophical writing — a place they were to occupy for more than a hundred years thereafter. It is to this part of the story that we now turn.

CHAPTER FOUR

Marvelous Particulars

While Oresme completed his *De causis mirabilium*, a well-known Italian physician and professor of medicine, Giovanni Dondi, was making preliminary observations for a treatise on the hot springs near his hometown of Padua. Like the petrifying lakes in Albertus Magnus' *De mineralibus*, springs with special therapeutic or rejuvenating properties were a familiar element in the canon of natural wonders and a staple of the literature of topography and romance.[1] Dondi drew on these associations in his treatise, composed c. 1382:

> When I first saw these waters and considered their properties, which seem outside [*extra*] the nature of other waters and other springs, I wondered not a little and, not finding causes for those properties that were wholly satisfactory, I was for a long time in doubt on many points. But now I have learned from passing years and gathered from long experience that there is nothing that is not wonderful, and that the saying of Aristotle in the first book of the *Parts of Animals* is true, that in every natural phenomenon there is something wonderful — rather, many wonders. Thus indeed it is, brother: among wonders are we born and placed and surrounded on all sides, so that to whatever thing we first turn our eyes, it is a wonder and full of wonders, if only we examine it for a little. But of many things which are equally wonderful, familiarity and daily use and frequency either remove or lessen our wonder. For this reason, therefore, I do not wonder as I used to, but finding everything wonderful and reflecting on it, I have told myself not to wonder at anything very much.[2]

This passage contains familiar elements: the wise man's search for causes, his eventual retreat from wonder. Nonetheless, the tone and content are new, for Dondi here abandoned the jaundiced attitude toward wonder and wonders that had marked the work of high medieval natural philosophers. In an earlier section of the same work, he had already cited Aristotle's description of wonder as the beginning of philosophy (from

135

the *Metaphysics*) without any of the usual qualifications.[3] The passage quoted above goes even further, echoing not only Aristotle's portrayal of the natural world as a congeries of wonders—a text from *De partibus ani-malium* studiously ignored by thirteenth-century natural philosophers— but also Avicenna's view that careful consideration multiplies instead of decreases the number of wonders, "so that to whatever object the eye first turns, the same is a wonder and full of wonder, *if only we examine it for a little*."[4] Dondi saw this effect as only temporary, lessened by familiar-ity and, presumably, understanding—but the fact that he invoked wonder with such force was itself remarkable.

Dondi may have been one of the first Latin naturalists to make a place for wonder in natural inquiry, but he was by no means the last. In this chapter, we will trace the stages by which wonder and wonders began to enter natural philosophical writing during the two centuries between about 1370 and 1590. As Dondi's example suggests, the principal agents in this process were not academic philosophers but writers with natural philosophical training who were working in related fields of inquiry, most notably medicine, which also had links to alchemy, *materia medica* (or pharmacology), and magic. Unlike natural philosophy, all of these fields had a strong practical component. Unlike natural philosophers, who concentrated for the most part on developing universal causal arguments, men engaged in these enterprises had to come to grips with particular natural phenomena in the animal, plant, and mineral worlds.

For a number of interrelated reasons, natural marvels played an im-portant part in the elaboration of forms of natural inquiry based on the study of particulars. First, most marvels were either singular events or substances that derived their wonderful properties from occult qualities; according to contemporary ideas, both could therefore be known only through empirical investigation and were therefore amenable only to empirical investigation.[5] Second, their intrinsic fascination and charisma set them apart from more mundane phenomena, motivating and ennobl-ing their investigation. But the single most important factor was probably the fifteenth- and sixteenth-century European voyages of exploration, to Africa, to Asia, and ultimately to the "New World" of America, which yielded wonder on top of wonder. Many of these new marvels had never appeared in ancient or medieval texts; whether reported, depicted, or physically collected, they quickly overflowed the traditional confines of erudition and of medical and pharmacological inquiry to demand empiri-cal study in their own right.

Initially, medical writers and others involved in this enterprise made relatively modest claims. Their tone was often defensive, revealing the continuing hegemony of traditional natural philosophical styles of explanation, and they looked for legitimation to illustrious patrons and to the courtly associations of the marvelous. Beginning in the middle of the sixteenth century, however, a new ambition and a new confidence marked the works of a number of medical and philosophical writers on marvels. These men not only reclaimed wonder as a philosophical emotion, but also rehabilitated wonders as useful objects of philosophical reflection. In the process they dramatically expanded the purview of natural philosophy to include marvelous effects of all sorts, and indeed eventually to privilege the previously excluded realms of the empirical and the magical, neither of which were amenable to demonstration. The result was a new kind of philosophy — we will call it "preternatural philosophy" — which rehearsed new empirical methods of inquiry and new types of physical explanation.

Italian physicians and natural philosophers of the Renaissance period were most active in this area. Their work reflected a new social and cultural environment for natural inquiry. In earlier chapters we have stressed the differences between the visions of nature and discourses of the marvelous developed at court, in the context of Christian religious orders, and in the university. Even in the medieval period, these environments overlapped and intersected. Over the course of the period described in this chapter, however, those intersections became ever denser and the barriers between them progressively eroded — the result of rapid urbanization, the spread of printing, and many other important social and technological changes. The relatively fragmented cultural world of medieval Europe yielded to a more fluid situation in which intellectual life was increasingly dominated by an urban elite, both social and educational, made up for the most part of men with professional training (physicians, some surgeons and apothecaries, lawyers, some notaries) and educated patricians, some engaged in mercantile activities, who embraced learning as a leisure activity. Some of these men taught in universities, or served at court, or both. Less commonly they belonged to religious orders or, in the sixteenth century, the Protestant clergy. As a group, they were mostly urban in outlook, secular in orientation, and inspired by the courtly and aristocratic values of their most important patrons, even when not themselves courtiers or members of local aristocracies.[6] All of these qualities shaped the appearance of wonder and wonders in their work.

Marvelous Therapeutics
While marvels had no place in traditional Latin natural philosophy, learned physicians prized them for their healing powers. Wonders like Dondi's

healing springs, or plants and animals with other remarkable properties, had long been a part of therapeutics; this fell under the branch of medical learning known as *practica*, which dealt with the diagnosis, description, and treatment of individual diseases in the bodies of individual patients.[7] Although lower in prestige than its sister discipline of *theorica*, which considered general principles of disease and physiology and therefore more closely approached the demonstrative ideal of natural philosophy,[8] *practica* flowered in the fifteenth century, spurred on by an explosion in the marketplace for professional medical services. The result of progressive urbanization, this explosion appeared first in northern and central Italy, while other areas of Europe followed with more or less delay.[9]

Practica was a university discipline, and most of the authors who composed treatises on practical medicine were either physicians trained at the university or university professors themselves. They were well versed in natural philosophy, for medical faculties required a degree in natural philosophy as a prerequisite for admission. Some wrote primarily for their students or for other practicing physicians, while others, increasingly, addressed treatises of medical advice and information to the wealthy patrons on whom their fortunes depended.[10] It is in the latter treatises that one finds a special emphasis on marvels both domestic and exotic, such as comets or exotic substances with powers to heal or harm. Prominent among these were therapeutic springs of the sort praised by Giovanni Dondi.

Although mineral springs had long been a staple of Italian therapeutics, it was only after about 1350 that they captured the sustained attention of professional medical writers in central and northern Italy — the result of a marked revival of lay interest in thermal medicine, which had flourished among the Etruscans and ancient Romans.[11] Springs with special properties were among the most remarkable of natural wonders, since each was considered to be not just rare but unique. As Albertus Magnus had argued in *De mineralibus*, they owed their irreproducible character to the particular subterranean arrangement of mineral deposits and heat sources that gave to each its particular composition and temperature, as well as to the specificity of the celestial rays and influences received by each place on earth.[12] (This last property sparked a learned debate as to whether the water of such springs lost its virtue when bottled and removed.)[13]

In writing on medicinal springs, Italian physicians did not aim to provide demonstrative and necessary causal explanations for their individual properties — this continued to be patently impossible — but rather to fashion a set of methods and principles by which these properties could be rationally investigated, thereby bringing them within the purview of the naturalist. The most important and influential fourteenth- and fifteenth-century writers on this topic, in addition to Giovanni Dondi, were Ugolino of Montecatini, who composed his *Tractatus de balneis* in 1417, and Michele Savonarola, whose *De balneis et thermis naturalibus omnibus Italiae* dates from 1448–49.[14] All three men had taught in medical faculties, but these works reflected their treatment of princely patrons. Dondi's treatise grew out of his attendance on Duke Galeazzo Visconti of Milan, while Ugolino's was inspired, at least in part, by his long association with Pietro Gambacorta, lord of Pisa, and Malatesta de' Malatesta, lord of Pesaro. Savonarola dedicated his own work to Borso, son of the Marquis Niccolò d'Este, of Ferrara, who had hired him as court physician in 1440.[15] Furthermore, all three authors presented their subjects as marvels, although this was often something of a stretch. Thus Dondi not only evoked the emotion of wonder in the passage with which we began this chapter, but he also enumerated the "marvelous accidents" of the waters of the Paduan hot springs: their even temperature, the small animals that lived in them, the gypsumlike deposits they left in metal pipes.[16] In adopting this terminology he was probably modeling himself on an early thirteenth-century writer, Peter of Eboli, who had dedicated a long poem on the baths of Pozzuoli, traditionally one of the wonders of Naples (fig. 1.1), to Emperor Frederick II.[17] In the later fourteenth and fifteenth centuries Peter's poem enjoyed a significant revival among Italian princes, as the number of beautifully illustrated manuscripts of this work testifies (fig. 4.1).[18]

The uniqueness of individual mineral springs had important epistemological consequences, for if each spring was unique — so that even directly adjacent springs could have, say, wildly different temperatures — then their properties could not be deduced from first principles, but had to be derived from experience of the individual case. As Savonarola put it, "all these things are probable, lacking logical demonstration. But experience is the mistress of all these discords."[19] In this passage, Savonarola explicitly used the language of probability and opinion to underscore the fact that, although the properties of individual springs had natural causes, the particularities of place meant that those causes could not be known with certainty and those properties were not amenable to demonstrative or "scientific" knowledge. Thus each spring had to be studied individually and with the utmost attention, using all the information available to the senses: the color and temperature of the water, its smell and taste, the nature of the illnesses it cured. At the same time, Savonarola realized that this situation put him on epistemologically shaky ground: "I have described

Figure 4.1. The baths of Pozzuoli
[Peter of Eboli, *De balneis puteolanis*], MS 838 (G. 2396), fol. 13r, Biblioteca General de la Universidad, Valencia (Naples, third quarter of 15th century).

The bath of S. Anastasia was one of the many therapeutic baths at Pozzuoli, which qualified as one of the wonders of Naples and its environs (see fig. 1.1). According to Peter, this spring was especially good for pains in the joints: "Indeed, it is a wonderful thing, that as long as one digs in the sand in the middle of its trench, the water remains hot. It takes away symptoms, as long as it has just been taken from its spring, but if it is carried away and loses its heat, it is of no use." This manuscript (in which the work is erroneously attributed to Arnald of Villanova) once belonged to Alfonso of Aragon, Duke of Calabria.[15]

140

in positive terms this way of investigating the cause of the heat of thermal baths," he wrote,

> [although] I judge that this material is not conducive to demonstration and cannot be defended from contradiction. But it has seemed to me the most expeditious [mode of investigation] and the most consonant with human minds. On account of this let no one bite me [*me quisquam non mordeat*], since I have thus [at least] supplied [the basis] for investigating another and perhaps truer cause.[20]

These defensive statements make sense in the context of the competitive nature of elite medical practice.[21] But they illuminate even more vividly the more general difficulties of naturalists who, in attempting to construct a paradigm for the philosophical knowledge of wonders, were forced to abandon the limpid certainty of *scientia* for the muddy waters of sensory experience and probable opinion. As we argued in the preceding chapter, both of the latter labored under the epistemological stigma of uncertainty and the sociological stigma of the mechanical arts — a stigma that did not threaten more theoretically oriented types of medical writing. As an erstwhile university professor himself, fully versed in Aristotelian philosophy, Savonarola was well aware of the degree to which his project left him open to intellectual attack.

Encouraged by their aristocratic patients, Savonarola and his colleagues chose to navigate these waters, but it is not surprising to find them repeatedly emphasizing their courtly connections as a source of intellectual and social legitimation. Ugolino and Savonarola referred repeatedly to their noble and princely patrons (as well as to their own enormous salaries as court physicians), and both stressed the interest of those patrons in their work. Thus Ugolino noted that Malatesta de' Malatesta had pressed him to investigate the water of Bagno ad Aqua, near Siena; in so doing, he was only following his patron's own (presumably impeccable) example, for Malatesta had already made some preliminary distillations, and, in Ugolino's words, "his reverence begged me to perform experiments [*experientias*] on it and to amplify and improve [his findings]."[22] Savonarola recorded his discussions with the Captain of Carmagnola concerning the relative heat of the baths of Abano and Saint Helena, and he cited the involvement of another noble, named Gelasio, in the study of a recently discovered spring in Carpi: when Gelasio heard that Savonarola was writing a book on baths, he sent him a sample of the water, which had cured a whole herd of cows of a urinary disorder in 1448.[23] By such remarks, these physicians may have hoped to discourage philosophical purists from "biting" them for venturing into the uncertain territory of marvelous phenomena and probable opinion and for embracing a very different epistemological model of natural inquiry from the demonstrative ideal advocated by natural philosophers.

The work of the Italian balneologists was also informed by a sensibility very different from the impassive and distanced stance of the professor of *scientia*, engaged in transmitting to his students the certain causal knowledge that he had received from his own teachers. Like some contemporary alchemical literature — and in marked contrast to contemporary writing on natural philosophy and theoretical medicine — the language of the treatises on springs was autobiographical — at times confessional, as in the passage from Giovanni Dondi that we cited above.[24] (Indeed, the connections between alchemy and this kind of writing were direct; alchemy flourished at a number of Italian courts, and both Michele Savonarola and Ugolino da Montecatini had strong alchemical interests.)[25] Dondi's and Savonarola's references to wonder and wonders signal their assimilation of courtly language and values, and may also have been intended to pique the interest of courtly patrons. No doubt the primary readers of their treatises continued to be physicians; it is unlikely that many princes labored through the technical Latin, despite any glamorous veneer. But the importance of natural wonders in courtly literature and recreational reading contributed to the appeal of the topic and made it at the very least a plausible instrument for ambitious physicians aiming to consolidate an aristocratic clientele.

What then were the characteristics of an emergent study of wonders as it appeared, however tentatively, in fourteenth- and fifteenth-century Italian treatises on healing springs? First, as Savonarola emphasized, it was of necessity empirical: anomalous by definition, the causes and properties of springs could not be deduced from first principles, but had to be inferred from sensible "signs" — odor, color, taste, sound — which "are never found to be wholly effective or infallible, but give knowledge approaching the truth."[26] Second, such a study might have a collaborative component, since many of its objects were newly discovered and needed to be described for the first time. As more and more springs came to light in the fifteenth century, it became increasingly clear that this task lay beyond the abilities of any single individual. Writing in the 1370s, Giovanni Dondi relied primarily on previous textual evidence and his own (and his father's) observations. But Ugolino of Montecatini and Michele Savonarola also collected information orally from local naturalists who had studied the phenomena firsthand. Thus Ugolino noted that because he had never visited the springs at Siena, he was relying on the testimony of two Sienese physicians, Marco and Francesco; when he went to Viterbo to inspect the

Figure 4.2. Ficino's *De vita*
Attavanti workshop (attr.), title page, Marsilio Ficino, *De vita*, MS LXXIII, 39, fol. 4r, Biblioteca
Medicea Laurenziana, Florence (1489).

This page comes from the manuscript of Ficino's *De vita* given to Lorenzo de' Medici by Filippo
Valori in 1489. Like Valori, Ficino was a friend and client of Lorenzo, and he dedicated his treatise to the man he called "most generous." The sumptuous manuscript, decorated with a portrait
of Ficino in the initial of the proemium and the Medici arms at the bottom of the page, suggests
the courtly environment in which the work was produced.[16]

baths there, he wrote, "I wanted for the day that I was there to inform myself from the local doctors and others."[27] Such remarks hint at a nascent community of empirical inquirers, alongside the well-established textual community of readers and writers. However intermittently and unsystematically, these men consulted with each other in order to accumulate and collate their observation of previously unrecorded natural phenomena.

Finally, this new, quasi-philosophical approach to natural wonders differed from traditional natural philosophy in embracing the emotion of wonder itself. In addition to clothing the physicians' epistemologically shaky enterprise in borrowed courtly splendor, the discourse of wonders also seems to have served another purpose: to focus the attention of observers on the particular phenomenon at hand. It was all very well for Oresme to note that the causes of natural effects were divinable if one "paid sufficient attention"; but the habit of paying close attention to natural phenomena (particularly unprepossessing natural phenomena, such as pools of warm, stinking, muddy water) required a special discipline of both the senses and the mind. Like alchemists, medical men had long exercised this discipline in other contexts: assessing symptoms in the human body, following the course of a complicated reaction. The emotion of wonder helped them to transfer those skills to phenomena of the natural environment, since wonder, as Aristotle had suggested in *De partibus animalium*, fostered habits of concentration and meticulous attention.

Thermal springs were only one of the marvels in the physician's therapeutic arsenal. The many plant, animal, and mineral substances stocked by apothecaries and prescribed by physicians included some that qualified as wonders. Rare by definition, most of these were also exotic and had to be imported from Africa and Asia; their resulting cachet and expense associated them strongly with elite practice. For example, during Lorenzo de' Medici's last illness, in 1492, his physicians treated him (unsuccessfully) with a medicine composed of ground gems. Similar substances figured prominently in the astrological medicine developed by Lorenzo's friend and court philosopher, Marsilio Ficino, as described in Ficino's *De vita* of 1489 (fig. 4.2). In Book II of this work, Ficino singled out many exotica as "especially helpful" for the prolongation of life; in addition to the more mundane wine, mint, and plums, one typical passage recommended "musk, amber, fresh ginger, frankincense, aloes, jacinth [probably a form of sapphire], and similar stones, or similar herbs."[28] The drugs and objects described in Book III, which explained how to call down beneficent planetary influences, had an even more marvelous cast. Each "planet" had its retinue of terrestrial wonders, to which it was attracted; for example, the sun (a planet in the contemporary earth-centered cosmology)

144

resonated with "solar" substances such as gold, carbuncles, balsam, cantharides, and crocodiles, as well as carved gems described in meticulous detail.[29] Ficino described all such objects as possessed of "occult and wonderful powers."[30]

Neither the theory nor the practice of this kind of marvelous medicine was wholly new in the late fifteenth century. The doctrine of occult properties and specific forms was well established in academic medicine and natural philosophy, and it had been further elaborated, in the generation or so before Ficino, by a number of medical writers at the University of Padua.[31] But Ficino was the first to fuse this tradition with his reading of the newly discovered works of late Greek Neoplatonic philosophers and the treatises then ascribed to Hermes Trismegistus. The result was a truly magical therapeutics, which employed hymns and images as well as natural substances of the sort described above.[32] We will return below to the natural philosophical vision that underpinned Ficino's work. Here it is enough to say that it exemplified an approach to wonders very different from the researches of the Italian balneologists. Fundamentally bookish in character, it seems to have lacked any empirical component. Instead, it required its practitioners to master the properties of a world of wonderful plants, animals, and minerals as they appeared in encyclopedias, treatises of *materia medica*, and hermetic texts.

Such topics did not exhaust the interest of Renaissance physicians and their patrons in the marvelous aspects of medical practice. Medical writers also paid increasing attention to peculiar physical conditions and what they often called "marvelous" cures. The first and most idiosyncratic work on the topic was *De abditis nonnullis ac mirandis morborum et sanationum causis*, a collection of remarkable cases compiled by the Florentine physician Antonio Benivieni (a client of Lorenzo de' Medici and acquaintance of Marsilio Ficino).[33] The extract published posthumously in 1507 contains brief descriptions of more than a hundred noteworthy cases, beginning with the horrible new disease, syphilis, that had struck Italy in the 1490s and ending with conjoined twins displayed in Florence by their mother. In between, Benivieni described a host of medical peculiarities, mostly from his own experience: people who vomited worms, a girl born without a vulva, a man struck by lightning, a thief who survived hanging, and cases of chronic illnesses healed by prayer.

Benivieni's dedicatory epistle gave a sketchy rationale for this work. "In fact," he wrote, "one who treats many patients for many years knows many and various things worthy of wonder. Wishing therefore to take up this topic, as granted by my age and experience, I will briefly recount those things in our time that appear to be wonderful [*miranda*], or at least

not to be spurned. I think that my efforts will perhaps be of no little utility to many, who will be able to know the hidden causes of nature from
the inside."[34] Benivieni's professed agenda resembled that of the writers
on healing springs; like them, he found that finding causes for individual
marvels was trickier than it appeared. Unlike Savonarola, however, he
never confronted that problem directly, choosing instead mostly to compile and describe individual cases. When he attempted, very occasionally,
to reveal hidden causes, it was in a literal way: by means of post-mortems,
usually performed to satisfy his own curiosity, albeit with the consent of
the next of kin.[35]

Works devoted to remarkable medical cases continued to flourish in
the next century, spurred on by the revival of Hippocratism, with its emphasis on case histories, as well as a growing interest in marvels. Later and
better developed examples of the genre included Girolamo Cardano's *De
admirandis curationibus et praedictionibus morborum*, published in 1565;[36]
Marcello Donati's voluminous and much-cited *De medica historia mirabili*,
which first appeared in 1586;[37] and even, in some respects, the vernacular
Des monstres et prodiges (1573) of the French surgeon Ambroise Paré.[38]
These works and many others like them exemplify the ramification of medical writing on the marvelous over the course of the sixteenth century,
not only in Italy but also in the rest of Europe. They signal, moreover, the
emergence of a large and enthusiastic audience, lay and professional, for
published accounts of natural wonders of all kinds.

Preternatural History

While Benivieni was busy with his remarkable cases, Europeans marveled
at an event that would catapult wonders into unparalleled visibility, both
inside and outside the discipline of medicine. In 1492, Christopher Columbus arrived at what he thought were the eastern reaches of Asia, the islands
so vividly described in Marco Polo's *Travels*, which Columbus had read
and annotated with care.[39] The language of Columbus's journal, written
for his royal patrons, reveals the degree to which he was influenced by the
discourse of marvels in travel and topographical writing. His entry for
October 21, for example, described the island of Cuba in language that
recalls Polo's account of Quilon:

> If the [other islands] already seen are very beautiful and green and fertile, this
> is much more so, with large and very green groves of trees. Here there are
> some very big lakes, and over and around them the groves are marvelous [*en
> maravilla*]. And here and in all of the island the groves are all green and the
> verdure like that in April of Andalusia. And the singing of the small birds [is

so marvelous] that it seems that a man would never want to leave this place. And [there are] flocks of parrots that obscure the sun; and there are birds of so many kinds and sizes, and so different from ours, that it is a marvel [*maravilla*]. And also there are trees of a thousand kinds and all with their own kinds of fruits and all smell so that it is a marvel. I am the most sorrowful man in the world, not being acquainted with them. I am quite certain that all are things of value, and I am bringing samples of them and likewise of the plants.[40]

According to the cosmographer Sebastian Münster, this passage and many others like it led the king of Spain to remark that Columbus should have been known not as the Admiral (*Almirante*) but as the Wonderer (*Admirans*).[41] Indeed, the "discovery" of the land that came to be known as America proved to be an epochal event that not only yielded a host of exotic new naturalia for study, but also prompted a reconsideration of how nature herself might best be explored. The fifteenth-century Portuguese voyages of exploration had revealed small groups of new islands and expanded European knowledge of already familiar continents, but the revelation that the world included whole new continents undreamed of by the ancients opened up a fissure in time as well as space. Compared to earlier writing on travel, the works from the decades after 1492 demonstrate a heightened sense of novelty and possibility – of just how new and different things were able to be.[42] Partly as a result of these changes, sixteenth-century Europeans increasingly began comprehensively to describe and catalogue natural phenomena, domestic and exotic, for their own sake rather than primarily for their therapeutic powers, as had the Italian balneologists.

In the sixteenth century, as in the fourteenth, writers used the rhetoric of wonder to express this sense of novelty and possibility. As Stephen Greenblatt has argued, this rhetoric supported the conquest and subjection of the American population, just as it had three centuries earlier in Gerald of Wales's *Topographia Hibernie*, which contrasted the moral depravity of the Irish with the marvels of their conquered land.[43] At the same time, it gave new life to the long tradition of writing on natural wonders. Some thought that it rehabilitated Pliny, Solinus, and other ancient writers on marvels, whose veracity Gerald of Wales and many others had doubted. In an account of his voyage to Brazil, written in the 1560s, the French writer Jean de Léry noted,

> I am not ashamed to confess that since I have been in this land of America, where everything to be seen – the way of life of its inhabitants, the form of the animals, what the earth produces – is so unlike what we have in Europe, Asia, and Africa that it may very well be called a "New World" with respect

to us, I have revised the opinion that I formerly had of Pliny and others when they describe foreign lands, because I have seen things as fantastic and prodigious as any of those — once thought incredible — that they mention.[44]

By contrast, others used the Wonders of the West to illustrate the inadequacies of ancient knowledge, noting that Greek and Roman writers, far from being authoritative sources, were wholly ignorant of much of the natural world. In the words of Léry's contemporary Girolamo Cardano, "among natural prodigies, the first and rarest is that I was born in this century in which the whole world became known, although the ancients were familiar with little more than a third."[45]

Cardano was a physician and professor of medicine. As his comment suggests, the late fifteenth- and sixteenth-century European enterprises of exploration and colonization bore fruits not only for monarchs, conquistadors, colonists, and traders, but also for medical men. Along with marvelous birds and animals, Columbus and other travelers documented novel mineral substances and plant species, at whose therapeutic powers they could only guess. Viewed in this light, the newly accessible portions of the world presented themselves as a vast repository of potential wonder drugs — exotic remedies as powerful as the Asian bezoar, opium, or unicorn's horn.

The voyages of exploration and conquest coincided with, and partly fueled, an increasingly focused and programmatic movement to reform the field of *materia medica*, which gathered momentum from the 1490s on. This movement, which began in Italy but quickly caught on elsewhere in Europe, initially centered around editing, assimilating, and criticizing the works of Greek and Roman writers on medicine and the natural world like Aristotle, Dioscorides, and Pliny. But the reformers quickly realized that they needed to supplement these texts with their own experience in order to render them of any practical use.[46] Accordingly, professors of *materia medica* and their students and colleagues went on field trips, collected plants, and visited fish markets, looking for both new and familiar species.[47] While most of these men, for eminently practical reasons, concentrated on domestic flora and fauna, others increasingly looked farther afield, to the opening horizons and novelties of the non-European world. Some, like Garcia de Orta, Nicolas Monardes, and the influential Gonzalo Ferrando de Oviedo, catalogued the pharmaceutical riches of the East and West Indies.[48] Others, like Pierre Belon, Prospero Alpino, and Leonhard Rauwolf, traveled to the eastern Mediterranean, in the footsteps of Galen, to bring back news of what Belon called the "singularities" he found there.[49]

Ulisse Aldrovandi, professor of natural history at the University of Bologna, underscored the scholarly importance of these efforts. While applauding the work of physicians like Guillaume Rondelet and Conrad Gesner, who focused on Old World plants and animals, he emphasized how much else there was still to learn: "And although all these men have increased the knowledge of natural things," he wrote, "nonetheless a great field remains for the scholars of this and future centuries, since Christopher Columbus and others, with the greatest effort, expense and danger, have discovered as it were another world, in which such remarkable wonders have been disclosed and discovered and described and brought back to us at many princes' behest."[50]

The language used by Belon and Aldrovandi, with its invocation of "wonders" and "singularities," shows the continuing force of the discourse of the marvelous, now revivified by the exploration of previously unimagined lands. But the language of wonder also served another purpose: to attract the attention of wealthy patrons and lay readers who might find little to engage them in volumes of more matter-of-fact prose. The rise of printing created a large and growing audience for literature of this sort. Increasingly, interest in exotic and domestic natural history was not confined to scholars like Aldrovandi; texts on this topic quickly became a fixture in the broader market for large and lavishly printed books.[51] The expanded audience accounts for the growing importance of woodcut illustrations in such works (fig. 4.3), as well as the growing number of vernacular editions and translations. In 1577, Monardes's natural history of the Indies came out in English under the appealing title, *Joyfull Newes out of the Newe Founde Worlde, wherein is declared the Rare and Singuler Vertues of Diverse and Soundrie Hearbes, Trees, Oyles, Plants, and Stones, with thier Applications, as well for Physicke as Chirurgierie*.[52] Ambroise Paré's vernacular treatise on monsters and prodigies, first published in 1573, exemplifies the attempt to appeal to a broader audience; uniting two popular topics, exotica and portents, the volume was copiously illustrated with woodcuts lifted unapologetically from the works of a host of authors ranging from Conrad Lycosthenes, an influential compiler of prodigies, to the cosmographer André Thevet.[53]

Closely connected with this new surge of interest in natural wonders was the emergence of collecting as an activity not just of patricians and princes, as in the High and later Middle Ages, but of scholars and medical men as well.[54] Unlike princely collectors, who continued to prize precious materials and elaborate workmanship, physicians and apothecaries collected mainly naturalia, which reflected their own interests in therapeutics and were also relatively affordable (fig. 4.4).[55] Like contemporary

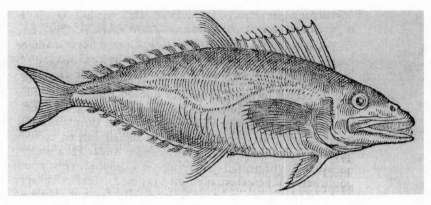

4.3.2

4.3.1

4.3.3

Figure 4.3. Wonders of the sea
4.3.1. *Ortus sanitatis* (Mainz: Jacob Meydenbach, 1491), sig. z viii v.
4.3.2–3. Pierre Belon, *L'Histoire naturelle des estranges poissons marins* (Paris: Reynaud Chaudière, 1551), pp. 14, 53.

The *Ortus sanitatis* is perhaps the best-known illustrated book of natural history from the fifteenth century and has separate sections devoted to plants, animals, birds, fish, and precious stones. The frontispiece to the section on fish (fig. 4.3.1) shows two scholars discussing, among other marvelous sea creatures, a monkfish and a siren. The illustrations in Belon's work on "strange fish," published sixty years later, are far more naturalistic, as in the case of the tuna (fig. 4.3.2), though the text continued to emphasize the wonderful properties of its subject. Regarding the nautilus, for example (fig. 4.3.3), Belon wrote, "besides being rare, it is of a strange and marvelous nature," and he noted that he had personally seen specimens belonging to the physician Jean de Rochefort and the captain Guischard, general of the Sicilian galleys.

4.4.1

4.4.2

Figure 4.4. Apothecaries as collectors
4.4.1. Ferrante Imperato, *Dell'historia naturale* (Naples: Costantino Vitale, 1599).
4.4.2. Benedetto Ceruti and Andrea Chiocco, *Musaeum Francisci Calceolari Iunioris Veronensis* (Verona: A. Tamus, 1622).

The frontispiece of the treatise on natural history by the Neapolitan apothecary Ferrante Impe-rato (fig. 4.4.1) is the first image of an early modern collection. It recalls both the hanging croco-diles in contemporary churches (see fig. 2.6) and the armoires in which contemporary ecclesias-tical collections were stored (fig. 2.3.1). The rather stilted inscription that accompanies this etching (fig. 4.4.2) of the collection of the apothecary Francesco Calzolari as it appeared in the early seventeenth century exhorts the reader: "Viewer, insert your eyes. Contemplate the wonders of Calzolari's museum and pleasurably serve your mind."

works of natural history—and like the princely *Wunderkammern*—these collections were often presented and experienced as wonders. The most famous and extensive such collection in sixteenth-century Europe belonged to Aldrovandi. Hailed by contemporaries as a "second Pliny" and a "Bolognese Aristotle," Aldrovandi described his collection in 1595, toward the end of his distinguished career:

> Today in my microcosm, you can see more than 18,000 different things, among which 7000 plants in fifteen volumes, dried and pasted, 3000 of which I had painted as if alive. The rest—animals terrestrial, aerial and aquatic, and other subterranean things such as earths, petrified sap, stones, marbles, rocks, and metals—amount to as many pieces again. I have had paintings made of a further 5000 natural objects—such as plants, various sorts of animals, and stones—some of which have been made into woodcuts. These can be seen in fourteen cupboards, which I call the Pinacotheca. I also have sixty-six armoires, divided into 4500 pigeonholes, where there are 7000 things from beneath the earth, together with various fruits, gums, and other very beautiful things from the Indies, marked with their names, so that they can be found.[56]

Given its dimensions, Aldrovandi might well boast that his collection was sought out by "many different gentlemen passing through this city, who visit my *Pandechio di natura*, like an eighth wonder of the world."[57]

The marvelous associations of the natural history collection, full of rare and exotic objects with unknown therapeutic potential, appear particularly clearly in a small volume issued in 1584 to promote the collection (and the business) of a prosperous and erudite apothecary, Francesco Calzolari of Verona. Compiled by the physician Giovanni Battista Olivi, this first published catalogue of a natural history collection made intense and repeated use of the rhetoric of wonder. Even in his initial complimentary epistle, Antonio Passieno (another physician) described Calzolari's "Museum" (fig. 4.4.2) as

> a most abundant repository and true treasury of all remarkable medicinal things, in which I observed each one placed in wonderful order in most decorative and elegant compartments and cases. First, [Calzolari] sought exceptional herbs and then the rest from their own distant places and regions, sent to him as gifts from the greatest princes and rulers; here it is pleasing to see not a few whole plants and plant roots, rinds, hardened or liquid saps, gums, flowers, leaves, fruits, and rare seeds and to recognize them as authentic. Also many metals. I omit how many dried terrestrial and aquatic animals I was astounded to find that I had never seen before.

He underscored this last point, remarking that "everything I saw there I observed and reflected on with the greatest wonder."[58]

What specific aspects of Calzolari's collection evoked wonder in its visitors? As in the case of Aldrovandi's *Pandechio*, the number and variety of its contents topped the list, together with its elaborate organization and special furnishings. Important too were the beauty, rarity, and exoticism of many of its objects; in his own description, Olivi repeatedly used words such as *rara* and *peregrina* ("strange"). Some objects were remarkable for their unusual form and behavior, such as the "Plinian nautilus," which Olivi "counted among outstanding marvels [*miracula*]" because it moved by shooting out water behind it as well as by setting up its own sail (fig. 4.3.3).[59] Others were more mysterious, like the fossils, with their uncanny imitation of animals and plants. But Olivi reserved his strongest expressions of wonder for objects with occult powers. These ranged from the traditional (for example, the lodestone, "a marvelous thing, which attracts a man in greatest wonder") to the novel ("liquid amber" from Mexico, which Monardes and Garcia de Orta recommended for wounds) — from the baleful (the "marine hare," a poisonous fish with no known antidote, which provokes miscarriages on sight) to the panacea (the bezoar, which "acts by its total substance, and not by its manifest qualities, which are corrosive").[60] This wealth of marvels struck even a connoisseur like Aldrovandi with amazement: "Among all the most erudite Italian apothecaries," he wrote in an appendix, "I have examined [Calzolari's] theater of nature with pleasure, rapt in wonder at so many natural things."[61]

Like most contemporary collectors of naturalia, Calzolari focused on therapeutic mirabilia; these could be obtained from local herbalists and merchants who dealt in drugs and spices or from travelers and other collectors, who traded objects back and forth with great enthusiasm.[62] Much rarer (and presumably more expensive) were anatomical and pathological rarities of the sort described by Benivieni, though these appeared in contemporary collections as well. Calzolari himself owned a mummified head, clearly visible in a seventeenth-century engraving (fig. 4.4.2). According to Aldrovandi, the physician Giovanni Battista Luchini possessed "a skeleton — or rather a completely dried body, with the flesh — of a famous man."[63] In Germany, the physician Johannes Kentmann collected stones found in the bodies of his own patients and of his medical colleagues — in their brains, their lungs, their intestines, their kidneys, and their bladders — and described these stones as "provoking great wonder in the learned and unlearned alike"[64] (fig. 4.5). Ambroise Paré owned the body of a child with two heads, two arms, and four legs, which he had dissected in 1546; Marcello Donati, author of the influential *De medica historia mirabili*

4.5.1

ARCA RERVM FOSSI-lium Ioan.Kentmani.			
1 TERRAE	* *	2	SVCCI NA-TIVI.
3 EFFLORE-SCENTES	* *	4	PINGVES
5 LAPIDES	* *	6	LAPID. IN A-NIMALIBVS
7 FLVORES	*	8	SILICES
9 GEMMAE	*	10	MARMORA
11 SAXA	* *	12	LIGNA IN Saxa corporata.
13 ARENAE	*	14	AVRVM
15 ARGENTVM	* *	16	ARGENTVM VIVVM
17 AES SEV CV-PRVM	* *	18	CADMIA MET. PLVMBAGO
19 PYRITES	* *	20	PLVMBVM NIGRVM
21 CINEREVM	*	22	CANDIDVM
23 STIBI	*	24	FERRVM
25 STOMOMA	* *	26	MARINA VARIA

Quicquid terra finu, venufq; recondidit imis,
Thefauros orbis hæc breuis arca tegit.
Laus magna eſt tacitas naturæ inquirere vires,
Maior in hoc ipfum munere noſſe Deum.
Georg.Fabricius.C.

4.5.2

Figure 4.5. The collection of Johannes Kentmann

4.5.1–2. Johannes Kentmann, *Nomenclatura rerum fossilium* (Zurich: Jacob Gesner, 1565), sigs. a5v–6r.

4.5.3. Johannes Kentmann, *Calculorum qui in corpore ac membris hominum innascuntur ... cum historiis singulis admirandis* (Zurich: Jacob Gesner, 1565), fol. 13v.

Kentmann, a German physician, was one of the earliest collectors of rocks and stones of all sorts, including fossils. Here he shows the armoire with drawers in which he kept his specimens (fig. 4.5.1), according to a classification in twenty-six parts (fig. 4.5.2). He had many of these specimens carefully drawn (fig. 4.5.3), including these four intestinal stones — three small ones (labeled A) passed by Mauritz von Thimen, a "nobleman in our diocese," and one large one (B in view, C in section) from the gut of an old man in Torgau, where Kentmann had his practice.

4.5.3

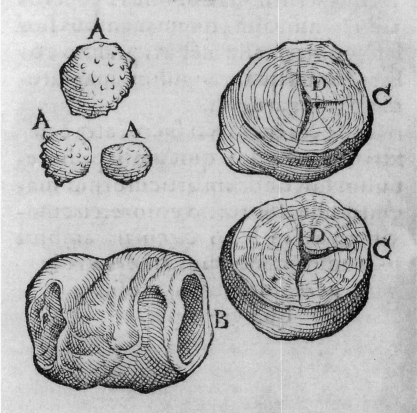

DE CALCVLIS
CALCVLI IN INTESTI-
nis generati.

I X.

A.
Calculi Mauritij à Thimen, quales singu
lis mensibus sine dolore excernit.
B.
Lapis integer M. Balthasaris Arnoldi.
Lapis

(1586), boasted that he kept a skeleton in his house "for the convenience of scholars, in which all the bones, that is, all the joints from the head to the furthest toes appear fused into one."[65]

In the hands of physicians, apothecaries, and professors, these collections served multiple purposes. On the one hand, as Donati indicated, they were places for research, where medical scholars could study the range of variation possible in human and animal anatomy, or where they could explore the healing properties of natural substances. Thereby, as Olivi put it, they might help "restore the medicines of the ancients to their splendor and bring them back into the light."[66] On the other hand, such collections operated as tools in professional and social self-fashioning, allowing their proprietors to build reputations, careers, and networks of clients and patrons through visits and the exchange of objects as well as through their written works.[67] In this sense, the aim of the naturalist's collection of marvels, like the collections of princes from the dukes of Burgundy to Rudolf II, was to transfer the emotion of wonder from the objects themselves to their erudite and discriminating owner.

In the end it proved difficult to reconcile the function of the collection as a repository of natural wonders with its function as a site of medical research. This was particularly true in the case of its most exotic items: objects brought to Europe from the New World. While naturalists who focused on domestic objects might build impressive, comprehensive museums, those who collected exotica found themselves in the possession of scattered plants and animal parts of which they had no firsthand knowledge and no information from ancient or modern texts. This led Aldrovandi to complain that "although many historians of these places have described many plants and animals that are born there, nonetheless, they have not written of them *ex professo*; rather, of the profusion of natural things they found there, they have been forced to write and refer to their variety as it were incidentally and in passing, not even judging what they are or what utility they have to man."[68] Despite a few notable exceptions — quinine, guaiac wood, tobacco, sarsparilla — the effects of these products were uncertain, so that Aldrovandi, Olivi, and others rejected them as impractical objects for collection and occasions for falsification and price-gouging by unscrupulous dealers. Writing in the 1580s, Olivi argued that "it is not thought to be expedient to use remedies from such distant places, except in order to show the beauty of their nature."[69] Such things became instead either objects of pure curiosity and display or subjects of a nascent discipline of natural history, which gradually separated itself from medicine and came to include natural objects regardless of their therapeutic use.[70] Formerly identified with *materia medica*, a functional and subordinate branch of medical studies, natural history first established its independence with the work of men like Aldrovandi and then was presented by Francis Bacon in the seventeenth century as foundational to a reformed natural philosophy — a story we tell in Chapter Six.

As embodied in the sixteenth-century collections of naturalia, however, natural history declared its autonomy from traditional natural philosophy as well. The principles of natural inquiry that informed these collections differed dramatically from those that shaped the work of those few medieval natural philosophers with an interest in natural history, most notably Albertus Magnus. For Albertus, the description of particular substances and their particular properties was always, at least in theory, subordinated to the goals of natural philosophy, with its emphasis on necessary and universal causal explanation. By contrast, sixteenth-century collectors of naturalia showed relatively little interest in such universals, or in regularities of any sort. They emphasized diversity more than uniformity, and the breaking of classificatory boundaries more than the rigors of taxonomy; thus the great interest of Kentmann's collection of stones found inside human beings (fig. 4.5.3) lay in the spectacular interpenetration of the animal and mineral realms.[71]

By the same token the empiricism of the collectors, like that of the natural history writers and the earlier Italian balneologists, did not emphasize the induction of universal principles from particular phenomena. They subscribed rather to what William Eamon has called the "epistemology of the hunt" — particulars to point the way to other particulars, just as the hunter extrapolates the presence of his quarry from its footprints and other clues.[72] In this way, they remained close to the discipline of medical *practica*, which had long reasoned from particular symptoms to the illness of a particular body. Despite their distance from the traditional agenda of natural philosophy, however, the sixteenth-century natural history texts and collections were not completely disconnected from philosophical investigation; in their focus on wonder and on wonders, they reflected and echoed the contemporary appearance of a new strand of natural philosophy that also took preternatural phenomena as its primary objects.

Preternatural Philosophy

Between the commonplace and the miraculous lay the large and nebulous domain of the marvelous, which had long resisted philosophical explanation. Medieval natural philosophers had never disputed the existence of anomalous and occult phenomena, nor doubted that these arose from natural causes. With a few notable exceptions, however, they had largely excluded such marvels from the purview of natural philosophy as neither regular nor demonstrable. Many Aristotelians and neo-Aristotelians continued to hold this position, but they were increasingly challenged by a new type of philosopher, whom we will call the "preternatural philosopher." As a group, the preternatural philosophers resembled the preternatural historians, albeit with a somewhat more academic cast. Some enjoyed the patronage of secular or ecclesiastical princes, intermittently or continuously — Ficino; Henricus Cornelius Agrippa; Francis Bacon and Scipion Dupleix in the next century — and one, Giovanni Battista della Porta, was himself of noble birth. Pietro Pomponazzi was an academic philosopher who taught at the University of Padua. The majority, however, were physicians, including both professors of medicine (Girolamo Cardano; Fortunio Liceti) and men who lived primarily from their own practices (Levinus Lemnius; William Gilbert; in some respects, the idiosyncratic Paracelsus). Academically and professionally trained, they wrote for the most part in Latin for learned audiences, although their works often appeared later in vernacular translation. These authors shifted the marvels of nature from the periphery to the center. In the process, they reclaimed for natural philosophy not only wonderful phenomena but also the emotion of wonder itself.

The objects of preternatural philosophy coincided with the traditional canon of marvels. They included both the results of occult action, such as magnetic attraction or the reputed power of the amethyst to repel hail and locusts, and rare individual phenomena, such as bearded grape vines, celestial apparitions, and rains of frogs or blood. Each of these two types of marvel posed a different philosophical challenge. Although occult properties were in principle as regular in their operation as manifest ones, they were opaque to reason and resistant to explanation except by a general appeal to substantial or specific form. Rare and anomalous phenomena similarly resisted philosophical explanation, and natural philosophers typically ascribed them to chance (that is, to a tangled knot of accidents exceptionally conjoined). Because of their complexity, chance events also escaped the limits of human understanding; as Oresme had put it in his *De causis mirabilium*, "these things are not known point by point [*punctualiter*] except by God."[73] Preternatural philosophers, in contrast, aimed to unravel such phenomena on a case-by-case basis. No longer content, as Oresme had been, to offer vague assurances that most marvels had natural causes even if these could not humanly be determined, they sought specific explanations for individual instances.

In order to accomplish their aims, preternatural philosophers expanded the range of explanations as well as the explananda. Rarely, however, did they invoke causal mechanisms that were entirely new. Rather, they brought together and developed a set of miscellaneous causes that had long been familiar to natural philosophy — Albertus Magnus had invoked many of them — although some were controversial within it: "spirits" (in the form of tenuous vapors, both inside and outside the human body); occult qualities; sympathies and antipathies; and the power to shape the external world attributed to the human intellect, celestial intelligences, and, especially, the human imagination. Most of these causes had some sort of an Aristotelian pedigree, but they had been greatly elaborated by medieval writers in both the Islamic and the Christian worlds.[74] What these causes had in common was their imperceptibility to the senses, unlike the manifest qualities of hot, cold, wet, and dry, or the grosser processes of motion by contact.

The first Renaissance philosopher to place this set of causal mechanisms in the foreground and to elaborate them as a group — in this sense, the first of the preternatural philosophers — was Ficino, whose use of marvelous therapeutics we have discussed above. Ficino was a key figure in the history of late fifteenth- and sixteenth-century European philosophy. As the principal Latin translator of the newly recovered texts of Plato and his late antique followers, including the authors of the hermetic corpus, he inaugurated a new strand of Latin philosophy that synthesized Neoplatonic ideas with the Christian Aristotelianism of Thomas Aquinas and other medieval Latin writers. While Ficino retained basic features of traditional cosmology, he treated the marvelous properties of natural substances as products of — and signposts to — the fundamental metaphysical structure of the universe, conceived in terms of correspondences and emanation. For Ficino the world was, in the words of Alfonso Ingegno, a "secret web of hidden links" that connected all earthly phenomena to the realm of the celestial intelligences and spheres.[75]

In many respects, the elements of Ficino's discussion of wonders were traditional: thus he attributed the therapeutic powers of the bezoar or unicorn horn to occult properties impressed by the heavens, and he invoked the doctrine of specific forms.[76] But he went beyond contemporary medical theory in invoking a second kind of marvelous action, one that involved not natural substances, but the human soul itself. As he described in his *Theologia platonica*, completed in 1471, the human mind "not only claims a divine right to shape and form matter through the means of art, but also to transmute and dominate the species of things, which work is indeed called a marvel [*miraculum*], not because it is beyond the nature [*praeter naturam*] of our soul, when it becomes the instrument of God, but because whatever is great and happens rarely gives birth to wonder."[77] Even without divine support the human soul can produce marvelous effects through its own faculty of imagination.[78] Elevated by God, however, the souls of "holy men" are even more powerful, particularly when reinforced by the devotion of those around them, allowing them to create peculiar meteorological disturbances, for example, or perform marvelous cures.[79] Here the outlines of Avicenna's controversial teachings concerning the powers of sub-celestial intelligences can be discerned.[80] Although Ficino, like other Latin writers, tended to materialize the soul's action by supplying an intermediary in the form of those vapors known as "spirits," the thrust of his work was to use late Platonic ideas to revivify elements in the thirteenth-century magical tradition that had fallen under philosophical and theological disapproval.

Ficino influenced other late fifteenth- and early sixteenth-century Italian writers interested in exploring marvels. The physician Andrea Cattani's *De intellectu et de causis mirabilium effectuum* (c. 1504) contained similar ideas,[81] as did the treatise on incantations (1520) of Pietro Pomponazzi, professor of philosophy at the University of Bologna. Indeed, Pomponazzi took Ficino's work a step further, attempting not only to explore the general operation of occult causes, subtle spirits, and the powers of the imagination, but also to explain reports of particular anomalous phenomena. He took many of these *historiae*, as he called them, from ancient and medieval sources (as being more credible), but also included several recent examples, such as the remarkable appearance of an image of St. Celestine in the skies over Aquila after the inhabitants had prayed to their patron for relief from torrential rains. According to Pomponazzi, the apparition did not require a supernatural explanation but could have been produced by the vehement imagination of the inhabitants and certain "vapors" they emitted, which impressed the image of the saint on the humid air.[82]

Much more than Ficino, whose relationship to demons was ambiguous, Pomponazzi expressed what came to be a fundamental trait of preternatural philosophy: its refusal to use demons to explain puzzling phenomena.[83] Demonology was in some ways the alter ego of preternatural philosophy, for demons also worked marvels.[84] Of sharper intelligence, fleeter foot, and lighter touch than humans, they were constrained to act through natural causes, but because their action was so swift and so subtle, their works were often erroneously construed as supernatural. The early sixteenth-century theologian Johannes Trithemius warned that the apparent miracles performed by infidels were in fact demonic impostures: "Who then can deny that the demons, of a still subtler nature and with greater experience [than human natural magicians], are able to effect many marvelous things naturally which are not understood by any ordinary man?"[85]

It was still trickier to distinguish a demonic marvel from a natural one, since the secondary causes involved were identical. The only distinction between the works of demons, on the one hand, and the pretergenerations of unassisted nature, on the other, was the agency of a free will. This is why early modern demonologists, intent on fixing the boundaries between the natural and the demonic, became authorities on the preternatural in the all-too-concrete context of witchcraft trials. Preternatural philosophers, on the other hand, expelled demons from their treatises — whatever their personal beliefs regarding the existence and operation of demons — and generally confined themselves to more natural explanations (though Agrippa's invocation of celestial intelligences came perilously close to the demonic). In this they reflected not only a particular set of intellectual commitments, but an awareness of theological dangers as well. Those who dabbled in magical ideas and practices, like Cattani and Ficino, took care explicitly to assert their Christian orthodoxy. As Ficino noted regarding the marvels performed by the elevated human soul, "the particular things concerning marvels [*miraculis*] that we have taken from the opinions of the Platonists, we assert only so far as Christian theologians do approve."[86]

Ficino's contribution to the philosophy of the preternatural was threefold. First, he emphasized and elaborated many of the causal mechanisms that were to become staples of the sixteenth-century philosophy of the preternatural: spirits, imagination, celestial influences, elevated souls. Second, he embedded these in an influential metaphysics and cosmology. Through his work the rather limited idea of specific form, used by high and late medieval writers on medicine and philosophy primarily to account for the mysterious properties of a relatively small class of substances, was generalized and expanded, becoming the linchpin of an increasingly influential view of the universe. Ficino and his followers interpreted the physical order as a network of invisible correspondences — sympathies, antipathies, and astral relationships of domination and submission — that provided the deep structure of the natural world.

Finally, although Ficino reached back to earlier magical theory, he reclothed the ideas he found there in the language of wonder, which had been signally lacking in the discourse of the academic writers on philosophy and medicine. Following Ficino, magical action, rooted in the powers of the soul or the occult properties conferred on terrestrial objects by the heavens, was increasingly referred to as "wonderful" (*mirabilis*) and its effects as *mira*, *miracula*, or *mirabilia*. This language was immediately taken up by writers like Cattani and Cornelius Agrippa von Nettesheim, whose treatise on occult philosophy (heavily dependent on the pseudo-

Albertine *De mirabilibus mundi* as well as on Ficino) invokes wonder with such regularity that it begins to lose its rhetorical effect.[87] Ficino's relative ease with the rhetoric of the marvelous grew out of the environment in which he worked. He was not an academic philosopher: he had not even completed his medical degree, and he never held a university post. Like Agrippa, who served the Emperor Maximilian and Margaret of Austria, Ficino moved in courtly circles, addressing his work to highly placed patricians including Lorenzo de' Medici himself (fig. 4.2).[88] In this context his invocation of the marvelous, with all of its romance and courtly associations, was no longer the intellectual and social liability it had been in the context of thirteenth- and fourteenth-century schools and universities; rather, as court physicians like Savonarola had already discovered, the language of wonder wrapped its often recondite material in a glamour that enhanced its appeal.

Ficino's work was important not only in its rhetorical and philosophical details, but also in the example it offered for a new kind of philosophical writing, oriented toward producing overarching, speculative, and synthetic accounts of nature rather than debating technical questions and refining Aristotelian distinctions; this last form still dominated most treatises produced in a university context for a university audience.[89] Much of the work in this newer vein was in fact critical of certain aspects of Aristotelian philosophy and advanced a view of nature and natural philosophy that emphasized the power of human knowledge to transform the material world.

One of the most broadly influential exponents of these ideas in the mid-sixteenth century was Girolamo Cardano, practicing physician, polymath, and occasional professor of medicine at the universities of Pavia and Bologna, who produced by his own account over two hundred works. In Cardano's hands, wonders became not merely an important part of philosophy, as they had been for Ficino, but in many ways the center of the entire philosophical enterprise. Drawing on contemporary travel writing, collecting practices, and natural history, Cardano used wonders as the foundation for his inquiry into the nature of the universe, which he interpreted as a network of occult interactions, shot through with the effects of contingency and chance. In such a world, wonders — rare, occult, and chance phenomena — became the key explananda in the physical world.[90]

The outlines of this enterprise appear clearly in Cardano's two encyclopedic works of natural philosophy, *De subtilitate* and *De rerum varietate*, first published in the 1550s.[91] *De rerum varietate* exemplifies the key role Cardano gave to wonders. It began with a survey of various "places and regions," from India to Norway to Peru, followed by what Cardano called the "wonders of the earth" (the colored mountains of Quahumetallau, for example, or earthquakes in southern Italy), "wonders of water" (a petrifying Irish lake, the finless fish of Loch Lomond), "wonders of air" (the irresistible winds of Assyria, or — an episode witnessed by Cardano himself — the deaths of four men and a dog after they climbed down into a newly dug sewer), and "wonders of the heavens" (the birth of a monster or a mute). Cardano supplemented these with a profusion of marvels taken from other natural realms: metals, stones, plants, animals, and humans. Throughout he employed the language of wonder. As he reflected at one point, "our age has seen nothing that is not marvelous, nor does nature play less wonderfully in small things than in large."[92]

In his appreciation of the inexhaustible variety and beauty of nature, Cardano reflected the sensibility and the interests we described in the preceding section. He repeatedly invoked the work of contemporary writers on topography and natural history — Rondelet, Münster, and Gesner, for example — as well as the vast horizons of previously unknown natural phenomena revealed by the voyages of European discovery. He praised Gonzalo Oviedo as "an outstanding author, and in my judgment truthful and learned, so that I count him alone equal to the ancients among the historians of our age."[93] And he was himself a collector and a connoisseur of others' collections, including the treasure of Saint-Denis, which he visited and by which he was impressed.

At the same time, however, *De rerum varietate* presented Cardano in the traditional guise of the natural philosopher: the man who, by debunking their rarity and elucidating their causes, was able to make wonders cease. Thus he remarked about a meteorite venerated in Emesa, "It will appear less wonderful if I tell you about a similar stone owned by my friend Guglielmo Casanato."[94] He also gave simple natural explanations of other preternatural phenomena by invoking celestial influences, elemental composition, or climatic change. If one year the English crows all laid their eggs in winter, for example, it was only because that particular winter was very warm, and, as he put it, "what's so wonderful about that?"[95] In such passages Cardano strongly recalls Oresme in the *De causis mirabilium*: pushing aside demonic and supernatural explanations, he proceeded to demystify the world. Unlike Oresme, however, in case after case Cardano invoked particular phenomena and provided particular explanations for them. If maize grows without rain in the high valleys of the Andes, it is because the heat of the sun is very weak there and dries only the surface of the soil. If there are no snakes in Ireland, it is because there is a great deal of bitumen in the earth, which kills them by its dryness and its noxious smell.[96]

Why did Cardano dare to propose particular explanations for particular natural wonders, where Oresme had thrown up his hands in despair? It was not just that his standards for a satisfactory explanation were lower than Oresme's, as is evident from these examples. Far more important was the vastly expanded material at his fingertips — the result not only of the voyages of exploration, but also of the gradually increasing density of European cultural and commercial contacts with the wonder-bearing margins of the world. As Cardano rejoiced, with regard to his own personal wonderfulness, "I was born in a rare century, which has come to know the whole world."[97] Circumstances had provided Cardano with a new and expanded natural history that he could use, in the original Aristotelian manner, as a basis for (in at least some respects) a new natural philosophy. More specifically, it allowed him to develop explanations for specific cases by increasing the number of examples at his disposal, which allowed him to test and refine possible explanations for wonderful phenomena by comparing them to others of similar type. Consider, for example, the case of petrifying lakes. After describing the experiments made by Hector Boethius on a lake in Ireland, Cardano hypothesized that its properties resulted from the presence of bitumen. On the other hand, Palestine and Iceland also had large amounts of bitumen and water, but no petrifying lakes at all. Cardano concluded that water and bitumen were not themselves sufficient, but had to be accompanied by exactly the right degree of cold and the right quality of water.[98] Thus by comparing instances of a particular kind of wonder, Cardano found himself able to sort through an intricate combination of causes to arrive at a satisfactory conclusion.

Despite these explanatory successes, Cardano did not attempt to eliminate wonders from natural philosophy, or wonder from the informed contemplation of nature, in the manner of thirteenth- and fourteenth-century Aristotelians. Rather, he wanted to rehabilitate wonder for philosophy by presenting it as a highly differentiated emotion, its intensity carefully calibrated to its object in the psyche of the wise. Correspondingly, he wished to establish a scale of natural wonders, based on information culled from the four corners of the earth, that would allow him to distinguish between the extraordinarily wonderful (unique phenomena like the blue clouds of the Strait of Magellan),[99] "things worthy of wonder, but not great wonder" (Mexican foot jugglers, or the fact that when Delcano navigated the globe from west to east he gained a day),[100] and the emphatically "not marvelous" (the appearance of a red cross in the sky over Switzerland).[101] In this way he would purge philosophy of false wonders — many of these, like the Swiss cross, would generally have been read as prodigies — and prevent the devaluation of true ones in the face of a daily avalanche of new candidates.[102]

Cardano adopted the same agenda in the final section of *De rerum varietate*, which treated *thesauri*, or collections. Here he laid out guidelines for a collection worthy of a philosopher — indeed, a wonder cabinet — based on (of all things) the treasure of Saint-Denis. Here, too, he described how to differentiate true wonders from the inevitable interlopers. He included detailed instructions on how to identify a genuine bezoar (it will make you vomit a *small* amount of poison, as Cardano knew from personal experience) or how to spot a faked mermaid (scrutinize the join). But there is no doubt that he, like the Abbot Suger, welcomed and honored the pleasure that accompanied the contemplation of such wonders, based, as he put it, on "the enjoyment of the thing."[103]

Far more explicitly than Ficino, Cardano elaborated a new notion of wonder appropriate to a natural philosopher. This wonder was neither the fearful wonder of the vulgar, crippled by ignorance and superstition, nor the Augustinian wonder of the devout, in awe before the marvelousness of the creator. Still less was it Aristotle's wonder at the regular and the functional. Instead it was a philosophical version of the pleasurable wonder that informed the medieval literature of topography and travel — an emotion now passed through a professional lens. It was a secular version of Suger's wonder before the Crista:[104] the wonder of the connoisseur, so familiar with a multiplicity of extraordinary phenomena that he knew which truly deserved his amazement. This wonder was a finely graduated register of response that only the best-informed and the most philosophically sophisticated could deploy.

This emphasis on connoisseurship and virtuosity formed part of an aristocratic model of knowledge that marked the work of many of the preternatural philosophers. Contrasting with both the academic model of thirteenth- and fourteenth-century natural philosophers and the monastic model of the medieval encyclopedists, this approach emphasized the exclusive nature of its material, which it identified with the exclusive nature of its audience, an elite at once intellectual and social. This theme was a staple in the introductory matter of treatises on preternatural philosophy. When Cardano described wonders as the aristocracy of natural phenomena — "causes, powers and properties of things that are varied and not vulgar, but difficult, occult, and very beautiful," according to the title page of the first edition of *De subtilitate* (1550) — his words recalled the letter from Trithemius that introduced Agrippa's *De occulta philosophia*: "I warn you to keep this one precept, that you communicate vulgar things to the vulgar and arcana only to your highest and secret friends."[105] In

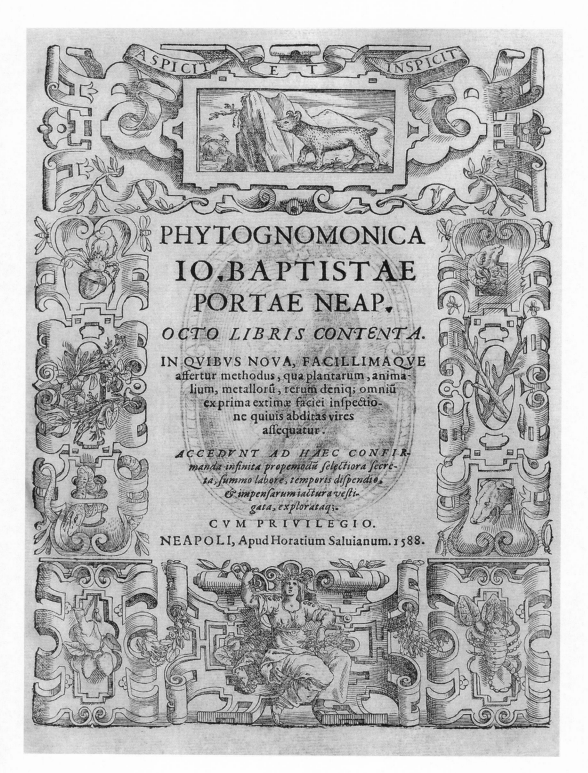

4.6.1

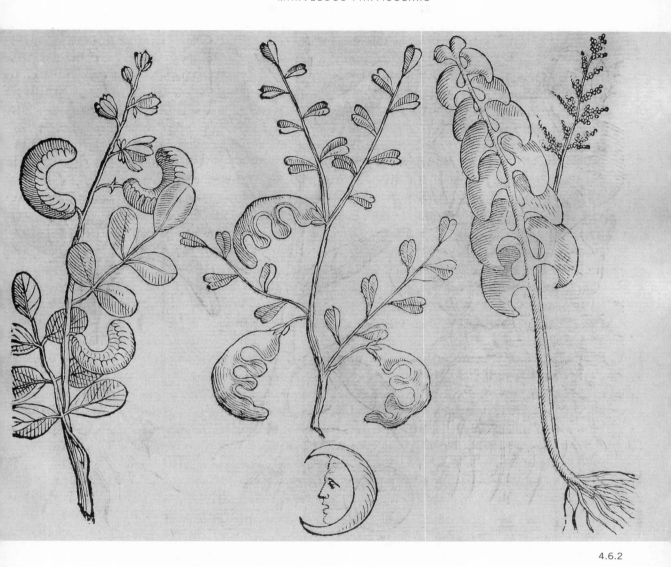

4.6.2

Figure 4.6. Della Porta on the secret structure of nature
4.6.1–2. Giambattista della Porta, *Phytognomonica* (Naples: Orazio Salviani, 1588), title page and p. 317.

In the *Phytognomonica*, della Porta explained how to look at the external forms of plants in order to "investigate their internal secrets": their manifest and occult powers and the planets that governed them. At the top of the title page (fig. 4.6.1), a lynx (reputed to be able to see through solid objects) stands as an emblem of the preternatural philosopher's exploration of secrets and hidden truth, while some of the images below show applications of della Porta's theory; the heart-shaped fruits in the lower left-hand corner, for example, are juxtaposed with a human heart, for which they were a specific remedy. Similarly, certain plants (fig. 4.6.2) reveal by their lunar signature that they are bound by bonds of sympathy and influence to the moon.

his dedicatory epistle, to Eric XIV of Swabia (1564), Levinus Lemnius described the subject of his *De miraculis occultis naturae* as royal, "so that these marvels [*miracula*] of nature can be consecrated to no one more aptly and more appropriately than to King Eric."[106]

The marvels of preternatural philosophy demanded more, however, than just a selective audience. The most difficult of phenomena to explain (on account of what Cardano called their "subtlety"), they yielded their secrets to only the highest and most skillful of interpreters, thereby defining a philosophical elite. Itself an esoteric quality — "that reason, by which things sensible to the senses and intelligible to the intellect are to be comprehended with difficulty"[107] — subtlety was associated with natural causes and processes that lay at the very edge of human perception and understanding. Wonders were a gauntlet thrown down to the man who could explain not only particulars, but extraordinary particulars, thus proving himself a wonder in his own right. In this way, wonder became a reflection not of ignorance but of virtuosity and connoisseurship: the product not only of great experience and erudition, but also of impeccable taste.

These aristocratic associations of wonder received full expression in the works of Giovanni Battista della Porta, a younger contemporary of Cardano and, unlike him, a noble by birth as well as sensibility. In his hands, the explanation and manipulation of wonders became, in the words of William Eamon, "courtly science par excellence. The magus's essential characteristics — his passionate quest for secrets, his craving for rarities, his cultivation of wonder, and his tendency to view science as a theatrical performance designed to delight and astonish spectators — perfectly fit the courtly manner."[108] The preface to della Porta's *Magia naturalis* (first edition 1558) contained the obligatory exclusionary rhetoric. But it reflected an aristocratic ontology as well. Like Cardano, della Porta aimed to uncover the secret workings of nature, and he emphasized occult forces, described in terms that recalled the human social hierarchy. According to della Porta, the invisible sympathies and antipathies linking the universe reflected the obedience of lower beings (terrestrial phenomena, such as plants, stones, and animals) to higher ones (the planets and the celestial spheres). God had "arranged, by his providence, that all these lower things are ruled by the higher, according to a certain natural law."[109] Della Porta's magical universe in fact replicated the structures of Italian patronage, with each planet presiding over a hierarchy of subordinates, each of which bore its master's emblem in the form of what were known as signatures, and to each of which it conferred power in the appropriate degree (fig. 4.6).[110]

Della Porta identified as the most elevated aspect of his philosophy the fact that it conferred power as well as understanding. As he argued in his preface,

> This sublime science undertakes the knowledge of things and causes, and while it considers and investigates the arcana of nature, it brings forth not only vulgar works,... but certain marvels and monsters of nature [*Naturae monstra quaedam, et miracula*]. Thus it excels all others (with the exception of theology), so that all other arts and sciences appear to serve it like a queen. Therefore I set it above the others by right, as by far the most difficult, and thus also exalted and royal.[111]

In della Porta's world, wonders were not just symbolic, confirming the hierarchical order as intended by God. They allowed the man who understood them to control forces that went far beyond those attributed to bezoars and unicorn horns in the treasures of the dukes of Burgundy. Della Porta's preternatural philosophy was also a preternatural technology; much more than simple anti-poison amulets, his wonders became the keys to an invisible network of sympathies and antipathies that gave its master nearly unlimited power.

Preternatural philosophers like Pomponazzi, Cardano, and della Porta became the self-declared virtuosi of natural philosophy, taking on the phenomena most difficult to explain and therefore the most wondrous. Their bravado attracted some criticism from more traditional philosophers. A generation later, the neo-Aristotelian Julius Caesar Scaliger attacked Cardano by ridiculing the arbitrariness of the idea of subtlety, underscoring the relativity of wonder to circumstances, and castigating Cardano's finely shaded scale of the marvelous as hopelessly confused.[112] But the field belonged to the preternatural philosophers, at least in terms of the influence and diffusion of their work. Cardano's two treatises appeared in German and French as well as multiple Latin editions, while della Porta's *Magia naturalis*, issued more than twenty times in Latin, was translated into Italian, English, and Dutch as well.[113] The French translation of Cardano's *De subtilitate* (1556) influenced the vernacular works of Ambroise Paré and the potter Bernard Palissy. The wide reception of Cardano's and della Porta's treatises signals the growing variety and vitality of late sixteenth-century philosophical culture, which flourished outside the universities, embracing not only physicians, professors, and their aristocratic or princely patrons, but ambitious artisans as well. This development was largely dependent on the technology of printing, which greatly expanded the potential audience for treatises of all sorts.

We would like to end by stressing the centrality of men with medical training in the developments described in this chapter — the elaboration of new forms of empiricism, the creation of natural history collections, the development of new approaches to natural philosophy, and the shaping of a sensibility of cultivated wonder toward phenomena in the physical world. Most of the men whose work we have discussed in this chapter had been educated as surgeons, apothecaries, or (the vast majority) physicians. As Harold Cook has emphasized, physicians were the principal group of early modern European intellectuals with advanced training in the study of nature.[114] They were numerous and prosperous — much more so than professional natural philosophers — and their work as both medical practitioners and scholars allowed them to move freely through the various social environments we have described: academic, professional, and courtly. In the process, they acted as the principal cultural mediators in the arena of natural inquiry — the men who assimilated, developed, and disseminated new approaches to the study of nature in both the Latin and the vernacular realms.

Spurred on by the efforts of these medical men (apothecaries and surgeons as well as physicians), books of marvels poured off the printing presses of Europe during the second half of the sixteenth and early seventeenth centuries. These books included not only the works of della Porta, Cardano, and many of their contemporaries,[115] but also medieval treatises on wonders, such as the pseudo-Albertine *De mirabilibus mundi* and even a version of Oresme's *De causis mirabilium*.[116] This period can be called with justice an "age of wonder." Wonder and wonders commanded attention — as objects of philosophical analysis, as the focus of a self-conscious sensibility, and as a nexus of cultural symbols — not only in the natural philosophy and medicine of the age, but also in its literature and art.[117] The following chapters describe both the hegemony of wonder and its ultimate decline from favor among European elites. We begin with one particular object of wonder, the monstrous birth. Perhaps more than any other kind of marvel, this phenomenon aroused passions and mobilized interests among Europeans of every social class; hence monsters will be our guides in a broad survey of the early modern topography of wonder.

CHAPTER FIVE

Monsters: A Case Study

Many Europeans in the late fifteenth and sixteenth centuries would have agreed with Girolamo Cardano that the period in which they lived was both quantitatively and qualitatively more wondrous than earlier times.[1] After issuing six illustrated Latin and German broadsides on a series of monstrous births during the period between September 1495 and April 1496, for example (fig. 5.1), the humanist and imperial publicist Sebastian Brant refused to comment publicly on the birth in June of a child with two heads: "Some people have pressed me to write," he noted, "[and I would do so,] except for the fact that monsters have become so frequent. Rather than a wonder [*miraculum*], they appear to me to represent the common course of nature in our time."[2]

Brant's words indicate a shift in the place and nature of wonders in fifteenth- and sixteenth-century European culture. During the Middle Ages, Europeans associated natural wonders above all with the margins of the world, most particularly with the plants, animals, and minerals of Ireland, Africa, and Asia. In the Renaissance, these wonders began to migrate palpably toward what had been the Mediterranean and European center, and not just in the packs and cargoes of traders, explorers, and collectors. Thus Benvenuto Cellini claimed to have seen a salamander in a blazing fire during his youth in Florence, while the basilisk, once a fearsome Eastern lizard, began to crop up in European settings: in *De rerum varietate* Cardano reported a possible Italian sighting.[3] Similarly, various late medieval European maps showed representatives of the monstrous races in places as close as Norway. The character of these newly proximate marvels changed, furthermore, as they began to move inward. The archetypal medieval wonder was the Blemmy or basilisk, member of an exotic race or species, whereas late fifteenth- and sixteenth-century writers increasingly privileged individual monsters like Brant's two-headed baby or the apparently inexhaustible supply of human-animal hybrids and conjoined twins that graced popular broadsides as well as learned books. In a treatise

Figure 5.1. Sebastian Brant's monsters
Sebastian Brant, *De monstroso ansere atque porcellis in villa Gugenheim* (Basel: Bergmann von Olpe, 1496).

In 1496 Brant published two broadsides, one in German and one in Latin (fig. 5.1), concerning the birth in April of conjoined geese and pigs. He associated these with two other recent monstrous births, both of which he had discussed in other broadsides: the conjoined human twins born in Worms in September 1495 and the Landser sow, born in March 1496. (The latter was the subject of a famous print by Albrecht Dürer.) In births of this sort, according to Brant, "kind, creating nature warns us with prodigies and portents."

published in 1575, Cornelius Gemma, professor of medicine at Louvain, enumerated a conventional list of monstrous races, "men competely wild in appearance and way of life: Fauns, Satyrs, Androgynes, Ichthyophages, Hippopodes, Sciopodes, Himantipodes, Cyclops," and so on. But, he noted, "it is not necessary to go to the New World to find beings of this sort; most of them and others still more hideous can still be found here and there among us, now that the rules of justice are trampled underfoot, all humanity flouted, and all religion torn to bits."[4]

As the last part of Gemma's comment emphasizes, this new geographical distribution of wonders did not spring exclusively from the voyages of exploration. Certainly, as Europeans began truly to experience (rather than merely to imagine) the world as a sphere instead of a circle, the categories of margins and center lost coherence. Rather than being confined to the exotic edges of the world, wonders might crop up in Europe as easily as in any other place on earth. But this explanation is only partial; as Brant's corpus of broadsides indicates, the shift was already quite pronounced by the mid-1490s, when the age of European exploration was still in its infancy. Furthermore, it does not account for the increasing emphasis on individual wonders: not only monsters, but earthquakes, volcanic eruptions, comets, and the host of celestial apparitions that increasingly preoccupied writers like Brant.

Individual marvels of this sort, as we argued in our first chapter, belonged to a different tradition from the exotic species, and had as a result a different meaning.[5] Typically interpreted as prodigies, as Gemma emphasized, individual wonders were seen as signs of human sin and the righteous wrath of God. Augustine's attempt to portray both individual wonders and exotic species as equivalent signs of divine omnipotence had found relatively few enthusiasts except among some medieval encyclopedists, so that the two traditions had developed separately until the sixteenth century, when both forms of marvels gravitated toward one other in the works of authors as diverse as Conrad Lycosthenes and Ambroise Paré. While the tradition of exotic species evolved and changed over the course of the fifteenth and sixteenth centuries, in response to the cultural contexts and pressures described in the preceding chapter, it was enriched and complicated by the dramatic emergence of the prodigy tradition, in the years around 1500, as a matter of urgent and nearly universal concern.

Why did marvels of all kinds, individual prodigies as well as exotic species, thrust themselves with such force into the consciousness of early modern Europeans? In this chapter we trace the trajectory of the early modern preoccupation with wonders through a single type of marvel, the monstrous birth. In an earlier article, we described the evolution of sixteenth- and seventeenth-century attitudes toward monsters in linear terms: originally part of the prodigy canon, with its ominous religious resonances, monsters shifted over the course of the sixteenth century to become natural wonders — sources of delight and pleasure — and then to become objects of scientific inquiry. At this last stage, they finally shed their associations with earthquakes and comets, finding a home in the medical fields of physiology and comparative anatomy.[6] Viewed in the light of more recent research, our own and that of others, we now reject this teleological model, organized as a progress toward rationalization and naturalization. Rather, naturalization — the explanation of marvels by natural causes — had its advocates even among medieval writers, while examples of monsters read as divine signs or enjoyed as *lusus naturae* can be found until the late seventeenth century.

Instead of three successive stages, we now see three separate complexes of interpretations and associated emotions — horror, pleasure, and repugnance — which overlapped and coexisted during much of the early modern period, although each had its own rhythm and dynamic. Like everything else having to do with wonders, these complexes cannot be detached from the particular audiences, historical circumstances, and cultural meanings that shaped and nourished each of them. Rather than canonizing the view that equates a religious response to monsters as prior to and less advanced than a naturalistic reading of them (an interpretation that characterized not only our earlier work, but also that of Georges Canguilhem and Jean Céard[7]), the last section of this chapter historicizes the origins of that very view. We argue that the march toward the naturalization of marvels was an illusion, created by a new unanimity among intellectuals in the late seventeenth century. Earlier, learned treatises offering ominous interpretations of monsters had flourished alongside those that furnished strictly natural explanations. But by the 1670s theologians as well as physicians and natural philosophers increasingly rejected portents as politically and religiously volatile. It is this emergent unison among elites, rather than a coherent and novel movement to naturalize monsters and other prodigies, that requires explanation.

These complexes of reactions to monsters were at once cognitive and emotional. If the category of wonders cohered around the emotion of wonder, then that emotion was itself protean, sliding on one side toward pleasure (as in the courtly literature of romance and secrets) and on the other side toward fear (as in the works of thirteenth-century natural philosophers). We will begin by exploring what we will call the prodigy complex, in which monsters functioned as signs of divine wrath and evoked the emotion of horror or terror. We will then turn to the pleasure of monsters, treated as the sports of a benign nature and ornaments of a benevolent creator. In the last section, we will examine a largely new emotional complex — of impassivity or downright repugnance — as it appeared in medicine, philosophy, and theology.

Horror: Monsters as Prodigies

In March of 1512, the Florentine apothecary Luca Landucci made the following entry in his diary:

> We heard that a monster had been born at Ravenna, of which a drawing was sent here; it had a horn on its head, straight up like a sword, and instead of arms it had two wings like a bat's, and at the height of the breasts it had a *fio* [Y-shaped mark] on one side and a cross on the other, and lower down at the waist, two serpents, and it was hermaphrodite, and on the right knee it had an eye, and its left foot was like an eagle's. I saw it painted, and anyone who wished could see this painting in Florence (fig. 5.2).[8]

Eighteen days later, Landucci wrote, a coalition of papal, Spanish and French troops

> took Ravenna and sacked it, being guilty of many cruelties.... It was evident what evil the monster had meant for them! It seems as if some great misfortune always befalls the city where such things are born; the same thing happened at Volterra, which was sacked a short time after a similar monster had been born there.[9]

Landucci's entry concerning the Ravenna monster echoed his compatriots' responses to the conjoined twins born two hundred years earlier, as described in Chapter One.[10] That monster too was taken as an evil omen: according to Giovanni Villani, the priors refused to allow it inside the city hall, deeming it, "according to the opinion of the ancients, a sign of future harm to the place it is born."[11] News of both monsters also circulated quickly through letters, drawings, and word of mouth; not unlike Landucci, Francesco Petrarch had described seeing in his youth a "painted image" of the twins, sent by Tuscan friends to his father, who was traveling in France.[12]

But there were also significant new elements in Landucci's response. For one thing, news of the Ravenna monster spread much faster and more widely than reports of the earlier monster, which concerned only local writers and Tuscan expatriates. For another, the Ravenna monster was not seen as an isolated case, like the early fourteenth-century twins, but as

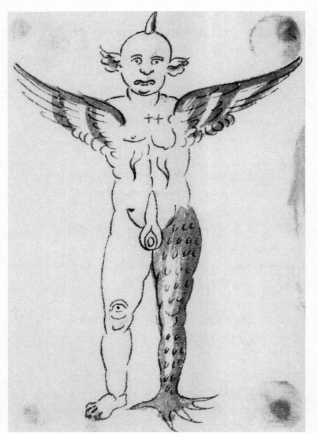 

5.2.1   5.2.2

Figure 5.2. The monster of Ravenna (originally Florence)
5.2.1. Diary of Marino Sanudo, MS Marc. it. VII.234 (= 9221), fol. 179v, Biblioteca Marciana, Venice.
5.2.2. German broadside, Einblatt VIII, 18, Bayerische Staatsbibliothek, Munich (1506).
5.2.3. *Ars memorandi* (Pforzheim: Thomas Anshelm, 1502), sig. a iiir.

The Ravenna monster had an unsually complicated iconographical and textual history. The details of its conformation can be traced back to a monster reportedly born in Florence in 1506. News of this creature's birth circulated quickly in the form of drawings, like this one that the Venetian patrician Marin Sanudo pasted into his manuscript diary in August (fig. 5.2.1), and woodcuts, like this German broadside from the same year (fig. 5.2.2), clearly based on a closely related image. According to the text of the broadside, the child was reported to the Pope, who decreed that it be starved to death. Like the Pope-ass (see fig. 5.5.1), the monster of Ravenna/Florence is portrayed as a composite, each element of its monstrous body pointing to a particular sin.[17] For this reason it resembles the composite memory images used to learn texts, like the beginning of the Gospel of John in this early sixteenth-century *Ars memorandi* (fig. 5.2.3).

5.2.3

one of an accelerating series of monstrous births and other prodigies that were documented in contemporary texts from the late fifteenth century on. Landucci himself recorded several other instances: the Volterra monster of 1474, which he learned about through a letter from a friend of his father's; three monstrous births that took place in the Venetian territories in 1489; and a thirteen-year-old boy with a parasitic twin, displayed for money in Florence in 1513.[13] This multiplication of monsters sprang at least in part from the new technology of printing, which greatly facilitated the spread of news through pamphlets and broadsides. But it also testifies to a heightened and more urgent interest in prodigies in the later fifteenth and early sixteenth centuries. While Petrarch merely noted his father's injunction to tell his own future sons about the conjoined twins, relating them to the slow unfolding of generational time, and while the Florentine priors could only invoke classical authority for their own discomfort, Landucci saw the Ravenna monster as presaging immediate and specific evil, based on his personal knowledge of at least one other instance (the Volterra monster). This expectation was quickly confirmed by the disastrous sack of Ravenna, which took place less than a month after that monster's birth.

In fact, the culture of monsters and prodigies exploded in late fifteenth-century Germany and Italy, in connection with specific political, religious, and military events. In Italy, it fed on the apocalyptic and reformist teachings of spiritual leaders like Savonarola, as well as on the sense of crisis and anxiety prompted by the invasions of the French and imperial armies and the disastrous military defeats suffered by the forces of the papal state.[14] In Germany it was a response to the perceived Turkish threat, as well as to internal political developments: continuing tension between the Emperor and the German princes, temporarily assuaged at the Diet of Worms in 1495.[15] In both contexts, imperial and papal publicists produced a spate of pamphlets and broadsides that drew on, and in turn fueled, the widespread culture of prodigies, while turning it to particular political ends.

In neither society was this culture of prodigies and prophecy exclusive to a single class or group. It was shared by erudite humanist scholars such as Brant, who appealed to the ancient tradition of divination; literate urban merchants and artisans such as the apothecary Landucci; and by peasants, laborers, and others without direct access to the written word. As Ottavia Niccoli has meticulously documented, the information and assumptions that structured this protean "system of prophetic signs" circulated in a variety of social spaces, from the church and the marketplace to the lecture hall and the papal court itself.[16]

The printed pamphlets and broadsides that were the most widely disseminated vectors of this culture varied in length and elaboration. Some
were simply descriptive: an image accompanied by a few lines of text noting the date and time of birth or names of parents and witnesses (fig.
5.2.2). Others, like Brant's, were highly moralized and allegorical, the
image overshadowed by extended reflections on the monster's political or
spiritual meaning (fig. 5.1). But very few left any doubt as to the emotion
such births were expected to evoke: they were described as *orrendi*, *orre-*
voli, *horribili*, *spaventevoli*, or *stupendi* in Italian; *espouventables*, *terribles*,
horribles in French; *erschroeckliche*, *grausame*, *grewliche* in German; "dreadful," "horrible," "terrible" in English. As these semantic clusters indicate,
monsters, like such other prodigies as earthquakes, comets, or celestial apparitions, aroused the intense fear identified by contemporaries as horror.

This horror did not spring simply from the confusion of categories —
animal and human, for example, or male and female — that anthropologists have placed at the heart of ideas of pollution; its roots lay rather in
the perceived violation of moral norms.[17] European Christians took a still
more somber view, treating prodigies as almost invariable harbingers of
locally targeted divine retribution in response to human sin. For Christians, in other words, the monster or prodigy was a sign of God's just
wrath. Important only as the bearer of a divine message — the reason so
many monsters died hours after birth, as soon as this had been delivered —
it pointed both to events that came before it and to events that were to
follow: the sin or sins that had prompted divine punishment and the
punishment itself, which could take the form of plague, famine, war, or
the like. The monster itself was a paradoxical product of God's mercy,
an alert and a warning issued to allow sinners one last chance to reform
themselves and avert the catastrophe to come. Because such catastrophes
were communal, Christians usually interpreted monsters as signaling not
individual but collective sin; it is for this reason that they rarely blamed
the monster's parents, still less the monster itself.[18] The horror of the
monster was thus manifold, directed at the precipitating state of sin, its
impending punishment, and the monster that served as a sign of both.

Many contemporary commentators on monstrous births confined themselves, like Landucci, to signaling (or retrodicting) approaching disaster.
Some, however, went further, speculating on the sins that had prompted
the divine warning and linking them to the monster's own configuration,
in a kind of point-by-point hermeneutics that treated the monster itself as
a revealed text. Writing on the monster of Ravenna, for example, the contemporary French chronicler Johannes Multivallis related its deformities
to particular moral failings:

The horn [indicates] pride; the wings, mental frivolity and inconstancy; the lack of arms, a lack of good works; the raptor's foot, rapaciousness, usury and every sort of avarice; the eye on the knee, a mental orientation solely toward earthly things; the double sex, sodomy. And on account of these vices, Italy is shattered by the sufferings of war, which the king of France has not accomplished by his own power, but only as the scourge of God.[19]

The significance of the monster's conformation helps to explain why so many reports of monstrous births contained images, not only in printed broadsides, whose woodcuts served as a come-on to potential buyers, but also in private chronicles and diaries.[20] Many such figures resembled contemporary images used in the art of memory, each pictorial element relating to a different item to be recalled (fig. 5.2.3). Such images could be used as objects of religious meditation, serving as a vivid reminder of sins to be avoided and the consequences to follow if they were not. In their composite nature, they also recalled the iconography of pagan idols and, especially, of demons, which were frequently represented as hybrid figures, constructed of many different animal and human parts (fig. 5.3). These visual references strengthened their associations with temptation, punishment, and sin.[21]

At the same time, particularly in the early days of printing, the images of monsters and other prodigies conduced to belief, just as photographs in tabloid newspapers do today, underscoring the authenticity of the report and heightening its emotional effect. Through illustrated broadsides, even non-eyewitnesses could experience the horror of the birth. Such considerations also help to explain the frequent references to witnesses and the emphasis, even in the most abbreviated broadsides, on the crucial circumstantial details of place, date, and time (fig. 5.4). Not only did these lend credibility to the account, but they were also integral to the cultural meaning of the monster as prodigy, since they allowed the audience to determine for whom the warning was intended and when the threatened disaster might occur.

Although pamphleteers, chroniclers, and publicists invoked the general symbolic and causal order represented by such monsters, writers of the numerous sixteenth-century treatises on prodigies explored it in much more detail.[22] One of the most influential was the Alsatian humanist and Protestant scholar Konrad Wolffhart, usually known by the Greek form of his name, Lycosthenes, whose *Prodigiorum ac ostentorum chronicon* appeared in print in 1557.[23] The influence of this work went far beyond the learned readers and writers who could afford and appreciate the large and highly illustrated Latin volume. Versions of it appeared in German and in English (the latter under the apt title, *The Doome, calling all Men to the Judgment*), and its examples and images were quickly absorbed and disseminated by countless vernacular publications.

The title page of Lycosthenes' book described it as a "chronicle of prodigies and portents that have occurred outside [*praeter*] the order, movement, and operations of nature, in both the upper and lower regions of the world, from the beginning to our own times—which sort of portent does not happen by chance, but, displayed to the human race, announces the severity and the wrath of God against its crimes, as well as great changes in the world." Lycosthenes presented creation as a unity, in which the physical mirrored the moral world. Just as human sin was a rupture in the moral order, so prodigies were a rupture in the physical order: the first of Lycosthenes' almost 1500 prodigies was Eve's encounter with a talking snake.

Like most other writers on prodigies, both Catholic and Protestant, Lycosthenes did not deny that many (though not all) monsters and other portents had natural causes; he was well aware of the arguments outlined above in Chapter Three. But he noted that those causes were often difficult to determine, and he deemphasized them in order to maximize the horror, in a strategy that was the obverse of Nicole Oresme's.[24] "We do not condemn natural explanations," he wrote, "and we greatly respect astrology too. But we know that nature is God's minister in matters both favorable and unfavorable, and that through her agency he aids the pious and punishes the impious, according to their different conditions." Even given a plausible natural explanation, "it is nonetheless impossible to deny that a monster is an imposing sign of divine wrath and malediction."[25]

Although most sixteenth-century writers on prodigies agreed on these basic tenets—not only Protestants like Lycosthenes, but also Catholics like Friedrich Nausea, bishop of Vienna[26]—there was nonetheless a distinct shift in the tenor of their analyses over time. Where Multivallis, writing on the Ravenna monster in 1512, read it as an emblem of the traditional sins of sodomy, avarice, pride, and worldliness, later authors like Lycosthenes were much more likely to stress quite different failings: blasphemy, religious error, heresy, conspiracy, and sedition.[27] This shift in emphasis reflected the religiously and politically charged atmosphere of Reformation society, riven by confessional rivalries and civil war. Mid-sixteenth-century writers, particularly Protestants, were also much more likely to place the accelerating frequency of monsters and other prodigies in an eschatological framework, as signs of the imminent end of the world.[28] In the dedicatory epistle to his *Chronicon*, Lycosthenes cited the biblical text most often invoked in this context, from the apocryphal book

5.3.1

5.3.2

Figure 5.3. The monster of Krakow

5.3.1. Pierre Boaistuau, *Histoires prodigieuses*, Wellcome MS 136, fol. [29v], Wellcome Institute for the History of Medicine, London (1559).

5.3.2. Jacopo Palladino of Teramo, *Belial* (Augsburg: Günther Zainer, 1472), fol. 86v.

The famous Krakow monster, reportedly born in the 1540s, was widely viewed as demonic rather than natural in origin; in his influential *Histoires prodigieuses*, Pierre Boaistuau used it to explore the problem of whether demons could engender offspring. This miniature (fig. 5.3.1), taken from the splendid dedication copy Boaistuau had prepared for England's Elizabeth I, shows that the standard depiction of this monster recalls the conventional iconography of demons—often portrayed with heads on their joints, as in the woodcut illustration to James of Teramo's satirical *Belial* (fig. 5.3.2).[18]

185

ff den Achten tag des monats Apprillen von der geburt Jhesu Christi als man zalt/fünffzehenhundert vnd Sechzehen Jare/inn der halben stunnd als nach mit nacht die glocf ains geschlagenn hat ist in des wolgebornenn herren herrn Ulrich Graue zü Muntfort vnnd stat Tettnang vonn ainer frawen mit namen Anna Bingerin / Conradenn Millers da selbs Eewirtin ain solch künd wie ob siet/mit dem fießlin vnnd schwenzel an seiner prüst auch grossem plafarben gewechs an seinem bauch aussen her vñ Ain Retin vñ Rotfarbe straumen darüber wie ain geschwer geborn worden/So des/selbig künd wachend/seine Bain von ain ander/vnnd so es schlaffend auch der massen gelegen/aber alwegen das fießlin in der hand haltend vnd ist ain döchterlin vnnd biß an/ den Neünden tag lebendig gewesen/solich künd hat der ob bestimbt Herr vnnd Grauff seinen maller Maister Matheyssen miller Maler burger zü Lindaw mit fleyß haysen verzaychnen oder konterfenn vnd zü trucken ver ordnen wie oben gesehen wirt.

Figure 5.4. Conjoined twins documented
Hans Burgkmair the Elder, *Disz künd ist geboren worden zu Tettnang* (Munich, 1516).

The text of this broadside emphasizes the role of the artist not only as disseminator of prodigious events, but also as witness and guarantor of their truth: "On the eighth of April, in the year 1516 after the birth of Christ, in the half hour after the clock had rung once after midnight, a child such as you see above was born to the noble lord Ulrich Graue of Muntfort and the town of Tettnang by a woman named Anna Bingerin, wife of Conrad Miller.... When the child was awake its legs were apart and when it slept it held its little foot in its hand. And it was a little girl and lived nine days. The said lord and count called his artist, Master Matheysen Miller, citizen of Lindau, to draw or portray the child and ordered it printed, as is seen above." Burgkmair may have produced his woodcut on the basis of Miller's circulated sketch, as in fig. 5.2.

186

of 2 Esdras 5: "And the sun shall suddenly begin to shine at night, and the moon during the day. Blood shall drip from wood, and the stone shall utter its voice...the wild animals shall roam beyond their haunts, and menstruous women shall bring forth monsters...."[29] The structure of Lycosthenes' work bore out this apocalyptic concern: although his *Chronicon* covered almost four-and-a-half millennia, he devoted almost a tenth of it to prodigies that appeared between 1550 and the book's publication in 1557. Although this avalanche of recent prodigies was largely an artifact of printing, which made so many reports available, to contemporaries it also suggested the imminence of the final reckoning. Assuming that the pace was likely to continue, if not accelerate, Lycosthenes included eighteen blank pages at the end of his 1552 edition of Julius Obsequens's fourth-century book on prodigies, so that readers could register portents as they occurred.

Early modern Christians' interpretation of prodigies was closely tied to external events. For this reason, it is difficult to identify a clear pattern of "naturalization" or "rationalization" of monsters and other prodigies before the late seventeenth century: as we have stressed in Chapter Three, medical writers and natural philosophers had supplied natural explanations for monsters and other wonders from at least the thirteenth century. The portentous interpretation of monsters as objects of horror did not slowly fade or disappear, but reasserted itself in waves according to local circumstance. Feeding on the anxieties and aspirations of the moment, it drew its power from conditions of acute instability: foreign invasion, religious conflict, civil strife. Thus the high-water mark for Italian interest in prodigies lay in the years between 1494 and 1530, during the most destructive phase of the Italian wars.[30]

In Germany, which had shared the late fifteenth-century Italian preoccupation with monsters, the concern with prodigies continued well into the seventeenth century, fueled by the political and religious struggles of the Reformation, which had emerged as the principal threat to the stability of German society. Even during this period, the specific cultural meanings given to monsters altered according to circumstance. In the early 1520s, they became staples of the propaganda war between supporters and opponents of Martin Luther. The most influential pamphlet in this vein was Luther's and Philipp Melanchthon's famous *Deuttung der czwo grewlichen Figuren* (1523), illustrated with woodcuts by Lucas Cranach (fig. 5.5): this read two recent monsters, the "Pope-ass" and the "monk-calf," as divine reproofs of monastic and papal corruption.[31] The time of the Schmalkaldic War (1547–48), between the Emperor and the Protestant Schmalkaldic League, was also extremely fertile in prodigy reports

5.5.1

5.5.2

Figure 5.5. The monk-calf and the Pope-ass
5.5.1–2. Lucas Cranach the Elder, woodcut illustrations for Martin Luther and Philipp Melanchthon, *Deuttung der czwo grewlichen Figuren, Bapstesels czu Rom und Munchkalbs zu Freijberg ijnn Meijsszen funden* (Wittenberg, 1523).

In this influential pamphlet, Luther and Melanchthon interpreted two recent monsters as prodigies and signs of divine wrath. According to Melanchthon, the "Pope-ass" (fig. 5.5.1), reportedly found dead in the Tiber in 1496, referred to the multiple and monstrous corruptions of the Roman papacy; Cranach underscored the connection by juxtaposing the creature with the Pope's Castel Sant'Angelo, identified by its banner with the keys of St. Peter. Unlike the Pope-ass, which belonged to the genre of allegorized composites like the monster of Ravenna/Florence (fig. 5.2), the "monk-calf" (fig. 5.5.2) was a calf born in Freiberg with a mantle of skin resembling a cowl. Luther interpreted this as a sign that the monastic state was "nothing other than a false and lying appearance and outward display of a holy, godly life."[19]

188

and broadsides, as was the period between 1618 and 1630, which marked the most acute phase of the Thirty Years' War.[32] In France and England, on the other hand, the culture of prodigies flowered only later, in the 1560s and 1570s, in the context of the French wars of religion and the accession of Elizabeth — herself part of John Knox's "monstrous regiment of women."[33] It died down in England in the early seventeenth century, to draw new energy from the political and religious upheavals of the 1640s and the English Civil War; only in the years after around 1670 did it begin to move into eclipse.[34]

Perhaps the clearest index of the sensitivity of the literature of monsters to external political and religious circumstances was the much reprinted six-volume French series of *Histoires prodigieuses*. Although the first two volumes, by Pierre Boaistuau (1560) and Claude Tesserant (1567), linked monsters and other prodigies to divine punishment, they suggested other, less ominous interpretations as well. According to Boaistuau, the main cause of monstrous births was divine judgment, swift and terrible, visited upon the sexually incontinent or bestial as a visible sign of "the horror of their sin" (fig. 5.3.1).[35] Yet he mentioned in the next breath the natural causes of the maternal imagination, excess or deficiency of seed, and indisposition of the uterus. Thus a child with four arms and legs, born on the day the Genoan and Venetian forces made peace, was at once the divine sign of brotherly reconciliation and the result of a narrow womb.[36] Tesserant thought monsters born in 1487 in Padua and Venice might have been the fearful presages of the misfortunes soon to be visited upon Italy, but he remarked concerning conjoined twins born near Heidelberg in 1486 ("very wondrous, for the rarity of the example") that no misfortunes and indeed "almost nothing memorable" had happened in Germany during that year.[37]

The *Histoires prodigeuses'* third, fourth, and fifth volumes, on the other hand, appeared between 1575 and 1582, at the height of the French wars of religion; they insisted unequivocally that all monsters were prodigies, sent directly by God to admonish Christians to "repentance and penitence."[38] In contrast, the anonymous author of the sixth volume, published in 1594, during a lull in the hostilities, worried that earlier volumes might have bored their audience and promised to "give more pleasure to readers for the most part curious about stories of wondrous things." Although he indicated that the years between 1567 and 1573 had been particularly fertile in monsters, because of God's righteous indignation against Protestant heretics who had risen up against the true faith, he construed the word "prodigious" in his title to mean not only portents but also all things that are "not ordinary and ... have caused great wonder [*grande admiration*]."[39]

Thus, in addition to signaling the sensitivity of the horror complex to political and religious circumstances, the volumes of the *Histoires prodigieuses* indicate that monsters could excite pleasure as well.

Pleasure: Monsters as Sports

Monsters had long been an object of spectacle. The thirteenth-century English chronicler Matthew Paris recorded the discovery in 1249 of an extraordinary young man in the Isle of Wight. "He was not a dwarf," wrote Matthew, "for his limbs were of just proportions; he was hardly three feet tall but had ceased to grow. The queen ordered him to be taken around with her as a freak of nature to arouse the astonishment of onlookers. The length of his tiny body is sixteen times that of this line."[40] By the later fifteenth century, at least in Italy, parents of monstrous children regularly showed them for money. According to the Florentine physician Antonio Benivieni, "a certain woman by the name of Alessandra, from the Milanese countryside, came to Florence and showed her [conjoined] twin boys for profit,"[41] and Landucci described a similar case in a diary entry from 1513:

> A Spaniard came to Florence, who had with him a boy of about thirteen, a kind of monstrosity, whom he went round showing everywhere, gaining much money. He had another creature coming out of his body, who had his head inside the boy's body, with his legs and his genitals and part of his body hanging outside. [The boy] grew together with his smaller brother and urinated with him, and he did not seem greatly bothered by him.[42]

By 1531, such displays were being officially licensed. In that year, Tommasino Lancellotti, a chronicler from Modena, recorded the arrival in Ficarola of a woman who "has a permit from the vicar of the bishop of Ferrara, and whoever wants to see her pays."[43] Over the course of the sixteenth century, similar references to the public display of monsters became increasingly common, not only in Italy, but also in Germany, England, and France.

It is hard to determine from chroniclers' and diarists' fragmentary and often laconic references whether the people who viewed such monsters responded to them with pleasure or fright. The fact that Landucci, who underscored the prodigious nature of all the other monsters whose birth he registered, made no such reference to the boy displayed in Florence suggests that he considered this case different and less threatening than the rest — perhaps because the boy had accommodated well to his condition and because it was generally believed that prodigious monsters died shortly after birth. But even dead infants elicited interest and appreciation: as a nun from Modena recorded in her diary in 1550, "Messer my father, that is Messer Giovanni Lodovico Pioppi, paid a foreigner from Germany or France one *bolognino* for each of us in the nunnery. He showed us an embalmed dead male baby in a box, who appeared to have two child's faces, and for the rest a single body, very beautiful to see and of good complexion — a wonderful thing."[44] At very least such passages, like the multiple messages of the *Histoires prodigieuses*, indicate that early modern spectators might register both pleasure and horror, depending on circumstances.

The multiplicity and lability of the meanings that early modern writers assigned to monsters is reflected in the wide variety of texts in which they appeared: from broadsides to Latin medical treatises to a whole new genre of books devoted entirely to the pleasures of reading about natural wonders. Despite their differences, these works shared certain features in their presentation of monsters. First, the same examples and images circulated endlessly back and forth between learned and popular works. The authors of vernacular wonder books plundered Herodotus and Livy as well as the latest broadsides for examples; readers might well encounter not only the same monster but also the same illustration in Lycosthenes' chronicle of portents, Fortunio Liceti's medical treatise on the types and causes of monsters, and Paré's vernacular *Des monstres et prodiges*.[45] Second, almost all accounts of monsters, from diary entries to the proceedings of academies, shared implicit narrative conventions that ultimately derived from the portentous interpretation of monsters. Details of place and date and often names of parents and witnesses unfailingly appeared (fig. 5.4). What had begun as an urgent concern of locals to decipher why this particular monster was born in a particular place at a particular time gradually developed into the canons of verisimilar reporting inherited by novelists and journalists.[46] All of these various accounts of monsters insisted ad nauseam on their truth and reliability, apparently anticipating skepticism on the part of readers. This literary convention served several purposes. Not only did it uphold the integrity of the author, but a true monster reverberated more loudly in the imagination than a made-up one; neither the natural explanations of philosophers nor the portentous readings of broadsides would have held much interest had the reality of the monster in question been in doubt.

What determined whether a monster elicited pleasure or horror? In part, response depended on external political and religious circumstances: in times of war, civil conflict, and confessional upheaval, almost anything was grist for the prodigious mill. In part, it had to do with the actual configuration of the monster in question, and how easy it was to assign it a natural cause. By the last quarter of the sixteenth century, there existed a specialized body of medical writing on the causes of monsters. This initially was made up of sections of treatises on topics like the wonders of nature or human generation and came later to include whole monographs devoted to the topic, both Latin and vernacular.[47] Readers and writers seem to have found it easier to come to terms with monsters that fell into the standard causal categories of this literature. Knowing that such beings had a simple natural explanation devoid of any moral component, they could view them with pleasure, as manifestations of the playfulness of nature or, at worst, the vagaries of chance. Prominent among these were three kinds of monsters: those interpreted as the result of an excess or defect of matter (for example, giants, dwarfs, conjoined twins, people with missing or supernumerary limbs); those produced by the mother's imagination (for example, hairy children); and those caused when the contributions of mother and father were almost evenly balanced (hermaphrodites and people of unstable sex). Hybrid monsters seen as springing from the intercourse of humans and animals – though there was debate as to whether this was possible – had a somewhat different status, as the behavior that gave rise to them was itself abhorrent. The resulting monster, even if not the product of special divine intervention, was nonetheless a sign of sin.[48]

As this last type suggests, the fact that a monster could be explained by natural causes did not always disqualify it as a prodigy. Monsters had a kind of twilight status. They were not really miracles (or Protestants who claimed that miracles had ceased would not have made so much of them); nor were they natural events, in the sense of regularities. Instead they provided a paradigm for the cooperation of primary and secondary causes.[49] The apparent contradiction stems from the comparative coarseness of the modern distinction between the natural and supernatural. The more nuanced ontology of the early modern period did not treat natural and supernatural causes as mutually exclusive. As a 1560 German broadside insisted concerning parasitic twins from Spain, "with such signs God warns us, even though [they are] naturally born."[50] But monsters for which a natural causal mechanism was available were not *necessarily* prodigies, in the absence of good reasons to interpret them as such – good reasons being, for example, that a child was conceived out of wedlock, the likely product of witchcraft, or born to parents of the wrong confession.[51] Very different were monsters so unusual and outlandish that it was hard to fathom how they could have been produced by any natural process, like the monster of Krakow (fig. 5.3.1) and the monster of Ravenna (fig. 5.2),

which sixteenth-century writers universally pronounced to be super-
natural in origin and which were reported to have been killed at birth.[52]
Regarding the former, for example, the German medical writer Jakob
Rueff wrote in his influential *De conceptu et generatione hominis* of 1554,
"I will ascribe the causes of this hideous monster to God alone."[53]

Caspar Peucer, a Protestant scholar and son-in-law of Melanchthon,
took up the question of how to distinguish prodigious from non-prodi-
gious monsters in his treatise on divination published in 1560.[54] In a section on *teratoscopia* (divination by monsters), he asked how one could distinguish a monster in the strict and etymological sense — something that
shows (*monstrat*) the wrath of God — from a misbirth that was merely rare.
The former differed from the natural and the expected not in small ways,
he answered, but by its whole shape and being, and the best way to identify one was through the immediate emotions it elicited. Because their
natural causes were invariably obscure, true monsters "have always terrified human minds, overcome by presages of sad and calamitous events, and
affected them with wonder and fear."[55]

Most monsters, however, did not fall into this extreme category, which
meant that they were at least potentially sources of pleasure: wonder shading into delight. This pleasurable dimension of monsters can be inferred
from the contexts in which they were encountered and the company they
kept. The presence of monsters in contemporary collections, both princely
and professional, shows that they were potential objects of aesthetic appreciation, rubbing shoulders with less ambiguous wonders of nature and art.
In Mantua, for example, Isabella d'Este's two-bodied puppy found a place
alongside cameos, medals, antique vases, corals, nautilus shells, and her
precious unicorn horn.[56] By the late sixteenth century, the Gonzaga collection also included a human fetus with four eyes and two mouths,[57]
while the collection of Ferdinand II of Tyrol, at Ambras, contained portraits of a giant and a hairy man from Teneriffe, together with the latter's
entire family, all in elaborate court dress (fig. 5.6).[58] Monsters also appeared
in the collections of doctors and apothecaries:[59] the image of Ferrante
Imperato's museum in Naples clearly shows a two-headed snake and a
lizard with two bodies joined to a single head (fig. 4.4.1).

The display of monsters as edifying and pleasurable spectacles was
not confined to princes and medical men but had long been a staple of
marketplaces and fairs. This was the case throughout Europe, though the
documentation for seventeenth-century London is particularly rich. The
diaries of John Evelyn and Samuel Pepys were peppered with appreciative
references to four-legged geese, hairy men and women, dwarfs and giants,
and conjoined twins. Converging on London from all over England, if not

5.6.1   5.6.2

Figure 5.6. Portraits of monsters from Schloss Ambras
5.6.1. Portrait of Petrus Gonsalvus, Kunsthistorisches Museum, Vienna (German, c. 1580).
5.6.2. Portrait of Gonsalvus's daughter, Kunsthistorisches Museum, Vienna (German, c. 1580).
5.6.3. Portrait of giant and dwarf, Kunsthistorisches Museum, Vienna (German, late 16th century).

Gonsalvus was born in Teneriffe in 1556. Raised at the court of Henry II of France, he also spent time at the court of Margaret of Parma, together with his wife and children. The artist of the Ambras portraits (figs. 5.6.1–2) is unknown. The giant and dwarf in another painting in the same collection (fig. 5.6.3) have been tentatively identified with Giovanni Bona, court giant of Ferdinand II, and the dwarf Thomerle.[20] The painting illustrates one of the display strategies common to many collections, which exaggerated the effect of their contents by surprising juxtapositions.

194

5.6.3

5.7.2

Figure 5.7. Duplessis's monsters
5.7.1–2. James Duplessis, *A Short History of Human Prodigious and Monstrous Births*, Sloane MS 5246, British Library, London (c. 1680).

Although most of the monsters recorded by Duplessis could be seen on display in London, like this hermaphrodite (fig. 5.7.1) — shown here with its modest paper flap raised — one "prodigious birth" belonged to his own family. This baby (fig. 5.7.2) was borne by his own mother-in-law. According to Duplessis, "the Occasion of this Monstrous birth was Caused by her Loosing her Longing, for a very Large Lobster which she had Seen in Leadenhall Market for which she had been Asked ane Exorbitant Price." After her husband brought the lobster home for her, she fainted, and "when Recovered she Could not Endure the Sight of it, the Meschief was done when her Time of being Delivered she Brought forth this Monster which was in all Respects like a Lobster Boyld and Red Excepting that instead of a Hard Shell or crust it was all a Deep Red Flech with all its Claws and Jonts it Died as Soon as Born. I James Paris her Son in Law had this Picture Drawn according to her Directions."

197

all over Europe, these paraded in taverns and coffeehouses as well as fairs. On December 21, 1688, for example, Pepys confided to his diary that he had seen a bearded woman at Holborn: "It was a strange sight to me, I confess, and what pleased me mightily."[60] Others collected the ballads, broadsides, and handbills that advertised such shows. The most assiduous included the Parisian diarist Pierre de l'Estoile, in the last decade of the sixteenth and first decade of the seventeenth century, and James Duplessis, Pepys's one-time servant, who in 1730 sold his memoir and collection to the naturalist and president of the Royal Society of London, Sir Hans Sloane (fig. 5.7).[61]

The protests of the pious afford the most telling evidence that monsters could produce pleasure, thereby dramatically undercutting their effectiveness as omens. François de Belleforest, author of the third volume of the *Histoires prodigieuses*, which appeared in 1575 in the midst of the French wars of religion, worried that if the term "monstrous" was applied indiscriminately to everything that was merely rare, it would lose its primary signification as a portent.[62] Similarly, the preacher who presided at the 1635 burial of a pair of conjoined twins in Plymouth chided his congregation for their unlawful "delight" in monsters. Urging them "to look higher, and to take notice of the special hand of God," he complained that the "common sort make no further use of these prodigies and strange births, than as a matter of wonder and table talk."[63]

Monsters seem indeed to have been the subject of much gossip and amusing discussion, for they figured prominently in conversation manuals, a new genre of literature directed toward vernacular readers who wished to better their social position by increasing their store of instructive and entertaining conversation. Thomas Lupton's brief but engaging *A Thousand Notable Things* (1586) includes accounts of many monsters among its other wonders (the loyalty of dogs, ways to clean amber and ivory, women who after many years of marriage turned into men).[64] Guillaume Bouchet's *Serées* (1584), dedicated to the merchants of Poitiers, recounted the honest pleasures of meals among family and friends, both male and female, enlivened by songs, clowns, and discussion of topics ranging from "Wine" to "Hunchbacks, Counterfeits, and Monsters." Exceptionally for this last, Bouchet reported that the men closeted themselves without the women, lest talk of monsters work upon the maternal imagination. The discussion itself revolved around the medical causes of monsters, their legal and theological status, where they were to be found (abundant in Africa, scarce in France), whether they were portents (probably not), and some individual cases.[65] The discussions held between 1633 and 1645 under the auspices of Théophraste Renaudot's Parisian Bureau d'Adresse — part employment agency, part medical and legal aid society, part academy — had a similar flavor. They ranged over an equally broad and varied list of topics, including "Of Embalming and Mummies" and "Of the Motion, or Rest of the Earth," as well as "Of the Little Hairy Girl Lately Seen in the City" and "Of Hermaphrodites." In the last, discussants touched upon the Ovidian associations of hermaphrodites, conflicting medical opinion as to the causes and possibility of perfect hermaphrodites, and stories of sex change.[66]

Whether or not the discussions printed by Bouchet and Renaudot took place precisely in the format and with the content ultimately made available to readers, both the *Serées* and the *Conférences* were presented as new, self-consciously open and egalitarian models of intellectual sociability. Both authors attacked what they presented as the wolfish solitude of clerics and scholars. Both emphasized social and intellectual openness; Bouchet remarked that "the common table . . . moderated and lowered the haughty," and Renaudot invited all comers to participate in his Monday discussions, assuring them that they would remain anonymous in the published proceedings. Both identified civil discourse with a panoramic survey of all viewpoints on a subject and with the pointed absence of any final conclusion.[67] That monsters became situated in these conversational experiments suggests that they were objects of lively interest and enjoyment for a socially heterogeneous audience, and perhaps also that they were associated with efforts to reform both the substance and the style of early modern intellectual life.[68]

There is further evidence for the delights of monsters in both the genre and tone of their textual presentation. In the seventeenth century, monsters sometimes became grist for satire, not only in plays like *The Winter's Tale* and *The Tempest*, but also in the traditionally omen-conscious broadside. A 1640 English tract about Mistress Tannakin Skinker, a "Hog-faced Gentlewoman" born to a respectable Dutch couple, reads as clearly tongue-in-cheek. Describing Tannakin's speech as a mixture of "the Dutch Hoggish Houghs, and the Piggs French Owee, Owee," the author interspersed fairy tales of enchanted princesses with the story of the hog-girl's ill-starred courtship.[69] Later in the century *The Athenian Gazette*, a London journal that purported to be the organ of "the Athenian Society for the Resolving [of] all Nice and Curious Questions" but was in fact a parody of the *Philosophical Transactions of the Royal Society*, had no difficulty in accommodating queries on monsters alongside ones such as "How shall a *Man* know when a *Lady loves* him?" and "Why are Angels painted in Petticoats?"[70] In such company, monsters could hardly horrify.

But it was not only satirists who exploited the pleasure of monsters in the later sixteenth and seventeenth centuries: physicians and medical professors acknowledged it as well. Realdo Colombo, one of Vesalius' successors in the chair of anatomy at the University of Padua, devoted the fifteenth and final book of his *De re anatomica* (*On Anatomy*), published in 1559, to what he called "anatomical rarities," including such wonders as hermaphrodites, a skeleton with fused joints, and a man without any sense of taste.[71] Colombo regularly invoked the word "monster" to describe these rarities, but he made no mention of their possible significance as signs and portents. Instead, like Cardano in *De rerum varietate*, he approached these matters as a connoisseur able to distinguish the truly wondrous from the not-so-wondrous on the basis of fifteen years of dissecting during which, he boasted, he had seen everything except the body of someone mute from birth.[72] Scorning neophyte anatomists who believed that every variable organ they saw indicated a "monster,"[73] Colombo flaunted his own erudition by instructing the reader in the degree of wonder appropriate to each oddity: an extra finger merited only mild wonder, the man without taste — a Venetian named Lazarus who earned his living by eating anything for pay — was worthy of much wonder, and hermaphrodites were the most wonderful of all.

Other medical writers seconded Colombo's sophisticated sense of appreciation and enjoyment, referring to monsters as nature's "sports" or jokes (*lusus*). The Paduan physician Liceti claimed in his 1616 treatise on the causes of monsters that there was nothing under the sun more "rare and wondrous" than monsters, and he suggested an alternative to the "vulgar etymology" that linked monsters to divine signs: "monsters" derived from *monstrare* ("to show") not because God uses them to demonstrate his wrath to sinners, but because their "novelty and enormity provoking as much wonder as surprise and astonishment, everyone shows them to one another."[74] Contesting the "vulgar" opinion that identified monsters with errors or failures in the course of nature, Liceti likened nature to an artist who, faced with some imperfection in the materials to be shaped, ingeniously creates another form still more admirable.[75] On this view, monsters revealed nature not as frustrated in her aims, but as rising to the challenge of recalcitrant matter, a constricted womb, or even a mixture of animal and human seed. "It is in this that I see the convergence of both Nature and Art," wrote Liceti, "because one or the other not being able to make what they want, they at least make what they can."[76]

We will discuss the image of nature as virtuoso artisan at length in Chapter Seven. Here it is enough to note that many other seventeenth-century writers, philosophers as well as doctors, hastened to express their appreciation of creatures that the vulgar might find disturbing but which finer minds recognized as tokens of nature's ingenuity and fecundity rather than of God's wrath. In his *Religio medici* (1642), the English physician Thomas Browne declared "there are no *Grotesques* in nature," for even in monstrosity "there is a kind of beauty, Nature so ingeniously contriving the irregular parts, as they become sometimes more remarkable than the principall Fabrick." For Gottfried Wilhelm Leibniz, monsters exemplified the pleasure nature took in variety, akin to the pleasure cultivators of tulips and carnations took in unusual colors and shapes.[77] A flagrant decoupling of form and function, as in the case of the enchanting but useless variety of seashells or flowers, was widely celebrated as evidence of nature at play.[78]

Although one can find expressions of this attitude into the eighteenth century, it did not go unchallenged by other learned writers, who in increasing numbers rejected monsters as objects of both pleasure and horror—deeming them signs of neither natural ingenuity nor divine wrath. Proponents of a neo-Aristotelian tradition that emphasized the fit of form to function, they regarded monsters as nature's errors. In their writings we see the emergence of a third complex of reactions to monsters, organized around a feeling of distaste or repugnance.

Repugnance: Monsters as Errors

In 1560 Benedetto Varchi, the Florentine philosophical writer, published two lectures he had presented twelve years earlier to the Florentine Academy, one on human generation and the other on monsters. In the prologue to the first, he sang the praises of nature as maker of wonders: "There is no one of wit [*ingegno*] so high or low that he does not sometimes contemplate the wonders [*miracoli*] of Nature, no less a pleasurable marvel than a marvelous pleasure. Rather, the more witty someone is, and the greater his understanding, with so much greater wonder and enjoyment does he labor to understand the causes of [the marvel]."[79] In the second lecture, however, Varchi excepted monsters from this happy set of associations, describing them as a "foul and guilty thing" and attributing them to the "errors and sins of whoever makes them."[80] In referring to "sin," Varchi was not invoking the prodigy tradition: for him, the errors in question were those of nature rather than humankind. His distaste for monsters rested on a neo-Aristotelianism similar to that of his contemporary, Julius Caesar Scaliger, who had attacked the wonder-mongering in Cardano's *De subtilitate* of 1550.[81] Like Scaliger—and Aristotle before him—Varchi considered that nature's wonder lay in her marvelous regularities, such as sunlight, moonlight, and the stars, rather than in the accidents and chance missteps that occasionally marred her work.[82] Thus

Varchi rejected both pleasure and horror as appropriate responses to monstrous births. For him, monsters were deformed and ugly, and they therefore evoked rather repugnance and distaste.

In Varchi, we find one of the earliest expressions of the attitude that was increasingly to dominate the reaction to monsters announced by the self-consciously learned in the seventeenth century — not only medical theorists and natural philosophers, but also theologians, humanists, and other men of letters. Monsters inspired repugnance because they violated the standards of regularity and decorum not only in nature, but also in society and the arts. A monstrous birth undermined the uniform laws God had imposed upon nature; the "monstrous regiment of women" rulers threatened the order of civil society; the intrusion of marvels into poems and plays destroyed literary verisimilitude. These standards were at once cognitive, moral, and aesthetic. The horror and pleasure of monsters did not wholly disappear in either learned or popular circles, but both responses gradually became disreputable among the intellectual elite. As the involvement of theologians suggests, the rejection of monsters as occasions of horror and pleasure cannot be ascribed to secularization or the "rise of science" — the "triumph of rational thought over monstrosity," in the words of Canguilhem.[83] The repugnance of monsters was not so much the consequence of making nature autonomous of God as it was of enslaving nature entirely to God's will. Nature was no longer permitted to play. In this section we trace the rise of the repugnance of monsters in three contexts: anatomy, theology as it intersected with natural philosophy, and aesthetics.

Describing conjoined twins born in Florence twelve years earlier, depicted by Bronzino, and finally dissected in the Rucellai gardens in the presence of a group of "most excellent physicians and painters,"[84] Varchi himself placed monsters in an anatomical framework. His reaction of distaste, which so contrasted with Colombo's frank wonder at nature's sports, was magnified toward the end of the sixteenth century in the works of northern European anatomists such as Martin Weinrich and Jean Riolan the younger. Proponents of an Aristotelian and Galenic tradition that emphasized the fit of anatomical form to physiological function, these anatomists regarded monsters as organisms that had failed to achieve their telos, their perfect final form. They not only insisted upon the restrictive Aristotelian definition of monsters as offspring that did not resemble their parents (thereby excluding the monstrous races of Pliny and Augustine, as well as the sea serpents and ostriches of Paré and other wonder books),[85] they also asserted the ugliness of monsters from the standpoint of Aristotelian final causes. As Weinrich wrote in his *De ortu monstrorum commentarius* of 1596, "All that is imperfect is ugly, and monsters are full of imperfections."[86]

Comparing the views of Colombo and Riolan on hermaphrodites illuminates the contrasting sensibilities of wonder and repugnance. The difference is not one between natural and supernatural explanations, nor even between rival explanatory traditions within medicine and natural philosophy. Both anatomists wrote in the purely naturalistic vein of the medical tradition, and neither so much as mentioned a prodigious interpretation. Moreover, both wrote as academic anatomists, interested in the internal as well as external conformation of bodily organs. But whereas Colombo saw hermaphrodites as the wonder of anatomical wonders, male and female combined in one body, Riolan, half a century later in 1614, described them as merely deformed men or (mostly) women. Chiding Colombo for mistaking an enlarged clitoris for a penis, Riolan insisted upon the anatomical impossibility of the true hermaphrodite, that is, one with functional reproductive organs of both sexes.[87] He castigated even more severely the provincial physician Jacques Duval, who had described the recent case of the Rouen hermaphrodite Marie/Marin le Marcis as an example of nature's ceaseless variety.[88] Riolan accused Duval of anatomical incompetence and insinuated that he had distorted the facts of the case to increase the wonder of the tale.[89] Riolan reclassified Colombo's and Duval's wondrous hermaphrodites as deformed women, who ran grave legal risks of being accused of "abusing their sex ... scandalous crimes which brought prejudice to the honor and the life of the persons accused."[90] For Riolan, monsters of all sorts represented a full-scale violation of the rule of law, both natural and civil—the "perversion of the order of natural causes, the health of the people, and the authority of the king."

Monsters continued to figure in anatomical literature throughout the seventeenth and early eighteenth centuries, and dissections were regularly reported in learned journals like the *Journal des Sçavans*, the *Philosophical Transactions of the Royal Society of London*, the *Histoire et Mémoires de l'Académie Royale des Sciences* in Paris, and the *Miscellanea curiosa medicophysica Academiae Naturae Curiosorum* in Leipzig and other German cities. But the final decades of the seventeenth century witnessed a slow shift in the questions that motivated these dissections. When Robert Boyle reported to the Royal Society of London on the dissection of the monstrous head of a colt in 1665, or the Besançon correspondent of the *Journal des Sçavans* sent an account of a dissection of conjoined human twins in 1682, the dissections were detailed and matter-of-fact, certified by medical men and other eyewitnesses in the form already established in prodigy broadsides.[91] The reports focused exclusively on the individual monster at hand, in all of its singularity. By the early eighteenth century, however, the anatomical study of monsters increasingly drew its justification from the knowledge it could provide, by contrast, about the functions of the normal organism, rather than from the wonder to be gleaned by examining singular cases in great detail. Bernard de Fontenelle, writing in his capacity as Perpetual Secretary of the Paris Académie Royale des Sciences, defended the comparative anatomical study of both animals and monsters on grounds that they furthered the understanding of the normal human body. "Even monsters are not to be neglected," he wrote in his history of the Académie: "The mechanism hidden in a certain species or in a common structure develops in another species, or in an extraordinary structure, and one could almost say that Nature in multiplying and varying its works cannot avoid sometimes betraying its secret."[92]

The early eighteenth-century annals of the Académie bore out Fontenelle's program. Anatomical interest in monsters did not disappear; indeed, it intensified.[93] Monsters became embedded in a larger embryological context, enlisted as evidence for one or another ontogenetic theory. Benignus Winslow and Louis Leméry debated the divine versus accidental origins of monsters in the early decades of the eighteenth century, and Albrecht von Haller and Caspar Friederich Wolff later battled over preformationism and epigenesis, using monsters to argue their cases. When anatomists dissected individual monsters, they related them back to the normal organism. An eight-month-old human fetus without a brain provided the occasion for reflections on the seat of the soul and animal spirits, for example,[94] while a stillborn with only "a single little hole placed between the two ears" supported the hypothesis that the fetus is nourished by the umbilical cord rather than the mouth.[95]

These anatomical observations did not converge in any single theoretical explanation for monsters. In the early eighteenth century, anatomists could not even agree on the definition of a monster, much less on a classification.[96] No explanation was without its critics; debates on the causes of monsters raged for decades in the annals of scientific and medical societies and in learned treatises.[97] What united the early eighteenth-century anatomies of monsters was not any particular theory, but the general framework of inquiry. Although each report began with an account of the historical particulars that could have been taken from one of the more laconic broadsides, the emphasis throughout was upon matching form to function — a normal function either disrupted by malformation (emphasized by those who, like Leméry, believed monsters to have accidental causes), or served by extraordinary means (the favorite examples of

Winslow and his allies, who insisted on divinely guided preformationism).

Anatomists no longer exclaimed over the rarity of a malformation, but rather over its perverse functionalism. Thus a report to the Royal Society of London on a woman with a double womb praised the "admirable contrivance of Nature" by which each womb had conceived "perfectly well-formed" children, all brought to term.[98] Jacques-François-Marie Du Verney, surgeon and anatomist at the Paris Jardin du Roi, argued that twins joined at the hips had a "supple ligament" in place of the usual hard cartilage joining the pubic bones, in order to allow each twin some degree of independent movement; this testified to "the richness of the Mechanics of the Creator, at least as much as the more regular productions [of nature]" (fig. 5.8).[99] Function was paramount, and anatomists usually reserved their scant wonder, or rather admiration, for the foresight and wisdom literally embodied in the design of what the Paris anatomist Jean Mery called the "machine of the human body."[100]

Thus by the early eighteenth century, at least in the elevated ranks of the Paris Académie, monsters had been normalized in the sense that they were habitually related to a functional standard; the irregular, in Canguilhem's words, had been "subjected to the rule."[101] But this cannot be described as the advent of naturalization; physicians and natural philosophers had long attributed monsters to natural causes.[102] Rather, the new anatomical view of monsters corresponded to a particular view of the natural order as absolutely uniform and not subject to exceptions, even in the name of sport. Fontenelle took a severe line with marvel-mongers on the occasion of the dissection of the deformed fetus of a lamb lacking head, chest, vertebrae, and tail. "One commonly regards monsters as jests of nature [*jeux de la nature*]," he wrote, "but philosophers are quite persuaded that nature does not play, that she always inviolably follows the same rules, and that all her works are, so to speak, equally serious. There may be extraordinary ones among them, but not irregular ones; and it is even often the most extraordinary, which give the most opening to discover the general rules which comprehend all of them."[103] In principle, monsters interested the anatomist not because of their rarity or singularity and the concomitant wonder evoked by nature's sports, but because they revealed still more encompassing and rigid regularities.

Over the course of the sixteenth and seventeenth centuries, natural philosophers and theologians traced a parallel trajectory, although few adopted the strict naturalism of the anatomists. In creating an inviolable order of nature, natural philosophers and theologians did not aim to make nature independent of God. On the contrary, not since Augustine had natural philosophy and theology so thoroughly bent nature to the will of

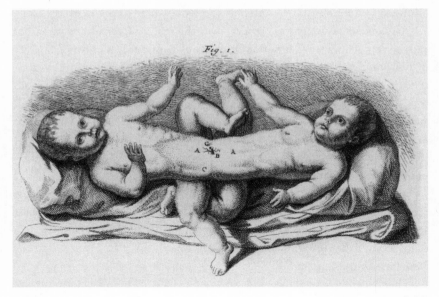

5.8.1

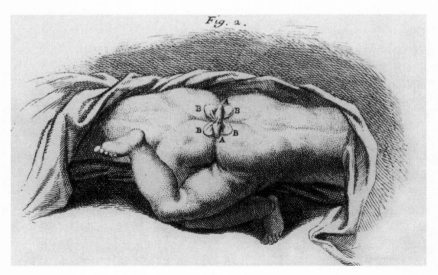

5.8.2

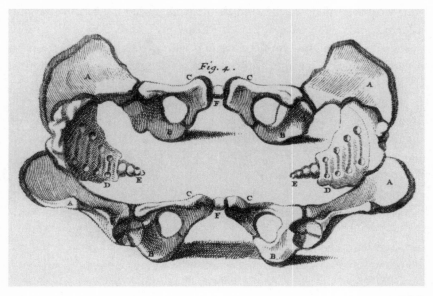

5.8.3

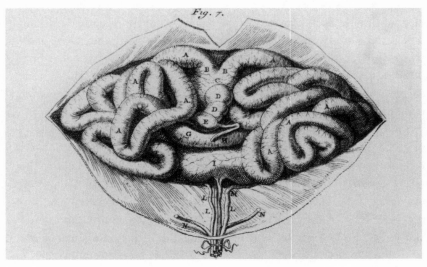

5.8.4

Figure 5.8. Conjoined twins dissected
5.8.1–4. Jacques-François-Marie Du Verney, "Observations sur deux enfans joints ensemble,"
Mémoires de l'Académie Royale des Sciences, année 1706 (Paris, 1707), figs. 1, 3, 4, 7.

Born without difficulty in September 1706, thanks to the ministrations of a skillful midwife, these twins had a single placenta and a single umbilical cord. They died after seven days, in part — according to Du Verney — "because they were uncovered too often, to satisfy the curiosity of numerous people."[21] The twins were dissected and the particulars of their internal and external conformation meticulously recorded, taking the process of documentation well beyond the Florentine relief of c. 1317 (fig. 1.8.2) or Burgkmair's broadside of 1516 (fig. 5.4).

God. Moreover, philosophers and theologians excluded from the natural order not only supernatural but also many previously accepted natural causes (such as the astral influences and plastic virtues beloved of the preternatural philosophers), because they granted nature too much intelligence and independence of action.[104] Only a nature consisting solely of "brute, passive, stupid matter" would not usurp divine prerogatives. The essence of the new attitudes toward nature among natural philosophers was not so much naturalization as subordination: the subordination of anomalies to watertight natural laws, of nature to God, and of citizens and Christians to established authority.

Reacting to religious polemics that fanned schism into civil war across Europe, as well as to heterodox forms of piety that undermined ecclesiastical and political authority by a direct appeal to God, many natural philosophers and theologians closed ranks in a campaign to detach monsters and other marvels from divine signs and warnings. Lengthening and loosening the chain of causes connecting God to monstrous births, comets, and other phenomena in the prodigy canon weakened the power of religious sectarians and political dissidents, who enlisted God in their cause by appealing to such breaches in the ordinary course of nature. Although few early modern natural philosophers, much less theologians, denied God's ability to intervene directly in the natural and social order, many doubted publicly that God's aims were best served by constantly overturning his own laws. Disorderly nature could be used by disorderly people, as Jean Riolan had already argued in an anatomical context.

Because this transformation was largely triggered by political and religious conflicts, which unfolded at different paces across Europe, it had no uniform chronology. Themes that emerged in mid-sixteenth-century Italy — in the work, for example, of Varchi and Pietro Pomponazzi — appeared in France only decades later, and in England and Germany only in the mid-seventeenth century, as local wars of religion ignited and cooled, and political upheavals swelled and subsided. But the gradual pacification of religious war and the stabilization of political and ecclesiastical authority throughout Europe in the latter half of the seventeenth century did harden a consensus among the learned that marvels should evoke neither horror nor wonder, only distaste. No longer portents of impending doom or ornaments of God's creation, monsters became rather problems in theodicy.

Appealing to the voluminous medical and natural philosophical literature on monsters, Boyle mocked nature's inability to mold even "so small, and soft, and tractable a portion of matter" as an embryo: "How these gross aberrations will agree with that great uniformity and exquisite skill, that is ascribed to nature in her seminal productions, I leave the naturists to make out."[105] Of course, if nature could be reproached for monsters, so could God. Boyle's solution to this problem contrasted maternal nature to patriarchal God, opposing benevolent concern for the welfare of individual creatures to the uniformity and regularity of the universe. Nature, Boyle implied, could be accused of irresponsibility in producing monsters and other aberrations because she was "but a nursing mother to her creatures"; God, as stern judge and creator, was under no such compassionate obligation and could punish or neglect his creatures with impunity.[106] Moreover, God subordinated the welfare of particular creatures to "his care of maintaining the universal system and primitive scheme or contrivance of his works."[107] Either anomalies were part of that universal system and therefore only apparently irregular, or they were the unfortunate but rare consequences of an order that worked most of the time in the best interests of most creatures. Hence monsters were "preternatural" only in a restricted sense, for although they deviated from the form of their species, "the causes, that produce that deviation, act but according to the general laws whereby things corporeal are guided."[108] For Boyle, therefore, irregularities like monsters were no longer divine suspensions of the regular order of nature, but rather the unintended consequences of God's refusal to amend his ordained regularities in particular cases.

The meanderings of monsters in early modern natural philosophy and theology illustrate the intimate links between cognitive and emotional responses to such anomalies. As portents signifying divine wrath and imminent catastrophe, monsters evoked horror: they were *contra naturam*, violations of both the natural and moral orders. As marvels, they elicited wondering pleasure: they were *praeter naturam*, rare but not menacing, reflecting an aesthetic of variety and ingenuity in nature as well as art. As deformities or natural errors, monsters inspired repugnance: they were neither ominous nor admirable but regrettable, the occasional price to be paid for the very simplicity and regularity in nature from which they so shockingly deviated.

In natural philosophy, what had made monsters wondrous rather than revolting was not recourse to God, but differing concepts of nature and, especially, of nature's relationship to God. Nature, as viewed by Cardano, Liceti, and other preternatural philosophers, was an active and even inventive artisan who fulfilled final causes and who had her settled customs from which she occasionally deviated out of necessity or whim. She worked through a panoply of mechanisms, from the astral influences and the imagination of preternatural philosophy to the collisions of matter in motion of mechanical philosophy. She was the servant of God: subservient to divine providence, but capable of a certain playful originality.

In Boyle's contrasting view, nature was no more than the sum total of dead, passive matter and the "general and standing laws of motion" ordained by God. Although God could in principle suspend or alter these laws, in fact he sustained them inexorably, as Fontenelle had emphasized, even at the cost of aberrations that awakened only distaste. The voluntarist rhetoric of much late seventeenth-century natural philosophy appeared to echo Augustine's when it claimed that nature was but the manifestation of unfettered divine will — "as easie to be *altered* at any time, as to be *preserved*," as Samuel Clarke put it in his Boyle Lectures of 1705.[109] Yet even militant voluntarists like Boyle assumed that God in fact almost never changed his mind. They enslaved nature to God, and God to his own laws. Many theologians, acutely aware of how people mobilized disruptions of the natural order to disrupt the civil order, were as chary of prodigies and miracles as the natural philosophers.[110] In the name of simplicity, uniformity, and universality, not only nature but also God lost the spontaneity that made for surprises in the established order of things.

The emotional shift from wonder to repugnance that occurred in anatomical and natural philosophical accounts of monsters found a parallel in late seventeenth-century aesthetics. Neoplatonizing theorists of the fifteenth and sixteenth centuries had granted an ever-greater and more positive role to the artist's *fantasia* as a creative power that imitated or even rivaled that of God.[111] After the rediscovery of Aristotle's lost treatise on *Poetics*, literary critics, especially in Italy, also began to elevate the poet's inventive imagination above the slavish imitation of nature.[112] Within the aesthetic of Mannerism, which emphasized invention and technical virtuosity, artists (like poets) provoked audiences to wonder not only by depicting marvels in their works, but also by the extraordinary force of imagination with which they did so: the artist or poet himself became the marvel, not the work. In his treatise on painting of 1591, Gregorio Comanini praised Giuseppe Arcimboldo's portrait of Emperor Rudolf II as Vertumnus because "the application of such fruits to the parts of the body is so ingenious that wonder is forced to turn to downright astonishment" (fig. 5.9).[113] John Donne's contemporaries professed themselves wonderstruck at his poetry, "hovering highly in the aire of Wit."[114] Such an aesthetic saw even monsters as works of art, awakening wonder in onlookers by their rarity and oddity, as well as by the ingenuity of their maker. Human monsters who survived to adulthood were often commemorated in admiring woodcuts, copper engravings, and verse, and

Figure 5.9. A wonderful portrait
Giuseppe Arcimboldo, *Rudolf II as Vertumnus*, Skokloster Palace, Skokloster (1591).

Arcimboldo's portrait of Rudolf II was intended both as a display of wit and as an allegorical comment on the eternity and fruitfulness of his reign. The fruits and vegetables that make up the emperor's head come from various times of the year, illustrating his identification with Vertumnus, god of the seasons. The effect is to emphasize the victory of Rudolf's rule over time and to associate it with the eternal spring of the mythical Golden Age.[22]

211

some of those who ended up at court were subjects of official portraiture, as in the case of the hairy man from Teneriffe (fig. 5.6.1).[115]

By the late seventeenth century, however, this marvelous aesthetic had fallen into disfavor among many patrons, theorists, and artists, and monsters in particular excited disgust rather than admiration. William Drummond scorned the poetry of Donne and other "metaphysicals" in the same breath as he impugned the wonder of monsters: "It is no more Poesy than a Monster is a Man," he wrote. "Monsters breed Admiration at the First, but have ever some strange Loathsomeness in them at last."[116] Neoclassical critics appealed to the opening lines of Horace's *Ars poetica*, which ridiculed the monstrous in painting — hybrids, for example, between man and horse or woman and fish — and by analogy exaggeration and impropriety in poetry.[117] The articles on literary and artistic subjects that appeared in the mid-eighteenth-century *Encyclopédie* of Denis Diderot and Jean d'Alembert gave full expression to these ideas. The unsigned essay on the marvelous in literature noted that the "fictions and the allegories which are the parts of the marvelous system please enlightened readers only to the extent that they are drawn from nature, sustained with verisimilitude and accuracy, in sum conform to received ideas."[118]

Enlightenment critics admitted that imagination was essential to art, but it was an uneasy admission. They emphasized that the regulated imagination restricted itself to objects similar to those given by sensation; the more the mind "departs from these objects," according to d'Alembert, "the more bizarre and unpleasant are the beings it forms."[119] Not that art strived for the mirrorlike imitation of nature: the task of the imagination was to perfect and embellish nature, to make what French playwright and critic Jean-François Marmontel, in his *Encyclopédie* article on "Fiction," described as "a new world... a world such as it should be, had it been made only for our pleasures."[120] This new world had no room for the marvelous, the monstrous, the fantastic, the all-too-creative products of an unruly imagination. For Marmontel, the emblem for the deranged artistic imagination was the monstrous "mixture of neighboring species"[121] — what the Chevalier de Jaucourt, writing on "Verisimilitude," referred to as "mingl[ing] incompatible things, coupling birds with serpents, tigers with lambs." These "fictions without verisimilitude, and events prodigious to excess," he wrote, "disgust readers of mature judgment."[122]

At stake was not truth to fact, but rather verisimilitude — not truth itself, but the appearance of truth, which relied on conventions of plausibility, decorum, and seemliness. Monsters affronted not truth but taste; Marmontel permitted artists an occasional Pegasus, as long as the wings were in proportion.[123] Because monsters were defined by distaste, and because distaste in turn depended on custom and appearance, Voltaire, writing on monsters in his *Dictionnaire philosophique*, was hard put to say exactly what one was. He had seen a woman at a fair with four breasts and a cow's tail on her chest: "She was a monster, without difficulty, when she let her bosom be seen, and a pleasant enough looking woman when she did not." Neither excess nor deficiency of parts defined a monster any longer, for monstrosity was a matter of beholding and beholder. Monsters were "animals whose deformities horrify," but the horror sprang from violated convention, not violated nature: "The first negro was however a monster for white women, and the first of our beauties was a monster in the eyes of negroes."[124] This moment of symmetric shock was comically captured in the libretto of Emmanuel Schikaneder's and Wolfgang Amadeus Mozart's *Die Zauberflöte* (1791): at their first encounter the Moorish slave Monostatos and the featherclad birdcatcher Papageno recoil in mutual horror from one another, shrieking in unison "Surely that's the devil!"[125]

As the irony of these symmetric monstrosities makes clear, Voltaire's *horreur* differed in emotional texture from the horror of portents: monsters were unseemly, not unnatural or supernatural. Liceti had also pointed to the relativity of monsters, arguing that the legendary races of satyrs and sirens would hardly regard themselves as monstrous. But for Liceti true monstrosity by definition inspired wonder, not horror or revulsion, and it did so universally, in every beholder. Thus the exotic races could never qualify as monsters, for neither they nor their neighbors were wonderstruck, surprised, or horrified at their appearance.[126] Voltaire's view on the relativity of horror echoes that of medieval travel writers like John Mandeville on the wonder of the monstrous races. But Voltaire went even further. He emphasized the impossibility of a universal response even to individual monsters, who became such only because aesthetic judgment was formed by visual habit, and visual habits reflected willy-nilly the diversity of culture and custom.

Why did Liceti's monsters inspire wonder and Voltaire's, repugnance? Both definitions hinged on emotional response; both presupposed rarity and surprise; both insisted upon natural explanations. But Liceti's monsters—always individual cases, never races—trespassed upon the norms of nature, of like reproducing like. Voltaire's monsters, of whatever sort, offended the norms of custom. For Liceti, monsters showed nature at her most inventive and ingenious; for Voltaire, they defied conventions of beauty and decorum. Both the wonder and the repugnance of monsters were strongly felt: Voltaire's "horror" may seem exaggerated when compared to the horror of divine punishment, but it does register an intensity

213

of response. The shift from norms of nature to norms of custom did not weaken the emotional charge of monsters. Rather, both norms of nature and custom became more rigid in the early eighteenth century. What had once been nature's habits hardened into inviolable laws; what had once been irregular and unpredictable public conduct hardened into a regimen of propriety and social rules. Monsters did not — could not — violate nature's laws, but in infringing upon society's customs, they cast doubt on the stability of both orders.

CHAPTER SIX

Strange Facts

In 1675 Gottfried Wilhelm Leibniz proposed a privately financed "academy of sciences." There would be exhibitions of "all sorts of optical wonders...unusual and rare animals...extraordinary rope-dancer...artificial meteors...ballets of horses...museums of rarities" alongside displays of calculating machines, the air pump, telescopic observations, an anatomical theater, and the equality of the oscillations of the pendulum: "All respectable people would want to see the curiosities in order to be able to talk about them;...Princes and distinguished persons would contribute some of their wealth for the public satisfaction and the growth of the sciences. In short, everybody would be aroused and, so to speak, awakened...."[1] In Leibniz's imagined academy, the marvels of the fair, the court, and the cabinet of curiosities merged seamlessly with the marvels of natural philosophy, and all served to stimulate a sense of expanded possibility in art and in nature.

Although this particular Leibnizian fantasy was never realized, it nonetheless captured the atmosphere of wonderstruck novelty that suffused natural philosophy and natural history throughout the seventeenth century. The marvels of natural philosophy shaded imperceptibly into the larger cultural category of marvels: air pumps performed next to rope-dancers; optical wonders could be admired alongside museums of rarities. The familiar canon of sixteenth-century marvels — monstrous births from broadsides and fairs, optical illusions from how-to books on natural magic, rarities imported from exotic locales for the collections of the curious — expanded in the seventeenth century to include new instruments and observations. Natural philosophy and natural history reciprocated by studying monsters, exotica, and other wonders with unprecedented intensity. The sciences of nature during this period produced and consumed marvels as never before or since. That princes would pay for, and the public flock to, such attractions had been true for centuries; what was unprecedented was Leibniz's conviction, shared by many of his contemporaries, that the sciences would thereby grow. What had begun in the late fifteenth and sixteenth centuries as a primarily medical inquiry into topographical wonders and bodily abnormalities expanded first into a series of dogged efforts by a few natural philosophers (often with medical training) to explain these oddities, and later yielded a cascade of strange facts that commanded the attention of the most prominent seventeenth-century naturalists. This chapter is about how and why marvels entered natural philosophy after centuries of exclusion, in the context of a new epistemology of facts and a new sociability of collective empirical inquiry.

Marvels made their way into natural philosophy over an extended period and by several routes. A long arc of some two centuries spans the composition of the treatises on healing springs and Leibniz's fantasy of an academy of sciences. During this period marvels spread from courts to private academies sponsored by princes and noblemen to universities to state-chartered scientific societies. Geographically, this institutional diffusion began in northern Italy in the late fifteenth century, reaching France, the Low Countries, and England by the early seventeenth century, and the German principalities by the mid-seventeenth century. These different sites — court, private academy, university, scientific society — at which naturalists encountered marvels were sometimes encompassed within a single career. Galileo, for example, held chairs at the universities of Pisa and Padua, served as court philosopher of the grand duke of Tuscany, and was a member of the Roman Accademia dei Lincei. Through this enlarged circle of institutional contacts, early modern naturalists were more likely to have firsthand experience of wonders than their medieval predecessors. Marvels could also be a vehicle for social mobility, bridging university and court for the natural philosopher on the make: Ulisse Aldrovandi hoped to parlay an anatomy of a dragon whose appearance had coincided with the investment of Pope Gregory XIII into papal favors;[2] Galileo touted both lodestones and "the Medicean stars" as *meraviglie* to potential patrons at the Tuscan court.[3] There were very few scientific careers in the seventeenth century untouched by an encounter with the marvelous: Galileo pondered the luminescent Bolonian stone, Christiaan Huygens studied double refraction in Iceland spar, Robert Boyle wrote reports of monstrous births, Isaac Newton speculated about a "certain secret principle" of chemical sociability, Leibniz trafficked in artificial phosphors.

Marvels also spanned Latin and vernacular natural philosophy and natural history. The French translation (1556) of Girolamo Cardano's *De subtilitate* (1550), for example, went through seven editions and exerted considerable influence on the vernacular but philosophically ambitious works of surgeon Ambroise Paré and of artisan Bernard Palissy.[4] The latter, who took vehement exception to Cardano's explanation of petrified shells found on mountain peaks, vented his annoyance that "the books of other philosophers were not translated into French," as *De subtilitate* had been, so that "I could refute [them] as I refute Cardano."[5] As Palissy's riposte illustrates, the relationship between Latin and vernacular texts on the natural history and natural philosophy of marvels was by no means restricted to one of transmission and meek reception.

The case of Palissy suggests that marvels circulated freely not only among the elite contexts of university, court, and academy, but also between these locales and more public and popular discourses, like those of Paré and the innovative Bureau d'Adresse[6] in early seventeenth-century Paris. Examples of marvels came from the latest broadside as well as from Pliny or Cardano, and explanations were exchanged between Latin and vernacular texts. For example, the explanations of monsters offered by Paré's French work (aimed at an audience evidently broad enough to include "des Dames et Demoiselles"[7] as well as fellow surgeons) and by Fortunio Liceti's Latin treatise *De monstrorum natura, caussis et differentiis* (1616) overlapped to a considerable extent. Moreover, the many articles devoted to monsters in the annals of the scientific societies established in the latter half of the seventeenth century did not significantly alter the repertoire of stock explanations, although they multiplied descriptions and anatomies of specific cases. Just as the marvels of the fair mingled with the marvels of the new philosophy in Leibniz's imagined academy, so no sharp line separated lay and learned discourse on marvels during much of the seventeenth century.

The people who investigated marvels in the seventeenth century were a larger and more motley group than the medieval community of natural philosophers, who had largely been bound to the universities and church schools, or that of the sixteenth-century physicians, who had shuttled between university chairs and princely courts. By the early decades of the seventeenth century, professors like Aldrovandi and physicians like Liceti who inquired into the wonders of nature were joined by erudite Jesuits like Athanasius Kircher, gentlemen *virtuosi* like John Evelyn, and members of academies such as the Accademia dei Lincei in Rome or the Academia Naturae Curiosorum, founded in Schweinfurt in 1652. Not all marvel mongers in the seventeenth century concerned themselves with natural philosophy; nor did all natural philosophers and natural historians attend to marvels. But there was an unprecedented (and never-to-be-repeated) overlap between the two groups. This was in part because marvels, described in words and displayed as things, saturated early modern

European culture, thrusting themselves into the consciousness of nearly everyone, from prince to pauper to philosopher. The overlap also reflected important changes in the community and activities of naturalists. The printing press and the vogue for collecting stimulated contacts between naturalists from Bologna to Leiden to Uppsala, who eagerly exchanged letters, learned favors, and specimens with one another.[8] Books and collections also served to widen participation in natural philosophical discussions: the artisan Palissy had no Latin, but he could read and write works in the vernacular, as well as assemble his own cabinet, and thereby enter learned debates about the origins of figured stones and the strange properties of springs and fountains.

In the middle decades of the seventeenth century, title pages of works in natural history and natural philosophy began to address themselves to "the curious" or "the ingenious" of Europe.[9] These terms defined a new community of inquirers primarily by sensibility and object, and only secondarily by university training or social status. Disposable time and income as well as education were to some extent assumed by the avocations of empirical inquiry, voluminous (and usually polyglot) correspondence, and collecting, but they were not the core qualifications of "the curious." More central was a highly distinctive affect attached to equally distinctive objects: a state of painstaking attention trained on new, rare, or unusual things and events.[10] To count as one of the "curious" was hence to combine a thirst to know with an appetite for marvels, which also came to be known as "curiosities" in this period. The "curious" or "ingenious" constituted themselves as a self-declared, cosmopolitan elite, one which spanned national and confessional boundaries, and which was the immediate ancestor of the Republic of Letters of the Enlightenment. The new community of the curious was nearly as socially diverse as it was geographically far-flung, embracing aristocrats and merchants, physicians and apothecaries, lawyers and clergymen of all denominations; but it was united in its preoccupation with the marvels of art and nature.

Late sixteenth- and seventeenth-century naturalists were not only a larger and more diverse group, intellectually and socially, than medieval natural philosophers; they also organized themselves in new and different ways. It is significant that Leibniz chose to call his grand entertainment an "academy of sciences," for by 1675 the academy, rather than the court or university, had become an important institutional locus for innovative natural philosophy and natural history. Leibniz himself had extensive dealings with two of the most prominent, the Royal Society of London and the Paris Académie Royale des Sciences (both founded in the 1660s), and he became the first president of the Berlin Akademie der Wissenschaften some forty years later. In the latter half of the seventeenth century, scientific academies became the focal point for the inquiries of naturalists into marvels — or, in contemporary parlance, for curiosities investigated by the curious. The annals of these academies document scores of marvels, sent in by correspondents at home and abroad.

Academies in principle weighed the credibility of these marvelous reports by strict criteria: Were there eyewitness accounts or only hearsay? Were the witnesses men of stature and character? Were they professionally qualified to observe the phenomena in question? These criteria applied to the extrinsic credibility of testimony and derived from legal doctrine and practice. More difficult to assess was the intrinsic credibility of things. The prominence of wonders in the sixteenth and seventeenth centuries broadened the sense of the possible in natural history and natural philosophy. Although naturalists of both the Old and New Worlds were increasingly skeptical about certain specific claims made by Pliny and other ancient authorities,[11] the most immediate impact of the torrent of new discoveries was to lower the scientific threshold of credibility. Lands where "the winter and summer as touching cold and heate differ not," inhabited by birds of paradise and armadillos, swept by hurricanes and tornadoes, and peopled by tribes arrayed in exquisite feather mantles or nothing at all, seemed to many Europeans at least as strange as anything in Pliny. As Sir Walter Ralegh reminded readers inclined to doubt his stories about the headless Blemmyes in Guyana, "Such a nation was written of by *Maundeuile*, whose reportes were held for fables many yeares, and yet since the *East Indies* were discovered, wee finde his relations true of such thinges as heretofore were held incredible."[12] With an intensity that approached that of the great religious debates of the day, naturalists queried themselves and each other as to what and whom to believe, and why.

Marvels posed these ontological and evidentiary questions in their sharpest form for sixteenth- and seventeenth-century natural philosophers. It is perhaps not surprising that speculations about natural magic and books of secrets should abound with wonders both natural and artificial,[13] but it is noteworthy that reformers as hostile to arcana as René Descartes should also have felt obliged to insist that "there was nothing so strange in nature" as to defy explanation by his mechanical philosophy.[14] Francis Bacon scorned the pretensions of natural magic but proposed a collection of all that was "new, rare, and unusual" as a propadeutic to his reformed natural philosophy.[15] By the middle decades of the seventeenth century, no natural philosopher with systematic aspirations could wholly ignore marvels. The several late-sixteenth- and early-seventeenth-century challenges to ancient and medieval authorities in natural history and natural philosophy created an intellectual free-for-all that spilled out of the university amphitheaters into courtly academies like the Accademia del Cimento in Florence,[16] private salons like the Montmor circle in Paris,[17] and even into the streets, as when Palissy set up placards at Paris intersections inviting "the most learned doctors [*médecins*] and others" to three lessons on the peculiar properties of fountains.[18] Everywhere marvels figured prominently as both the stuff and proof of the new philosophies, as in Leibniz's imaginary academy of sciences.

In this chapter we examine the impact of this preoccupation with the marvelous on seventeenth-century natural history and natural philosophy. We argue that marvels played a brief but key role in forging a new category of scientific experience: the fact detached from explanation, illustration, or inference. We will first set forth Bacon's influential vision of a natural philosophy reformed in part by a natural history of marvels, and then examine how two of the most prominent new scientific institutions of the seventeenth century — the Royal Society of London and the Paris Académie Royale des Sciences, both established in the early 1660s to promote natural history and natural philosophy — went about gathering, sifting, and pondering the strange facts Bacon had recommended so urgently. We will describe how the sociability of the late-seventeenth-century scientific societies subverted the philosophical mission Bacon had envisioned for strange facts, and conclude with an analysis of the problems of belief raised by strange facts. Throughout we will be concerned with how marvels raised problems of evidence, explanation, and experience central to the study of nature during this period.

Baconian Reforms

Sir Francis Bacon, Lord Chancellor of England under James I and prolific writer on topics ranging from the common law to friendship to horticulture, mapped between 1605 and 1620 a program for the reform of natural history and natural philosophy that reverberated loudly throughout the seventeenth and eighteenth centuries. Although almost no one adopted in its entirety his plan for a "Great Instauration," or new beginning in natural philosophy, almost everyone — Descartes, Boyle, Leibniz, Hooke, Huygens, Newton — incorporated fragments of his vision into their own work. His influence on seventeenth-century views of scientific method and organization was vast. In this section we discuss Bacon as a preternatural philosopher in the tradition described in Chapter Four, albeit a preternatural philosopher who ultimately broke with that tradition. Bacon's injunctions to catalogue and explain wonders transformed preternatural philosophy by making it indispensable to a reformed natural philosophy.

In the early decades of the seventeenth century what we have called preternatural philosophy flourished.[19] Its practitioners went so far as to tackle phenomena usually identified as divine portents. The French historian and natural philosopher Scipion Dupleix, for example, instructed his readers that if comets presaged the death of princes, it was because the same dry exhalations that fed the comet's flame afflicted the high and mighty, whose delicate constitutions and luxurious tastes made them susceptible to acute diseases.[20] Liceti explained that women sometimes bore children with horns and tails not because they had slept with demons but because their overwrought imaginations had imprinted a diabolical shape upon the soft matter of the fetus.[21] Presses all over Europe steadily issued treatises in both Latin and the vernacular, dedicated to explaining phenomena like the appearance of three suns in the sky, the attractions and repulsions of the magnet, stones figured with the images of organisms or landscapes, earthquakes, the power of the imagination to imprint soft matter, monsters — in short, all that happens "extraordinarily (as to the ordinary course of nature) though not lesse naturally."[22] The proviso "though not lesse naturally" was key to the enterprise, for however strange and even incredible the objects of preternatural philosophy might seem, the working premise was that all such anomalies might be ultimately explained by natural causes. Hence its claim to be "philosophy," the repository of causal explanations, as opposed to mere "history," an assemblage of disconnected particulars.[23] Indeed, preternatural philosophy in some ways set the most demanding standards for scientific explanation in the late sixteenth and early seventeenth centuries.

Bacon's vision of a reformed natural history and natural philosophy marked a turning point in preternatural philosophy, at once its culmination and its downfall. In Bacon's plans for an enlarged natural history that would embrace "nature erring" and "nature wrought" as well as the familiar regularities of "nature in course," in his insistence that such a reformed natural history of particulars must precede and correct natural philosophical generalizations, in his conviction that the causes of natural things were "secret," and in his prediction that a close study of the wonders of nature would promote the invention of the wonders of art, Bacon was the most ambitious of the preternatural philosophers. But precisely because he made preternatural phenomena so central to a renovated study of nature, he thereby undermined preternatural philosophy. In the hands of Pietro Pomponazzi, Cardano, Liceti, and others, the explanation of marvels supplemented Aristotelian natural philosophy by boldly taking on phenomena that were neither commonplace nor regular, and whose causes were neither manifest nor constant. However, no wholesale remaking of standard natural philosophy was expected or desired. On the contrary: preternatural philosophy assumed all the categories of conventional natural philosophy, and simply defined itself by their negation. In contrast, Bacon believed that "a substantial and severe collection of the heteroclites or irregulars of nature, well examined and described,"[24] would shake Aristotelian natural philosophy at its foundations, shattering its axioms and discrediting its logic of syllogisms. Bacon's deployment of preternatural philosophy was hence key to his program to reform natural philosophy as a whole.

That program was launched with *The Advancement of Learning* (1605), in which the recently knighted Bacon addressed to King James I of Scotland and England a "general and faithful perambulation of learning, with an inquiry what parts thereof lie fresh and waste, and not improved and converted by the industry of man." He hoped thereby to inspire a "kingly work" to cultivate the heretofore fallow or barren territories of knowledge.[25] Among knowledge's "unmanured" plots, Bacon numbered natural history, which he divided into the history "of nature in course, of nature erring or varying, and of nature altered or wrought; that is, history of creatures, history of marvels, and history of arts."[26] The first he found in "good perfection," but the second and third had scarcely been begun. Waving aside the wonder books written for "pleasure and strangeness," Bacon called for a complete collection of exceptions, deviations, singularities, strange events, and other staple phenomena of preternatural philosophy. Although this collection would have to be purged of "fables and popular errors," Bacon insisted that it should nonetheless include well-attested accounts of witchcraft, divination, and sorcery: "For it is not yet known in what cases, and how far, effects attributed to superstition do participate of natural causes."[27] Only genuinely supernatural phenomena, "prodigies and miracles of religion," were to be excluded.

The uses of the history of marvels would be twofold, according to Bacon. First, it would serve to correct natural philosophical axioms derived only from commonplace phenomena; and second, it would pave the way for innovations in the mechanical arts. Although Bacon claimed that art (in the broad sense of all that is manmade, including technology) and nature differed in neither essence nor form but only in "efficients"[28] (that is, the causes that move each), he nonetheless described art as nature under constraint. Nature under the compulsion of art resembled nature erring in the variability of effects visible in both cases, revealing possibilities hardly glimpsed under ordinary conditions. When nature wandered from its wonted paths without the prodding of art, the marvels thereby produced mimicked the variability induced by art — or rather, marvels were proto-art, nature anticipating art.[29] This is why Bacon contended that "from the wonders of nature is the nearest intelligence and passage towards the wonders of art: for it is no more but by following and as it were hounding Nature in her wanderings, to be able to lead her afterwards to the same place again."[30] Nature and art met in marvels, because marvels of both kinds forced nature out of its ordinary course.[31]

In this rapprochement of nature and art by way of marvels, Bacon echoed the views of the earlier natural magicians like Marsilio Ficino — and of some demonologists. Marvels occurred naturally through a chance conjunction of causes; natural magicians (and demons) manufactured marvels by engineering such conjunctions. The natural magician delved into the hidden and secret properties of things, tapped the invisible but powerful forces of the imagination and the stars, and above all imitated the incessant matchmaking of nature ("so desirous to marry and couple her parts together,"[32] as Giovanni Battista della Porta put it) in knitting together causes ordinarily found apart. Bacon therefore called natural magic the "operative" counterpart to speculative natural philosophy, and defined it as "that great liberty and latitude of operation which dependeth upon the knowledge of forms." Once speculative natural philosophy had revealed these forms, operative natural magic could manipulate them, just as demons were believed to do. Not that Bacon had any truck with demons: he warned that certain "darksome authors of magic" had strayed from the "clean and pure natural."[33] But with or without demons, natural magic could manufacture marvels aplenty in Bacon's view. Although he ridiculed extant books of natural magic as credulous, frivolous, and secretive, he did not doubt the potential of this art to counterfeit natural marvels as astonishing as any recounted in fabulous romances.[34]

Bacon intended his tripartite natural history to serve the traditional function of supplying natural philosophy with raw materials, as well as the avowedly untraditional function of providing what he called "the stuff and matter of true and lawful induction" that would reveal the hidden forms and genuine axioms of natural philosophy.[35] Bacon made clear that he did not mean by induction an enumeration of cases, which was endless and vulnerable to counterexample, but rather "a form of induction which shall analyse experience and take it to pieces, and by a due process of exclusion and rejection lead to an inevitable conclusion."[36] With this new-style induction Bacon attempted to counter the open-endedness of empiricism based on particulars, including the empiricism of preternatural philosophy. Whereas Aristotelian demonstrations had allegedly produced closure in natural philosophy, the work of collecting and accounting for even the rarities of nature would "never come to an end," as Cardano sighed at the end of his four-hundred-page treatise on the subject. As the title "New Organon" of Bacon's methodological treatise of 1620 indicates, he intended his work to replace Aristotle's old logic of syllogisms (set forth in the *Prior Analytics* and called, together with the *Posterior Analytics*, the "organon," or tool) with a new logic of things equal to the "subtlety of nature."[37] In this new logic the natural history of particulars would play an unprecedented role not only as the permanent foundation of natural philosophy,[38] but also as a constant check upon abstraction and generalization. Bacon sometimes referred to natural history as a "warehouse," one that must be constantly replenished and drawn upon if natural philosophy were ever to fathom the secrets of nature.[39] He emphasized the novelty of using natural history as the foundation of natural philosophy,[40] and he was correct to do so. Institutionally, natural history was an emergent discipline, the first university chairs in the subject having been created in the sixteenth century.[41] Bacon placed the fledgling discipline in the intellectual vanguard by making it at once the source and the safeguard of his new logic of nature.

The safeguard was needed to curb the inborn tendency of the human understanding to premature generalization. Of all the cardinal philosophical sins, Bacon dreaded intellectual impatience most. Both in the formation of axioms and the abstraction of notions, he claimed that the understanding naturally leaps from a scant handful of particulars to the loftiest level of generalization, for it is more at home in the manipulation of words than of things.[42] Baconian empiricism — the careful scrutiny of particulars — was hard work, and the understanding shunned it. To counteract the unseemly haste of the understanding, Bacon recommended that it be "hung with weights": the mind would be disciplined by a method that shuttled perpetually back and forth between particulars and axioms.[43] Neither keenness of intellect nor steadiness of character could cure the mind of its restless aversion to particulars: "We must lead men to the particulars themselves, and their series and order; while men on their side must force themselves for awhile to lay their notions by and begin to familiarise themselves with facts [*cum rebus*]." Nor would attention to particulars by itself correct the errors and vacuity of natural philosophy: the senses were themselves infirm and prey to deceptions, and so "great a number and army of particulars" would overwhelm the understanding. Only a strict method of ordering, digesting, and distilling particulars could reveal both axioms and new particulars.[44] By thus zigzagging between the universal and the particular, Bacon's "new organon" would yield the knowledge of underlying forms that he called the "interpretation of nature."

Bacon's "interpretation" of nature, as opposed to the over-hasty "antici-pations" of the understanding,[45] was to consist of tables of agreement, dif-ference, degree, and rejection among simple natures. The Baconian tables to some extent formalized the patterns of reasoning already employed by Cardano in his search for hidden causes of unusual phenomena.[46] Bacon not only systematized the kind of comparative empiricism practiced by earlier preternatural philosophers; he advanced an elaborate taxonomy of the special types of particulars that would compose the tables: the twenty-seven "Prerogative Instances." Bacon intended both tables and instances as aids to the discovery of the "laws" or "forms" of simple natures such as whiteness or heat.[47] The "Prerogative Instances" were privileged em-pirical particulars, which opened up shortcuts for the senses and under-standing to the sought-for forms, and which abridged the tedious work of induction over all particulars. At least five of the prerogative instances were drawn directly, albeit not uncritically, from preternatural philoso-phy: "similar instances," "singular instances," "deviating instances," "bor-dering instances," and "instances of power." Bacon himself singled out this set of five instances as a coherent group with a particularly urgent and general role to perform in the renovation of natural philosophy.

Similar instances were resemblances of form — between the mirror and the eye, between tree gums and rock gems, between the beaks of birds and the teeth of animals, between the rhetorical trope of surprise and a descending musical cadence. These "real and substantial resem-blances" Bacon distinguished sharply from the sympathies of natural magic, but he nonetheless imagined nature to be criss-crossed by analo-gies, which would supply the axioms of first philosophy (*philosophia prima*).[48] Singular instances were exceptional species within a genus — the elephant among quadrupeds, the letter *s* among consonants, the magnet among minerals — which "sharpen and quicken investigation, and help to cure the understanding depraved by custom, and the common course of things."[49] Closely allied in substance and function to singular instances were deviating instances, which were marvelous individuals rather than species, "errors, vagaries and prodigies" that break intellectual habits and also point the way to the wonders of art. Bacon thought these "prodigies and monstrous births" too abundant to require any examples; he only reiterated his call for a credible natural history of them (fig. 6.1).[50] Bor-dering instances apparently combined two species — bats, flying fishes, a "biformed foetus" — and although Bacon admitted that they might be sub-sumed under singular or deviating instances as "rare and extraordinary," he deemed them worth a separate class.[51] Fifth and last, instances of power collected the wonders of art, by direct analogy with the wonders

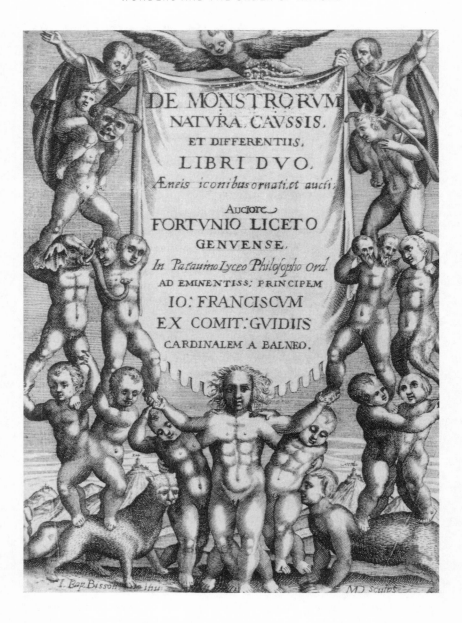

Figure 6.1. Compilations of "monsters and prodigious births"
Fortunio Liceti, *De monstrorum natura, caussis et differentiis libri duo* [1616], 2nd ed. (Padua, 1634), frontispiece.

Francis Bacon assumed that readers of his *Novum organum* (1620) were so familiar with "strange and monstrous objects, in which nature deviates and turns from her ordinary course" that actual examples would be superfluous. Among the works in which Bacon and his readers might have encountered such wonders was a richly illustrated compendium of monsters first published by the Italian physician Fortunio Liceti in 1616; its engraved title page provides an enticing sense of its contents.

of nature described in the four prerogative instances immediately preceding. Their function was also similar, namely to challenge the understanding to find forms general enough to include the anomalous as well as the commonplace: "But as by rare and extraordinary works of nature the understanding is excited and raised to the investigation and discovery of Forms capable of including them; so also is this done by excellent and wonderful works of art."[52]

Bacon considered these five prerogative instances concerning the wonders of art and nature to form a cluster, the utility of which was not restricted to the investigation of any single simple nature, in contrast to the other twenty-two prerogative instances. This cluster formed the core of a natural history that ought to be collected immediately, in order to prepare the understanding to receive the "dry and pure light of true ideas" by ridding it of "daily and habitual impressions."[53] The function of the preternatural cluster was as much destructive as constructive: marvels would be the battering ram that broke down the axioms of Aristotelian natural philosophy and would clear the way for the new axioms of Bacon's interpretation of nature. A mind awakened by wonders would reject syllogisms based solely on the familiar and the commonplace.

Here Bacon and other preternatural philosophers parted ways. By naturalizing marvels the preternatural philosophers had aimed to fortify, not subvert, natural philosophy. Their explanations were drawn either from Aristotelian natural philosophy — for example, recalcitrant matter resists form — or from supplementary sources, such as the Avicennnian and Neoplatonic theory of the imagination, which could peacefully coexist with Aristotelian natural philosophy. Even at its strangest, preternatural philosophy was an annex rather than a contradiction of traditional natural philosophy: non-standard phenomena given non-standard explanations. Bacon treated many of the same phenomena that had interested the preternatural philosophers — divination, imagination, sympathies and antipathies, extraordinary weather — but he did so with the aim of finding causes broad enough to encompass both the ordinary and the extraordinary. Notably, these unified and comprehensive causes were hidden within "the more secret and remote parts of nature" and thus resembled the occult properties of preternatural philosophy more closely than the manifest causes of conventional natural philosophy. The reason for studying the marvels of art and nature was to show that these apparent exceptions to the rules could be explained by different, deeper rules:

For we are not to give up the investigation, until the properties and qualities found in such things as may be taken for miracles of nature be reduced and

227

comprehended under some Form or fixed Law [*sub aliquâ Formâ sive Lege certâ*]; so that all the irregularity or singularity shall be found to depend on some common Form, and the miracle shall turn out to be only in the exact specific differences, and the degree, and the rare concurrence; not in the species itself. Whereas now the thoughts of men go no further than to pronounce such things the secrets and mighty works of nature, things as it were causeless, and exceptions to general rules.[54]

Although Bacon still spoke the older language of nature occasionally straying from her wonted paths, nature's "particular and special habits [*consuetudines naturae particulares*]" were already giving way to "fundamental and universal laws [*leges fundamentales et communes*]" in his vision of a natural philosophy without exceptions.[55]

Bacon broke radically with the sensibility as well as with the metaphysics of preternatural philosophy. For Cardano, Liceti, Dupleix, Lemnius, and others, at least some of the wonders of art and nature remained wondrous even after explanation. Just which ones merited just how much wonder was a matter for educated judgment, but connoisseurship of wonders no more stifled delight than did connoisseurship of artworks. Both the dedicatory epistles of these works and the numerous citations to them in the frankly entertaining wonder books suggest that they were intended to please as well as to instruct their readers— "useful as well as pleasant and enjoyable," as Lemnius put it in his dedicatory epistle to Matthias Gallomontois de Heesuvüyck, Prelate of Metelbourg.[56] In contrast, Bacon seldom let an opportunity pass to reprimand those who pursued the study of nature for pleasure. He was particularly vehement on the possible abuses of his proposed natural history of marvels. He insisted upon short descriptions, "though no doubt this kind of chastity and brevity will give less pleasure both to the reader and the writer."[57] He complained that hitherto natural history had been too eager to record "the variety of things," which, though delightful, obscured "the unity of nature," which should be the foundation of natural philosophy.[58] Acknowledging that the "knowledge of things wonderful is indeed pleasant to us, if freed from the fabulous," Bacon nonetheless argued that the pleasure of nature's marvels derived "not from any delight that is in admiration itself, but because it frequently intimates to art its office."[59] That is, the wonders of nature deserved appreciation because of their utility in inspiring wonders of technology, not because of the pleasurable wonder they evoked.

Bacon knew how to praise the pleasures of learning, although he was considerably more comfortable with curiosity than with wonder, which he called "broken knowledge."[60] But neither "natural curiosity" nor "variety and delight," much less ambition, vainglory, or profit, were the proper ends of knowledge: learning should be "a rich storehouse, for the glory of the Creator, and the relief of man's estate."[61] Bacon intended his reformed natural philosophy to serve Christian charity in alleviating the material hardships of human life. Natural history, of whose pleasures Bacon was particularly wary, was also a "granary and storehouse of matters, not meant to be pleasant to stay or live in, but only to be entered as occasion requires."[62] The temptation to linger over the variety of nature, so richly indulged in the works of Cardano and Paré, must be resisted, according to Bacon. As his contemptuous references to books of wonders and natural magic make clear, Bacon was well aware of the pleasures of marvels, hawked at fairs, sung at inns, read *in camera*, and discussed at dinner. He wrote in a milieu so saturated with marvels that he thought it superfluous to give examples of "deviating instances," because these rarities were paradoxically so abundant.[63] The sheer cultural availability of marvels may well have thrust them into the central epistemological role they played in Bacon's plan for a reformed natural philosophy. But Bacon wanted wonders without the pleasure of wonder, as mere means to more solemn and useful ends.[64]

Bacon warned against premature theorizing even more sternly than against unseemly wonder. He feared a natural history adulterated with natural philosophy, much as he might have feared a suborned witness in court. He intended his refurbished natural history to correct, not simply enlarge, extant natural philosophy. Natural history had therefore to guard its independence from natural philosophy, lest Baconian innovations be "tried by a tribunal which is itself on trial." In his sketch for a natural history of heavenly bodies (apparently a compilation of astronomical observations), Bacon complained that astronomers and astrologers had contaminated their observations with "arbitrary dogmas." He sought instead a natural history of heavenly bodies wholly innocent of astronomy and astrology, "all theoretical doctrine being as it were suspended: a history embracing only themselves...pure and separate."[65] In his treatise on English common law, Bacon recommended exposition by concise rules and aphorisms on similar grounds, "because this delivering of knowledge in distinct and disioyned Aphorismes doth leave the wit of man more free to turne and tosse, and make use of that which is so delivered to more severall purposes and applications."[66] The "warehouse" of natural history should stand open to many possible uses and interpretations.

In his programmatic writings on natural philosophy, Bacon recalled the myth of Atalanta, the swift huntress who lost the race to her suitor Hippomenes by stooping to pick up the golden apples he had temptingly strewn in her way. She stood for the folly of sacrificing great long-term to small short-term gains, real gold to fool's gold. Atalanta symbolized rash impatience for Bacon, and he revealingly invoked her bad example to warn against "that unseasonable and puerile hurry to snatch by way of earnest at the first works which come within reach" of axioms and effects from natural history.[67] Natural history should be a discipline for the mind, a slow and meticulous exercise in self-restraint, as well as the warehouse for natural philosophy. Pleasurable variety and precocious theory, golden apples both, threatened that discipline.

Bacon's "Histories" — particulars scrubbed clean of conjecture and severed from theory — became the model for the "facts" in late seventeenth-century natural philosophy. In both Latin and in several European vernaculars, the word "fact" and its cognates (Latin *factum*, Italian *fatto*, French *fait*, German *Tatsache*, Dutch *feit*[68]) originally meant "deed" (cf. "feat"), "that which is done." This sense, shading from "deed" to "crime," is still preserved in English in legal phrases like "to confess the fact" or "after the fact." Lawyer that he was, Bacon used the word primarily in this sense: "All crimes have their conception in a corrupt intent, and have their consummation and issuing in some particular fact."[69] But Bacon's use of the phrase "matter of fact" (sometimes "matter in fact") hints at the new usage that emerged in English (somewhat later in French) in the middle decades of the seventeenth century.[70] "Matters of fact" (that is, those concerning the crime itself) were, Bacon advised, to be scrupulously distinguished from matters of judgment in court proceedings: "But the greatest doubt is where the Court doth determine of the veritie of the matter of fact; so that is rather a point of tryall than a point of iudgement, whether it shall bee re-examined in errour."[71] This distinction roughly approximated that between a particular of observation and the inferences drawn from it — a distinction central to Bacon's project of a purified natural history and essential to the epistemology of late seventeenth-century natural philosophy.

But Bacon's "Histories" were not identical to the "matters of fact" of the Royal Society of London and other early scientific societies, and neither corresponded exactly to our own "fact." All three are particulars as opposed to universals, and all three cordon off the data of experience from conjectures or arguments founded upon those "givens." There, however, the resemblance ends. Bacon thought it quite possible to be mistaken about the particulars of natural history and about the "matters of fact" of the courtroom. Moreover, his "Histories" were selective, covering "many things which no one who was not proceeding by a regular and certain way to the discovery of causes would have thought of inquiring

230

after."[72] The results were immediately put to work in the investigation of causes, digested into "tables" and classified according to prerogative instances to yield the underlying "forms" of simple natures. In contrast, the "matters of fact" of the early Royal Society and the Paris Académie Royale des Sciences were as often as not left to float free both of a motivating causal inquiry or a unifying causal explanation. Indeed, too bizarre or singular to be classified, much less theorized, they often seem to have been chosen with an eye to thwarting any explanation or generalization. Although these reports of extraordinary phenomena were collected on the model of Bacon's "History of Pretergenerations," they were seldom employed to the ends he had envisioned. Instead they became epistemological ends in themselves, defining a new kind of scientific experience: the strange fact.

Strange Facts in Learned Societies

"A Girl in Ireland, who has several Horns growing on her Body," "Description of an Extraordinary Mushroom," "Observations on a Monstrous Human Foetus," "Of Four Suns, which very lately appear'd in France," "Rare and Singular New Phenomenon of a Celestial Light:"[73] the learned journals of the latter half of the seventeenth century were filled with accounts of the new, the rare, the unusual, the astonishing. Members of newly established scientific societies such as the Royal Society of London and the Paris Académie Royale des Sciences studied extraordinary phenomena with nearly as much zeal as they did natural laws and regularities; the scientific correspondence of the likes of Boyle and Leibniz was studded with eager queries and reports concerning fluids whose volume waxed and waned with the moon, phosphors that glowed in the dark, and other examples of "the workings of Nature where she seems to be peculiar in her manner" (figs. 6.2, 6.3).[74]

The strangeness of the strange facts in the early scientific journals was underscored by language redolent of the exclamations of broadsides, prodigy books, and accounts of notable cabinets: "new," "remarkable," "singular," "unusual," "extraordinary," "uncommon," and "curious" were the stock adjectives that enlivened the otherwise terse entries. Because the novelty of the phenomenon shaded imperceptibly into the originality of the investigator, the language of novelty and astonishment spread to less outlandish topics. The Dutch mathematician and natural philosopher Huygens, for example, admired the "singular and remarkable properties" of the curve described by a hanging chain;[75] similarly, the French chemist Edmé Mariotte heralded his experiments on the color of wine as "new" and "curious."[76] For the lucky and the alert, scientific reputations stood to

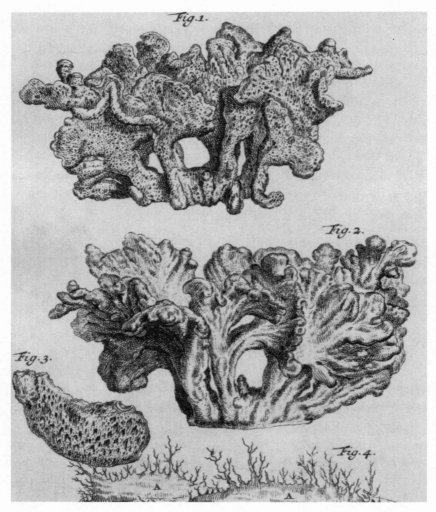

6.2.1

Figure 6.2. Strange facts in the Académie Royale des Sciences
6.2.1. Joseph Pitton de Tournefort, "Description d'un champignon extraordinaire," *Histoire de l'Académie Royale des Sciences* (read 3 April 1692), in *Mémoires de l'Académie Royale des Sciences* 1666/99 (Paris, 1733), vol. 10, pp. 69–70.
6.2.2. Gian Domenico Cassini, "Nouveau phenomene rare et singulier, d'une lumiere Celeste, qui a paru au commencement du Printemps de cette Année 1683," in ibid., vol. 8, pp. 182–84.

Descriptions of strange facts strained the resources of language and tended toward multiple analogies that decomposed the oddity into a mosaic of features, each to be mapped piecemeal onto a familiar element of experience, as in the case of the legendary chimera. Engravings helped fuse these composites into a single visual impression and lent additional credibility to the report. Regarding the object shown in fig. 6.2.1, Tournefort noted that "naturalists count more than 80 different kinds of mushrooms: But among all these species, there is none similar to the mushroom described here, nor so extraordinary." The astronomer Gian Domenico Cassini described the object in fig. 6.2.2, a whitish light stretching along the zodiac from Aries to Taurus, as "one of the rarest spectacles [ever] observed in the heavens."

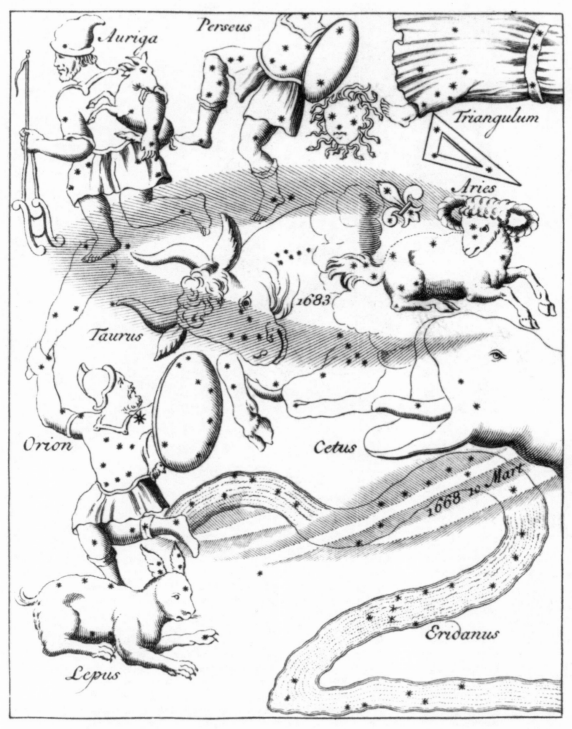

6.2.2

233

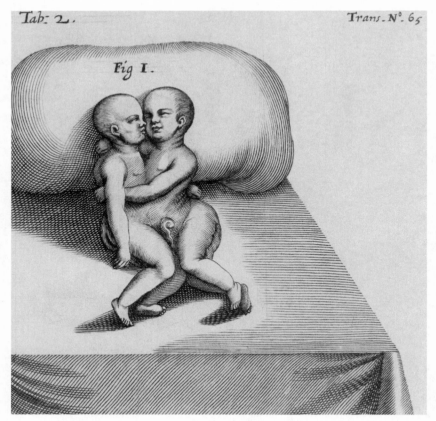

6.3.1

Fig. 6.3. Strange facts in the Royal Society
6.3.1. "A Narrative of a Monstrous Birth in *Plymouth*, Octob. 22. 1670," *Philosophical Transactions* 5 (1670), pp. 2096–98.
6.3.2. John Winthrop, "An Extract of a Letter . . . Concerning Some Natural Curiosities . . . , especially a Very Strange and Very Curiously Contrived Fish, Sent for the Repository of the *R. Society*," *Philosophical Transactions* 5 (1670), pp. 1151–53.

Of the children in fig. 6.3.1 physician William Durston observed: "These Twins were exactly like one another: very well featured, having also pretty neat and handsome Limbs. . . . We might have proceeded to further Observations, but time and the tumultuous concourse of people, and likewise the Father's importunity to hasten the Birth to the Grave, hindred us." In the same year the Royal Society published the report of John Winthrop, governor of Connecticut, who had sent the Royal Society a specimen of a "strange kind of *Fish*" hauled up outside of Massachusetts Bay (fig. 6.3.2): "Its Body (as was noted by M. *Hook*) resembling an *Echinus* or *Egg-Fish*, the main branches, a *Star*, and the dividing of the branches, the Plant *Missel-toe*."

234

6.3.2

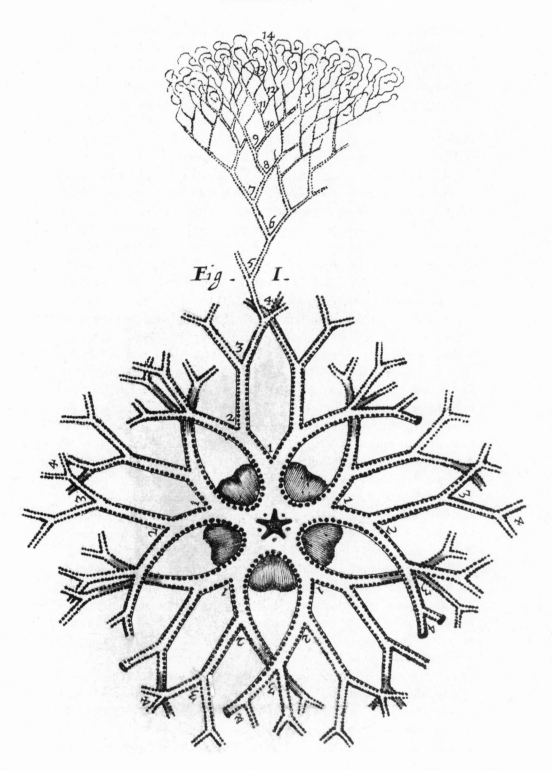

Fig. I.

be made overnight with a spectacular discovery that startled lay and learned reader alike, and during this period the spectacular was largely synonymous with the marvelous. The Paris Académie account of astronomer Gian Domenico Cassini's 1683 discovery of a mysterious light near the Milky Way (fig. 6.2.1) remarked upon "these dazzling novelties which serve no less to make a century illustrious than to give some savant the occasion to distinguish himself," noting that the appearance of a new star in the heavens had "hatched, so to speak, the famous Tycho Brahe into the learned world."[77]

This appetite for strange facts neither exhausts nor explains the whole of seventeenth-century natural philosophy, but it does illuminate much of what was distinctive about seventeenth-century scientific empiricism, particularly the collective empiricism promoted by the new scientific academies of the late seventeenth century. Moreover, although strange facts have largely disappeared from the annals of mainstream science, their epistemological legacy endures. The strange facts of seventeenth-century natural philosophy were the Ur-facts, the prototypes of the very category of the factual. Strange facts defined many (though not all) of the traits that have been the hallmarks of facticity ever since: the notorious stubborness of facts, inert and even resistant to interpretation and theory; their angular, fragmentary quality; their affinity with concrete things, rather than with relationships. The disparity between strange facts and modern matter-of-fact facts is equally revealing: strange facts were anything but robust, seldom public, and too singular to be amalgamated into sums or tallied in tables. The strange facts of early modern natural philosophy show that scientific facticity has a history, one that begins but by no means ends in the seventeenth century.

Because facts are so fundamental to the varieties of scientific empiricism that have emerged and entrenched themselves since the seventeenth century, the claim that facts have a history[78] that begins in the seventeenth century may sound paradoxical, as if one were to claim that experience itself began only in the seventeenth century. Facts are however only one subspecies of scientific experience, and by no means the most ancient. The treatises of Aristotle abounded with closely observed particulars about everything from autumn rains to the mating behavior of pigeons to dreams. But these observations almost always made their appearance in a demonstrative or pedagogical context: they were mustered in support or contradiction or illustration of a claim or conjecture. Aristotle had distinguished his "histories" of animals from a philosophical inquiry into causes,[79] but even the histories bundled particulars about this or that species into generalizations about classes — for example, that bloodless animals are all smaller than blooded animals, or that all furry animals are viviparous.[80] The texture of Aristotelian empiricism was smooth, fusing particulars into the universals that could serve as the premises and conclusions of syllogisms.

In contrast, the empiricism of the late seventeenth century was grainy with facts, full of experiential particulars conspicuously detached from explanatory or theoretical moorings. Peter Dear has described how the scholastic notion of "experience" was gradually modified in the course of the seventeenth century from "generalized statements about how things usually occur" to "statements describing specific events," particularly experiments.[81] Although many seventeenth-century experiments were sterling examples of the new facts, the factual was not simply a synonym for the experimental. What chiefly distinguished the new empiricism of facts from the old empiricism of experience was not experiment but the sharp distinction between a datum of experience, experimental or observational, and any inference drawn from it. The distinction between explanandum and explanans — *hoti* and *dioti* — was as old as Aristotle,[82] but the epistemological autonomy and centrality of matters of fact in seventeenth-century natural philosophy was unprecedented.

The prominence of facts in seventeenth-century natural philosophy is at first glance puzzling, for they offered neither the certainty nor the universality traditionally expected of demonstrative *scientia*. Thomas Hobbes' 1651 classification of the branches of knowledge drew the conventional distinction between *historia* and *scientia* clearly: "The Register of *Knowledge of Fact* is called *History*," and neither civil nor natural history could aspire to the "Registers of Science,...[which] contain the Demonstrations of Consequences."[83] The facts of civil and natural history could not be linked together into a certain demonstration of what must be the case in general; nor could they offer certainty as to what was the case in individual instances. Every fact of the matter could conceivably have happened otherwise. Hence the jurists and historians[84] obliged to rely upon mere facts often adopted an apologetic tone when they wrote about the solidity of their proofs vis-à-vis those of mathematicians and philosophers. "For in the sciences," wrote French jurist Jean Domat in his late-seventeenth-century treatise on civil law, "proofs consist in the linking of truths together, as they necessarily follow from one another. But in the facts [*faits*] that could happen or not happen, depending on causes whose effects are uncertain, it is not by sure and immutable principles that one knows what has happened."[85] Lacking principles and demonstrations, the jurists and naturalists who relied on facts had to fall back on the perception, memory, and hearsay of witnesses — all notoriously unreliable sources of truth.

The intrinsic uncertainty of facts, as opposed to demonstrations, helped erode absolute standards of certainty in seventeenth-century natural philosophy. In response to the revival of philosophical skepticism and the mutually annihilating theological strife of the late sixteenth century, some natural philosophers, such as Pierre Gassendi, John Wilkins, and Boyle, argued that certainty was not an absolute but rather came in degrees, descending from "metaphysical" to "physical" to "moral." Demonstrations enjoyed metaphysical certainty; some principles of natural philosophy might aspire to physical certainty; but proofs derived from facts could at best attain only to the moral certainty of judgments in daily life.[86] A natural philosophy grounded on facts hence sacrificed demonstrative certainty to a mere "concurrence of probabilities," as Boyle put it.[87]

If the facts of seventeenth-century natural philosophy were not certain, neither were they robust. In contrast to the "hard facts" of the Victorians, the problem was to make them stay, not to make them go away. This was particularly true of the strange facts observed in the field, which were rare and came unbidden. Who could say when and where and for how long the next aurora or monstrous birth would appear? Some local wonders — a well that emitted searing heat, ants from Surinam that marched in formation, a quarry rich in ammonites — were stable enough to permit leisurely and prolonged investigation. They strongly resembled the wonders of the medieval topographical tradition in being associated with an exotic locale and in performing with some regularity for curious travelers. But most of the strange facts that preoccupied seventeenth-century natural philosophers burst upon the scene as briefly and surprisingly as the portents that they so closely resembled.

Aristotelian experience had been both universal and commonplace; Baconian facts were often not only particular but also anomalous. Some seventeenth-century naturalists worried that extraordinary particulars would swamp ordinary ones, but this was a position seldom held consistently. Bishop Thomas Sprat, early Fellow of and apologist for the Royal Society, could in one breath reproach Pliny and other ancient naturalists for attending only to "the *greatest Curiosities*" rather than "the *least*, and the *plainest* things," and in the next exhort the experimenters at the Royal Society to observe not only what nature does "in a constant rode" but also "what with some kind of sport and extravagance: industriously marking all the various shapes into which it turns itself when it is persued, and by how many secret passages it at last obtains its end."[88] The instructions to guide travelers in collecting natural historical information from foreign parts drawn up by Boyle and printed in the *Philosophical Transactions* included queries about "unusual and remarkable" symptoms of diseases among inhabitants, "any thing that is peculiar" in the animals, and in general all matters "either peculiar to it [the foreign country], or at least uncommon elsewhere."[89]

Why were seventeenth-century natural philosophers willing to exchange the universal certainty and prosaic availability of traditional philosophical experience for the localized particularity and elusiveness of strange facts? We have already described the epistemological motives advanced by Bacon: destructively, collections of strange facts would unseat the home truths and bland axioms that had made Aristotelian natural philosophy at once so irresistibly plausible and so intellectually stagnant; constructively, these outlandish phenomena would reveal the hidden causes of all phenomena in nature, common and uncommon. As in medicine, seventeenth-century natural philosophers (many of whom had medical training) expected to understand the normal by a careful study of the pathological. Methodologically, the cluster of five preternatural prerogative instances would serve as an observational approximation of controlled experiments — or rather, as a record of the experiments nature performed on itself.

The strange facts collected by seventeenth-century scientific societies fulfilled their destructive function more successfully than they did their constructive one. Taken together, strange facts helped natural philosophers distinguish essential from incidental properties of the nature under investigation. For example, singular instances (such as the magnet among stones) or bordering instances (such as the bat between birds and quadrupeds) spurred natural philosophers to redraw the boundaries among natural kinds. Was immobility really a defining characteristic of stones? Was having wings really the sine qua non of bird-ness? Both John Locke and Leibniz invoked monsters to throw doubt on the very existence of natural kinds.[90] In general, strange facts served as a repository of counterexamples to the commonplaces of natural philosophy and even (as in the case of natural kinds) the tenets of metaphysics. However, the natural philosophers who gathered and read strange facts so eagerly were notably reluctant to seek the general forms that Bacon had hoped would unite ordinary and extraordinary phenomena under a single rule. The anatomists who studied monsters did indeed increasingly draw general embryological and physiological conclusions from the anomalies they dissected, although the rules of ontogeny eluded them. But they were the anomaly among the many *devotés* of the anomalous in the Royal Society and the Académie Royale des Sciences. Very few of those seventeenth-century naturalists who reported strange facts ventured to provide an explanation even for the case at hand, much less to relate that case to the ordinary

239

course of nature. Their reluctance contrasted sharply with the explanatory ambitions of the preternatural philosophers, especially those of Bacon, who had provided the most influential rationale for including strange facts in natural philosophy.

The Sociability of Strange Facts

No doubt one reason for the failure of the constructive part of the Baconian project was its sheer difficulty. Part of what made strange facts strange was that they baffled ready explanation. Moreover, the investigative strategy that worked for one class of wonders — for example, monsters as keys to ordinary embryological development — seldom worked for other classes of wonders — for example, rains of blood as keys to ordinary precipitation. The wonders of nature were a miscellany briefly forged into a coherent category by their cultural prominence as portents and pleasures and by their utility as natural philosophical counterexamples. Finally, the rarity, remoteness, brevity, or variability of wonders made them unpromising objects of sustained investigation.

Yet these difficulties had loomed just as large for the preternatural philosophers of the sixteenth and early seventeenth centuries, who nonetheless boldly advanced explanations for the oddest of oddities. Pomponazzi had come perilously close to explaining away miracles as well as marvels, and Cardano told his readers that the "subtlety" of his treatise *De subtilitate* (1550) referred to no particular kind of object but rather to "that reason, by which things sensible to the senses and intelligible to the intellect are to be comprehended with difficulty."[91] The preternatural philosophers had partly defined their task by its difficulty according to conventional philosophical lights: the preternatural was all that slipped through the meshes of traditional epistemology — the subsensible, the variable, the rare. Bacon's epistemological reflections echo this theme of difficulty by emphasizing the deviousness of nature, "full of deceitful imitations of things and their signs, winding and intricate folds and knots."[92] From this standpoint, difficulty was a challenge but not a block to philosophical inquiry into the causes of marvels. Indeed, the resources (many reports of oddities from all over, and a method for comparing them) that had encouraged Cardano to venture case-by-case explanations of marvels were available in enlarged and improved form to late seventeenth-century academicians.[93] Their journals amassed a Baconian natural history of marvels; Bacon himself had made Cardano's implicit methods explicit and systematic in the second book of the *Novum organum*. If the preternatural philosophers of the sixteenth and early seventeenth centuries had ventured explanations for marvels, why did the members of late seventeenth-century scientific academies so seldom attempt explanations of strange facts?

The answer does not lie in the kinds of explanation available to the late seventeenth-century academicians. It might be argued that, because of the influence of the mechanical philosophy and Copernicanism, they could not invoke astral influences and the power of the imagination with the same impunity that Pomponazzi had. Yet there were good precedents within the mechanical philosophy for explaining marvels. In his *Principia philosophiae* (1644) Descartes had boasted that there was no marvel in nature that his mechanical philosophy could not explain; in the French translation of 1647 he offered as an example a mechanical explanation of why a corpse bled in the presence of its murderer.[94] Descartes' explanations for such marvels strongly resembled the airy emanations Bacon had invoked to explain light, contagious infection, and fascination.[95] These sorts of subtle fluids had a distinguished career in late seventeenth-century natural philosophy, most famously in the work of Newton. Hence the reasons for the reluctance of the academicians to advance explanations for strange facts cannot be due solely to the range of permissible explanations after 1660, any more than it can be ascribed wholly to the difficulty of the undertaking.

Instead, the reasons must be sought in the context in which strange facts were studied and reported. In contrast to the preternatural philosophers, who recounted and explained marvels as individuals, the academicians investigated strange facts as a collective. Although the Fellows of the Royal Society might still be called "virtuosi," they were not the virtuoso philosophers and connoisseurs of nature's secrets that Cardano and della Porta had made themselves out to be. The strutting, aristocratic tone of sixteenth-century treatises on preternatural philosophy contrasts with official proclamations of self-effacing modesty and collective diligence made by members of the early Royal Society and Académie Royale des Sciences. The Royal Society of London and the Paris Académie Royale des Sciences were both established in the 1660s to promote natural philosophy, especially the new-style natural philosophy of empirical particulars, pursued as a group endeavor. These state-chartered[96] scientific academies grew out of the tradition of private humanist academies founded in Italy and France in the sixteenth and early seventeenth centuries, which promoted an ideal of civil discussion in opposition to the point-counterpoint wrangling of university disputations.[97] Humanist models were self-consciously courtly rather than monastic or scholastic, and the elegant dialogues of Galileo and Marin Mersenne, in which competing scientific and philosophical viewpoints confronted one another with wit and courtesy, reflected the influence of earlier courtly manuals such as Baldesar Castiglione's *The Courtier* (1528).[98] The regulations and protocols of the Royal Society and Académie Royale des Sciences dealt explicitly with comportment: for example, the 1699 regulations of the Paris Académie admonished members holding different opinions to refrain from "using any term of scorn or harshness" against one another;[99] Sprat in his *History of the Royal Society* (1667) explained that the Fellows rejected all dogmatic philosophy because the "warlike State of Nature" it fostered was inimical to cities, trades, civility, and all "these nobles productions [that] came from mens joyning in compacts, and entring into *Society*."[100]

The great advantage of facts, especially strange facts, was that they seemed to offer a way of pacifying the "warlike State of Nature" — a state only too familiar to seventeenth-century natural philosophers. Civility was often in woefully short supply in the new natural philosophy of the academies as well as in the old scholasticism of the universities, as the blistering controversies over sunspots between Galileo and Christopher Scheiner or between Newton and Hooke over the inverse square law of gravitational attraction amply testify. Although debates over alleged matters of fact, including experimental results, were hardly unknown, confrontations between rival theories and systems unleashed the most ruthless and relentless polemics. This graduated register of reactions showed how early modern scientific reputations were made. A spectacular discovery — for example, Galileo's telescope observations of 1609–10 — could bring sudden fame; but even the most accomplished observers and experimenters, like Brahe, Huygens, and Newton, rested their claims to glory in natural philosophy on their explanations of their results, not on the results themselves. There was a gradient in the preeminence of theories in the various disciplines, ranging from the relative indifference of the anatomists to the *esprit de système* of the astronomers, but the overall bias was clear. Whatever the causes of the bias — a lingering preference for head over hand; the link between experiment and observation and collective activities in which credit was divided; the unreliability and inconclusiveness of the merely empirical; the simple magnitude of the intellectual effort required to synthesize many and varied findings — the existence of such a bias was both beyond dispute and the basis of further disputes.

Therefore, in order to cool down the more hot-tempered natural philosophers and to protect the fragile sociability that made collective inquiry feasible and fruitful, academicians came to show a distinct preference for facts rather than explanations and theories as the subject matter of their discussions. In their notably collective preface to a natural history of animals and plants based on close anatomical investigation of individual specimens, the naturalists of the Paris Académie disavowed any desire to contradict ancient and modern authorities — if their results disagreed with those of their predecessors, it was because "nature was variable and inconstant." They would restrict themselves narrowly to the particulars of a given case, avoiding all generalization. Hence, instead of treating "the doctrine of animals like that of the Sciences, speaking always generally, we only describe things as being singular: and instead of claiming, for example, that bears have 54 loins on each side, we only say that a bear we have dissected had an entirely particular conformation."[101] In a summary report on this natural history project, which described dissections of everything from chameleons to ostriches, the anatomist and academician Claude Perrault claimed that research collectives were less susceptible than individuals to the temptations of sacrificing scrupulous observation to some elaborate system because the glory "would be a very small thing, being divided among so many people, who all contributed to this work."[102] Collective empirical research would rein in individual theoretical ambitions, or so it was hoped. Sprat offered a similar (if idealized) account of the amicable tone at Royal Society gatherings:

> It was in vain for any man amongst them to strive to preferr himself before another; or to seek for any greater glory from the subtility of his Wit; seeing as it was the inartificial process of Experiment, and not the Acuteness of any Commentary upon it, which they have had in veneration. There was no room left, for any attempt, to heat their own, or others minds, beyond a due temper; where they were not allow'd to expatiate, or amplifie, or connect specious arguments together.[103]

As Steven Shapin and Simon Schaffer have argued apropos of Boyle's controversy with Hobbes, "matters of fact were crucially differentiated from metaphysical theses or bold conjectures" in ways "highly functional for these disputes."[104]

This does not mean that members of the new scientific societies did not sometimes disagree publicly with one another, even about matters of fact, both ordinary and extraordinary. The very presence of prohibitions against acrimonious disputes in their charters and histories suggests that open controversies continued in natural philosophy, and more direct evidence can be readily gathered from their annals. What had changed was sensitivity to the risks of disagreement for collective inquiry. The distinction between facts and the inferences drawn from facts — a distinction that was always artificial, always precarious — was an attempt to dampen dispute within a specific context of inquiry, one that required the voluntary collaboration of many investigators bound by little else than a self-declared identity as one of "the Ingenious in Many Considerable Parts of the World," as the title page of the *Philosophical Transactions* addressed its readers.

What was true of experimental facts was still more true of the observational facts of strange phenomena. Although Bacon had intended them as windows onto hidden causes, their role in the early scientific academies was to fly in the face of all extant theory and to paralyze conjecture, frustrating explanation on all sides. Only a fraction of the strange facts reported in the annals of the Royal Society of London or the Paris Académie before 1700 were placed within any explanatory framework, be it mechanical, natural magical, or Neoplatonist. Even those few authors who did venture hypotheses did so tentatively and carefully separated the "history" of the oddity in question from its suggested causes.[105] Edmond Halley hoped readers of his account of an aurora borealis would rest content with a "good History of the Fact," for he hesitated to offer "the Etiology of a Matter so uncommon."[106] Boyle believed his investigations of a luminescent diamond would provide clues as to the nature of light, and he tried to "lessen the wonder" of the marvelous gem by briskly rubbing other diamonds into a dim glow.[107] Yet he scrupled to mine those clues for an explanation of this and the other instances of luminescence he investigated so diligently, protesting the need for still more observations: "It is not easy to know, what phaenomena may, and what cannot, be useful, to frame or verify an hypothesis of a subject new and singular, about which we have not as yet (that I know of) any good hypothesis settled."[108]

Boyle's own epistemological modesty and his concern to divorce "matters of fact" from "conjecture" no doubt contributed to his reluctance to formulate a hypothesis, but many contemporary naturalists who shared neither his probabilism, his suspicion of abstraction, nor his irenicism also balked at explaining strange facts.[109] Strange facts were stubborn unto obduracy, challenging not just this or that theory but flying in the face of all theories. Their opacity in the face of interpretation, natural philosophical as well as religious, splintered accounts into fragments, blocking all narrative connections of significance or sense. The accounts asked where and when and whom, but no longer why there, why then, why them (the queries posed to portents), or even the natural philosophical how. Strange facts were the prototypical facts because they were the most inert; they were impartial among competing systems and theories because equally immiscible with all of them.

By no means all seventeenth-century natural philosophers embraced the new faith of facts, strange or otherwise. Theory-neutral facts were as controversial in mid-seventeenth-century as they are in late-twentieth-century philosophy of science. Although Descartes admitted the necessity of experiments in natural philosophy, he trusted only his own, or those performed under his direction. Those made by others he dismissed as badly explained or false, for the results had been made to "conform to their principles."[110] Newton believed that his "new theory of light and colours" had been "deduced from the phenomena," and was outraged when Huygens, Hooke, and others attempted to drive a wedge between the experimental facts and Newton's own interpretation of them.[111] Even Bacon appreciated the difficulty of disentangling facts from interpretations. He recommended the strict discipline of method to check the mind's native tendency to refract all experience through a theoretical lens.[112] Avoiding abstract terms, weighting falsifying instances as heavily as confirming ones, meticulously compiling and refining tables of observations — all these techniques were meant to retrain minds in the thrall of the Idols of the Theater and thus apt to color observations with conjecture. Pure, unalloyed facts were hard-won; they were not the simple givens of sense or common sense.

Yet despite these doubts and disadvantages, the scientific academies embraced Baconian facts as the stuff of their investigations and discussions. Although no scientific academy wholly forsook the inquiry into causes, that inquiry was deliberately curtailed by epistemological caution and the desire to protect a fragile form of intellectual sociability perceived as vulnerable to dispute and controversy. In part, an empirical epistemology prescribed sociability. The sheer labor of collecting observations and performing experiments dictated a collaboration among far-flung researchers. Bacon and Descartes had both dreamed in vain of a patron who would play Alexander to their Aristotle, paying for the hands and materials needed for the investigations both thought crucial to a reformed natural philosophy. Both clearly had in mind underlings, not colleagues. In his utopian fragment *The New Atlantis* (1627), Bacon envisioned a society centered around a scientific research institute, the House of Solomon, which was organized hierarchically from the "depredators" on up to the "interpreters of nature" and employed "a great number of servants and attendants."[113] Descartes preferred servants to volunteers in performing experiments because the former "can be made to do exactly what they're told."[114] The academies of the late seventeenth century still made use of paid help, but the bulk of the empirical labor depended on a network of correspondents whose reports fattened the volumes of the *Philosophical Transactions* and the *Histoire et Mémoires*. Natural philosophy was sociable because it had become collaborative, and it had become collaborative not only because it had become factual, but also because no Maecenas had appeared to foot the bills for assistants and equipment.

Natural philosophy had also become sociable out of loyalty to an academic ideal of truth revealed through civilized conversation, in which all viewpoints were aired and communally judged. Sprat explained that the Royal Society was a society because in "*Assemblies*, the *Wits* of most men are *sharper*, their *Apprehensions readier*, their *Thoughts fuller*, than in their *Closets*." Conversely, many of the founding members of the Royal Society and the Académie Royale des Sciences had personal experience of how intellectual disputes could destroy this fruitful sociability: Samuel Sorbière, secretary to the private Montmor circle in Paris, found the bickering among savants in that group to be even worse than the disputations of the university scholastics; intellectual rivalry between members had more than once splintered the group into factions.[115] The academicians thought it worth the price of severe restrictions on theorizing (restrictions that were almost never imposed on individuals, only on assemblies) in order to reap the benefit of sustained, peaceable discussion. They must have set great store by this benefit, for it cost them dearly. As the *Histoire* of the Paris Académie confessed to its readers, here they would find no systems or theories, only "detached pieces" of knowledge, wrenched apart by a "kind of violence."[116]

The Credibility of Strange Facts

Knowledge parceled out in "detached pieces" lacked many of the marks by which naturalists might assess plausibility. By definition strange facts contradicted everyday expectations, fragmented the categories meant to contain them, and repelled explanation. All criteria of coherence with the familiar and well understood — the yardstick of probability and analogy — failed in the face of strange facts. So did more straightforward empirical criteria: in most instances, independent corroboration was impracticable or impossible. Moreover, scientific memoirs about strange facts shared both marvelous subject matter and the literary conventions for reporting it with genres of dubious authenticity, the novel and the broadside. Daniel Defoe's fabricated *Journal of the Plague Year* (1722), the Paris Académie's account of a strange celestial light, and the latest broadside about a monstrous birth all claimed, in good faith or bad, to be "true and certain relations" of recent events in real locales, with named witnesses. Hence how to verify strange facts, especially how to distinguish them from invented marvels, was an urgent problem for members of seventeenth-century scientific societies.

Both the urgency and the elaborate apparatus of verification (which could be and presumably often was faked, as in the case of Autolycus in Shakespeare's *The Winter's Tale*) may strike modern readers as the essence of cautious journalism, but early modern readers would have recognized in them the contemporary conventions for reporting portents and miracles. Because in the Christian tradition portents and miracles were not only wonders but also signs,[117] the specifics of who, what, where, and when were essential to the hermeneutic work of disclosing why God had so dramatically suspended the order of nature. Since miracles could be signs of sanctity as well as warnings of impending doom, they need not be so precisely dated and localized as portents. In the context of early modern religious strife, however, the verification of miracles became an acute problem for ecclesiastical authorities, both Catholic and Protestant. For Protestants, miracles had ceased sometime after the establishment of Christianity; their reformed faith required no new miracles to confirm revelations already divinely vouchsafed centuries ago.[118] Hence every alleged miracle was a weapon in the hands of their Catholic enemies. For Catholics, miracles were double-edged weapons: useful when wielded against the Protestants, but dangerous when turned against the church hierarchy by dissident sects like the Jansenists, or by popular piety. After the Council of Trent laid down strict guidelines for the investigation of miracles and Pope Urban VIII in 1625 transferred responsibility for the inquest to the local bishop,[119] Catholic reports on miracles from Bavaria to the Loire valley were crammed with circumstantial detail, the expert opinion of doctors and surgeons in the case of cures, and the testimony of sworn witnesses.[120] When natural philosophers attested to strange facts with well-documented testimony, they were following ecclesiastical precedents.

The strange facts of the scientific academies were redolent of other religious associations as well. Many of the phenomena were easily recognizable from the portent and prodigy literature of the sixteenth century: monstrous births, bizarre weather, celestial apparitions, and comets figured prominently in early scientific journals. But just as the angularity of the strange facts had thwarted causal explanations, so the same bald style of presentation helped to sever events still bristling with portentous associations from any such interpretation. In some cases these interpretations were rejected outright, as when the *Journal des Sçavans* branded the "thousand misfortunes" allegedly presaged by the comet of 1680 "a popular error";[121] in other cases, a studied silence had much the same effect, as when the *Philosophical Transactions* reported in severely neutral tones how a lightning bolt had struck a Pomeranian church in the midst of Sunday mass on 19 June 1670, overturning two full communion chalices and scattering the wafers.[122] Despite — or perhaps because of — the ominous associations of blasphemers struck down by lightning in mid-oath, compounded by rancorous Protestant-Catholic controversy over sacramental miracles like the Eucharist, the account remained resolutely mute concerning the religious meaning of the incident. In a similarly pointed deadpan, the Paris Académie reported in 1703 how a young man deaf from birth had suddenly regained his hearing upon the ringing of church bells:[123] to not remark upon the palpably miraculous overtones of such a story spoke volumes. Sprat made his skepticism about portents explicit when he defended the "Experimental Philosopher" against charges of religious incredulity: "He cannot suddenly conclude all extraordinary events to be the immediat Finger of *God*, because he familiarly beholds the inward workings of things: and thence perceives that many effects, which use to affright the *Ignorant*, are brought forth by the common *Instruments of Nature*."[124]

This diffidence on inflammatory theological and political topics might easily be traced to explicit bans on discussion of such divisive issues, particularly in the case of the Royal Society of London.[125] However, it would be a mistake to ascribe such restraint to a concerted effort to explain marvels by appeal to natural causes, *pace* Sprat. The late-seventeenth-century campaign against portents and religious enthusiasm was not waged in the pages of the *Philosophical Transactions* or the *Histoire et Mémoires*,[126] although some reports of strange facts did dismiss possible portentous interpretations. Halley, for example, went so far as to suggest apropos of an unusual appearance of the planet Venus in daylight that "shewing the genuine Causes of rare Appearances" would free the vulgar from "the vain apprehensions they are apt to entertain of what they call *Prodigies*."[127] The "genuine Causes" were however only exceptionally spelled out in reports of strange facts, so that the naturalization of portents remained at best a promissory note. Strange facts resisted causal explanation as well as portentous interpretation among the controversy-shy scientific societies.

Still rarer than attempts to explain strange facts were attempts to explain them away. Amidst the piles of oddities accumulated in the fledgling scientific journals of the late seventeenth century, very few doubts about the bare reality of the phenomena surfaced. This was not for want of caution or cautionary tales. Expressions of skepticism about this or that wondrous property of plants and animals were rife in natural history from the mid-sixteenth century onward: the French Jesuit Etienne Binet remarked that Pliny's claim that diamonds could be softened with the blood of a ram was nowadays "mocked in Paris;"[128] the English physician and naturalist Thomas Browne refused to believe that bears licked their formless cubs into shape.[129] Frauds were not unknown: in 1681 the *Journal des Sçavans* had to retract a report about a child who had sprouted a gold tooth in Vilnia when further inspection revealed that the tooth had been covered with thin gold leaf, and the editor took the occasion ruefully to warn readers about "how necessary it is to be circumspect about giving in to prodigies, and about believing everything that one is told."[130] In response to such doubts, reports of strange facts included the names and particulars of witnesses, and editors sometimes demanded expert testimony as well. Henry Oldenburg, Secretary to the Royal Society and editor of the *Philosophical Transactions*, thanked Boyle for sending in a detailed account of a monstrous birth, but requested additionally the "double attestation of the two physicians."[131] Apparently, high social standing and good will were not enough; specialist knowledge was also required. Echoing Bacon, natural philosophers repeatedly cautioned one another on the dangers of credulity and on the need to sift the evidence of witnesses with great care. Boyle, for example, delicately assayed the credibility of Benvenuto Cellini's claim to have seen a carbuncle (a gem that allegedly blazed in the dark) with his own eyes: Cellini was an expert on gems, employed by several princes as goldsmith and jeweler, and so unlikely to be fooled by a forgery; moreover, according to Boyle, he "appears wary of what he delivers, and is inclined to lessen, than increase the wonder of it."[132]

Yet for all of this circumspection, not to mention the sarcasm of wits who poked fun at naturalists "who greedily pursue/ Things wonderful, instead of true,"[133] only a vanishingly small percentage of the strange facts reported in learned journals of the late seventeenth century were ever challenged. The *Journal des Sçavans* exceptionally annotated a report about how in certain parts of the Pyrenees raindrops turn to stone as soon as they touch the ground, "[t]his remark requires a bit of caution;"[134] the *Histoire et Mémoires* rejected a report of a "liquid phosphor" in the sea near Cadiz.[135] But these flickers of doubt were few and far between until the early decades of the eighteenth century.[136] Part of the reluctance to contradict the testimony of a named witness may have been social: as Steven Shapin has noted, to gainsay the word of a gentleman — and most correspondents of the Royal Society counted themselves gentlemen — in Restoration England was a grave insult, even an invitation to a duel. A delicate economy of civility governed the reporting on wonders. On the one hand, "gentlemen were cautioned against acquiring the reputation of 'wonder-mongers' and, accordingly, boasters. It was uncivil to put demands upon auditors' belief." On the other hand, gentlemen who ran the "moral risk" of nonetheless reporting wonders were sometimes credited just because they had affronted good sense and good taste, presumably with good reason.[137]

Social grounds for belief were closely intertwined with epistemological ones. The empiricism professed by the seventeenth-century scientific societies depended crucially on a division of observational labor among a wide-cast network of mostly volunteers. The *pensionnaires* of the Paris Académie Royale des Sciences received a stipend, but they were as eager as the unpaid Fellows of the Royal Society for reliable accounts of natural phenomena and discoveries from provincial and foreign correspondents who were unpaid.[138] This held with double force for natural historical observations. Experiments were expensive, laborious, time-consuming, and capricious, but could in principle be performed at the place and time of one's choosing. Strange facts were more inconvenient to observe: the fossil stones peculiar to one Oxford quarry could not be found elsewhere; an aurora borealis might appear once in a lifetime; rarities like monstrous births were just that. It was neither practical nor in some cases even possible to repeat these observations; at best one might scour ancient chronicles for similar-sounding instances, as Cassini did with his strange celestial light.[139] But this was done to strengthen the induction, not to test the report. Trust was (and remains) essential to this form of collective empiricism,[140] and there is very little evidence that late-seventeenth-century natural philosophers found their credibility strained. As Domat noted apropos of legal testimony, it would be absurd to doubt all witnesses merely because a few lie, for "it was the natural order that men tell the truth as it is known to them."[141]

For Baconians there existed further epistemological and methodological grounds for crediting strange facts. These were to enlarge ordinary natural history, correct natural philosophical axioms with counterexamples, redefine natural kinds, point the way to inventions of art, and liberate the understanding from theoretical preconceptions. From this perspective, the very outlandishness of strange facts was a positive virtue. There were more things in heaven and earth than had been heard of in traditional natural philosophy, and strange facts were their heralds. The openmindedness required by Baconian empiricism predisposed natural philosophers to lend a sympathetic ear to marvelous tales that had been ignored by their predecessors and would be ridiculed by their successors. In reviewing the many ancient and modern accounts of carbuncles, Boyle trod a typically fine line between credulity and skepticism: "Though I be very backward to admit strange things for truths, yet I am not very forward to reject them as impossibilities."[142]

Finally, there were metaphysical grounds for lowering the threshold of belief for strange facts. Although phrases like "the laws of nature" had become common currency in late-seventeenth-century natural philosophy, such laws were seldom taken to imply strict, much less mathematical, regularity throughout nature. Interlocking "municipal" and "catholic" laws could create as much variability in nature as they did within the polity whence the metaphor was borrowed.[143] When the Paris academicians failed to reproduce Groningen professor Johann Bernoulli's experiments on glowing barometers, the Perpetual Secretary Bernard de Fontenelle diplomatically refused to doubt Bernoulli's account, only remarking that "when one knows nature a little, one also knows the peril of deciding about natural effects."[144] Natural philosophers were only too accustomed to the "bizarreries" of nature. Hooke thought precise measurements otiose in natural philosophy, since "Nature it self does not so exactly determine its operations, but allows a Latitude to almost all its Workings;"[145] Boyle also believed nature to be variable and natural kinds to have fuzzy boundaries.[146] Although the Parisian academicians stopped short of the extreme nominalism of some of their English colleagues, they were equally pessimistic about the possibility of an exact fit between theoretical rules and actual experience of nature. The architect François Blondel, recalling the "prodigious effects" of the hurricanes he had observed in America, concluded that the earth was subject to "continual mutations."[147]

All of these factors — social, epistemological, methodological, and metaphysical — combined to make strange facts credible to late-seventeenth-century members of scientific societies. This situation was to change dramatically by 1730, at least for leading academies like that in Paris: if late-seventeenth-century natural philosophers had sunk the threshold of belief unusually low for strange facts, their mid-eighteenth-century successors raised it unusually high.[148] At the same time that the learned journals were regaling their readers with all manner of natural marvels, philosophers were supplementing the traditional legal criteria for the evaluation of the plausibility of testimony with new criteria for the evaluation of the plausibility of events. The Port Royal *Logique* (1662) advised that historians evaluating the trustworthiness of traditional accounts attend to the intrinsic credibility of the event itself, as well as to the extrinsic credibility of the witnesses.[149] Locke extended the distinction to natural philosophy: the probability of an account varied not only according to "the number and credibility of testimonies" but also "as the conformity of our knowledge, as the certainty of observations, as the frequency and constancy of experience."[150]

As might be expected from a Fellow of the Royal Society, Locke used his criteria of probability to warn against excessive incredulity, relating the story of the King of Siam who had rashly dismissed the Dutch ambassador for tall tales about how water became hard in winter.[151] But some of his contemporaries turned this reasoning against excessive credulity, notably against credulity in the case of putative portents like comets. Rotterdam professor Pierre Bayle argued apropos of the comet of 1680 that the sheer bulk of testimony was insufficient to warrant belief, for the "fabulous opinions" recently discredited in natural history had been supported by the testimony of innumerable persons. "One may rest assured," he asserted, "that an intelligent man who pronounces only upon that which he has long pondered, and which he has found proof against all his doubts, gives greater weight to his belief, than one hundred thousand vulgar minds who only follow like sheep." Bayle also insisted that the content of testimony should be inspected before assenting; reports of marvels and miracles were particularly suspect.[152]

These new evidentiary criteria signaled a new metaphysics that favored nature's regularities over its variability, as well as a new epistemology that feared the acceptance of the false more than the exclusion of the true. The confrontation of old and new appeared in microcosm in a report concerning a monstrous birth sent in 1715 to the Paris Académie by the Montpellier physician Eustache Marcot. Marcot's autopsy on the baby, so deformed as to resemble "a toad," revealed no recognizable brain — a finding that greatly puzzled Marcot, for he could not understand how the infant had survived even for a few hours without an organ to supply animal spirits. Marcot transposed his own perplexity into the likely incredulity of his readers, remarking that he knew full well how reports on monsters were "sometimes accompanied by circumstances so bizarre, and so extraordinary, that it is difficult to credit them entirely." Either these strange facts were dismissed outright or blindly believed, both undesirable extremes in Marcot's opinion. But how to balance the rival claims of what Marcot called "reason" and "experience"? Without observation, reason itself would "give birth only to chimeras"; yet "[a]ll the facts one reports are not true." In the end even Marcot, observer of monsters and champion of singular cases, sided with suspicion over trust. He exhorted "disinterested persons to give themselves the trouble to perform contested experiments, to mark doubtful facts and suspect observations, so that one grounds nothing upon them."[153] Marcot hardly needed to add that strange facts were *prima facie* candidates for "doubtful facts and suspect observations."

Suspicion and trust, belief and skepticism, were rooted in metaphysical presuppositions about possibility and impossibility in nature. Leibniz's imagined academy of sciences, with its tightrope walkers and telescope, its rare plants and animals and its electrical globe, evoked the vigorous, if short-lived, presence of the marvelous in seventeenth-century natural philosophy. The strange facts that were the literary counterparts of Leibniz's kaleidoscopic display of wonders testified to new forms of scientific experience, which worked to loosen the constraints upon the possible in nature. Although Leibniz himself believed firmly in the constancy of natural laws, which even God could not or would not alter, he did not thereby exclude variable, rare, or unusual phenomena from natural philosophy. The avalanche of strange facts reported by naturalists, especially in the annals of scientific academies, may have stymied explanation of specific effects, but it stimulated metaphysical reflections on the reality of natural kinds and on the interplay of natural causes. For both Locke and Leibniz, the marvels they had seen and read about shook the conceptual deep structure of the world. Two other seventeenth-century philosophers, Bacon and Descartes, drew equally subversive conclusions from another kind of early modern collection of marvels in which strange things replaced strange facts: the *Wunderkammern* that challenged the metaphysical opposition of art and nature.

CHAPTER SEVEN

Wonders of Art, Wonders of Nature

On 22 April 1632 the Lutherans of the city of Augsburg presented their ally King Gustavus Adolphus of Sweden with an extraordinary cabinet (fig. 7.1.1). Constructed of oak and ebony, the cabinet was richly inlaid with medallions of Limoges enamel, beaten silver, marble, agate, lapis lazuli, and intarsia panels of multicolored woods, and crowned with a mound of crystals, corals, and shells surrounding a goblet fashioned from a Seychelles nut chased in gold and ornamented with the figures of Neptune and Thetis (fig. 7.1.2). Its secret compartments and drawers opened by means of hidden latches to reveal cunningly wrought artificialia and naturalia, including an anamorphic painting, an Italian spinet that played three tunes by an automatic mechanism, a pitcher made out of a nautilus shell worked with gilded silver, mathematical instruments, and a mummified monkey's claw. Laboriously assembled by master craftsmen over a period of six years (1625–31) under the direction of the Augsburg merchant and collector Philipp Hainhofer, the cabinet not only housed in its recesses but also embodied in its design wonders of both art and nature.[1]

These wonders were not simply juxtaposed. Repeatedly and deliberately, Hainhofer intertwined the realms of art and nature in single objects: the goldsmith H. Lencker turned the Seychelle nut into a luxurious goblet; on the agate that formed part of the cabinet's back panel, the painter Johann König incorporated the natural swirls and veins of the limestone into a painted scene of Moses leading the Israelites through the Red Sea (fig. 7.1.3). Even the cabinet's unadorned naturalia — the amethyst crystals, the coral branches, the inlays of landscape marble, the delicately curled and spiked seashells — suggested artistry by their striking forms. Nature the geometer had measured the regular crystals, nature the painter had sketched landscapes in stone, nature the architect had sculpted baroque curves and volutes in shells. Hainhofer made a business of procuring luxury items for princes and knew firsthand the princely *Wunderkammern* of

255

7.1.1   7.1.2-3

Figure 7.1. The Uppsala *Kunstschrank*
7.1.1–3. *Kunstschrank*, Uppsala University, Uppsala, Sweden (1632).

Officials of Augsburg paid merchant and collector Philipp Hainhofer the huge sum of 6,500 guilders for this cabinet, which they presented to their Protestant ally King Gustavus Adolphus of Sweden upon his visit to the city in the midst of the Thirty Years' War. Queen Christina of Sweden transferred the cabinet to the University of Uppsala in 1695. The *Kunstschrank* combined the finest craftmanship and the most extravagant productions of nature, with an emphasis upon rare and precious materials. In both its form and its contents the Uppsala *Kunstschrank* was a compact epitome of the princely *Wunderkammer*.

7.2.1

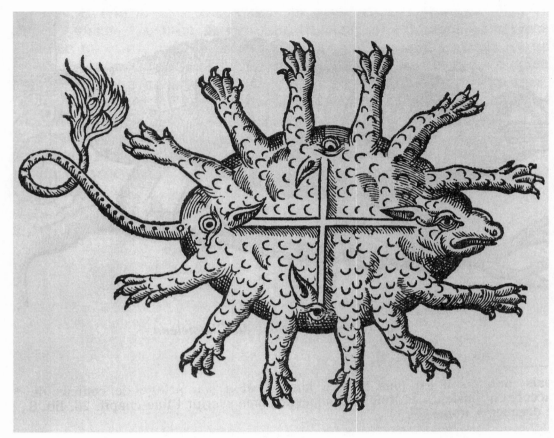

7.2.2

Figure 7.2. Art and Nature at play
7.2.1. Jakob Zellner, Ivory frigate, Grünes Gewölbe, Dresden (1620).
7.2.2. Ambroise Paré, *Des monstres et prodiges* (Paris, 1573), p. 139 in Céard edition.

One characteristic of *lusus* (sports) applicable to works of both nature and art was the flagrant disregard for function in material or form. Carved by a Dresden court artist in 1620, this ivory ship in full sail (fig. 7.2.1) bears the complete genealogy of then-reigning Kurfürsten Johann Georg I of Saxony incised on its hull; the intertwined coats of arms of Johann Georg and his wife Magdalena Sibylla von Brandenburg appear in relief on its mainsail. Whimsical craftmanship could turn hard, brittle ivory into wind-filled sails — or parody organic symmetries into a monster. How, wondered Ambroise Paré, did this African creature (fig. 7.2.2) decide which way to go? Overendowed with paws and eyes, it displayed nature's occasional propensity for forms as elaborate and superfluous as the delicate carvings on Zellner's ivory frigate.

259

Dresden, Ambras, and Munich. Thus informed about courtly tastes, he commented on how rock crystal vases in particular were "a great pleasure for princes, since they have Nature and Art side by side,"[2] and when in 1628 he delivered a cabinet made in Tuscany to the Archduke Leopold of Austria, he exclaimed over its inlaid painted agates, in which "art and nature played with one another."[3]

The cabinet of Gustavus Adolphus distilled the early modern *Wunderkammer* ("chamber of wonders") to an essence. In contrast to the more specialized sixteenth-century professional collections described in Chapter Four, the copious, various, and costly *Wunderkammern* contained precious materials, exotica and antiquities, specimens of exquisite workmanship, and natural and artificial oddities — all crammed together in order to dazzle the onlooker. If each object by itself elicited wonder, all of them densely arrayed floor to ceiling or drawer upon drawer could only amplify the visitor's gasp of mingled astonishment and admiration. And if the artificialia and naturalia of these collections were wondrous placed side by side in the studied miscellany of the typical cabinet, they were still more wondrous when fused with one another, obscuring the boundaries between the wonders of art and the wonders of nature. Despite wide divergences in who collected what and why, almost every early modern *Wunderkammer* in some way exploited the peculiarly intense wonder of crosses between art and nature.

During the heyday of the *Wunderkammern* in the late sixteenth and early seventeenth centuries, the ancient opposition between art and nature first blurred and then dissolved in natural philosophy, most notably in the works of Francis Bacon, René Descartes, and their many followers. We will argue for a link between the art-nature crosses of the *Wunderkammern* and the collapse of the art-nature opposition in the study of nature. *Wunderkammern* like the Hainhofer cabinet exploited the old opposition between art and nature to feign pleasant paradoxes and also hazarded new combinations of the two that subverted the distinction altogether. It was in such collections of rarities and marvels (and only later in natural history and natural philosophy) that art and nature first mingled and ultimately merged. Although the prophets of the reformed natural philosophy disdained the cult of the wondrous, the *Wunderkammern* nonetheless left telltale traces in their remarks on the unity of art and nature. Bacon invoked the "wonders of nature" to bridge the natural and the artificial; Descartes held up the automata of the cabinets and grottoes as models for the microscopic machines underlying all natural phenomena. Given their distrust of the sensibility of wonder in natural history and natural philosophy, it is all the more striking that both Bacon and

Descartes appealed to the stock wonders of the cabinets for examples of the unity of art and nature. Implicit in the typical objects of the *Wunderkammern* that drew nature and art together in mutual emulation — the landscape veined in marble, the mechanical duck that swam and quacked, the nautilus shell garlanded in gold — was a personification of nature as an elevated kind of artisan. She (for the personification of nature was traditionally and invariably feminine) was neither Aristotle's humble maker of mundane, functional objects like beds and ships, nor the creative, almost divine artist exalted by the Neoplatonic art theory of the Italian Renaissance. Rather, she was the creator of luxury items, as elaborate as they were useless, combining costly materials with fine craftmanship. Like the goldsmith, the ivory turner, and the painter of miniatures, she was freed from the demands of utility. The virtuoso artisan could play with form and matter, just as nature occasionally "sported" with her ordinary species and regularities.

The same aesthetic of variety, whimsy, and extravagance informed both a frigate of ivory turned to a lacy delicacy (fig. 7.2.1), and an African animal alleged to have four eyes, four ears, and six pairs of paws pointing in all compass directions (fig. 7.2.2). Ambroise Paré "marveled greatly" at how this creature's many eyes, ears, and feet could all perform their functions, concluding that "here Nature has played, to make [us] wonder at the greatness of her works."[4] The frigate's ivory sails and rigging were too fragile to be even touched; the African animal was Buridan's ass thrice over, lacking any sufficient reason to impel it in any one direction. Both defied function as well as the recalcitrance of matter, harnessing consummate skill to no useful end whatsoever.

This chapter explores how the wonders of art and nature displayed in the *Wunderkammern* helped transform the ontological categories of art and nature in early-seventeenth-century natural history and natural philosophy. What was a work of art and what a work of nature, and how could the naturalist tell? What was wondrous about the mutual emulations of art and nature? And was nature artist or art? These were the questions raised by the marvels displayed in the *Wunderkammern*, illustrated in natural history treatises, and debated among natural philosophers. Eventually these questions undermined the ancient ontology that opposed art and nature, with profound consequences for the early modern understanding of the natural order.

Art and Nature Opposed

In Shakespeare's *The Winter's Tale* the shepherdess Perdita scorns to include the "carnations and streaked gillyvors,/ Which some call nature's

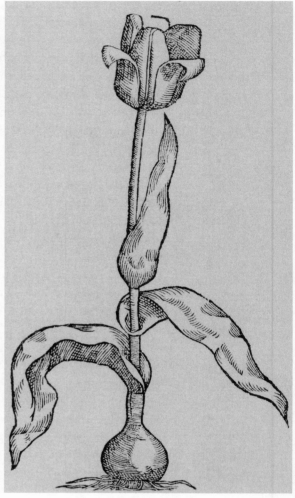

7.3.1   7.3.2

Figure 7.3. Art and Nature as rivals
7.3.1–2. Rembert Dodoens, *Historie of Plantes*, trans. H. Lyte (Antwerp, 1578), pp. 154, 213.

The gilly-flower (carnation) scorned by Perdita was one of several flowering plants early modern gardeners cultivated into showy varieties not originally found in nature. Dodoen's carnation (fig. 7.3.1) is a good example. His "smal Tulpia" (fig. 7.3.2) resembles the flower originally imported to the Low Countries from the Levant in the sixteenth century. Because tulips vary greatly even in an uncultivated state, florists were able to cross-breed them to produce the striking varieties avidly collected and painted in seventeenth-century still lifes. As in the case of grafting fruit trees, these horticultural practices provoked criticism as well as admiration: could art, should art aspire to outdo nature?

bastards" in her winter bouquet of rosemary and rue presented to Polixenes, King of Bohemia:

> Perdita: For I have heard it said
> There is an art which in their piedness shares
> With great creating nature.
> Polixenes: Say there be;
> Yet nature is made better by no mean
> But nature makes that mean. So, over that art
> Which you say adds to nature, is an art
> That nature makes. You see, sweet maid, we marry
> A gentler scion to the wildest stock,
> And make conceive a bark of baser kind
> By bud of nobler race. This is an art
> Which does mend nature — change it rather — but
> The art itself is nature. (4.4.86–97)

Here Polixenes pleads for art — in this case, the gardener's art of grafting — as the helpmate rather than as the upstart rival of nature. Art may "mend nature," but only by following nature's own "mean." Art does not usurp but rather extends "great creating nature" (fig. 7.3).

But Perdita remains stubbornly unconvinced ("I'll not put/ The dibble in earth to set one slip of them:" 4.4.99–100), and the rest of the play vindicates her position on the self-sufficiency and superiority of nature. Polixenes is appalled when his own "gentler scion" Florizel proposes to marry the "low-born lass" Perdita, relenting only when her true nature as long-lost princess reveals why her manner "smacks of something greater than herself/ Too noble for this place" (4.4.158–159). And the "statue" of Perdita's mother Hermione, allegedly "performed by that rare Italian master, Julio Romano, who, had he himself eternity and could put breath into his work, would beguile Nature of her custom, so perfectly he is her ape" (5.2.91–94), turns out to be none other than the living, breathing Hermione herself. Art can neither mend nor vye with nature; its apparent triumphs turn out in the end to be nature's own: the princessly shepherd-ess really is a princess; the lifelike statue really is alive.

Perdita's wisdom on the sovereignty of nature over art had acquired a long philosophical pedigree by the time *The Winter's Tale* was performed in 1611. The founding text in this lineage, still widely cited throughout the seventeenth century, was Aristotle's distinction between art and nature in his treatise on *Physics*:

Of the things that exist, some exist by nature, some from other causes. By nature the animals and their parts exist, and the plants and simple bodies (earth, fire, air, water) – for we say that these and the like exist by nature. All the things mentioned plainly differ from things which are *not* constituted by nature. For each of them has within itself a principle of motion and of stationariness (in respect of place, or of growth and decrease, or by way of alteration). On the other hand, a bed and a coat and anything else of that sort, *qua* receiving these designations – i.e. in so far as they are products of art – have no innate impulse to change.[5]

Only natural objects can constitute true species or kinds characterized by internal principles of change and faithful reproduction. In contrast, the matter of artifacts derives from that of nature, and the forms impressed upon them cannot generate similar forms: beds do not beget little beds.[6] Hence on the Aristotelian view, artifacts lack ontological identity, or "natures." No essence of either form or matter stamps them as what they are; they are therefore necessarily posterior to and parasitic upon natural objects. Art imitates nature, but even the most skillful artisan cannot equal, much less surpass, the model. This at least was the moral drawn by thirteenth-century scholastics, who argued that art could only act upon the accidents, not the substantial forms of bodies.[7] As Jean de Meun put it in the *Roman de la Rose*, "Art, no matter how hard she tries, with great study and great effort to make anything whatever ... will never make them go by themselves, love, move, feel, and talk – for she could kill herself before she could transmute the species, even if she didn't go to the extent of taking them back to their prime matter."[8]

Despite occasional attempts by earlier writers to dignify the status of art, this view still resonated at the turn of the seventeenth century, as in Perdita's disdainful speech on grafting. But by then other positions were possible as well: art might still ape nature, but it also might extend, assist, complete, contravene, or even surpass her.[9] Not all Elizabethans, for example, shared Perdita's dim view of grafting: the English philosopher John Case praised the purported grafting of a pear tree onto a cabbage as "a wonderful fact of art! The bloom of one plant thus grafted changes the whole tree into another species ... what can prevent me from concluding that something natural has really been done by art?"[10] Ferrante Imperato claimed that "art conduces to the perfection" of stones and metals;[11] Bernard Palissy wrote of how artificial fountains improved upon natural ones because "one has [here] helped nature, just as to sow grain, to prune and labor in the vineyards is nothing else but helping nature."[12] Although the majority of Renaissance writers might still have sided with Perdita on nature's superiority to art,[13] elite artists and artisans like Palissy and the collectors who patronized them could provide concrete counterexamples.[14]

However expanded and even daring these claims on behalf of art vis-à-vis nature may have been, all nonetheless still assumed the ancient opposition between the two. If the laurels were now sometimes awarded to art rather than to nature, rivalry nonetheless presupposed difference. Renaissance writers on topics ranging from moral education to cosmetics skipped nimbly from one commonplace to another, but they rarely escaped the oppositional logic of art versus nature. As a habit of the understanding, the Aristotelian opposition between art and nature still framed the mental world of early modern Europeans.[15] It is against this background that wonders moving between art and nature fascinated the proprietors of the collections variously known as cabinets of curiosities, museums, repositories, *studioli*, galleries, *thesauri*, and *Schatz-*, *Kunst-*, *Raritäten-* and *Wunderkammern*.[16]

The Wonders of Art and Nature Displayed

Princely and professional, large and small, institutional and individual, specialized and miscellaneous — all these kinds of *Wunderkammern* expressed most concretely and dramatically the early modern culture of wonders, serving as the nodes of a thickly cross-hatched network of commerce, correspondence, and tourism.[17] Hainhofer bought shells for his own collection from Dutch merchants at Frankfurt fairs and commissioned fine stonework from Florence and Milan; the port of Marseilles supplied southern French collectors with coral, shells, and handiwork from far west and east.[18] Naturalists and antiquarians exchanged specimens with their learned correspondents, and many traveled to see collections for themselves.[19] The collections were also the theaters in which old and new relationships between art and nature played off against one another, symbolized in the objects and their physical arrangement. In order to follow that play, we must first reconstruct the theater and its context.

The diaries of learned travelers bear witness to the number and diversity of the collections and the insatiable interest of their visitors. Michel de Montaigne confided his disappointment to his journal when he was turned away without a tour from the Ambras collection of Ferdinand II of Tyrol in 1580; John Locke paid an awkward visit to "old, morose, half-mad" Nicholas Grollier de Servière in Lyons, "he haveing not Latin nor I French," in order to see the "many sorts of clocks" and "many excellent pieces of turning in ivory."[20] By the latter half of the seventeenth century, published guides in several languages instructed visitors on the choicest rarities of the Leiden anatomy theater,[21] a testimony to the crush of tourists, and catalogues of other celebrated collections issued steadily from the press from the 1580s on.[22] Catalogues and treatises often included lists of the principal cabinets elsewhere in Europe or even the world.[23] Some even lectured readers on cabinet etiquette: make sure your hands are clean, follow the guide obediently, don't admire things that aren't particularly rare — you'll make yourself ridiculous.[24]

As this last injunction suggests, connoisseurship was expected of visitors to cabinets as well as of their proprietors. But connoisseurship of what? What properties linked coral, automata, unicorn horns, South American featherwork, coconut shell goblets, fossils, antique coins, turned ivory, monsters animal and human, Turkish weaponry, and polyhedral crystals? The disjointed lists typical of the travel journals and catalogues might serve as a nominalist's brief — one irreducibly individual object after another, the brute singularity of each resisting all attempts at generalization and categorization. Catalogues and inventories sometimes imposed a rudimentary classification upon the endless etcetera of the collections, but the affinities they picked out were of the weakest sort — most often objects were classified simply by the materials of which they were made — and they provide no clue as to why the objects were coveted, and moreover, coveted in common.[25]

Given the very different social identities and budgets of early modern collectors, there can be no single answer to the riddle of the contents of the collections. These form a spectrum ranging from the princely collection housed in the Hainhofer cabinet to the professional collection of the naturalist Olaus Worm. Princely collections of the Medici in Florence, Archduke Ferdinand II in Ambras, Emperor Rudolf II in Prague, Elector August in Dresden, Duke Albrecht V in Munich, King Frederick III in Copenhagen, or Czar Peter I in St. Petersburg traced their ancestry directly back to the medieval *Schatzkammern*; like them, these collections emphasized the lavish use of precious metals and gemstones, as well as expensive craftsmanship and paintings by recognized masters such as Dürer and Rubens (fig. 7.4.1). Even the princely collections varied considerably, according to individual tastes: August was a passionate collector of artisanal tools and machines; Rudolf II assembled a superb collection of paintings; Frederick III was keen on naturalia; Peter I favored human monsters, live as well as stuffed.[26] However, all these collections shared at least one important function, namely to display the prince's magnificence and taste before foreign dignitaries and potentates.[27] Like the dukes of Burgundy before them, early modern princes not only spent fortunes on elaborate festivals and theatrical spectacles on the occasion of royal entries and marriages,[28] they also vied with one another in the collection of marvels of art and nature that would impress visitors and trumpet their fame abroad. As at the Burgundian court, wonder could be easily transferred from objects and spectacles to the prince who commissioned them. It was a subtle but delicious exercise of one-upmanship for a prince to reduce his noble guests to speechless admiration by his collection of marvelous and costly objects, as when Giuseppe Arcimboldo's "various, ingenious, beautiful, and rare inventions, fill[ed] all the great princes present with great wonderment, and his lord [Emperor] Maximillian with great contentment."[29]

The collections of scholars, physicians, lawyers, and merchants also aspired to stupefy visitors with wonder, but their ostentation tended to be that of learning rather than of wealth. Humanists (often trained in Roman law) preserved the material fragments of the classical past in their collections of antiquities, especially coins and medals (fig. 7.4.2);[30] merchants used their foreign contacts to buy and exchange exotica (fig. 7.5.1); physicians, pharmacists, university medical faculties, and academies of natural philosophy were chiefly avid for naturalia (fig. 7.5.2). Medical interests strongly favored naturalia, for there was a thin line between pharmaceuticals and rarities such as unicorn horns and bezoars, or exotica such as quinine bark from South America.[31] Moreover, many naturalia could be purchased far more cheaply than works of art or craftmanship ingenious enough to attract the collector.[32] Throughout Europe in the sixteenth and seventeenth centuries, physicians and apothecaries dominated the collecting of naturalia, from Imperato in Naples to Worm in Copenhagen.[33] Similarly, medical faculties and societies established the earliest academic collections: a physic garden of medical simples established at the University of Pisa in 1543, was, by 1597, attached to a museum displaying amethysts, unicorn horns, a petrified human skull with coral growing out of it, Flemish landscape paintings, and Mexican idols.[34]

Despite clear patterns of specialist emphasis among the objects preferred by humanists versus medical men, it was the rule rather than the exception for these professional collections to embrace both artificialia and naturalia. The collection of the French antiquarian Boniface Borilly was, for example, dominated by Roman medals, but it also boasted "a head of a rat from the Indies," "a small picture made of the root of an olive tree upon which is naturally represented a human figure," "three well-polished coconuts, garnished with ivory, serving as flasks," and a celebrated "cyclops."[35] Conversely, naturalist Ulisse Aldrovandi also investigated antiquities.[36] The actual physical arrangement of many collections (in contrast to the more systematic classifications of catalogues and inventories) was often calculated to highlight this heterogeneity. As contempo-

7.4.1

7.4.2

Figure 7.4. *Wunderkammern*, princely and antiquarian
7.4.1. Jan Breughel and Peter Paul Rubens, *The Sense of Sight*, Museo del Prado, Madrid (1617–18).
7.4.2. Henrik van der Borcht, *Still Life with Collection Objects*, Historisches Museum, Frankfurt am Main (c. 1700).

Although the majority of early modern collections shared certain features — for example, the combination of artficialia and naturalia, with a pronounced appetite for the rare and the exotic in both categories — differences in social status, professional interests, and, above all, budget made for differences in what was collected and how it was displayed. Breughel's and Rubens's allegorical painting (fig. 7.4.1) depicts an imaginary collection, but one that contains examples of almost all the elements typical of a royal *Wunderkammer*, placed in a palatial setting: elaborate vessels fashioned from precious materials; showy scientific instruments; exotic artifacts and animals; rare and expensive flowers; and artistic masterpieces, including a garlanded Madonna painted by Breughel and Rubens, prominently displayed in the front right-hand corner. Van der Borcht's still life (fig. 7.4.2) also depicts an idealized collection, but one embodying the enthusiasm of the humanist scholar and antiquarian for the material remains of Greek and Roman antiquity. The ancient coins in the foreground are rendered so exactly that it is possible to identify them from extant collections.

7.5.1

7.5.2

Figure 7.5. *Wunderkammern*, mercantile and professional
7.5.1. Joseph Arnold, *Cabinet of Art and Rarities of the Regensburg Iron Dealer and Mining Family Dimpel* (1608), Ulmer Museum, Ulm, Germany.
7.5.2. Olaus Worm, *Museum Wormianum seu Historia rerum rariorum* (Leiden, 1655), title page.

In this watercolor (fig. 7.5.1), Joseph Arnold portrayed the *Cabinet of Art and Rarities of the Regensburg Iron Dealer and Mining Family Dimpel* (1608) as a smaller, tidier version of a princely collection. In addition to the cannons, emblems of the family trade, neatly arranged books and seashells play a more prominent role than in royal collections, and modest faience replaces urns of semiprecious stones worked in gold. This engraving of the collection of a professor of medicine in Copenhagen (fig. 7.5.2) shows a room stuffed with animals, plants, and minerals, many rare and exotic, including the crocodile and armadillo mounted on the right-hand wall. After Worm's death in 1633, his collection was bought by Duke Friederich III of Schleswig-Holstein, who merged its contents with his own princely collection of coins, exotica, porcelain, gold, and silver.[23]

271

rary engravings show, objects were positioned next to one another so as to maximize dissimilarity.[37]

The charitable description of this concerted diversity is encyclopedism; the uncharitable description is miscellany. But neither quite fits the variety peculiar to the cabinets of curiosities. "Encyclopedism"[38] pays tribute to the ampleness of the collections, which embraced so many and such wildly contrasting objects; it draws textual support from a few contemporary visions of the *Wunderkammer* as a microcosm of the universe, notably Flemish physician Samuel Quicchelberg's treatise on an ideal museum, which is to be organized as a "universal theater."[39] Yet the subtitle of Quicchelberg's work suggests that its contents were anything but a representative sample of the universe, for it featured "miracles of art," "all rare treasures," "precious stuffs," and "other singular things."[40] Even if Quicchelberg's treatise had exerted more influence on the actual practice of collecting — it enjoys a far greater reputation among modern historians of collections than it did among early modern collectors[41] — that influence could be described as "encyclopedic" neither in their sense nor in ours.[42] The *Wunderkammern*, especially those devoted mostly to naturalia, contrast dramatically with contemporary natural history treatises such as Konrad Gesner's *Historia animalium liber I. de quadrupedibus viviparis* (1551), which aspired to genuinely encyclopedic scope in both coverage of subject and sources consulted.[43]

Early modern accounts of museums occasionally use the word "microcosm," but in a special sense. The museum was not intended to represent the entire macrocosm in miniature, but rather, as French physician Pierre Borel had inscribed over the door of his own *cabinet*, "it is a microcosm or Compendium of all rare strange things."[44] Early modern collections excluded 99.9 percent of the known universe, both natural and artificial — namely, all that was ordinary, regular, or common. They therefore cannot qualify as a representative sample in the usual sense. If, however, "representative" is meant as representing nature at peak intensity or creativity, there is more warrant for the term in connection with the *Wunderkammern*.[45] Attempts to make *Wunderkammern* truly encyclopedic were foredoomed: English botanist Nehemiah Grew's call to include "not only Things strange and rare, but the most known and common amongst us"[46] in the Repository of the Royal Society of London was itself a great rarity among collectors. To label the *Wunderkammern* "encyclopedic" because they included so many different kinds of things arranged against the grain of familiar classifications is to mistake variety for universality.

Despite their variety, the *Wunderkammern* were not assembled and arranged by chance or caprice. Their objects belonged to recognizable

genres and were linked by hidden assumptions and aims. Some objects —
unicorn horns, bezoars, coral, birds of paradise — were practically ines-
capable and turn up in inventory after inventory.[47] But even less standard
objects could be grouped under relatively few rubrics: opulence (gigantic
emeralds, bowls formed of lapis lazuli); rarity (birds of paradise, Roman
coins); strangeness (a monstrous man covered with hair, a two-handled
fork); fine workmanship (nested polygons of turned ivory, clockwork);
and medical or magical properties (bezoars). Or they illustrated ambiguity and metamorphosis, bringing together what conventional classifications put asunder (coral, described by Ovid as a sea plant petrified
by the blood dripping from Medusa's severed head) (fig. 7.6). These criteria for collectibles often overlapped: a precious goblet fashioned from
a rhinoceros horn worked in gold was at once costly, rare, finely crafted,
ambiguous between art and nature, and undeniably strange. All converged in singling out the exceptional, the anomalous, and the bizarre.

The strategy of display piled one exception upon another, provoca-
tively subverting or straddling the boundaries of familiar categories. Was a
winged cat bird or animal? Was coral vegetable or mineral? Was a gilded
coconut shell nature or art? Distraction as well as disorientation amplified
the onlooker's wonder. Not only did individual objects subvert commonplaces or shatter categories; from every nook and cranny uncountable
rarities clamored simultaneously for attention. The cabinets paid visual
tribute to the variety and plenitude of nature, albeit very partially sam-
pled. Stuffed with singularities, they astonished by copiousness as well as
by oddity. Collectors did not savor paradoxes and surprises, they piled
them high in overflowing cupboards and hung them from the walls and
ceilings. The wonder they aimed at by the profusion of these heterogeneous particulars was neither contemplative nor inquiring, but rather
dumbstruck.

Although these "wonder chambers" were filled with "curiosities,"
they stimulated an emotional response that divorced wonder from curiosity.[48] Each object taken by itself could qualify as a "curiosity" in at least
one of three senses. First, it might embody meticulous workmanship, or
"care" (*cura*), hearkening back to the root meaning of the word. Robert
Hooke used the word in this sense when he described the appearance of a
blue fly under the microscope, observing that "the hinder part of its body
is cover'd with a most curious blue shining armour."[49] Second, it could
flaunt its lack of function. Bernard de Fontenelle, defending the activities
of the Paris Académie Royale des Sciences, admitted for example that
some parts of mathematics and physics "were only curious" rather than
"useful."[50] Third, a curiosity might excite a desire to know about the

7.6.1

7.6.2

Figure 7.6. Wonders on display
7.6.1. Writing set, Schloss Ambras, Kunsthistorisches Museum, Vienna (German, late 16th century).
7.6.2. Abraham Jamnitzer, goblet, Grünes Gewölbe, Dresden (late 16th century).
7.6.3. Antlers in tree trunk, Schloss Ambras, Kunsthistorisches Museum, Vienna.

Despite the private associations of the "cabinets" in which collections were housed, their objects were meant to be seen and admired, at least by a select audience. Highly wrought objects of extraordinary delicacy and intricacy exhibited the triumph of artisanal skill over recalcitrant matter — and drab utility. This ornate wooden writing set (fig. 7.6.1) in the form of a circular, turreted palace (44 cm high) was listed in the 1596 inventory of the collection of Archduke Ferdinand II of Tyrol in Schloss Ambras. Inventories of sixteenth-century European libraries testify to the great popularity of Ovid's *Metamorphoses*, which supplied themes for many *Wunderkammer* objects.[24] This goblet (68 cm high) by a renowned Nuremberg goldsmith (fig. 7.6.2) plays upon two Ovidian themes: the transformation of the nymph Daphne into a laurel tree, and the transformation of seaweed into stony coral by contact with blood dripping from Medusa's severed head. Marvels of nature overturned natural regularities and baffled explanation. In 1563, Ferdinand II purchased these stag antlers, mysteriously imbedded in a tree trunk (fig. 7.6.3), for his Ambras collection.

7.6.3

object in all its odd particularity, as when the *Philosophical Transactions of the Royal Society of London* praised an author for the "excellent researches of his Curiosity."[51] Curiosities heaped high in the *Wunderkammern* lent themselves easily to the first two senses, but not to the third. Workmanship and a fine contempt for base utility might be admired both individually and en masse, but focused inquiry required concentration on a single object to the exclusion of others.

The copiousness of the *Wunderkammern* also excluded certain kinds of wonder, as well as of curiosity. Individual novelties or rareties could move the spectator to the meditative, differentiated wonder of, for example, the Abbé Suger, reflecting one by one upon the beauty and allegorical significance of each gemstone. But the multiplication of novelties and rareties characteristic of the *Wunderkammern* stunned the spectator into the magnified wonder of astonishment. The accounts of travelers who made the rounds of European *Wunderkammern* reveal two recurring patterns. First, they often attempted to make a selection among the objects overflowing from cabinets and shelves by singling out a handful in a list — "a little organ made entirely out of glass from Barcelona," "one of the most perfect birds, which is called the king of the birds of paradise, that I have ever seen," "a cross very well worked, a foot in height, made of a natural branch of silver which vegetated in that form"[52] — in which the choicest wonders might be appreciated individually. Second, they became addicted to ever larger, more well-stocked collections as the jaded traveler's threshold of astonishment rose. The German traveler Zacharias von Uffenbach, for example, pooh-poohed the Royal Society of London Repository; he had already seen bigger and better elsewhere. John Evelyn had viewed so many marvels during his Continental grand tour that he yawned over the Oxford Anatomy School, "adorn'd with some rarities of natural things; but nothing extraordinary."[53] List and addiction worked at cross-purposes emotionally. Lists pulled in the direction of individuation; addiction, in the direction of amalgamation of impressions. Ultimately, both patterns dissipated in futility. The lists trailed off in fatigue, their compilers' eyes and judgment dazed by so much to look at and choose from. The addiction followed the inexorable logic of all addictions, with satiety hard on the heels of each new burst of stupefied astonishment.

The Wonders of Art and Nature Conjoined

Within this economy of astonishment the wonders that blurred the boundary between art and nature stood out, even in the company of so many competing wonders and paradoxes. Because the opposition of art and nature was still a conceptual reflex during the early modern period, its violation was especially startling. Bedrock assumptions were shaken, and the intensity of the wonder was correspondingly seismic. Words could and did trespass against the boundary between art and nature, but seldom with the impact of things seen with one's own eyes.[54] Not only did the *Wunderkammern* display artificialia and naturalia side by side; they featured objects that combined art and nature in form and matter, or that subverted the distinction by making art and nature indistinguishable. These wonders of art and nature juxtaposed, combined, and fused in the cabinets all illustrated an aesthetic of virtuosity. Both naturalia and artificialia embodied difficulties of material overcome seemingly without effort: hard, dense ivory turned into filigree; baroque pearls formed into a tiny jester by touches of gold and enamel; the brittle, porcelain-like material of a seashell curled and crimped into a murex. These objects crystallized painstaking labor, but labor cleansed of the sweat and toil of the workshop. Wit, delicacy, and extravagance were the leitmotifs of this kind of workmanship, as opposed to the simplicity and economy of useful things.

As many a naturalist since Aristotle had pointed out, the ordinary objects of nature, especially organic nature, hardly lacked variety and ingenuity. But these diverse forms differed from the *Wunderkammer* objects in serving equally diverse functions, which Aristotelians identified as the final causes of organisms and their parts. Philosophers traditionally measured nature's skill by the elegant economy with which she had fitted form to function: "Nature does nothing in vain," as the scholastic maxim put it. In contrast, the wonders of the cabinets gloried in superfluity, careless of function and extravagant in expenditure of labor and materials. It was precisely this pointless variety and studied uselessness that linked pure luxury and afunctional ornamentation to play. The *lusus naturae*, like the luxury object, multiplied form without function. Hence Paré thought that the endless variety of seashells showed nature sporting: "There are to be found in the sea such strange and diverse kinds of shells that one can say that Nature, chambermaid of great God, plays in fabricating them."[55] Nature approached art when her workmanship approached playfulness; wonder once again ensued from the convergence of opposites.

Certain classes of objects typical of the *Wunderkammern* threw into relief both convergences — nature to art, work to play. All of these exploited analogies of form between natural and artficial objects. Some were hybrids of art and nature that played with analogies of form and matter, such as the *lapides manuales* that created little scenes from suggestively shaped chunks of ore or crystals, or ornamented naturalia like the gilded Seychelles nut that topped the Hainhofer cabinet. Others

7.7.1

7.7.2   7.7.3

Figure 7.7. Hybrids of Art and Nature
7.7.1. Hans von Aachen, *Phaëthon's Fall*, Kunsthistorisches Museum, Vienna (c. 1600).
7.7.2. Bartel Jamnitzer, Decorated nautilus shell, Staatliche Museen, Kassel (1588).
7.7.3. *Handstein*, Schloss Ambras, Kunsthistorisches Museum, Vienna (mid 16th century).

Human artisans could force Nature into a collaboration by elaborating suggestive natural forms into artifacts. Crystals, seashells, and certain stones, such as agate and landscape marble, lent themselves to hybrid fantasies by their shape and variety. Hans von Aachen's oil painting on alabaster (37 by 45 cm) (fig. 7.7.1) was probably painted for the *Kunstkammer* of Rudolf II in Prague. The artist exploited the markings of the stone to form a background of clouds and landscape for Ovid's story of mortal hubris. Exotic naturalia such as coconuts, ostrich eggs, and nautilus shells attracted master goldsmiths like Bartel Jamnitzer of Nuremberg, who engraved and ornamented this nautilus shell (fig. 7.7.2). Glittering ores (including silver, quartz, and malachite) from Bohemian mines suggested Calvary to another goldsmith, who added silver figurines to create a *Handstein* of the Crucifixion (fig. 7.7.3); by 1596 it was part of the Schloss Ambras collection.

279

challenged the opposition by art imitating nature (or nature imitating art) so perfectly as to deceive the spectator about their provenance: automata, figured stones, bronze and ceramic casts, seashells, and trompe l'oeil paintings. Both hybrids and challenges implicitly likened nature to the virtuoso artisan rather than to the creative artist: ingenious without *ingegno*, fantastic without *fantasia*.

Phaëthon's fall from heaven painted upon alabaster "clouds" (fig. 7.7.1); a goblet formed from a nautilus shell (fig. 7.7.2); a Crucifixion scene constructed from evocatively shaped bits of quartz, marcasite, malachite, and various ores (fig. 7.7.3) — were these *Wunderkammer* showpieces naturalia or artificialia? Their raison d'être was to make such questions unanswerable. These hybrid objects undermined the nature-art opposition not only by transforming natural materials by craftmanship (the simplest piece of furniture or clothing did as much); they additionally exploited an analogy of form: between, for example, the shapes of chunks of ore and a hilltop fortress; between the markings on marble and crashing waves; between the lip of a nautilus shell and the lip of a pitcher. Nature had, as it were, already begun the work of art.

To detect a resemblance between a natural shape and an artistic scene or artifact was an exercise in artistic *fantasia* akin to Leonardo da Vinci's famous technique of taking inspiration from shapes in clouds or spots on walls.[56] Yet hybrids of art and nature in the *Wunderkammern* did not tax the inventive imagination in quite the same way as Leonardo's landscapes or even Arcimboldo's composite faces. Self-consciously imposed upon the most unlikely raw materials — clouds or ashes, fruits or flowers — these creations redounded to the glory of the artist's creativity. These were *objets projetés*, rather than *objets trouvés*. In contrast, the shapes discerned in alabaster or ore, albeit ingeniously arranged and supplemented, were legible to the onlooker as well as to the artisan. Art elaborated but did not invent these forms. Wonder flowed not from the artist's unfathomable *fantasia*, but rather from nature's anticipation of art.

The understanding of nature as artisan also suffused the naturalia ornamented with precious metals and stones, like the nautilus shell pitcher. Of course, part of the motivation for gilding the lily — or the coconut, nautilus shell, or ostrich egg — lay in pumping ostentatious value into the relatively cheap naturalia. This was especially true of the princely collections: certain cameos and carved stones imported from Milan could not even be classified in Rudolf II's Prague collection until they were set in the precious metals or encrusted with the gems that served as rubrics in the inventory.[57] But virtuosity also created a momentum of embellishment of its own: since all ornamentation was gratuitous from the strict standpoint of function, there was no logical stopping point for the heaping of costly materials upon bravura craftmanship.[58]

In the case of naturalia, the spiral of virtuosity had already begun before the artisan even touched the object. Naked and natural, the nautilus shell or rhinoceros horn were already marvels, rare and finely wrought. Nature's admirable workmanship was a gauntlet thrown down to the human artisan, who enriched the delicate, pearly shell of the nautilus with still more delicate carvings, burnished its luster with gold, outdid its rarity by adding fabulous figures of dragons, sea serpents, and satyrs, and finally threw in a branch of coral for good measure. Lest the point of the competition between nature and art be lost on spectators, cabinets often displayed ornamented and plain naturalia side by side.[59] In these hybrids art and nature competed rather than collaborated with one another, but in both cases nature tended to merge with art — or rather, with artisan.

How could the works of nature truly be assimilated to works of art? The hybrids of art and nature in the *Wunderkammern* flirted with such a union, but it was always clear where nature left off and art began. Another class of wondrous objects went further in erasing the boundary between works of art and nature. Although all of these objects could be unambiguously classified as art or nature if one knew how they were made, each mimicked the opposing category in its appearance. On the side of art, makers of automata and trompe l'oeil paintings and casts aimed to counterfeit nature so expertly as to deceive spectators; on the side of nature, figured stones and plants resembled paintings and sculptures.

From the side of art, the terms of the challenge — that is, to make art not just rival but duplicate nature — had been implicitly set by Aristotle and by Pliny. Automata reigned supreme among the ornate clocks and mathematical instruments of the *Wunderkammern* because they seemed to possess Aristotle's first and foremost hallmark of the natural, an internal principle of motion. They could not of course generate their own kind — an eighteenth-century German writer mocked the famous automatic duck that swam and quacked well enough to fool real ducks but could not reproduce.[60] But self-propelled motion was so hallowed a criterion in the medieval Aristotelian tradition that automata that walked, danced, or swam seemed to border on artificial life.

Automata, at least as literary inventions, had enlivened many medieval romances. But the early modern automata differed from their literary prototypes not only because they were real, but also because their makers aimed to imitate animals and humans, not to improve upon them. The four automata who danced, juggled, sang, and purveyed discreet advice on courtly etiquette in the twelfth-century *Roman de Troie* were far too

7.8.1   7.8.2

Figure 7.8. Challenges of Art to Nature
7.8.1. Jacques Vaucanson, *Mécanisme du flûteur* (Paris, 1738), frontispiece.
7.8.2. Bernard Palissy, Palissy-ware, Herzog Anton Ulrich-Museum, Braunschweig (late 16th century).

The ancient artistic ambition to imitate nature so perfectly as to deceive all viewers found its most successful expression not in painting or sculpture but in automata and nature casts. Grottoes and *Wunderkammern* were prime sites for artificial nature: Montaigne admired the automata at Pratolino; Palissy planned a grotto encrusted with casts of shells and reptiles for the Tuileries; Descartes' physiology may have been inspired by the automata in the Grotto of Perseus at Saint-Germain-en-Laye. The flutist, drummer, and duck made by French mechanic Jacques Vaucanson were the most celebrated of all automata, drawing paying crowds in Paris during the 1730s (fig. 7.8.1). The technique of casting snakes, frogs, lizards, and shells in bronze spread from Padua to Nuremberg to Paris in the early sixteenth century. French potter Palissy used casts in ceramic and enamel to decorate gardens and ornamental platters (fig. 7.8.2); his experience casting shells and animals may have inspired his theory of fossils.

282

accomplished to be confused with ordinary mortals, and cast in shiny metal to boot.[61] In contrast, the early modern tradition of automata that culminated in the celebrated flutist (unveiled in 1738) of French mechanic Jacques Vaucanson aspired to ever more exact replicas — Vaucanson called them "moving anatomies" — of living creatures.[62] (fig. 7.8.1) The automata that performed for visitors in the *Wunderkammern* of Prague and Dresden or in the gardens and grottoes of Pratolino — Neptunes brandishing tridents, schoolboys scribbling their lessons, maidens serving drinks — astonished by their verisimilitude.[63]

Two other kinds of mimetic objects, trompe l'oeil painting and casts from nature, hearkened back to Pliny's story of the contest between the Greek artists Zeuxis and Parrhasius: Zeuxis painted grapes so perfectly that birds pecked at them, but Zeuxis himself tried to draw back Parrhasius' painted curtain.[64] This anecdote was repeated innumerable times in early modern theoretical writings about painting and epitomized the critical view that the aim of art was an imitation of nature so complete as to border on deception. The Dutch art critic Cornelius de Bie praised those artists, beginning with Zeuxis and Parrhasius and ending with his own countryman the still-life painter Jan de Heem, who had "swindled" (*bedroghen*) beholders with the lifelikeness of their images. Whereas Zeuxis had competed with Parrhasius, de Heem competed with nature: "*D'Heem pingit, Natura stupet.*"[65]

Although some critics, especially in Italy, had advanced alternative theories exalting invention over imitation, *fantasia* over mimesis, narrative over description,[66] the *Wunderkammern* subordinated creative fantasy to the technical virtuosity of mimesis. This preference extended not only to paintings but also to casts from nature in ceramic or bronze, usually of reptiles and crustacea. In contrast to the exaltation of the artist's imaginative powers and individuality, the mimetic tradition remained closer to craft practices, emphasizing technical polish over soaring fantasy. The Netherlandish still lifes of natural rarities such as tulips and seashells testified to skill, labor, and the subordination of stylistic quirks to faithful description. Such paintings, products of the painstaking care that was the root meaning of "curiosity," were as much luxury items as the natural and artificial rarities they portrayed.[67] In contrast to works of Italian masters (vaunted as unique, priceless expressions of *ingegno* that could only be given, not bought), the Dutch paintings remained commodities, like the masterpieces of the goldsmith or the potter.[68] The nature casts of Wenzel Jamnitzer and Palissy were even more closely tied to the decorative arts of the workshop (fig. 7.8.2).[69] It is tempting to see here an implicit analogy in which the inspired artist is to creative God as the skilled artisan is to ingenious nature, with each pair joined in their own brand of rivalry. Particularly in their depictions of marvels, Netherlandish artists believed that their wondrous skill at once mimicked and paid tribute to the wondrous ingenuity of nature.[70]

Trompe l'oeil techniques evolved in the context of replacing objects with images — specifically, the memorabilia of pilgrimages, such as badges or pressed flowers, by illuminations of devotional manuscripts.[71] In the context of naturalia collections, such ersatz images served a special purpose, substituting for unavailable or ephemeral specimens. There were both princely and scholarly versions of these cabinets on paper: Joris Hoefnagel's presentation manuscript to the Habsburg Emperor Ferdinand I (including an exquisitely painted dragonfly with real wings)[72] belonged to the one category; the vast collections of woodcuts (fourteen cupboards full of blocks) and tempera illustrations (around eight thousand) commissioned by Aldrovandi for his Bologna collection belonged to the other.[73]

These cabinet images often differed significantly from the published woodcuts and engravings in contemporary natural history treatises such as Gesner's guide to fossils. Whereas most natural history images aimed for an idealized representation that could stand in for an entire species,[74] the cabinet images tended to capture all the idiosyncrasies of a particular specimen, especially if the specimen was a marvel and therefore *sui generis*. These images *contrafactum*, as they were called, enjoyed an almost notarial status as exact visual records, especially of preternatural phenomena, testifying to the minutest details as well as to the bare existence of the marvel.[75] In this particularizing, documentary spirit Aldrovandi's *Dendrologia* (1668) treated each aberrant lemon, like the leathery *Ore ferox Crocodile*, as a category unto itself, in sharp contrast to the tidy and frugal classifications of contemporary botanists, which seldom exceeded four types of citrus fruits.[76] Collectors wanted images so singular as to substitute for the actual object, as well as to certify possession.

Casts from nature in bronze, plaster, or ceramic also necessarily imitated nature in its brute particulars rather than species generality. Perfected in the bronze workshops of late medieval Padua, by the early sixteenth century the techniques had spread to the workshops of northern Europe, which turned out fountains, goblets, and inkpots around which bronze reptiles writhed.[77] The *Wunderkammer* of Albrecht V in Munich displayed "Animals, cast in metal, plaster, clay, or some other composition so that they appear alive, and which one can make still more deceptive by [adding] colors, lizards, snakes, frogs, crabs, and shells."[78] Palissy planned a wholly artificial grotto to be covered with ceramic figures, presumably cast, of every imaginable kind of plant, lizard, serpent, and shell, which would look entirely natural, "holding no appearance neither of form of art, nor of sculpture, nor the labor of the hand of man." Palissy imagined that the imitation of nature would be flawless, "so close to nature, that it will be impossible to describe," astonishing all who beheld it.[79] This wonder of art astonished not only because it was visually indistinguishable from nature, but also because it mimicked nature's own workings, the ceramic casts creating forms in the same way, Palissy believed, that nature impressed fossils into stone.[80]

In Palissy's projected grotto, art aimed at a perfect imitation of nature. Nature sometimes returned the favor with imitations of art. That art and nature might produce similar products by similar methods was hardly a new idea in the sixteenth century. Aristotle's explanatory framework of fourfold causation[81] applied equally to art and nature.[82] But the figured stones assembled in the early modern *Wunderkammern* posed a different kind of puzzle, one not so easily dispatched by Aristotelian reasoning on how artificial and natural processes converged. Aristotelian theories of art and nature matched form to function, and so long as some end could be discerned, there was no form in nature so intricate, so finely wrought that it could not be explained. Figured stones astonished not because of the perfection of their forms — the most commonplace organisms surpassed them in symmetry and complexity — but because their forms lacked any apparent function. In them nature seemed to imitate the useless artifacts of the *Wunderkammern*.[83]

Figured stones constituted a class of minerals whose sole unifying feature was striking form without apparent function. In the works of sixteenth- and seventeenth-century naturalists who described the objects they had seen or owned, readers could find many marvelous minerals distinguished by their strange forms. Gesner divided his stones into fourteen classes according to what they resembled, including (Class V) natural fossils similar to art; Aldrovandi collected marble fragments in which cats, dogs, fish, and humans were "sculpted by nature" (fig. 7.9.1); Athanasius Kircher described stones naturally lettered with the Greek and Roman alphabets, flowerlike crystals of topaz, and human figures found in marble; Oxford naturalist Robert Plot produced plate after full-page plate of the star-stones, scrotum stones, shell stones, and so on, that could be found in the quarries of Oxfordshire.[84] As these examples indicate, by no means all figured stones were fossils in our sense; the remarkable forms of polyhedral crystals, landscape marble, and moss agate could not be explained as the petrified remains of plants and animals. The coherence of the category did not depend upon a common explanation of their origins, but on the implicit analogy between the forms of nature and the forms of art.

286

There was no lack of explanations for these stones, including celestial influences impressed upon subterranean vapors, "gorgonizing" spirits that petrified animal and plant remains (as well as generating kidney stones in the human body), organic seed caught in rock fissures that nonetheless produced its usual form in the unusual medium, the vegetation of crystals like plants in stony matrices, plastic virtues shaping stones in accordance with divine archetypes, miraculous interventions, or simply chance.[85] In the context of the *Wunderkammern*, however, the figured stones counted among the sports of nature.[86] To categorize an object as a sport was not so much to explain it as to underline its wondrousness. The wonder of the *lusus* lay in the personification of nature not simply as artisan, but as virtuoso, unfettered by the requirements of function and utility. Just as goldsmiths and jewelers fashioned flagrantly useless cutlery with handles of branching coral, so nature painted certain stones, and also flowers and shells, for the sake of delightful variety — "it being," as Plot explained, "the wisdom and goodness of the *Supreme Nature*, by the *School-men* called *Naturans*, that governs and directs the *Natura naturata* here below, to beautifie the World with these varieties; which I take to be the end of such productions as well as most *Flowers*, such as *Tulips*, *Anemones*, &c. of which we know as little use as of *formed stones*."[87]

It was the mutual imitation of art and nature that was wondrous, not the objects in themselves. There was nothing wondrous about a natural duck that quacked and swam, but a duck automaton that imitated quacking and swimming to perfection was a marvel. Similarly, there was nothing remarkable about a painting of a cat or a landscape, but a landscape or a cat "drawn by nature" in marble was worthy of a *Wunderkammer*. When art and nature imitated one another, there was always an element of playfulness in the outcome. For nature to craft a swimming duck was to arrange all for the best, as the Aristotelian maxim had it. But for a human mechanic to craft a swimming duck was at once a triumph of ingenuity and a dismissal of utility. For an artist to paint a landscape was to ply his trade; for nature to do the same in stone was to indulge in a caprice as well as to master the most obdurate material in accomplished form.

Most wondrous of all were objects so ambiguous that spectators could not decide whether they were works of art or works of nature. Of the "great many other naturall & artificiall curiositys" Locke was shown on his visit to the Parisian abbey of Sainte-Geneviève in 1677, he singled out for special mention "a stone about 10 or 11 inches long & thicker then one's fist which was made up of severall stones soe locked into one an other that though they were perfectly distinct stones, yet they hung fast

7.9.1

Figure 7.9. Challenges of Nature to Art

7.9.1. Ulisse Aldrovandi, *Musaeum metallicum* (Bologna, 1648), p. 762.

7.9.2. Claude du Molinet, *Le Cabinet de la Bibliothèque de Sainte-Geneviève* (Paris, 1692), p. 218.

Pliny's *Historia naturalis* supplied early modern artists and naturalists with prodigious stories both of art imitating nature (the competition of Zeuxis and Parrhasius) and nature imitating art (the agate of King Pyrrhus). Figured stones — marble, crystals, agate, and fossils — constituted a coherent category in sixteenth- and seventeenth-century treatises on minerals and in *Wunderkammern*, because all displayed remarkably artful forms made by nature. Examples include this cat figured in marble, collected by Aldrovandi and published posthumously by his students (fig. 7.9.1), and the so-called stone of Hammon in the collection of the Bibliothèque Sainte-Geneviève (fig. 7.9.2). Hard-put to explain the hexagonal forms of crystals, the Bruges naturalist Anselm Boetius de Boodt could only wonder at nature's artifice: "Nature wishes that one should admire such things, but not that one should understand them."[25]

288

7.9.2

all together" (fig. 7.9.2).[88] In his catalogue of the collection, Locke's guide Claude du Molinet puzzled over the provenance of this "game of nature": "Most of those who have seen it have believed it [to be] artificial, but the most able Sculptors in Paris have judged it natural."[89] Positioned precisely at the crux between art and nature, the "Stone of Hammon" of the Sainte-Geneviève collection was a marvel of ambiguity and an emblem for all wonders that merged nature and art. It was precisely in these ambiguous wonders that Bacon and Descartes found their inspiration for an anti-Aristotelian metaphysics that rejected the opposition of art and nature altogether.

Nature as Artist, Nature as Art

In December 1594 the young men studying law at Gray's Inn, London staged a Christmas revel in which a "Prince of Purpoole" held mock court and received advice from several counselors on how best to reign. The Second Counselor encouraged the Prince to secure his fame through the cultivation of reason rather than force: for his glory and edification, the Prince should collect a library, a garden, and a menagerie of rare beasts and birds, a "still-house" equipped with instruments and furnaces, and "a goodly huge cabinet, wherein whatsoever the hand of man by exquisite art or engine hath made rare in stuff, form, or motion; whatsoever singularity chance and the shuffle of things hath produced; whatsoever Nature hath wrought in things that want life and may be kept; shall be sorted and included."[90]

The author of this fantasy was Francis Bacon, and its images and themes recurred in his later philosophical works. The Second Counselor held up to the Prince of Purpoole the examples of legendary learned kings such as Alexander and Solomon, and Bacon's later utopian fragment *New Atlantis* (1627) in fact described a "House of Solomon" that housed an enlarged and elaborated version of the garden, menagerie, still-house, and cabinet, containing, among other things, fruit trees made "by art greater much than their nature," mathematical instruments, "divers curious clocks," silks and "dainty works of feathers" of a fineness unknown in Europe, and "loadstones of prodigious virtue; and other rare stones, both natural and artificial."[91] Bacon intended these imagined "preparations and instruments" to dampen rather than to feed the wonder he elsewhere called "broken knowledge."[92] The Prince of Purpoole was told that "when all other miracles and wonders cease by reason that you shall have discovered their natural causes, yourself shall be left the only miracle and wonder of the world";[93] the sages of the House of Solomon were forbidden to use their knowledge to disguise natural effects "to make them seem more miraculous."[94] But Bacon nonetheless used the *Wunderkammern* both as inspiration for new establishments like the visionary House of Solomon for the investigation of nature and the perfection of the arts, and as intimations of a new ontological union of art and nature in wonders.

Throughout his programmatic writings on the reform of natural history and natural philosophy, Bacon reiterated that the opposition between art and nature was best overcome by attending to the wonders of each.[95] His two proposed supplements to the traditional natural history of "nature in course," "nature erring or varying," and "nature altered or wrought," were in Bacon's view closely related to one another. Both in the "sports and wantonings" of nature and in the productions of the mechanical arts, nature deviated from her normal course.[96] Natural marvels were nature's own works of art, and therefore exemplary for innovations of human art. Indeed, Bacon argued, if natural marvels delighted, it was chiefly because they furnished clues to new techniques and inventions:

> The knowledge of things wonderful is indeed pleasant to us, if freed from the fabulous, but on what account does it afford us pleasure? not from any delight that is in admiration itself, but because it frequently intimates to art its office, that from the knowledge of nature it may lead it whither, it sometimes preceded it by its own unassisted power.[97]

Bacon assimilated art to nature on notably different grounds than Aristotle had. For Aristotle, art and nature were alike in the regularities they produced, and therefore in the causes, especially formal and final causes, responsible for that order. Both nature and art might err, but this produced disorder and was peripheral to the analogy between them. For Bacon, in contrast, art and nature were most similar in their wonders, for nature's wanderings were creative, pregnant with hints for the mechanical arts rather than mere mistakes. Whereas for Aristotle monsters and other errors of nature destroyed order, for Bacon they created and inspired new orderings of things. As in the *Wunderkammern*, nature and art converged in marvels.

In making a full description of the mechanical arts the third division of his reformed natural history, Bacon firmly rejected the ancient opposition of art and nature. He complained of "the fashion to talk as if art were something different from nature, so that things artificial should be separated from things natural, as differing totally in kind; . . . Whereas men ought on the contrary to have a settled conviction, that things artificial differ from things natural not in form or essence, but only in the efficient."[98] This was a broad assertion of the identity of art and nature, in principle embracing ordinary as well as extraordinary phenomena. But when Bacon turned to practice, it was that "which is new, rare, and unusual in nature" that he held up as the model for discovery and invention in the arts.

Bacon's metaphors for nature's variations sometimes evoked the *Wunderkammer* — "sports and wantonings" — and sometimes the court of law — "errors into bonds." What remained constant in his writings on the unity of art and nature was the necessity of studying variation both to understand and command nature. Like the mythological Proteus, who "ever changed shapes till he was straitened and held fast; so the passage and variations of nature cannot appear so fully in the liberty of nature, as in the trials and vexations of art."[99] Intermediate between the "liberty of nature" expressed in regular species and "the trials and vexations of art" were nature's own variations, marvels halfway en route to art. Although Bacon may have disdained the sensibility of the *Wunderkammern*, in which astonishment stifled inquiry, he embraced the image of nature as virtuoso artisan, her highest workmanship displayed in her oddest productions.

Like Bacon, Descartes was wary of a sensibility of frozen astonishment and a science of nature that wallowed in secrets and rarities for their own sake.[100] Like Bacon, however, he appealed to stock objects of the *Wunderkammern* and grottoes to argue for the fundamental identity of art and nature. Whereas Bacon took "monsters and prodigious births of nature" as emblematic of nature's nearest approach to art, Descartes seized upon automata to make the same point. So closely did these artifacts resemble nature that one might, Descartes suggested, assume that nature was itself nothing but an assemblage of automata. If the severed heads of criminals still moved and grimaced for several moments after death, it showed that the bodily members can move without the exercise of will:

> This will not seem in any way strange to those who, knowing what diverse *automata*, or moving machines, can be made by human industry, using only very few parts, in comparison to the great multitude of bones, muscles, nerves, arteries, veins, and all the other parts which make up each animal, considering this body as a machine, which, having been made by the hand of God, is incomparably better ordered, and contains more wondrous movements [*mouuemens plus admirables*], than any that can be invented by men.[101]

Only complexity and scale differentiated automata made by human hands and those made by God: "for I do not recognize," wrote Descartes in his *Principia philosophiae* (1644), "any difference between the machines made by artisans and the various bodies which nature alone constructs, other than that machines that depend only on the effects of certain tubes, or springs, or other instruments, that, having necessarily some proportion to the hands that make them, are always large enough that their shapes and forms can be seen, instead of which the tubes and springs which cause the effects of natural bodies are ordinarily too small to be seen." In this more global claim, Descartes embraced all of nature, not just animals. All of the mechanical arts properly belonged to physics (in the ample Aristotelian sense of the study of all of nature); a clock that marks the hour by means of internal wheels is as natural as a tree that bears fruit.[102] Descartes merged natural and artificial in the grand clockwork of the organic and inorganic worlds.

The mimetic possibilities of automata worked strongly upon Descartes' theoretical and philosophical imagination. In his physiological speculations, he anatomized the human body as a clockmaker might disassemble a watch, speculating about the hydraulics of animal spirits flowing through tubular nerves and claiming that the motion of the heart "follows as necessarily from the simple disposition of the organs ... as that of a watch, from the force, placement, and configuration of its counterweight and wheels."[103] Descartes' thought-experiments about automata revolved around their real indistinguishability from animals and their apparent indistinguishability from humans. In the *Discours de la méthode* of 1637 he offered two criteria, language and flexible behavior adapted to circumstance, to tell real humans from impostor automata;[104] in the *Meditationes de prima philosophia* of 1641, skeptical doubts led him to question whether the passers-by he saw from his window were really men or merely "hats and cloaks which may cover automata?"[105] Both the claim that animals and human bodies can be understood as automata and the fear that automata in human form could deceive are a tribute to how striking Descartes found the resemblance between art and nature in these objects so characteristic of the *Wunderkammern*.

We do not claim that Bacon and Descartes integrated either the sensibility or the objects of the *Wunderkammer* into their visions of a reformed natural philosophy in any thoroughgoing fashion. On the contrary, they entertained strong reservations about both, albeit with somewhat different emphases: Bacon was more suspicious of the sensibility and more welcoming of the objects; Descartes just the reverse. Rather, we wish to draw attention to how their radical attempts to break down the opposition of art and nature repeatedly appealed to examples of the wonders of art and nature conjoined. Although in principle the identity of art and nature held for the most ordinary as well as the most extraordinary objects — and one can find claims of this breadth in the writings of both Bacon and Descartes — their habitual illustrations came from the cupboards of the

Figure 7.10. Art imitating Nature imitating Art
Still Life with Seashells, Kunsthistorisches Museum, Vienna (Italian, 17th century).

In this painting, the play between natural and artificial forms becomes a game of mirrors: a mimetic rendering of seashells, which, along with flowers and figured stones, counted as nature's most artful productions.

295

Wunderkammern. For early modern Europeans of education and means to travel, the *Wunderkammer* was the material site where the boundary between art and nature dissolved, and its wonders were the objects that displayed the closest resemblances between the two realms. For Bacon and Descartes, "pretergenerations" and automata were more than salient examples. In very different ways, they served as heuristics for imagining how nature and art might in theoretical and practical ways be bridged. For Bacon, "rare stones," monsters, and other marvels were nature's artifacts; for Descartes, automata were art's coarse-grained approximation of nature. Although both authors are remembered for making art natural after millennia of opposition, both could be read at least as easily as making nature artificial (fig. 7.10).

There were subtle but important differences in the imagery Bacon and Descartes enlisted to unite nature and art. While eschewing formal and final causes as explanations in natural philosophy, Bacon's richly anthropomorphized language of nature's "wanderings" and "wantonings" suggested an active nature capable of craft in both senses of the word, if not of deliberation. In contrast, Descartes' language wavered between a portrayal of nature as active artisan, constructing bodies just like the machines "made by artisans," and of nature as passive art, fashioned by "the hands of God." This ambiguity between nature as artisan and nature as art became a flashpoint of natural philosophical and theological controversy in the late seventeenth century. At issue were nature's autonomy, God's sovereignty, and the division of labor between God and nature.

These issues became urgent in the context of European monarchies threatened by insurrection in the middle decades of the seventeenth century, particularly during the English Civil War and Restoration. From the first treatises on physico-theology of the 1650s to the celebrated correspondence between Leibniz and the English theologian Samuel Clarke (whom Isaac Newton used as his mouthpiece) about natural order and divine voluntarism conducted in the 1680s, discussions of divine sovereignty often mirrored coeval debates about the temporal sovereignty of kings.[106] These debates turned on the relationship between God and nature: was nature God's "chambermaid," a personification that implied activity and some autonomy; or was nature "the Art of God," a metaphor of passivity and dependence? Servants would spare God the indignity of the menial labor needed to keep creation running, but servants could also rebel, as John Milton's account of Satan's mutiny against God in *Paradise Lost* (1667) made vivid.

Natural philosophers concerned about threats to God's sovereignty declared the variety, utility, and beauty of nature to be, in the words of

English natural philosopher and physician Walter Charleton, "*Artificiall, Nature being the Art of God.*" Charleton countenanced no division of labor between God and his "chambermaid" nature; not a sparrow fell but God attended. Writing as a Royalist in the midst of the English Civil War, Charleton expected instant chaos should God's solicitous attention to his creation falter for even a moment. Without this constant vigilance, the very plants would rebel against their "cold, dull, inactive life" and aspire to "motion and abilities for nobler actions."[107] Charleton's God must personally superintend everything, for in a world in which even vegetables (and Roundheads) might mutiny, to delegate authority was to invite usurpation. In a world of regicide — Charleton's royal patient Charles I had been executed in 1649 — no king, not even God, could afford the luxury of uppity servants.

These themes of usurpation and labor remained central to the debate over nature's agency carried on by English Neoplatonist philosophers Ralph Cudworth and Henry More, Robert Boyle, and Gottfried Wilhelm Leibniz after the Restoration of 1660. Boyle composed his *Free Inquiry into the Vulgarly Received Notion of Nature* in 1666, perhaps in reply to hylozoist views of nature advanced in More's treatise on *The Immortality of the Soul* (1662), and shortly after Charles II had ascended to the throne.[108] Boyle's tirade against the autonomy of nature bore the traces of this political context (in which the monarchy still seemed precarious) as well as of contemporary natural philosophical debates over the formation of figured stones and other strikingly "artificial" natural objects, and of theological conflicts over divine voluntarism.

However lowly the tasks assigned to nature as God's chambermaid, the very fact of a division of labor implied some measure of autonomy for nature. It was exactly this autonomy that was at issue in Boyle's rant against "the Vulgarly Received Notion of Nature." Although the mechanical philosophy of the seventeenth century has often been represented as a declaration of nature's independence from the meddling interventions of divine providence, some of its foremost spokesmen insisted vehemently on nature's absolute dependence and on God's equally absolute prerogative to alter his creation at will.[109] As England's foremost mechanical philosopher, Boyle protested loudly against granting nature the slightest discretion in its operations. Indeed, he went so far as to deny nature any real existence, declaring it to be a mere "notional" entity.[110] For Boyle, the central issue was usurpation: those who admired the works of nature stole praise and gratitude from God. It was disrespectful and even idolatrous to suggest that God needed an assistant, "an intelligent and Powerful Being, called nature."[111]

Although Boyle took too lofty a view of God's exalted station to countenance much divine labor, he was loathe to allow God servants, for this might lead to an ensouled and potentially usurping nature. Boyle's solution was to claim, like Descartes, that nature was artifact rather than artisan. Moreover, nature was an artifact of a peculiar kind, immediately recognizable from the inventories of the *Wunderkammern*: an "engine" or "automaton," words Boyle used interchangeably. Appealing over and over again to one of the most fanciful and intricate of early modern automata, the astronomical clock in the Strasbourg cathedral, Boyle imagined the entire world as nothing but a "great automaton," composed of still smaller automata, in the manner of Chinese nested boxes. God, as the most ingenious of engineers, had arranged for "all things to proceed, according to the artificer's first design, and the motions of the little statues [of the Strasbourg clock], that at such hours perform these or those things, [and] do not require, like those of puppets, the peculiar interposing of the artificer, or any intelligent agent employed by him."[112] Automata eliminated the need for servants, in particular one who might rival the deity, and at the same time kept God's hands free of demeaning labor.

Such extreme views about the passivity of nature were opposed by those who, like More, Cudworth, and Leibniz, judged the automata of the mechanical philosophy inadequate to the task of explaining natural regularities, especially regularities of form. None were satisfied that the mechanical philosophy had adequately explained how nature matched form to function, much less the afunctional forms of nature's sports and errors. All appealed to a standard set of counterexamples — sympathetic cures, musical instruments that vibrated in unison, the power of the maternal imagination to shape the fetus, the architecture of the spider's web or of geometric crystals — in order to justify their idea of a partially ensouled nature, variously described in terms of "plastic powers" (Cudworth), "spirit of nature" (More), or "indwelling active principles" (Leibniz).[113] More, Cudworth, and Leibniz worried less about the idolatry of nature than they did about the indignity of a God without servants, but they also feared usurpation, just as Charleton and Boyle had. Therefore they all insisted on the inferiority of the soul of nature to the rational soul of humans, not to mention to the mind of God. Although ensouled nature was elevated above the stupid matter of the mechanical philosophy, Leibniz warned that a fully anthropomorphized nature would revive "heathen polytheism;"[114] Cudworth admitted that human actions may lack the "Constancy, Eaveness and Uniformity" of natural operations, but that we nonetheless surpass nature in acting consciously.[115] Nature was no longer a virtuoso artisan, but she remained at least a "Manuary Optificer" to God's "Architect."[116]

In post-1660 natural philosophy, the debates that brought out the issues of nature's activity and artistry most sharply concerned the origins of highly elaborated forms. More admired the ingenuity of Descartes' explanation of magnetism in terms of tiny screws and of the descent of heavy bodies in terms of vortices, but thought the "Mechanical powers in *Matter*" inadequate to explain "sundry sorts of *Plants* and other things, that have farre more artifice and curiousity than the direct descent of a stone to the ground."[117] Unusual natural forms supplied his main counterexamples. More was much impressed by Kircher's account of a man with a birthmark on his arm in the shape of Pope Gregory XIII, seated on a throne with a dragon underfoot and an angel crowning him overhead. More made much of how nature here seemed to imitate art in these forms: "[Kircher] having viewed it with all possible care, does profess that the *Signature* was so perfect, that it seemed rather the work of Art than of exorbitating *Nature*; and yet by certain observations he made, that he was well assured it was the work of *Nature*, and not of *Art*, though it was an artificial piece that Nature imitated." Kircher ascribed the birthmark to the maternal imagination, a force which More saw as one manifestation of the darkling intelligence of "the Spirit of Nature … the great *Quartermaster-General* of Divine Providence."[118] For nature to imitate art, some minimal spark of soul and autonomy was necessary.

The "plastic powers" of active nature also gave rise to figured stones, according to some of the naturalists who had studied the specimens most closely. After a thorough study of trochites, ammonites, and "stone-plants," the English naturalist John Beaumont reported to the Royal Society that minerals must possess at least a "vegetative soul" in order to produce such "curious sports of Nature." Although willing to acknowledge that some figured stones were petrified organic remains, he protested that this could not account for all figured stones, referring skeptics to "those delicate Landskips which are very frequently (at least in this Country) found depicted on stones, carrying the resemblance of whole groves of Trees, Mountains and Vallies, & c."[119] In such explanations of typical *Wunderkammer* objects — the Royal Society Repository itself boasted a large collection of figured stones — nature preserved its identity as artisan rather than art, imprinting forms through "plastic powers" and "vegetable souls."

In this way, the controversy over the origins of figured stones and other "artificial" forms in nature overlapped with the debate over nature's autonomy. Commenting upon an ammonite he had viewed under the microscope, Hooke deemed it "quite contrary to the infinite prudence of Nature, which is observable in all its works and productions, to design everything to a determinate end, … that these prettily shaped Bodies should have all those curious figures and contrivances (which many of them are adorn'd and contriv'd with) generated or wrought by a Plastic Virtue, for no higher end than only to exhibit such a form."[120] Prudent nature supplanted playful nature, and prudent nature was only a short step from passive nature, what Sir Thomas Browne had called "the Art of God."[121] When Hooke discovered that under magnification the point of the finest needle was as rugged as a mountain range, he resuscitated the opposition between art and nature — or rather, between human and divine art: "Whereas in the works of Nature, the deepest Discoveries shew us the greatest Excellencies. An evident Argument, that he who was the Author of these things, was no other than the Omnipotent." Hooke thought it barely worth the trouble to examine manmade objects under the microscope, for the instrument revealed them to be "rude, misshapen things."[122] The automata of art had dazzled with their lifelike appearances, but the automata of God awed by their finished interiors as well.

By the final decades of the seventeenth century the boundary between art and nature had reemerged, as sharp and distinct as it had ever been. But it was now drawn on aesthetic and theological rather than ontological grounds. By the early eighteenth century, collectors had begun systematically to separate artificialia and naturalia.[123] In natural philosophy and natural theology, nature and art remained conjoined — but more because nature had become artificial, "the art of God," than because the artificial had been made natural. This was the departure point for the argument from design set forth by physico-theologians from William Derham to William Paley: the very artificiality of nature must imply a super-intelligent Artisan, as clock implies clockmaker. From Boyle Lectures to Bridgewater Treatises, naturalists presented the regularities of nature as evidence against an Epicurean universe ruled by chance, and for the existence and benevolence of a craftsman God. The third possibility, of nature designing rather than designed, was first reviled as hylozoism or Spinozan pantheism and then almost forgotten. David Hume's suggestion in his *Dialogues Concerning Natural Religion* (comp. 1751–5) that the regularities of nature might be immanent rather than imposed was not so much rebutted as roundly ignored.[124] Only poets could persist in personifying nature as artisan.

Was the *Wunderkammer* done in by the "rise of the new science"?[125] The attacks launched by Blaise Pascal, Boyle, and others against scholastic anthropomorphisms such as "Nature does nothing in vain" or "Nature abhors a vacuum" might suggest that the union of art and nature in the *Wunderkammern* fell victim to the mechanical philosophy. But if this was so, it was because the mechanical philosophy had established its own brand of anthropomorphic theology. As Hume pointed out, as far as anthropomorphism was concerned, there was little to choose between an artisan nature and an artisan God. Moreover, the monopolization of agency and intelligence by *res cogitans*, possessed only by God and man (Boyle added angels), was frankly anthropocentric at the expense of the rest of nature. If philosophers deprived nature of skill and autonomy, it was out of openly voiced fear that she might usurp the praise due to God. Boyle warned that admiration for the works of nature was disrespectful of the works of God, and that too great a veneration for nature hindered art by fostering "a kind of scruple of conscience to endeavour to emulate any of her works, as to excel them."[126] Even those who, like Cudworth, More, and Leibniz, allowed nature plastic powers or a *vis insita* were careful to describe her work as "drudging" and her intelligence as, in Cudworth's words, "either a *Lower* Faculty of some *Conscious Soul*, or else an Inferiour kind of Life or Soul by it self, but essentially depending on a Higher Intellect."[127]

It is ironic that the wonders that provided Bacon and Descartes with their best examples of how the works of nature and art converged became, by the early decades of the eighteenth century, the best examples of a confusion between art and nature. Enlightenment naturalists jeered at their predecessors for having assimilated figured stones to works of art; Enlightenment collectors mocked cabinets that jumbled artificialia and naturalia together. The ancient opposition of nature and art had been transformed into an opposition between the works of God and the works of man. Nature had become "the Art of God," no longer able to create art of her own.

CHAPTER EIGHT

The Passions of Inquiry

On 6 February 1672 Isaac Newton described his response to light shone through a prism in a letter to Henry Oldenburg, Secretary of the Royal Society of London:

> It was at first a pleasing divertissement to view the vivid and intense colours produced thereby; but after a while applying myself to consider them more circumspectly, I became surprised to see them in an oblong form; which, according to the received laws of refraction, I expected should have been circular.... Comparing the length of this colored spectrum with its breadth, I found it about five times greater, a disproportion so extravagant that it excited me to a more than ordinary curiosity of examining from whence it might proceed.[1]

No doubt this opening passage of Newton's celebrated "New Theory about Light and Colours" was as much a fictionalized reconstruction of actual events as were the accounts of the experiments that followed.[2] It is exactly the conventional character of Newton's introduction that bears weightiest witness to mid-seventeenth-century commonplaces about how the passions of wonder and curiosity interlocked in natural philosophy. Relaxed admiration — "a pleasing divertissement" — at the novel and beautiful spectacle of colors is cut short by a rush of "surprise" that the spectrum should be oblong rather than circular, which in turn rallies all the faculties into an alert and energetic state of more than ordinary curiosity to solve the puzzle.

Musing admiration, startled wonder, then bustling curiosity — these were the successive moments of seventeenth-century clichés describing how the passions impelled and guided natural philosophical investigations. The senses were first snared and lulled by delightful novelties; understanding snapped to attention as novelty deepened into philosophical anomaly; and body and mind mobilized to probe the hidden causes of the apparent marvel. Newton might have assembled the elements of this sequence from any number of mid-seventeenth-century sources — from Robert Hooke's promise that "Experimental Philosophy" would cater to "*material* and *sensible Pleasure*,"[3] René Descartes' account of wonder as the first of the passions, excited by the "rare and extraordinary" and useful in small doses to stimulate scientific inquiry,[4] or Thomas Hobbes' praise of curiosity as "the continuall and indefatigable generation of Knowledge."[5] Precisely because Newton trafficked here in platitudes, his specific models are difficult to pinpoint, for the list of possible suspects is too long.

Yet when Newton wrote in 1672, these platitudes about the interweaving of admiration, wonder, and curiosity into a sensibility of inquiry were of relatively recent coinage and to be of fleeting duration. For Newton and the generation before him, wonder sparked curiosity, shaking the philosopher out of idling reverie and riveting attention and will to a minute scrutiny of the phenomenon at hand. The passions of wonder and curiosity had, however, been traditionally remote from one another in medieval natural and moral philosophy,[6] and they were to separate once again by the mid-eighteenth century. Moreover, during the same period that wonder and curiosity first approached and then withdrew from one another, the trajectories of their valorization in natural philosophy also crossed, with curiosity ascending and wonder declining. On the one hand, the wonder that had once been hailed as the philosophical passion par excellence was by 1750 the hallmark of the ignorant and barbarous. On the other hand, curiosity, for centuries reviled as a form of lust or pride, became the badge of the disinterested and dedicated naturalist.

This intricate minuet of wonder and curiosity recast the collective sensibility of early modern natural philosophers. That sensibility in turn shaped their choice of what objects to study (or pointedly to ignore), of which methods and standards of evidence to apply, and of whom to believe. These shifts in sensibility resulted not only from a recombination of stable emotional elements — for example, wonder with or without curiosity — but also from changes in the felt experience of these emotions. We shall argue in this chapter that the affective content, as well as the epistemological import and moral standing, of wonder and curiosity shifted as these passions first drew near and then drifted away from one another. The realignment of wonder and curiosity within the fields of vices and virtues, passions and interests, emotionally restructured both.

Although there is a kinship of descent and, no doubt, some resemblance of feeling between the wonder praised by Augustine and that blamed by David Hume, or between the curiosity castigated by Bernard of Clairvaux and that celebrated by Hobbes, they are not of the same emotional species. We base our argument for their difference on the premise that the felt substance of an emotion depends to a significant degree on the company it keeps. An emotion classified next to, say, towering ambition and sexual jealousy differs in a significant sense from one bordering on envy and avarice, even though they may share a name and a host of other, more substantive features. What might be called the dynamic of an emotion changes with its neighbors — not beyond all recognition, but enough to create new possibilities for the objects and attitudes that give an emotion outlet and outline.

This is what happened to curiosity and wonder during the early modern period. Curiosity shifted from the neighborhood of lust and pride to that of avarice and greed, gaining a new respectability; wonder migrated from the pole of awed reverence to that of dull stupor, becoming the ruling passion of the vulgar mob rather than of the philosophical elite. We have divided our account of this transition into three parts: first, the seventeenth-century transformation of curiosity from a species of lust to one of greed, which had important consequences for the preferred objects of curiosity; second, the mid-seventeenth-century convergence of wonder and curiosity into a psychology of natural philosophical inquiry; and third, the divergence of curiosity and wonder in the first half of the eighteenth century, when wonder was demoted from premiere philosophical passion to its very opposite, and once-frivolous curiosity took on the virtuous trappings of hard work. Throughout, we shall be concerned with the impact of these changes on the objects and subjects of early modern natural history and natural philosophy.

Ravening Curiosity

In order to appreciate the magnitude of these changes in the nature and status of the passions of curiosity and wonder in the early modern period, we must first briefly review the long tradition these changes overturned. For almost two millennia the relationship between the passions of wonder and curiosity in the Western philosophical tradition was one of indifference and of opposition. Aristotle and his Latin commentators had made wonder (*thauma*) the beginning of philosophy and claimed that seeking knowledge was natural to humanity, but that quest had little to do with what they understood by curiosity (*periergia*).[7] As we have seen in Chapter Three, for medieval natural philosophers wonder was a necessary if uncomfortable (and therefore ideally short-lived) realization of ignorance; curiosity was the morally ambiguous desire to know that which did not concern one, be it the secrets of nature or of one's neighbors.[8]

Plutarch contrasted the search for truth in philosophy with the undisciplined curiosity of the vulgar;[9] Latin authors such as Seneca echoed this condemnation of aimless curiosity.[10]

However, the most vitriolic attacks on curiosity came from the patristic writers, none more influential than Augustine in entrenching the opposition between wholesome wonder and morbid curiosity, as we have already described in Chapter Three. At best such curiosity was perverted and futile; at worst it was a distraction from God and salvation. By dint of its proximity to lust and other appetites, curiosity was also closely akin to bodily incontinence. Still stronger was the repudiation of curiosity as allied with pride and therefore opposed to a seemly wonder at God's handiwork: this is Augustine's complaint against the astronomers, who suffer from an excess of curiosity and a deficiency of awe.[11]

These links with pride and lust were harmful and enduring ones for curiosity, echoed in the works of philosophers and moralists through the sixteenth century. Although Augustine was by no means the only Christian authority to castigate curiosity, his influence was the broadest and most long-lived, persisting well into the early modern period.[12] Both Desiderius Erasmus and Michel de Montaigne, for example, repeated many of Augustine's strictures against curiosity, stressing its affinity with malicious gossip and an impudent desire to know inessentials and secrets. Erasmus contrasted this to a pious curiosity that would admire the works of God in gratitude and wonder, without seeking to discover unknown causes, while Montaigne consistently opposed presumptuous curiosity to devout simplicity.[13] Yet as Hans Blumenberg and Carlo Ginzburg have shown, curiosity of a more audacious and frankly philosophical kind gradually took on ever more favorable colors during this period, particularly among travelers and students of nature.[14] By 1620, when Francis Bacon reassured his readers that Adam and Eve had sinned by seeking moral, not "pure and uncorrupted" natural knowledge, and that God benignly regarded the investigation of nature's secrets as an "innocent and kindly sport of children playing at hide-and-seek,"[15] the profound reevaluation of curiosity was well under way.

We do not aim here to chart this transformation of curiosity from grave vice to peccadillo to outright virtue.[16] Instead, we want to juxtapose the early modern brand of curiosity with the Augustinian one, in order to throw their differences into relief. Early modern curiosity was not simply Augustinian curiosity with a reversed moral charge; its emotional texture had also been altered. It had undergone two important changes: a shift from the dynamic of lust to that of greed, and, most important for natural philosophy, an alliance with wonder.

If there was a seventeenth-century spokesman on curiosity of stature comparable to Augustine's, it was perhaps Hobbes. Hobbes was by no means the only or even the first thinker of the period to relocate curiosity among the passions, but he was arguably the most voluble on the subject. Curiosity figures prominently in all of his major works dealing with human nature, for he judged it the quality that distinguishes man from beast, on a par with reason: "*Desire*, to know why, and how, CURIOSITY; such as in no living creature but *Man*; so that Man is distinguished, not onely by his Reason; but also by this singular Passion from other *Animals*."[17] He continued to subsume curiosity under the desires, but opposed it to the bodily appetites, including lust: "... this is a curiosity [seeking effects] hardly incident to the nature of any living creature that has no other Passion but sensuall, such as are hunger, thirst, lust, and anger."[18] Desire was of course an all-important and nearly all-embracing category in Hobbesian psychology, for only the ceaseless motions of mind and body, our appetites and aversions, keep us striving and, indeed, alive. Happiness lies in yearning, not in satisfaction. In this attraction-repulsion mechanics of the passions, curiosity was not merely one of a host of desires, but rather the archetypal desire, for it was a kind of *perpetuum mobile* of the soul. Always seeking, never satisfied, its pleasures were never exhausted. In contrast to the desires of the body, curiosity was pure desire, distinguished "by a perseverance of delight in the continuall and indefatigable generation of Knowledge, [which] exceedeth the short vehemence of any carnall Pleasure."[19]

The insatiability of curiosity, as pure *conatus* or endeavor, linked it to greed rather than lust in the early modern period. Lust slackens when its object is attained; curiosity is never quenched, never at rest. But whereas an eternally unsated bodily appetite would be the torture of Tantalus, from unslaked curiosity flows the most intense pleasure, according to Hobbes. Unlike lust, which aims at satisfaction, avidity aims at perpetuation of desire, darting from object to object, barely pausing to enjoy any of them.[20] Bacon (whom the young Hobbes briefly served as amanuensis) similarly celebrated the pleasures of learning because they alone were inexhaustible: "We see in all other pleasures there is satiety, and after they be used, their verdure departeth ... and therefore we see that voluptuous men turn friars, and ambitious princes turn melancholy. But of knowledge there is no satiety, but satisfaction and appetite are perpetually interchangeable."[21] The rhythms of curiosity were those of addiction or of consumption for its own sake, cut loose from need and satisfaction.

The psychology of endeavor was peculiar to Hobbes, but the new view of curiosity, as akin to the avarice of the miser, was not. Marin Mersenne, that scientific pen pal *extraordinaire* of the early seventeenth century, reached for the same analogies of movement and insatiability when he reflected upon curiosity:

> ... one could say there is a certain sort of current pleasure, not to be found in possession, due to the movement which accompanies it and which belongs to current life [*la vie actuelle*], instead of which enjoyment resembles habit and repose, which is almost imperceptible. And thus we always desire to go beyond, such that acquired truths only serve as means to arrive at others: this is why we no more take stock of those we have than a miser does of the treasures in his coffers.... [22]

This was not simply a variation on the ancient theme of the contemplative life of the scholar versus the active life of the citizen. In this passage Mersenne transformed the contemplative life from tranquil reflection — St. Jerome reading motionless in his study — to restless inquiry. In contrast to a culture of learning epitomized by commentary or exegesis, which dwells long and lovingly on the same passages read over and over, Mersenne's *curieux* do not linger over their accumulated store of truths. They do not digest what they have; they pursue endlessly what is just out of reach. Like Hobbes and Bacon, Mersenne linked learning to pleasure, and pleasure in turn to change. Repose may be a blessing, but an imperceptible one. Pleasure is keenest when the gradient of change is steepest.

Always on the move, avaricious curiosity never paused to savor any single experience, even the most perfect of its kind; this is "why the sequence of chords pleases us more than the continuation of the same chord, even if it were the most melting in all of music," as Mersenne put it. [23] Even those who disapproved of the cult of curiosity, as Descartes did, agreed that it was a restless emotion that kept the mind in a state of continual agitation. [24] Augustinian curiosity had also been of the flickering sort, but slack and aimless, in the guise of distraction; Aquinas had gone so far as to liken this wandering curiosity to sloth, so undirected was its affect. [25] In contrast, early modern curiosity replaced the earlier dynamic of self-dissipating passivity with one of self-disciplined activity, all faculties marshalled and bent to the quest. Avarice mobilizes means to ends with ruthless efficiency and demands considerable concentration and self-control. Curiosity had acquired a similarly keen edge, one that was always moving onward.

Because this new-style curiosity was voracious, it would be natural to assume that it was also omnivorous, indifferent to its objects so long as these were in steady supply. This, however, was not the case for the curious sensibility of early modern natural historians and natural philosophers, who had rather dainty tastes in this regard. Its typical objects closely resembled luxuries—"rare," "novel," "extravagant" were adjectives regularly paired with "curious"—and some actually were luxuries. Mersenne went so far as to define curiosities as luxuries, things not necessary to daily life, and likened the mind curious about nature to

> a king in his kingdom, who having been raised more splendidly and nourished more delicately, needs more things that his subjects and the rest of the people can do without; he has a number of officers, valets, and purveyors; thus the mind of man uses all the senses, and dispatches them to forage among all that nature has established here below, in order to serve not only for his necessities, but also for his pleasures and enjoyment.[26]

These "pleasures" gathered by the "valet" senses were most in evidence in the contemporary *Wunderkammern*, which displayed "Pictures, Books, Rings, Animals, Plants, Fruits, Metals, monstrous or Extravagant Productions, and Works of all Fashions; and, in a Word, all that can be imagin'd curious, or worth enquiry, whether for Antiquity or Rarity, or for Delicacy and Excellency of the Workmanship," in the words of a seventeenth-century tourist and connoisseur of cabinets.[27] These "curiosities" may have defied classification, but they were alike in being valuable by one criterion or another: value derived from precious materials (gold coins, jewel-studded caskets); value derived from rarity (a stuffed armadillo from Brazil, a doe with antlers); value derived from the labor required to produce them (Palissy-ware, a cherrystone carved with a hundred facets)—all of these luxuries were amply represented, indeed flaunted, in the *Wunderkammern*.

Luxuries are not only valuable; they are also otiose from the standpoint of utility. Early modern curiosity shared this lofty disdain for the useful. Just as political economy during this period morally rehabilitated the luxury trade,[28] so moralists reversed the traditional valuation of "vain" curiosity. The luxury trade allegedly transmuted the private vices of greed and voluptuousness into the public virtues of industry and prosperity; curiosity transmuted *vanitas* into *veritas*. Whereas earlier moralists in the Augustinian tradition had harped upon the frivolity and futility of curiosity, so irrelevant to the necessities of daily life and to salvation in the after-life, early modern writers on the passions associated the uselessness of curiosity with praiseworthy disinterestedness. French natural historian Pierre Belon contrasted the attitudes of the practical merchant with those of the *homme curieux* in a strange land: the one cares only about the best deployment of his money in buying and selling merchandise, while the other observes "infinite singularities" of no obvious utility.[29] Conversely, Bacon's frank concern that natural philosophy ultimately be made useful seemed to him to preclude the motivations of, *inter alia*, "a natural curiosity and inquisitive appetite."[30] When the *Philosophical Transactions of the Royal Society of London* bestowed the honorific "curious" on one of its correspondents, it was in recognition of his disinterestedness as well as of his erudition;[31] Bishop Thomas Sprat underscored this indifference of the curious to personal gain when he noted that gentlemen predominated amongst the members of the Royal Society.[32] In his history of the first decades of the Paris Académie Royale des Sciences, Bernard de Fontenelle attributed the failure of ancient Greek philosophers to investigate the secrets of the magnet to their disdain for "a curiosity apparently useless."[33] "Curious" in late-seventeenth-century natural philosophical usage, especially in French, came near to meaning the very opposite of "useful."[34]

Curiosity had thus become a highly refined form of consumerism, mimicking the luxury trade in its objects and its dynamic of insatiability. Krzysztof Pomian has aptly described the curiosity of early modern collectors as "the desire for totality," at once a desire and a passion.[35] Like the bodily appetites to which necessities minister, demand for the bare essentials cannot be indefinitely expanded. Only the market for inessentials holds out the prospect of unbounded consumption. To some early modern observers like the political economist Nicholas Barbon, the rising demand for luxury goods precisely paralleled the insatiable desires of curiosity: "The Wants of the Mind are infinite, Man naturally Aspires and as his Mind is elevated, his Senses grow more refined, and more capable of delight; his Desires are inlarged, and his Wants increase with his Wishes, which is for every thing that is rare."[36] Both the acquisitive and the inquisitive sought out the new, the rare, and the unusual, so that natural philosophy caught something of the prestissimo tempo of fashion. The Cartesian François Poulain de la Barre thought it natural that the sciences should change "like fashions;"[37] the Paris-based *Journal des Sçavans* ran a regular feature on the latest events of scientific interest under the heading "Novelties."[38] Some naturalists went so far as to suggest that nature had obligingly increased the supply of objects of curiosity in response, as it were, to increased demand: the astronomer Gian Domenico Cassini asserted smugly that no previous century could rival his own for "brilliant novelties" in the heavens and "marvels of all kinds."[39]

Wonder and Curiosity Allied

This affinity for the new, rare, and unusual — in short, for the marvelous — pushed curiosity closer to wonder in the middle decades of the seventeenth century. Ravening curiosity devoured novelties, and novelty — inflected as the rare, the singular, the exotic, or the extraordinary — had long awakened wonder in the literature of topography, romance, and travel.[40] However, as we have already pointed out, this was not the wonder of the Aristotelian natural philosophical tradition. Aristotle had made wonder the origin of philosophy, but it was a wonder engaged first and foremost by the ordinary, and a philosophy that aimed at universals.[41] Early modern wonder might still serve to "excit[e] the appetite of knowing the cause,"[42] as Hobbes wrote, but it was directed toward what "appears to us rare and extraordinary"[43]; natural philosophy now sought "a perfect Knowledge of all Particulars."[44] The transition from a natural philosophy based on universals to one based on particulars required a new economy of attention and the senses on the part of naturalists. As Bacon complained, the unimproved understanding "flies from the senses and particulars to the most general axioms.... For the mind longs to spring up to positions of higher generality, that it may find rest there, and so after a little while wearies of experiment."[45] The union forged between wonder and curiosity in the mid-seventeenth century was meant to cure the understanding of its lazy hankering for generalities by making at least certain particulars irresistible. Once conjoined, neither wonder nor curiosity looked quite the same.

But conjoined they were, and, in the minds of seventeenth-century natural philosophers, necessarily so. It was wonder that would capture curiosity and enlist attention for "the diligent, private, and severe examination of those little and almost infinite curiosities, on which the true Philosophy must be founded,"[46] as Sprat put it in his 1667 apology for the Royal Society. Although both Bacon and Descartes were uneasy about the excesses of wonderstruck curiosity in natural philosophy, they acknowledged its essential role as bait and motivation for intense efforts of attention. Bacon claimed that "by rare and extraordinary works of nature the understanding is excited and raised to the investigation and discovery of Forms capable of including them,"[47] while Descartes believed that those deficient in wonder were "ordinarily very ignorant."[48] Wonder caught the attention; curiosity riveted it.

Consider, for example, Robert Boyle's observations on natural and artificial phosphors, which to his "wonder and delight" glowed in the dark with a cold light. Not only diamonds but also rotten wood, stinking fish and meat, and a "noctiluca" distilled from human urine were the objects of his minute and intense scrutiny. Over and over again, "partly for greater certainty, and partly to enjoy so delightful a spectacle," he and his assistants (working in total darkness and clothed in black so as to catch the slightest gleam) observed the flickering light of wood and fish in the receiver of Boyle's air pump.[49] When his servant discovered a shining veal shank in the larder late one night, Boyle forgot his bad cold (caught while testing a new telescope a few nights before) and immediately launched into a series of eighteen observations, noting, *inter alia*, that the color and intensity of the light varied from patch to patch, that the wind that night came from the southwest and where the barometer stood, and that neither compression nor cold water would douse the pale flames.[50] Boyle spent another late night, with "no-body to assist me but a foot-boy," investigating a borrowed diamond that shone in the dark, plunging the gem into oil and acid, spitting on it, and "taking it into bed with me, and holding it a good while upon a warm part of my naked body."[51]

What is striking about these observations is their urgency, amplitude, and detail. Wonders were notoriously ephemeral, and the annals of seventeenth-century natural philosophy abound with stories of interrupted meals, forsaken guests, and missed bedtimes as observers dropped everything to devote themselves to a fleeting, fascinating phenomenon. Since Boyle seldom conducted his investigations without assistance, his servants must have dreaded those midnight orders to prepare the air pump to receive yet another morsel of putrefying fish.[52] Nor did the observations stop with the glowing wonder itself. Boyle registered the phase of the moon when the veal shank began to shine, the colors of the cloth with which he rubbed the diamond, and the emotional responses of the spectators to his noctiluca. The observer's focus of attention spread to encompass an indefinite number of particulars, all potentially hints as to the nature of light. At dead center of the spotlight of attention lay the phosphors themselves, described in fine-grained detail: the precise greenish hue of the brightest patch of veal; whether the diamond's light was a faint glimmering or little sparks of fire; how an artificial phosphor smelled of sulphur and onions. Boyle himself realized that this plethora of detail and the attention it presupposed magnified the wonder of any object that could stimulate such obsessive curiosity: "So many particulars taken notice of in one night, may make this stone [the diamond] appear a kind of prodigy."[53]

The feats of attention cultivated by the natural philosophers only superficially resembled the meditative disciplines preached by the Catholic founder of the Jesuit order, Ignatius Loyola, or by Protestant divines such as the English bishop Joseph Hall. In both its Catholic and Protestant forms, sixteenth- and seventeenth-century meditations started from the mundane — "Upon the Sight of Flies" or "Upon the Sight of a Drunken Man" are typical titles — and ascended quickly to lofty reflections on (in these examples) the unqualified gratitude owed to God, or the faculty of human reason.[54] The goal of meditation was not to dwell upon the images of memory and sense, but rather to cast particular impressions "in the moulds of resolution into some forme of words or actions."[55] In contrast, natural philosophical attention ideally lingered over the particulars, cherishing every detail. Some natural philosophers hesitated to generalize or conjecture about the particulars they recorded so fastidiously: although Boyle was, for example, prepared to stay up half the night to investigate a phosphor from every conceivable angle, he pleaded "ill health" when it came to conjecturing about the nature of light from his results.[56] They preferred to linger over and multiply the particulars. Hence scientific curiosity demanded stamina as well as a flash of interest. Curiosity had to keep the gaze glued to the object of observation, when boredom or distraction might have lured it elsewhere.

This was particularly true of objects unprepossessing either by their ordinariness or their base associations. Boyle studied phosphors because of their "nobility," light being "the first corporeal thing the great creator of the universe was pleased to make."[57] Even rotting fish and urine might acquire some glamour in this august connection. But Hooke inspected under the microscope a louse, a blue fly, a cork, and other objects that were at best prosaic and at worst disgusting. Some contemporaries found such tastes distinctly odd, if not reprehensible: Hooke was furious to recognize himself as the model for playwright Thomas Shadwell's character Sir Nicholas Gimcrack, a minutiae-crazed natural philosopher who has spent a fortune on microscopes and is made to remark, "'Tis below a *Virtuoso*, to trouble himself with Men and Manners. I study Insects."[58] Hooke could indeed wax eloquent over the beauties of a magnified fly, which he described in prolix and admiring detail:

> All the hinder part of its body is cover'd with a most curious blue shining armour, looking exactly like a polish'd piece of steel brought to that blue colour of annealing.... Nor was the inside of the body less beautifull than its outside, for in cutting off a part of the belly, and then viewing it,... I found, much beyond my expectation, that there were abundance of branchings of Milk-white vessels.[59]

In order to rivet the attention upon a common fly, Hooke had to transform it into a marvel by means of the microscope. Unmagnified, the fly barely registered in the observer's consciousness. Magnified and mar-velous, it became the quarry of avid curiosity, pursued with a "*material and sensible Pleasure.*"[60]

Wonder could provoke such intense and painstaking curiosity, but usu-ally at the cost of restricting its objects, or the ways in which the objects were viewed. Just what excited wonder was a matter of some debate, but novelty, rarity, variety, strangeness, and ignorance of causes turned up on almost everybody's list. Wonder did not have to be confined to the natural; much of mannerist art and Renaissance poetry aimed to evoke the same gasp of admiration and surprise by enlisting the rare, the singular, and the richly various.[61] Exquisite workmanship, both natural and artifi-cial, also commanded wonder, especially if the construction was intricate or miniature. Such objects were often termed "curious" by extension — for, as Hobbes remarked, the dissection of an object into its minute parts, be it by the eye of the body or the eye of the mind, prolongs the pleasur-able state of curiosity, by disassembling a single object into many.[62] The analogy between a curious work of art and a curious work of nature was often made quite explicit, as when the naturalist John Ray compared the tiny animals observed by the Dutch microscopist Antoni van Leeuwen-hoek to "[a]ny work of Art of extraordinary Fineness and Subtlety, be it but a small Engine or Movement, or a curious carved or turned Work of Ivory or Metals . . . beheld with admiration, and purchased at a great Rate, and treasured up as a singular Rarity in the *Museums* and Cabinets of the Curious."[63]

Fine workmanship on a small scale specified one class of objects wor-thy of wonder and curiosity; hidden causes and secret properties of nature specified another. Ignorance of causes was traditionally a spur to wonder; and curiosity, in both its natural and social forms, had a powerful affinity for delving and prying into secrets. Patristic and medieval literature hos-tile to curiosity associated this taste for secrets with demonic and natural magic,[64] damaging associations that persisted through the sixteenth cen-tury.[65] Although the social curiosity of gossips, busybodies, and jealous spouses remained a matter for censure well into the Enlightenment,[66] the revelation of nature's secrets — Bacon's game of hide-and-seek with God — became ever more praiseworthy in seventeenth-century natural philoso-phy. Given the epistemological gloom that settled over many seventeenth-century philosophers concerning the infirmities of the human senses and intellect, it is startling to find the very same pessimists recommending that "the curious Sight" follow nature "Into the privatest recess, / Of her imperceptible Littleness."[67] Hooke was prepared to disqualify logic from a significant role in natural philosophy because it could not reveal "what are the Kinds of secret and subtle Actors, and what the abstruse and hidden Instruments there made use of" in natural processes.[68] The intense natural philosophical interest in the invisible and subvisible, from Bacon's "latent configurations" to Descartes' microscopic mechanisms to Boyle's "hidden springs and principles," was no doubt overdetermined,[69] but one of the several contributing causes was that the hidden or secret was the ideal object of admiration and curiosity. The causes of the hidden and the secret, by definition obscure, thereby set in motion the sequence of wonder-curiosity-attention.

Thus the interlocking of the passions of wonder and curiosity privileged certain phenomena and things above others as objects of scientific investigation — namely, the new, rare, unusual, or secret. The contents of the *Journal des Sçavans*, the *Philosophical Transactions*, and the *Histoire et Mémoires de l'Académie Royale des Sciences* during the latter half of the seventeenth century bear loud witness to this preference.[70] If other, more ordinary and obvious objects were to stimulate curiosity and the attendant effort of attention, they, too, would have to be tricked out as wonders. A correspondent to the *Philosophical Transactions* felt obliged to describe the slow encroachments of sand upon arable land in Suffolk as "prodigious *Sands*";[71] another correspondent, John Winthrop of New England, worried apologetically "whether I may recommend some of the productions of the wilderness as rarities or novelties, but they are such as the place affords."[72]

Even those natural philosophers who wearied of the exotic and the anomalous subscribed to a psychology that led them to treat the prosaic and common as if it were extraordinary. Hooke, convinced that reason unprovoked by curiosity would never bestir itself, tried to make the common rare and the domestic exotic, in the hope of sharpening attention by engaging first wonder and then curiosity:

> In the making of all kinds of Observations or Experiments there ought to be a huge deal of Circumspection, to take notice of every least perceivable Circumstance.... And an Observer should endeavour to look upon such Experiments and Observations that are more common, and to which he has been more accustom'd, as if they were the greatest Rarity, and to imagine himself a Person of some other Country or Calling, that he had never heard of, or seen the like before: And to this end, to consider over those Phenomena and Effects, which being accustom'd to, he would be very apt to run over and slight, to see whether a more serious considering of them will not discover a significancy in those things which because usual were neglected.[73]

In short, philosophers should struggle to see the banal — for example, the fly — as marvelous. Only through deliberate self-estrangement could curiosity be inflamed and attention thereby heightened to the level deemed necessary to a scientific inquiry of particulars. No detail was too miniscule, no variation too subtle to be without potential philosophical significance to the observer suspicious of Aristotelian axioms, universals, and natural kinds. In the Dürer-like crispness of vision of naturalists who cultivated this strenuous curiosity, everything was foreground, nothing was background. Fearful of excluding anything, they strained each nerve to catch everything. The feats of concentrated attention this required were herculean, and it took rarity, novelty, and other sparks of wonder to sustain them.

The French philosopher Nicholas Malebranche gave distilled expression to the seventeenth-century alliance of wonder, curiosity, and attention in his influential treatise *De la Recherche de la vérité* (1674–75). Because wonder hardly affected the heart, it was the least corrupting to reason of all the passions. Without wonder, the efforts of attention required for the pursuit of truth would be drudgery:

> There is nothing so difficult than to apply oneself to a thing for a long time without wonder, the animal spirits not carrying themselves easily to the necessary places in order to represent it.... It is necessary that we deceive our imagination in order to awaken our spirits, and that we represent the subject upon which we wish to meditate in a new way, so as to excite in us some movement of wonder.

But it was not sufficient that wonder refresh attention; in order to be intellectually respectable, it must also awaken the curiosity necessary "to examine things in the greatest exactitude [*dernière exactitude*]." Wonder that failed to engage "a reasonable curiosity" stalled in the pathologically attentive state of astonishment. Malebranche insisted that wonder should lead to inquiry, but he admitted that wonder for wonder's sake was seductively sweet to the soul. Bathed in all the animal spirits set abubbling by the strange or marvelous object, the soul might be tempted to "enjoy its riches rather than to dissipate them" in energetic investigation.[74]

Gawking Wonder

The very attractions of wonder threatened to decouple it from curiosity and attention. Seventeenth-century analysts of the passions were ambivalent about the role of wonder in natural philosophy, acknowledging its utility as lure, but warning against its excesses. Descartes was perhaps clearest on the delicate balance to be struck between just enough and too much wonder. He recognized the utility of wonder "in making us learn and hold in memory things we have previously been ignorant of."[75] But this serviceable "wonder" (*admiration*) was to be distinguished from a stupefying "astonishment" (*estonnement*), which "makes the whole body remain immobile like a statue, such that one cannot perceive any more of the object beyond the first face presented, and therefore cannot acquire any more particular knowledge." Astonishment differed only in degree from wonder — "astonishment is an excess of wonder" — but their cognitive effects were diametrically opposed. Whereas wonder stimulated attentive inquiry, astonishment inhibited it and was therefore, Descartes asserted, always bad.[76] Hence extreme caution must be exercised lest wonder become habitual and indiscriminate, "the malady of those who are blindly curious, that is to say, who seek out rarities only to wonder at them, and not to understand them."[77]

Descartes' pernicious astonishment bore a strong resemblance to Augustinian reverential wonder. It stunned rather than tantalized the faculties and the senses, dousing rather than fanning curiosity. This form of paralyzing wonder still had its admirers among seventeenth-century writers, but these underscored its religious significance, often at the expense of natural philosophical investigation. Bishop Hall, upon hearing a peal of thunder, called upon the devout to "have their thoughts swallowed up with an adoring wonder" and rebuked "the sauciness of vain men that will be circumscribing the powerful acts of the Almighty within the compass of natural causes, forbearing to wonder at what they profess to know."[78] There was nothing dreadful or ominous in the objects that provoked Cartesian astonishment — on the contrary, Descartes dismissed them as "things of no importance" — but they, too, froze mind and body into inactivity (fig. 8.1).

Bacon similarly diagnosed excessive wonder as a thwarting of knowledge, scorning the aimless trials of empiricists whose research "ever breaketh off in wondering and not in knowing."[79] He believed that the "broken knowledge" of wonder also resulted from futile attempts to plumb the mysteries of God through the study of nature: "It is true that the contemplation of the creatures of God hath for end (as to the nature of the creatures themselves) knowledge, but as to the nature of God, no knowledge, but wonder; which is nothing but contemplation broken off, or losing itself."[80] Wonder at greatness was perhaps permissible in religion, but "vain admiration" of novelty bespoke embarrassingly meager learning in a philosopher.[81] In contrast to the studied neglect of marvels by academic Aristotelians, Bacon treated them as challenges flung down to his reformed

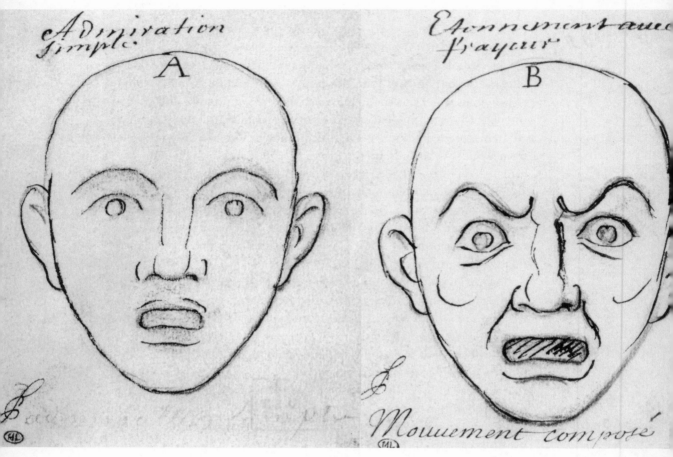

8.1.1   8.1.2

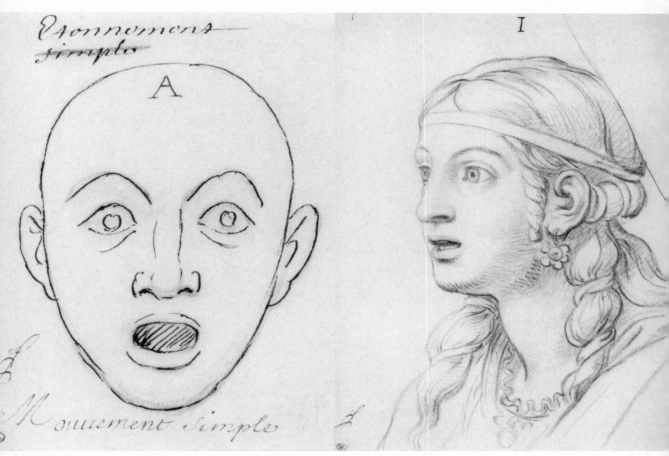

8.1.3   8.1.4

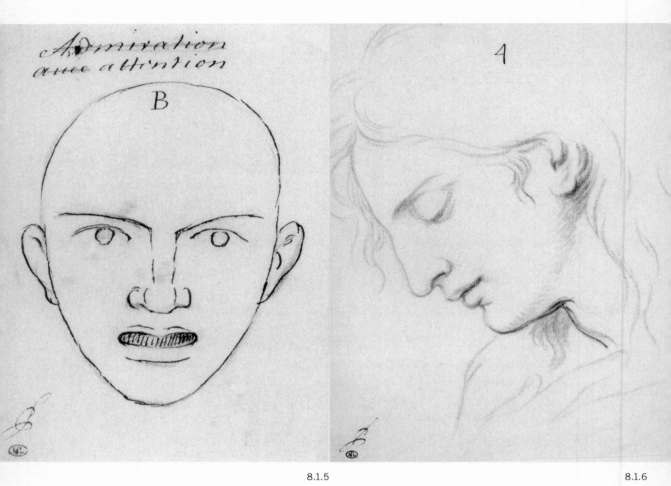

8.1.5   8.1.6

Figure 8.1. "Let us begin with wonder. . . ."
8.1.1. Charles Le Brun, *Simple Wonder*, G.M.6461.
8.1.2. Charles Le Brun, *Astonishment with Terror*, G.M.6498.
8.1.3. Charles Le Brun, *Astonishment — Simple Movement*, G.M.6465.
8.1.4. Charles Le Brun, *Wonder*, G.M.6464.
8.1.5. Charles Le Brun, *Attention*, G.M.6449.
8.1.6. Charles Le Brun, *Veneration*, G.M.6467.
All Musée de Louvre, Paris (c. 1668).

In 1668 the French artist Charles Le Brun lectured the Académie de Peinture in Paris on how to depict the human passions, as illustrated by his own engravings.[26] Le Brun's *Conférences sur l'expression générale et particulière* (posthumous 1696) drew heavily on Descartes' *Les Passions de l'âme* (1649) for his description of the individual passions and of their interrelationships. Hence wonder (*admiration*) was for Le Brun, as for Descartes, the "first of the passions." A combinatorics of the passions, mirrored in physiognomy, displays the nuances of wonder, variously allied with terror, reverence, and curiosity.

natural philosophy, which would no longer tolerate wonder born of ignorance.[82] In small doses wonder whetted the edge of curiosity, but in larger
amounts it both betokened and prolonged ignorance.

Late-seventeenth-century natural philosophers increasingly emphasized the vices rather than the virtues of wonder in scientific inquiry.
Wonder was ever more closely associated not only with ignorance of
causes but with ignorance *tout court*. Joseph Glanvill belittled the "rude
wonder of the ignorant," who were blind to "the chief wonders of Divine
art and goodness [which] are not on the surface of things laid open to
every careless eye."[83] The Anglo-French theologian Meric Casaubon disdained those who wondered at reports of a flying cat or of lobsters drag-
ging men into the sea, describing them as provincials "who were never
out of their own country, nor ever had the curiosity to read the travels of
others" and who were hence insufficiently exposed to the variety of the
world.[84] The wonder of the natural philosopher had become that of a
connoisseur, more discriminating and refined than the dull, effusive wonder of the vulgar stay-at-home. Curiosity and the knowledge it procured
had become more a cure for wonder than its product.

In the context of religious and political polemics carried out by portent and counter-portent, the wonder of the vulgar could become dan-
gerously volatile. The Cambridge Hebraist John Spencer was a *devoté* of
the pleasures of wonder and curiosity: "Objects as yet not fully known
(as objects rare and strange are) keep the Soul in a state of hope and expectation of some huge satisfaction in a greater intimacy and acquaintance
with them." His compilation of wonders—volcanoes, new stars, monstrous births, landscape marble—aimed to feed these appetites of the soul,
for "the Understanding is continually calling for a new scene of contem-
plations; *the eye is never satisfied with seeing*." Yet Spencer, writing shortly
after the restoration of Charles II to the throne, was acutely aware of the
more menacing prodigies abroad in the land and of their potential for
political and religious subversion. If one were to believe the broadside
press, "England is grown Afrika, and presents us every year since the
Return of His Majesty, with a scene of Monstrous and strange sights; and
all held forth to the people, like black clouds before a storm, the harbingers of some strange and unusual plagues approaching in the State." In
Spencer's opinion such prodigies fomented dangerous wonder, admixed
with awe, fear, and enthusiasm, and were all too capable of bringing down
established church and civil authorities.[85]

Wonder could betray ignorance, freeze inquiry, undermine authority,
and, Boyle feared, usurp the prerogatives of man and God over nature. In
his *Free Inquiry into the Vulgarly Received Notion of Nature*, also composed

during the early years of the Restoration, Boyle warned that admiration paid to nature was admiration stolen from God, and verged upon idolatry. Moreover, veneration for nature had, in Boyle's opinion, hindered human exploitation of other creatures assigned an inferior status by God.[86] In a 1685 treatise on worship Boyle argued that it was undignified for humans, the lords of creation, to admire "corporeal things, how noble and precious soever they be, as stars and gems, [for] the contentment that accompanies our wonder, is allayed by a kind of secret reproach grounded on that very wonder; since it argues a great imperfection in our understandings, to be posed by things, that are but creatures, as well as we, and, which is worse, of a nature very much inferior to ours."[87] Therefore only God could be a fit object for wonder and, Boyle believed, also for "not only an excusable, but a laudable curiosity."[88] *Pace* both those Augustinians who had championed reverential wonder over "saucy curiosity" and the Baconians who had feared that religious wonder would halt inquiry, Boyle did not scruple to link the two passions in the standard Hobbesian sequence — but now in the context of piety rather than of inquiry. Admiration for God engaged rather than rebuffed curiosity, moreover focusing it on the only object capable of sating the insatiable. This devout curiosity aspired to know ever more of divine perfection, and since "the more knowledge we obtain of [God], the more reason we find to admire him," so "there may be a perpetual vicissitude of our happy acquests of farther degrees of knowledge, and our eager desires for new ones."[89]

Boyle had carried to its extreme the Hobbesian logic of delightful wonder endlessly stoking inexhaustible curiosity. But in so doing he had wholly removed legitimate wonder from the sphere of natural philosophy. Aristotelian wonder at the ignorance of causes had been fleeting and perhaps uncomfortable; insofar as applied to nature alone, wonder for Boyle was humiliating toward man and idolatrous toward God. Boyle's wonder was in its texture the delightful wonder evoked by "masks, and other pompous and surprising shews or spectacles," rather than the awed wonder that Hall had marshalled for thunderclaps or that Spencer had deplored in broadside portents. But Boyle had rechanneled this pleasurable wonder, trailing curiosity in its wake, away from nature toward God and God alone.

Exclamations of wonder did not disappear from natural philosophy, but by the turn of the eighteenth century they were almost invariably lodged in passages glorifying God through his works. Moreover, the works glorified — the geometry of snowflakes, the anatomy of the human eye, the celestial mechanics of the solar system, the ratio of male to female births — were of a more homespun sort than the two-headed cats, petrifying springs, gold-toothed infants, and multiple suns that had been the objects of wonder in earlier natural philosophy. Wonder of a kind persisted in natural theology, nowhere more than in the lectures Boyle himself endowed,[90] but lavished on the one object judged worthy to receive it. Wonder was no longer a goad to curiosity, but to praise, for its ultimate object was in principle not a concrete individual in all its particularity but mind-numbing God in all his perfections. In practice, the wonder of the natural theologians did sometimes descend to raptures over particulars, especially the particulars of the intricate and the hidden. But wonder was now displaced almost entirely to commonplace objects praised as marvels of divine handiwork. Late-seventeenth- and early-eighteenth-century entomology was particularly rich in such natural theological expressions of wonder at the ordinary. The Dutch naturalist Jan Swammerdam, for example, thought the humble ant deserved as much admiration as God's largest and gaudiest creations, on account of "its care and diligence, its marvelous force, its unsurpassed zeal, and extraordinary and inconceivable love for its young."[91]

Swammerdam's ant, like Hooke's fly, was meant to be an object of wonder, but wonder in a different vein. Swammerdam was struck at least as much by the instincts and intelligence of insects as he was by their beauty. For Hooke, the magnified fly had aroused sensual pleasure; for Swammerdam the ant evoked moral admiration. There is some evidence that entomologists following in Swammerdam's footsteps had to work hard to inspire their readers with a proper wonder for such traditionally unwondrous objects. The French naturalist René Antoine Ferchault de Réaumur also recommended, in six fat volumes, the study of insects—in order to glorify God, but also as a source of marvels: not even in fairy tales was to be found "so much of the marvelous, and of the true marvelous as in the natural history of insects."[92] But Réaumur's marvels, like Swammerdam's, were those of insect habits (how they care for their young, or defend themselves against enemies), which presupposed careful observation before they could appear wondrous. Hooke or Malebranche would have seen this as putting the cart before the horse: instead of wonder snaring curiosity, curiosity had first to work to deliver the revelations that aroused wonder for the maternal tenderness of the spider or the ingenuity of the bee. But for Réaumur, only once the eyes have "become curious and attentive to observe" in the most ordinary settings did the insects that once "appeared fearsome, or least disgusting then offer a spectacle that attracts the attention."[93] This was a new, reversed dynamic of wonder, a sensibility of inquiry based on the principle of delayed gratification. Wonder was the reward rather than the bait for curiosity.

Wonder no longer set in motion feverish investigation but rather the argument from design. Design implied different stimuli as well as a different dynamic for wonder in natural philosophy and natural history. English physician and natural theologian John Arbuthnot thought regularities such as the slight but constant surplus of male to female births more remarkable than irregularities like monsters.[94] Comets were more wondrous for their newly discovered periodicity than for their notorious aberrations. The natural theologians assumed that the natural state of a universe without God would be formless chaos; hence, every expression of order, from insect anatomy to planetary astronomy, was taken as a proof of God's existence, power, and beneficence. Around the turn of the eighteenth century portents and prodigies, those spectacular divine interventions in the ordinary course of nature, lost their plausibility in theology as quickly as in natural philosophy. Although most theologians would have agreed that God always reserved the prerogative of a miracle, they were increasingly cool toward showy marvels. Natural theologians, natural philosophers, and natural historians glorified the meanest of God's works as the true wonders, bestowing their full measure of wonder on the Workman rather than on the works.

So it was that in the late seventeenth and early eighteenth centuries wonder came unhinged from curiosity, moving its home from natural philosophy to natural theology. This was one solution to the indignity of excessive or misplaced wonder. Another solution, however, was proposed to the problem: displace wonder not to God but to the tidy regularity of nature. Wonder should be pried apart from its venerable companions novelty, rarity, and ignorance of causes, and joined instead to parsimony, order, and simplicity. So rehabilitated, wonder might continue to reside in natural philosophy, but would direct the attention of researchers to quite different phenomena.

No one campaigned more tirelessly for this second solution than Fontenelle, Perpetual Secretary of the Paris Académie des Sciences from 1697 to 1740. Although Fontenelle often attacked the old objects of wonder such as *lusus naturae* in his reports for the Académie's annals, he addressed his most vigorous pleas for new objects of wonder to an elite lay audience. The children in Fontenelle's island utopia of Ajoia are made to chant an "Ode to the Marvels of Nature," with the refrain "the same Nature, always similar to herself;"[95] the narrator of Fontenelle's urbane dialogue on the implications of the new astronomy attacks the *devotés* of the "false marvelous ... [who] only admire nature, because they believe it to be a kind of magic of which they understand nothing."[96] It is not so much the variety of nature but the simplicity and economy of its underlying principles that should command our admiration: "[nature] has the honor of this great diversity, without having gone to great expense."[97] In defiance of the ancient dictum that wonder was the beginning — not the outcome — of philosophy, Fontenelle remonstrated with those who rejected the scientific study of nature and instead flung themselves into "admiration of nature, which one supposes absolutely incomprehensible. Nature is however never so wondrous [admirable], nor so wondered at [admirée] as when she is known."[98]

These were sentiments designed to enlist the sympathies of the enthusiasts of old-style wonder and its pleasures, who evidently suspected the new natural philosophy of turning baroque nature, profligate in variety and surprise, into a dully dressed matron of plain speech and regular habits. As if to compensate for slighting the traditional delights of wonder, Fontenelle emphasized all the more the pleasures of bottomless curiosity in the natural sciences, which afforded the observer "an extreme pleasure [in] the prodigious diversity of the structure of different species of animals... the trees of silver, the almost magical games of the magnet, and an infinity of secrets that art has discovered in observing closely, and in spying on nature," as well as an endless opportunity for observations in a science that would never be complete.[99] Here were all the familiar enticements to curiosity, along with promises that these delectable rarities and novelties would be available in inexhaustible supply. Fontenelle's curiosity was as ravenous, his attention as concentrated, as any recorded in the treatises of the mid-seventeenth century. But now wonder was to be, as in Réaumur's natural history, bestowed on the knowledge won rather than on the puzzle posed — the fruit rather than the seed of curiosity.

Fontenelle failed in his attempts to rescue wonder from sterility among the learned and to promulgate a reversed dynamic of curiosity and wonder among lay readers. There is some evidence of a growing split between lay and learned sensibilities with respect to the dynamic of wonder and curiosity in the early decades of the eighteenth century. A 1736 Parisian sale catalogue of a new shipment of exotic seashells distinguished between two classes of customers: the naturalists (Physiciens) who exercised the "recreation of the mind" by discovering the causes of the various forms of shells, and the curious (Curieux) who sought the "recreation of the eye" from the "variety of forms and colors with which they [the shells] are ornamented." Neither motive excluded the other, but there were immediate consequences for the retailer, naturalists wanting their shells "brutes" and the curious preferring them polished. It was the Curieux, not the savants, who left the shop dazzled by wonder: "the eye is struck so marvelously, that one can hardly fix it: the difficulty is to know that which one should wonder at the most, the perfection of the work of this one, or the vivacity of the colors of that one, or the marvelous symmetry of another, or the harmonious irregularity of yet another."[100] Arrested by the pleasing surfaces of things, this wonder bordered on stunned astonishment and did not trigger inquiry. Even in this sympathetic description of the lay response, the wonder of the *Curieux* was ironically decoupled from curiosity.

By the mid-eighteenth century, wonder had sunk among the learned to the level of the gawk, and kept disreputable company to boot. As far as philosophers were concerned, all of wonder (except for stylized effusions in natural theology) had been swallowed up by Cartesian astonishment and largely severed from curiosity. David Hume thought "the strong propensity of mankind to the extraordinary and the marvellous" injurious to truthful testimony and characteristic of "ignorant and barbarous nations." The "wise and judicious" dismissed all tales of marvels out of hand.[101] Wonder was conspicuously absent in Hume's analysis of the passions in *A Treatise of Human Nature* (1739–40). Curiosity figured primarily as "the love of truth" and secondarily as the desire to know one's neighbors' affairs. In neither case was curiosity sparked by so much as surprise, much less by wonder and astonishment: for Hume curiosity was an inborn desire excited by some idea that is at once forceful and "concerns us so nearly, as to give us an uneasiness in its unstability and inconstancy."[102] According to Hume, the curious seek to avoid the mental pain of uncertainty. In contrast, Hume linked surprise to pride and vanity, because all were pleasurable, self-regarding passions. The connection was not a flattering one: "Hence the origin of vulgar lying; where men without any interest, and merely out of vanity, heap up a number of extraordinary events, which are either fictions of their brain, or if true, have at least no connexion with themselves."[103] For Hume the love of truth and the love of the marvelous had become incompatible.

Almost alone among eighteenth-century theorists of the passions, Hume's friend and fellow Scot Adam Smith preserved wonder's role as the beginning of philosophy, but jettisoned all of its pleasurable associations. Smith composed a history of astronomy in which one cosmological system succeeded another by allaying or exciting philosophical wonder, so that wonder became the engine of progress in that science. Yet he found the passion to be an uncomfortable one, inducing "confusion and giddiness" in small doses and "lunacy and distraction" in large. The naturalist confronted with "a singular plant, or a singular fossil" must somehow classify it "before he can get rid of that Wonder, that uncertainty and anxious curiosity excited by its singular appearance, and by its dissimilitude with all the objects he had hitherto observed." Gone was Malebranche's *sentiment de douceur*, which barely stirred the heart and pulse. In a state of wonder, Smith claimed, imagination and memory come unmoored and "fluctuate to no purpose from thought to thought"; the eyes roll, the breath is bated, the heart swells. Smith's wonder resembled not so much fear, as it had for scholastic philosophers, as a nasty hybrid of seasickness and toothache. Naturally Smith's astronomers were as eager to rid themselves of this condition as Roger Bacon had been to flee the thirteenth-century brand of disagreeable wonder. Like musicians acutely sensitive to the slightest dissonance, Smith's philosopher winced at the tiniest discontinuity in nature, and as the musician sought to restore harmony, so the philosopher sought to restore the imagination to "tranquility and composure" by eliminating wonder.[104]

Smith's wonder was at least still a philosophical passion, albeit a fleeting and distressing one. But his was a minority position in the mid-eighteenth century. Hume was more typical in ascribing wonder and a "love of the marvelous" to the unlettered masses. It was nothing new for philosophers to cultivate a sensibility in self-conscious distinction to that of the vulgar, but the lines between learned and lay reactions had here been redrawn. In the late seventeenth century natural philosophers had piqued themselves on registering a thrill of wonder rather than a chill of fear in the face of strange phenomena. Boyle had witnessed the demonstration of the word "Domini" traced on paper with an artificial phosphor, which "shone so briskly, and looked so oddly, that the sight was extreamly pleasing, having in it a mixture of strangeness, beauty, and frightfulness, wherein yet the last of those qualities was far from being predominant."[105] Even after he had transferred all wonder from the name of God in strangely shining letters to God himself, Boyle retained his sense of wonder as "extreamly pleasing," a passion both philosophical and delightful. Like Malebranche,[106] he appreciated its power to please and for that very reason was stern about its proper objects and economy. Yet by the mid-eighteenth century, pleasurable wonder had become the indulgence of the folk, while natural philosophers took refuge in Hume's dour skepticism or in Smith's malaise.

If natural philosophers had steeled themselves against the blandishments of wonder, they had not put themselves beyond the reach of the cognitive passions altogether. Curiosity still impelled them, albeit a curiosity shorn of its hedonistic associations with desire. Since the early seventeenth century, curiosity had taxed attention to its limits, and eighteenth-century writers increasingly emphasized the arduous side of curiosity. The *Encyclopédie* praised the "noble curiosity" that demanded "continuous work and application"; it opined that only a select few were capable of such sustained attention. This laborious curiosity was sharply distinguished not only from the meddling curiosity of busybodies and the vain curiosity of astrologers, but also from the "curiosity for all kinds of novelties, [which] is the portion of the lazy."[107] Curiosity was now respectable, even laudable; but once severed from wonder, it was no longer quite the same curiosity of ravening desire and its attendant pleasures. The passions of inquiry had entered into a relationship rather like that of the ant and the grasshopper in Aesop's fable. Noble curiosity worked hard and shunned enticing novelties; vulgar wonder wallowed in the pleasures of novelty and obstinately refused to remedy the ignorance that aroused it.

Thus by the mid-eighteenth century wonder and curiosity were once again opposed, as they had been for Augustine. But the emotional status quo had not thereby been restored. Neither in their objects nor in their dynamics did the wonder and curiosity of the *Encyclopédie* resemble those of Augustine — or even those of Hobbes. Wonder was no longer reverential, tinged with awe and fear, but rather a low, bumptious form of pleasure. Although still excited by novelty and rarity, it quenched rather than inflamed the desire to probe hidden causes. Curiosity was neither Augustinian lust nor Hobbesian greed, but rather earnest application, unmotivated by the pleasures of wonder and unrewarded by those of inexhaustible desire.

These shifts in the status and dynamics of the cognitive passions of wonder and curiosity had important implications for the objects of natural philosophy. The striking mid-seventeenth-century preoccupation with the secrets of nature and strange phenomena went hand in hand with a psychology of investigation that used wonder as a fuse with which to ignite curiosity. Even natural philosophers as hostile to marvel-mongering as Bacon and Descartes, Hooke and Newton, subscribed to this psychology, suggestive evidence of its felt reality as well as of its philosophical acceptance. The attempts of Hooke, Fontenelle, and others to redirect wonder toward the ordinary and the orderly show the strains between the psychology of inquiry and its changing objects. In the end, wonder proved intractable to such a dramatic reorientation and ceased to be a philosophical passion. Or if it still claimed that status, as for Smith, it did so at the price of pleasure. At best wonder had become a meditative passion, in the form of natural theological admiration for God through his works. Curiosity survived the exclusion of wonder, but much altered in its objects and dynamic. Neither meditative wonder nor earnest curiosity dwelt on particulars, the stuff of mid-seventeenth-century inquiry, but rather ascended swiftly to universals and generalizations. The texture of experience had changed along with the sensibility of natural philosophers.

CHAPTER NINE

The Enlightenment and the

Anti-Marvelous

In the great *Encyclopédie* of Jean d'Alembert and Denis Diderot, the author of the unsigned article entitled "Merveilleux" doubted whether the cultivated French public could ever again stomach an epic on Homeric or even Miltonian scale, filled with the wondrous feats of gods, angels, and devils: "Whatever one says, the marvelous is not made for us."[1] Suitable or not, marvels had hardly disappeared from mid-eighteenth-century Europe, not even from Paris. Fairs and coffeehouses still showed monsters; cabinets still displayed curiosities; the almanacs and gazettes reported bizarre weather, talking dogs, and balloon flights; savants demonstrated the wonders of electricity and luminescence; provincial academies pursued preternatural history; and the popular French series of volumes in the Bibliothèque Bleue entertained their readers with wondrous stories.[2] Marvels persisted in eighteenth-century Europe, both as words and as things. Yet among intellectuals, among somber theologians as well as gadfly *philosophes*, the star of the marvelous had indeed waned, if not completely vanished. Whereas a scant century earlier John Spencer could write zestily of the pleasure and delight taken in the contemplation of "things rare and unusual,"[3] the Encyclopedists and their fellow "gens de lettres" beheld marvels with a mixture of irony and distaste. This chapter is about how and why the vanguard of European intellectuals, strongly oriented toward if not resident in metropolises like London, Paris, and Amsterdam, came to disdain both wonder and wonders in the first half of the eighteenth century.

The explanation that lies ready — all too ready — to hand is "the new science" of the late seventeenth century. If comets and monsters no longer terrified, if strange facts no longer fascinated, if sports of nature no longer amused, if wonders of art and nature no longer blurred together, then it was because, so runs the story, "the rise of the new science and its objective and rational approach to the study of nature" took "much of the wonder ... out of the observation of the physical world."[4] This account of the demise of wonder is simply a variation on a grander theme in eighteenth-century historiography, in which the entire Enlightenment movement catches fire from "the new science." Exponents of neither theme nor variation attempt to untangle the Gordian complexity of diverse and uneven developments in mathematics, astronomy, mechanics, optics, anatomy, and natural philosophy over some 150 years, agglomerated as "the new science." The new science of Andreas Vesalius or of Galileo? Of the universities or the academies? Of the infinitesimal calculus or the microscope? By what specific mechanism did "the new science" sweep away wonder and illuminate the Republic of Letters? Insofar as more detailed explanations are ventured at all, they are too many and too vague to persuade: what exactly have Cartesian mechanical philosophy, Harveyan comparative anatomy, Newtonian celestial mechanics, and Keplerian optics to do with one another, much less with the decline of wonders and the rise of Enlightenment?

The puzzles only multiply when the specifics of timing and content are examined more closely. Comets had ceased to be portents for the learned at least two decades before the mathematical calculations of Edmond Halley were published in 1705; his predictions were not confirmed until 1758.[5] The occult properties of gems may have struck most natural philosophers as ludicrous by around 1700, but the active principles of matter and gravitation so central to Newtonian natural philosophy were every bit as occult or even miraculous, as contemporaries like Gottfried Wilhelm Leibniz repeatedly pointed out.[6] The subtle spirits of preternatural philosophy were rechanneled as the imponderable fluids of eighteenth-century physics and chemistry, with the celebrated final Queries to Isaac Newton's *Opticks* serving as conduit.[7] Voltaire still upheld the power of the maternal imagination to form or deform the fetus, citing his own eyewitness experience in its support.[8] If sympathies and antipathies had faded from accounts of electricity and magnetism by the late seventeenth century, it was not because these phenomena had been subjected either to mathematization or instrumental measurement, techniques applied only a century later.[9] Finally, if the new science of Francis Bacon and René Descartes had destroyed the ancient distinction between art and nature, it is hard to see how the same new science can explain why collections of naturalia and artificialia split apart in the late seventeenth century.[10]

Nor did naturalists' mounting wariness of the wondrous stem from empirical investigation of individual cases. Although some wondrous objects and forces were summarily evicted from natural philosophy and natural history as illusions or frauds, a great many others endured well into the eighteenth century. Astral influences became the butt of satire (fig. 9.1.1); chameleons were shown to take nourishment from insects rather than air (fig. 9.1.2); collectors relabeled their unicorn horns as narwhal tusks (fig. 9.1.3).[11] But the Royal Society of London reported on Sir Robert Moray's observations of barnacle geese in Scotland; the Paris Académie Royale des Sciences printed Leibniz's account of a dog that could bark out some thirty words; fluids fully as invisible and impalpable as the emanations of celestial intelligences or the *vis imaginativa* became Newton's subtle aether, "Nature's universall agent, her secret fire."[12] For each wonder debunked, the learned journals of the latter half of the seventeenth century supplied a dozen new ones, duly certified by trustworthy witnesses and circumstantial details.[13] The labor of examining carefully each of these cases, not to mention the social risks of contradicting the worthies who had reported them,[14] would have been immense, and no such systematic effort to discredit wonders one by one was ever undertaken. Rather, the wonders of preternatural philosophy and the strange facts of the early scientific societies were discredited wholesale, by appeal to a new metaphysics and a new sensibility. Nature abandoned loose customs for inviolable laws; the naturalist abandoned open-mouthed wonder for skeptical sangfroid.

Both metaphysics and sensibility were part and parcel of broader transformations of intellectual life following the pacification of much of Western Europe after the protracted and devastating civil strife of the sixteenth and seventeenth centuries.[15] In positive terms, the regularity of the new natural order mirrored the decorum of the new social order; in negative terms, wonder in natural philosophy smacked of the disruptive forces of enthusiasm and superstition in religion and politics. Naturalists did not destroy the culture of the marvelous in the Enlightenment, any more than they had created it in the High Middle Ages. As we have argued in the earlier chapters of this book, wonders and wonder were not native to medieval natural philosophy, natural history, and medicine, but rather were absorbed from courtly, literary, and theological sources. These milieux infused wonders with value and nobility by association with princely power, civilized refinement, exotic luxuries, and divine artistry, and they eventually turned insignificant oddities into objects worthy of intense philosophical attention. Analogously, eighteenth-century naturalists abandoned wonders as part of a more global reaction against the political, religious, and aesthetic abuses of prodigies and marvels. Out of this reaction emerged a new cultural opposition between the enlightened and the vulgar, which turned on contrasting valuations of wonder and wonders. Central to the new, secular meaning of enlightenment as a state of mind and a way of life was the rejection of the marvelous.

331

9.1.1

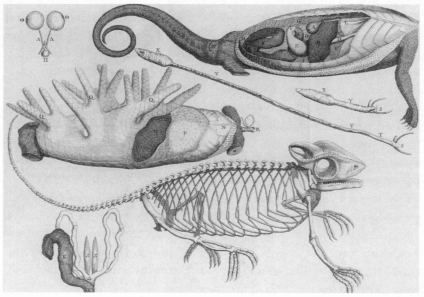

9.1.2

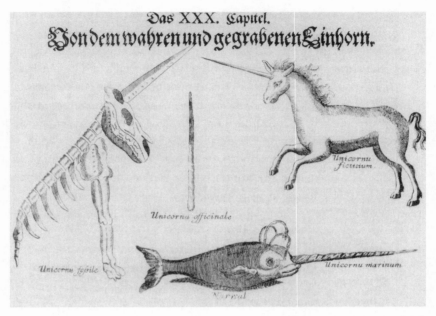

Das XXX. Capitel.
Von dem wahren und gegrabenen Einhorn.

9.1.3

Figure 9.1. Wonders investigated

9.1.1 Gian Domenico Cassini, "Description de l'apparence de trois soleils vûs en même temps sur l'horizon," *Mémoires de l'Académie Royale des Sciences* 1666/99 (Paris, 1733), vol. 11, pp. 234–40.

9.1.2. Académie Royale des Sciences, "Description anatomique de trois caméleons," in Perrault, *Mémoires pour servir à l'histoire naturelle des animaux* (Paris, 1733), pp. 35–68.

9.1.3. D. Michael Bernhard Valentini, "Von dem wahren und gegrabenen Einhorn," *Museum Museorum*, 2nd ed. (Frankfurt am Main, 1714).

Because ignorance of causes produced wonder, one way to neutralize a wonder was to explain it. Comparisons, anatomies, and conjectured explanations situated anomalies in a context of other, more humdrum phenomena, thus blunting their wonder. When three suns had appeared on the birthday of King Charles I of England in 1644, the parahelion had been the occasion for feverish astrological interpretations; almost fifty years later, when the astronomer Gian Domenico Cassini observed three solar images at sunrise in 1693 (fig. 9.1.1), he conjectured that such multiple images were due to the refraction of sunlight through ice crystals in the air. Because of their alleged abilities to change color to match their surroundings and to live on nothing but air, chameleons (fig. 9.1.2) qualified as minor marvels. As part of their major project in comparative anatomy, Claude Perrault and other Paris academicians submitted such claims to systematic investigation, using both live and dead chameleons; Perrault himself was more impressed by the discovery that the chameleon could swivel each eye independently. Unicorn horns, once the most precious objects in princely collections, plunged in value in the late seventeenth century as narwhal tusks flooded the market. In his 1704 compendium on natural history collections, Valentini attempted to sort out the various sources, fossil and otherwise, for "true and false" accounts of unicorns (fig. 9.1.3).

The Unholy Trinity: Enthusiasm, Superstition, Imagination

In December 1680 a comet visible to the naked eye was observed through-out Europe, northern Africa, and America, occasioning agitated debate about its nature and significance. Noting that "never had a comet been observed by such a large number of astronomers," the *Journal des Sçavans* reported on every shade of opinion, from those who maintained that comets were merely "natural things...presaging neither good nor evil" to those who counted them among the terrifying "signs and prodigies" sent by God as warnings, and all points in between — including one Monsieur Mallement de Messange, who advanced a Cartesian vortex theory of comets but also maintained that they could influence sub-lunary events.[16] Pierre Bayle addressed a book-length harangue to a doctor of the Sorbonne who "imagines, with the rest of the world, despite the reasons of a small number of select [persons], that comets are like the heralds of armies who come to declare war on the human race on the part of God." Bayle avowed himself surprised and indignant that a scholar could hold such opinions, fit only for a preacher aiming "to persuade the people."[17]

Bayle's association of prodigies with the religious manipulation of the folk had been forged in the learned critiques of enthusiasm, superstition, and imagination published by theologians, physicians, and philosophers in the late seventeenth century. Although these treatises were inflected by nationality and confession — "enthusiasm" (*Schwärmerei* in German) being the preferred target of Protestant writers and "superstition" that of their Catholic counterparts — all shared an anxious preoccupation with the nefarious role played by popular fear and wonder in religious and political subversion. Rabble-rousers who claimed divine inspiration (or even, in some cases, demonic possession) could topple prelates and princes by whipping up the mob into destructive frenzy. Meric Casaubon, stirred by an account of the demonic possession of a French nun, went so far as to define enthusiasm in political terms, as "a real, not barely pretended counterfeit [of divine inspiration], and simulatory, for politick ends...the nature of the common people being such, that neither force, nor reason, nor any other means, or considerations whatsoever, have that power with them to make them plyable and obedient, as holy pretensions and inter-ests, though grounded (to more discerning eyes) upon very little probabil-ity."[18] Although Casaubon later criticized the Royal Society for enthusiast tendencies, he and Bishop Thomas Sprat, early apologist for the Royal Society, were fully agreed on just why and how enthusiasm and prodigies were a dangerous mix.[19] As Sprat observed, "to hearken to every *Prodigy*, that men frame against their Enemies, or for themselves, is not to reverence the *Power* of *God*, but to make that serve the Passions, and interests, and revenges of men."[20]

In seventeenth-century English usage "enthusiasm" came to cover all individuals and sects — Puritans, Quakers, Anabaptists, millennarists, the Cévennes prophets — claiming supernatural authority, which they often pitted against the established authority of church and state.[21] In the minds of its critics, enthusiasm resembled prodigies and portents in fictitious cause and deplorable effect. Both allegedly came from God; both therefore aroused emotions of fear and wonder befitting divine emissaries. Because directives from God overrode all temporal and ecclesiastical obligations, both could inflame awed and terrified crowds to insurrection and heresy. Henry More deplored the "stupid reverence and admiration which surprises the ignorant" and enthralls them to "the dazeling and glorious plausibilities of bold Enthusiasts."[22] Edward Stillingfleet, chaplain to Charles II of England, denied perpetual revelation for fear it would breed "an innumerable company of *croaking Enthusiasts* [who] would be continually pretending *commissions* from heaven, by which the minds of men would be left in continual *distraction*."[23] Some fifty years later, the more conservative William Warburton, Bishop of Gloucester and defender of revealed religion against Lord Bolingbroke and David Hume, still worried about "[a]dmiration ... one of the most bewitching enthusiastic passions of the mind," which "arises from novelty and surprise, the inseparable attendants of imposture."[24] By the early decades of the eighteenth century the campaign against religious enthusiasm had gained so much momentum at least in England that some theologians felt obliged to exonerate Jesus from suspicions on this score.[25]

For late-seventeenth-century writers, the kinship between enthusiasm and prodigies lay in the wonder and fear both inspired, powerful levers by which usurpers could move multitudes against crown and church. Spencer warned that prodigies could undermine the state:

> How mean a regard shall the issues of the severest debates, and the commands of Authority find, if every pitiful Prodigy-monger have credit enough with the People to blast them, by telling them that heaven frowns upon the laws, and that God writes his displeasure against them in black and visible Characters when some sad accident befals the complyers with them?[26]

Alarmed by these threats to precarious public order, the enemies of enthusiasm and of prodigies wielded the same weapons against volatile fear and wonder: naturalization and pleasure.

The project of finding natural causes for enthusiasm and for prodigies

335

was explicitly strategic, a way of dissolving the fear of divine wrath and the wonder of divine intervention in the course of nature and human affairs, rather than an autonomous medical or natural philosophical inquiry. Casaubon and More invoked melancholy humors, epilepsy, and hysteria to explain enthusiasm, confident that "reverence and admiration" would cease once "the naturall causes of things are laid open."[27] Joseph Glanvill viewed prodigy-mongering as blasphemy against God, "whereby foolish men attribute every trivial event that may serve their turns against those they hate, to his immediate, extraordinary interposal." He also recommended a strong dose of natural philosophy to cure "causeless fear of some extraordinaries, in accident, or nature," citing the example of comets.[28] Neither the originality nor the naturalizing ambitions of these writers should be exaggerated. Nicole Oresme and Pietro Pomponazzi had also undertaken to naturalize (or, more precisely, to de-demonize) marvels in order to calm popular fears,[29] and they were notably more thoroughgoing in that enterprise than Casaubon, More, and Glanvill, all three of whom insisted elsewhere upon the reality of spirits and witches.[30] What was new among the late-seventeenth-century critics of enthusiasm and prodigies was their vivid sense, based on personal experience of religious conflict and civil war, of the urgent political dangers lurking in the emotions of wonder and fear when paired.

A complementary strategy was to decouple wonder and fear, emphasizing the delights of natural wonders at the expense of awe-inspiring divine interventions. Thus domesticated, wonder excited the soul to the contemplation and admiration of God's works rather than to terror at his wrath. This state of mind was not only more conducive to a religion based on faith and love; according to Spencer, it also furthered philosophy by a Baconian investigation of "Nature's voluntary errors and steppings out of her common road of Operation."[31] The Royal Society of London saw fit to review not only the cometary observations of the Danzig astronomer Johannes Hevelius, but also his views that comets should be "rather admired than feared; there appearing indeed no cogent reason, why the Author of Nature may not intend them rather as Monitors of his Glory and Greatness, then of his Anger or Displeasure."[32] Retracing the arc that had transmuted the horror of monsters into the pleasure of monsters,[33] these authors sought to realign wonder with a tame admiration rather than with terrified awe. Spencer argued that admiring wonder helped to undergird public order by banishing fears that "every strange accident [was] ... some sword of vengeance."[34]

To feel admiring rather than fearful wonder in the face of a marvel, particularly one fraught with prodigious associations, became the self-

conscious mark of the natural philosopher freed from the yoke of ignorance and enthusiasm. Halley was distraught that he had missed the
opportunity to observe the beginning of an aurora borealis, pointedly
described in terms highly reminiscent of the ominous battling armies in
the clouds pictured in broadside woodcuts, for "however frightful and
amazing it might seem to the vulgar Beholder, [it] would have been to me
a most agreeable and wish'd for Spectacle."[35] Christian Wolff, professor
of philosophy at the University of Halle, delivered a public lecture to an
overflow crowd on the occasion of an unusual light that appeared in the
sky on 17 March 1716, in which he exhorted his anxious audience to
regard the phenomenon as evidence of "God's might and majesty" rather
than as presaging "future misfortune."[36] Fear had never been a passion
becoming to a philosopher — it was the only passion for which Descartes
could find no use whatsoever.[37] But wonder had also traditionally been
taboo for philosophers, a badge of shameful ignorance.[38] The brief reha-
bilitation of wonder as a philosophical passion in the seventeenth century
provided a weapon in the battle against enthusiasm. Natural philosophers
fought fire with fire, pitting the calm wonder of admiration against the
fearful wonder of awe.

Although French ecclesiastical and civil authorities also struggled to
subdue the subversive forces of wonder and fear, particularly in the wake
of the celebrated Jansenist miracles (first of the Holy Thorn and later of
Saint-Médard),[39] the negative meanings of the English Protestant term
"enthusiasm" never took root in Catholic France. The cognate *enthousiasme* remained close to its classical meaning of poetic furor; the *Encyclopédie* article on the subject went so far as to praise it as "the masterpiece
of reason," with the aim of rescuing artists and poets from injurious associations with madmen.[40] When French writers castigated the dangerous
excesses of wonder and fear, their preferred term was rather "superstition," a word whose meanings underwent dramatic and revealing shifts in
the early modern period. Originally, superstition had referred to excessive or superfluous religious practices, zeal impelled by fear. Augustine
and the church fathers had associated superstition more particularly with
pagan religion and idolatry, and had identified false gods with real demons,
who had duped the Greeks and Romans into worshipping them. In the
sixteenth and early seventeenth centuries, superstition came to refer primarily to the worship of demons in the form of witchcraft or necromancy
or paganism — demons that were conceived to be evil and all too real. By
the late seventeenth century, however, superstition had evolved to mean
irrational fears of unreal entities.[41] Although superstition had always been
surrounded by a penumbra of fear, only in the late-seventeenth and

337

early-eighteenth centuries did it also become a disease of the imagination.

Like enthusiasm, superstition was cause for both alarm and pity. Terrified by spectres and visions of their own making, the superstitious forsook both right reason and sound religion. Writing in the *Encyclopédie*, the Chevalier de Jaucourt considered superstition more dangerous to the established order than atheism, for the atheist "is interested in public tranquillity, out of love for his own peace and quiet; but fanatical superstition, born of troubled imagination, overturns empires."[42] Theologians evidently agreed, for they led the campaign against superstition, arm in arm with the *philosophes*.[43] Like the English anti-enthusiasts, the French critics of superstition made circumscribed use of naturalization. The superstitious were afflicted by febrile imaginations; the apparitions and prognostications that haunted them could be explained away by natural causes. If certain persons, for example, could foretell the future, it was because they relied on natural auguries from the "elements, meteors, plants, and animals," just as mariners and farmers had for centuries. Monks who girded pregnant women with the belt of St. Margaret in order to insure easy childbirth "exposed themselves to the ridicule of the learned world." Yet demons still lurked in the background, for every such superstition, no matter how absurd, was "of necessity a pact with the Devil."[44] In a 1702 treatise commended by the Paris Académie Royale des Sciences, the Oratorian Pierre Lebrun used the mechanical philosophy of Descartes not only to discredit marvels like the alleged power of coral to ward off thunderstorms, but also to demonstrate that demons were responsible for the powers of divining rods.[45]

Critiques of enthusiasm and superstition eroded the cultural credit of the marvelous in diverse and complex ways. Fear of fear, especially of fear tinged with misappropriated wonder, dominated both critiques, and for much the same reasons. A population amazed by a false prodigy or a counterfeit miracle might judge insubordination to civil and ecclesiastical authorities to be a lesser risk than insubordination to an angry God. After nearly two centuries of prodigy-fueled strife, marvels that bordered on miracles had to be handled like explosives. Hence, both Catholic and Protestant theologians also controlled miracles ever more tightly in principle and practice.[46]

Certain branches of preternatural philosophy came to be tarred with the same brush as enthusiasm and superstition: More attacked the Paracelsians for "philosophical enthusiasm";[47] Lebrun branded occult properties and secrets of nature as "superstitions."[48] These critical associations are at first glance puzzling, for although some Paracelsians had claimed divine inspiration,[49] preternatural philosophers concerned with the hidden virtues of things (like Pomponazzi and Fortunio Liceti) had made a principle out of excluding demonic explanations. In this they differed sharply from More and Lebrun, who never doubted the agency of good and evil spirits in the world. What connected preternatural philosophy with superstition and enthusiasm in the minds of such writers was not in the first instance a view about the autonomy and sufficiency of natural causes — on the contrary. Rather, the implicit analogy was psychological: the marvels of preternatural philosophy, like the excesses of enthusiasm and superstition, provoked wonder, and could therefore be manipulated to instill that peculiar and peculiarly destructive form of fear linked with the demonic or the divine. Reporting on a luminescent veal neck in a 1676 letter to the Royal Society, Dr. J. Beal of Somersetshire mused over how easily he might have frightened his servants and neighbors, had he "had a mind to act Pageantries, or to spread a story of Goblins" by smearing his "hands and face with the tincture of light."[50] When Robert Hooke regaled the Royal Society with a kind of magic lantern show devised from a camera obscura, several lenses, and a lot of candles, he speculated on how "Heathen Priests of old" might have made use of such a wonder of art to counterfeit the miracles "of their Imaginary Deities."[51] Wherever wonder was, lurked the possibility of deception and manipulation; and the reputation of wonders, even demonstrably natural ones, was tarnished by association.

Underlying almost all of these critiques of wonder and wonders was a new understanding of the pathological imagination as a breeding ground for enthusiasm, superstition, and marvels. For the preternatural philosophers of the sixteenth and early-seventeenth centuries, the imagination could produce genuine marvels — apparitions, monsters, sudden cures — by the emanations of subtle effluvia imprinted upon soft matter. More, writing in 1662, still had no doubts about the power of the imagination to produce "real and sensible effects," although the issue had been debated since the sixteenth century.[52] By the early decades of the eighteenth century, however, the powers of the imagination had contracted to the mind and, among the highly susceptible, the body of individuals. External effects like the formation of the saint's image on humid air were eliminated. The imagination — or more precisely, the pathological imagination — became an almost bottomless reservoir for the explanation of bodily anomalies among the impressionable. Bizarrely deformed monsters, the transports of the *miraculés* of Saint-Médard (fig. 9.2), the astonishing Mesmeric cures (fig. 9.3) — all the phenomena that eluded medical and natural philosophical explanation were chalked up to febrile imagination, including many previously ascribed to demons.[53] In politically volatile cases like the inves-

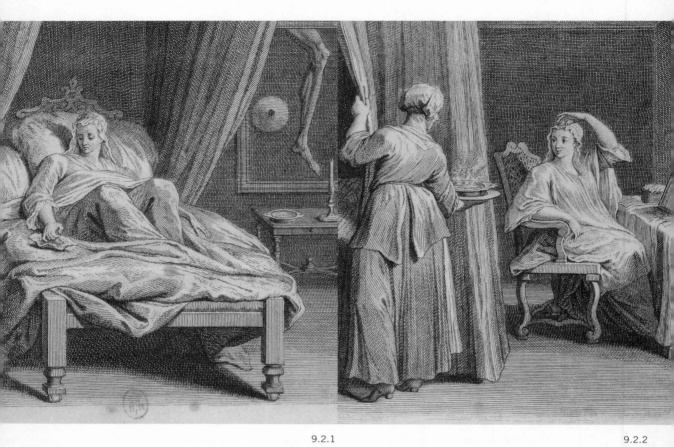

9.2.1   9.2.2

Figure 9.2. Inflamed imagination: The miracles of Saint-Médard
9.2.1–2. Louis-Basile Carré de Mongeron, *La Vérité des miracles opérés à l'intercession de M. de Pâris et autres appelans* (Utrecht, 1737), vol. 1, pp. 1–40.

The explanation of last resort for marvelous and even miraculous events among Enlightenment skeptics was the pathological imagination. Although David Hume dismissed all miracle reports on principled grounds, he was impressed by the evidence assembled by Louis-Basile Carré de Mongeron in support of the authenticity of the miracles worked at the tomb of François de Pâris in the Parisian church of Saint-Médard in the early 1730s. In a typical report, Carré de Mongeron produced eighteen notarized accounts testifying that Mademoiselle Louise de Coirin of Nanterre was cured in August 1731 of paralysis and breast cancer by a bit of earth brought from the grave of Pâris; his illustrations show Mademoiselle de Coirin both before (fig. 9.2.1) and after (fig. 9.2.2) the miracle. "And what," queried Hume, "have we to oppose to such a cloud of witnesses but the absolute impossibility or miraculous nature of the events which they relate?"[27]

tigation of the alleged Jansenist miracles or the efficacy of animal magnetism, the imagination served as a blank check by which to naturalize well-documented but potentially disruptive marvels. Critics of the wondrous believed further that the imagination could also manufacture false marvels, including many objects of preternatural philosophy as well as the delusions of enthusiasm and superstition. If naturalists had once seen the forms of landscapes and animals in marble, it was because they had imaginatively projected them onto the random swirls and whorls of the stone, just as unlettered peasants had imaginatively projected battling armies onto an aurora.[54]

Distrust of the imagination was an ancient theme, particularly among the Stoics, who had routinely opposed it to the faculty of reason.[55] The novelty of late-seventeenth-century diatribes against the imagination lay in transporting this ancient and global opposition into the specific context of prodigies, marvels, and miracles. Benedict Spinoza was the most daring in his claim that the biblical prophets had been possessed by their own imaginations rather than by divine afflatus.[56] But those who would not go so far as to reduce revelation to a malady of the imagination had no hesitation in doing so in cases of enthusiasm and superstition. John Locke ascribed enthusiasm to "the conceits of a warmed or overweening brain";[57] Anthony Cooper, third Earl of Shaftesbury, warned against the enthusiasm of crowds, in whom "the Evidence of the senses [is] lost, as in a Dream; and the Imagination so inflam'd, as in a moment to have burnt up every Particle of Judgment and Reason."[58] French philosopher Etienne de Condillac explained how imagination joined forces with hope and fear to cement the superstitions of animism and divination.[59] Almost no one denied the positive creativity of the imagination — Voltaire exalted its function in art; Condillac insisted upon its contribution to philosophy[60] — but almost everyone who wrote on the subject also singled out for censure what even Voltaire called "fantastic imaginations, always deprived of order and good sense."[61]

The fantastic imagination played a double role with respect to marvels. On the one hand, it served as the explanatory resource of last resort for extraordinary but well-attested phenomena that resisted all other attempts at natural explanation. The members of the commission appointed by the Paris Académie Royale des Sciences to investigate the marvels of animal magnetism here followed in the footsteps of Marsilio Ficino and Pomponazzi.[62] On the other hand, the imagination could be invoked to explain away marvels like the sympathies and antipathies of plants and animals, the patterns displayed in stones, or the occult properties of gems. A correspondent of the Paris Académie, reporting on a monstrous birth, ack-

Figure 9.3. Inflamed imagination: Mesmerism exposed
Le Magnétisme dévoilé (Paris, 1784).

In 1784 the royal commission of inquiry from the Académie Royale des Sciences and the Paris medical faculty concluded that there was no evidence for the invisible magnetic fluid Anton Mesmer and his disciples claimed to manipulate.[28] They attributed the cures and convulsions worked by Mesmerism to touching, imitation, and, above all, the imagination, "that active and terrible power, which produces the great effects that one observes with astonishment in the public treatments."[29]

342

nowledged that prejudiced observation and imagination could lead the naturalist into the fantasies of astral influences, enchantments, sympathetic powders, and alchemy.[63] These latter were now the illusions of natural philosophy, just as enthusiasm and superstition were the illusions of religion. In both cases the imagination betokened a lapse of self-awareness, self-control, or self *tout court*. In order to succumb to the power of the imagination, the disciplines of sober judgment, strict observation, and even (in the case of the convulsions at Saint-Médard or the crises of the Mesmeric tub) bodily control had first to give way. The commissioners from the Paris faculty of medicine and Académie Royale des Sciences appointed by the king in 1784 to investigate animal magnetism contrasted their own composure when magnetized with the violent convulsions of the regular patients undergoing identical treatment: "calm and silence in the one case, motion and agitation in the other; there multiple effects, violent crises, mind and body in a habitual state of distraction and disturbancy, nature exalted; here, the body without pain, the mind undisturbed, nature preserving both its equilibrium and its ordinary course."[64] Mesmeric patients, enthusiasts, demoniacs, and preternatural philosophers might well be sincere, but they were nonetheless self-deceived, victims of mutinous imagination. To be in the grip of fantastic imagination was to lose hold of one's true self, to surrender to an insurrection from within.

Vulgarity and the Love of the Marvelous

Because a self fortified by reason and will could resist the onslaughts of the unholy trinity of enthusiasm, superstition, and imagination, not everyone was believed to be equally at risk. Particularly susceptible were women, the very young, the very old, primitive peoples, and the uneducated masses, a motley group collectively designated as "the vulgar." In the works of the learned, the vulgar stood as the antonym of enlightenment; they were barbarous, ignorant, and unruly. When, in the early eighteenth century, the "love of the marvelous" also came to be seen as a hallmark of the vulgar, it was a sure sign that enlightenment and the marvelous were no longer compatible.

Like "enthusiasm" and "superstition," "vulgarity" was more a floating epithet than a precise term of reference in the seventeenth and early-eighteenth centuries. An insult long hurled by the learned, it gained wide currency in the early modern books of popular errors.[65] Two influential exemplars of the genre, Laurent Joubert's *Erreurs populaires* (1578) and Thomas Browne's *Pseudodoxia Epidemica* (1646), illustrate the lability of the term. The French physician Joubert dedicated his book to the queen and explicitly addressed "all qualities of people." When he upbraided "the vulgar" for, for example, failing to follow medical advice with respect to diet, he indicted common, especially lay opinion, rather than that of any particular class. Ancient sages and church fathers could also be accommodated within this capacious sense of "the vulgar," as when Joubert attacked the "vulgar" opinion of Herodotus and Augustine concerning the natural language of humanity. When he debunked the bestiary lore about pelicans and beavers, Joubert excused common folk for believing these tales, "since several great philosophers and ancient doctors held such opinions."[66] "Popular" and "vulgar" were pejorative but non-specific terms, rough synonyms for conventional (but mistaken) wisdom.

Seventy-five years later, Browne's inventory of errors identified the vulgar more narrowly with the people, whom he called "the most deceptible part of mankind, and ready with open armes to receive the encroachments of Error." Literal-minded and untutored, they lay at the mercy of their senses and appetites. Their individual imperfections were magnified in the aggregate, "for being a confusion of knaves and fooles, and a farraginous concurrence of all conditions, tempers, sex, and ages, it is but naturall if their determinations be monstrous, and many wayes inconsistent with truth." Self-deceived and duped by priests, politicians, and charlatans, "they must needs be stuffed with errors." Yet Browne also allowed that anyone, regardless of status or condition, could lapse into vulgarity by relinquishing reason, "although their condition and fortunes may place them many Spheres above the multitude, yet are they still within the line of vulgarities, and Democraticall enemies of truth."[67] Like Joubert, Browne traced the errors of the people back to Pliny, Strabo, Ctesias, and other ancient authorities, although he strongly implied that the lion's share of the fault lay with those who "swallowed at large" the accounts of these worthy authors.[68] For Browne, vulgarity was a state of intellectual lassitude, in principle applicable to anyone, but in fact most frequently found among the folk.

Learned worries about enthusiasm and superstition in the late seventeenth century tightened the association between vulgarity and the unlettered people. Browne had thought them particularly prone to deception by demonic marvels, especially prodigies like comets and celestial apparitions:

> Thus hath he [Satan] also made the ignorant sort beleeve that naturall effects immediatly and commonly proceed from supernaturall powers, and these he usually derives from heaven, his own principality the ayre, and meteors therein, which being of themselves, the effects of naturall and created causes, and such as upon a due conjunction of actives and passives, without a miracle must arise unto what they appeare, are alwayes looked upon by ignorant

spectators, and made the causes or signs of most succeeding contingencies. To behold a Rain-bow in the night, is no prodigie unto a Philosopher.[69]

By 1680 naturalists made scant reference to demonic deceptions, but the polarization of learned and lay opinion on such matters had become a cliché in reporting about the comet of that year: "The Astronomers observe its course, and the People make it foretell a thousand misfortunes."[70] Bayle found it understandable that "the People were carried of their own accord to error and superstition" in the face of rare events like comets, for they lacked philosophical instruction in the course of nature.[71] Ignorance was the badge of vulgarity, and ignorance in the face of marvels bred the fear upon which enthusiasm and superstition allegedly fed.

Thanks to their knowledge of natural causes, seventeenth-century philosophers in theory rose above the fear of the vulgar, but they were not above wonder. Interlocked with curiosity, wonder was central to the mid-seventeenth-century psychology and epistemology of empirical investigation.[72] Moreover, pleasurable wonder could quench the inflammatory fears of enthusiasm and superstition. Yet wonder was also relative to knowledge, "the most ignorant being most prone to wonder."[73] Late-seventeenth-century natural philosophers were much exercised by the problem of distinguishing learned from vulgar wonder. Glanvill disdained "the *rude wonder* of the ignorant" and suggested that "the chief wonders of divine art, and goodness are not on the surface of things, layd open to every careless eye."[74] Hard work and exacting observation dignified the better sort of wonder. Fontenelle claimed that learned wonder waxed rather than waned with knowledge, and found the universe "ever more marvelous, in the measure that it is better known."[75] Learned and vulgar wonder thus differed not only in object, but also in quality.

But the distinction between learned and vulgar wonder proved precarious. By the early decades of the eighteenth century, wonder had followed fear in becoming part and parcel of vulgarity. Fontenelle's essay *De l'origine des fables* (1724) repeated the commonplaces about the links between ignorance and belief in prodigies, but added the novel claim that all of humanity loved marvels and that it was "principally the false marvelous which is the most pleasurable." Far from inspiring fear, marvels so intoxicated ancient and savage peoples that they could barely narrate the simplest episodes without inserting all manner of extraordinary embellishments. The people had always been gullible, but, according to Fontenelle, even philosophers of earlier times had occupied themselves in explaining "facts imagined to please." It required "a kind of effort and a particular attention in order to relate exactly only the truth." On this account, it was the slow work of centuries of civilization to sift out true facts from false imaginings in fables, and to achieve the taste and discipline for the unadorned truths of history: "Ignorance declined little by little, and consequently one saw fewer prodigies, one made fewer false systems in philosophy, histories were less fabulous; for all of these are linked."[76]

In the grip of enthusiasm and superstition, the vulgar had been terrified of marvels; possessed by the love of the marvelous, they now craved them. This turnabout tells us more about the learned who defined themselves against vulgarity than it does about actual shifts in popular *mentalités*. Jacques Revel has observed that the equation of ignorance or false belief with an inferior social group was part of the wide-reaching "cultural normalization" in the late seventeenth century through restored monarchical and ecclesiastical control.[77] However critical *philosophes* like Fontenelle and Voltaire might have been about established religion and political repression, they were firmly on the side of social order and social hierarchy enforced by an absolutist monarch. The philosopher was now wary of wonder, for it could distort or even fabricate observations, as well as undermine civil and religious order. Previously, the folk had been "deceptible"; now they were deceptive, albeit without guile. In the context of enthusiasm and superstition, vulgar fear and its political manipulation had obliged the learned to insist upon the natural causes of wonders, not to deny their existence. Hence learned wonder could allegedly correct and soothe vulgar wonder. But if there existed, in David Hume's words, a "usual propensity of mankind towards the marvellous" that flourished best among the ignorant and uncivilized,[78] then all wonder and wonders were contaminated with vulgarity. Not only putative prodigies and miracles with subversive political and religious overtones, but also natural wonders became automatic objects of learned suspicion.

This new tone of mingled superiority and suspicion toward vulgar wonders appeared first in reports of individual strange phenomena and eventually drummed the marvelous in general out of natural philosophy. In the early decades of the seventeenth century, Parisian professors of philosophy had not scrupled to stuff their lectures on medicine and natural philosophy with traditional mirabilia;[79] early volumes of the *Philosophical Transactions of the Royal Society of London* and the *Histoire et Mémoires de l'Académie Royale des Sciences* had bulged with accounts of monstrous mushrooms, bizarre echoes, and odd lights in the sky.[80] But by the early eighteenth century, learned societies were on their guard against false marvels, however natural.

As Perpetual Secretary of the Académie Royale des Sciences, Fontenelle used the annual *Histoire* of that body to preach a new skepticism. In

1703 he published a retraction of a 1700 report about "a supposedly inaccessible mountain" near Grenoble that turned out to be nothing more than an ordinary rock, and he blamed "the fabulous genius of mankind" for the original error.[81] In the same vein, he debunked a report of a liquid phosphor at sea, adding that it was as much the duty of the Académie to "disabuse the Public of false marvels as to report on true ones."[82] Fears of fraud were not entirely groundless: in 1721 Louis Fremin, minister in Geneva, had to retract his report of the supposed monstrous offspring of a mating between a cock and a cat, which had "attracted a huge crowd from Geneva," when the hoax was exposed (though Fremin protested weakly that at least the "cohabitation of the cock and cat" was true). The effusively embarrassed and apologetic letter to the Académie from Fremin's local sponsor Bouquin, a lawyer at the Paris parliament, reveals how socially and morally charged such errors had become: "I was deceived...by an excess of credulity; that which redoubles my ill humor towards him [Fremin] is the just resentment you Sir could have against me to have engaged the Gentlemen of the Academy of Sciences in advances [*avances*] which did not hold up."[83] By forwarding Fremin's letter, Bouquin had become an accomplice not so much in deception as in credulity, but his consternation nonetheless indicates a degree of moral responsibility. Although Fremin himself had been an innocent dupe, Bouquin clearly felt that the Genevan's gullibility had been culpable, an abuse of trust among the learned occasioned by too much trust in the vulgar.

The language of Bouquin's apology played upon the double sense of credit, the medium of trust in finance and in testimony. He had tendered "advances" to the Académie that could not be backed; his correspondent Fremin had drawn upon unreliable reserves of credit. The economy of trust essential to collective empiricism shifted its foundations in the early decades of the eighteenth century. Beginning with Bacon, seventeenth-century natural philosophers had devised intricate codes for gauging the reliability of reports filed by a network of farflung correspondents. As Steven Shapin has shown in the case of the early Royal Society, criteria of knowledge, skill, character, social status, and, above all, "integrity and disinterestedness" were applied to testimony about things natural.[84] The extreme case of testimony about marvels had pushed these criteria to their limits, but marvels had not thereby been excluded. The system of evaluation had been solely about testimony, and unimpeachable eyewitnesses could secure the credibility of even the strangest facts. By the late seventeenth century, however, a new criterion, intrinsic plausibility, had emerged as a counterweight to testimony. Hume's essay on miracles (1748) provided this double system of the internal probability of things and the external probability of testimony with its classic formulation,[85] but natural philosophers contemptuous of the vulgar love of the marvelous had already been playing by the new rules of evidence for decades.

The new construction of vulgarity around wonder and wonders also reoriented the early modern debate over learned incredulity and vulgar credulity. Demonology and the witchcraft trials supplied the original context of debate, in which numerous authors simultaneously attacked vulgar credulity and learned incredulity concerning the existence and activities of demons. There is ample evidence of the increasing reluctance of magistrates to hand down convictions in sorcery and witchcraft trials, evidently because they distrusted the motives of plaintiffs and the reliability of witnesses.[86] As in the case of enthusiasm and superstition, considerable intellectual energy, especially on the part of physicians, went into explaining how putative cases of diabolical malice might instead be due to natural causes.[87] Indeed, an over-readiness on the part of illiterate country folk to press charges of witchcraft on the occasion of bad luck or odd happenings became a paradigm case of "superstition" in the seventeenth and early eighteenth centuries. Although the balance of evidence for and against demons fluctuated from author to author, almost every writer preached a course of moderation. Jacques de Chavanes was typical in deploring the hysteria of peasants in Burgundy who had reacted to a hailstorm in 1644 by accusing village idiots of sorcery, yet in the next breath castigating the magistrates, physicians, and philosophers who claimed access to "all the secrets of nature" and thereby explained away all witchcraft.[88]

In the context of demonology, vulgar credulity meant confusing natural phenomena with diabolical mischief, just as vulgar credulity confused natural phenomena with divine inspiration and prodigies in the case of enthusiasm and superstition. In the context of natural history and natural philosophy, credulity had a different valence. Since the late sixteenth century, naturalists had campaigned for a strict review of the natural history transmitted by ancient authorities, especially Pliny—a review implementing, as Bacon put it, "due rejection of fables and popular errors."[89] This purification of natural history contrasted with the principled skepticism of early-eighteenth-century savants in at least three ways. First, it was mainly directed against ancient and some modern authorities, rather than against ignorant contemporaries. Second, it proceeded piecemeal rather than by wholesale elimination; and third, it did not exclude wonders, either in precept or in practice. The authors of Bacon's "popular errors" turned out to be "Plinius, Cardanus, Albertus [Magnus], and divers of the Arabians."[90] Konrad Gesner, Pierre Belon, Joubert, Browne, and a host of other early modern naturalists discarded this or that bit of fabulous natural history, only to wholeheartedly embrace another: Joubert dismissed the claim that salamanders could live in fire, but thought it quite possible that certain people with cold complexions could go for months on end without eating;[91] Browne ridiculed the belief that elephants lacked knee joints, but did not "conceive impossible" that they could speak and write.[92]

Some learned authors even went so far as to reproach the vulgar for incredulity with respect to wonders. The English naturalist and antiquary Joshua Childrey hoped that his natural history of the rarities of Britain would teach the vulgar "not to *mis-believe* or *condemn* for *untruths* all that seems strange, and above their wit to give a reason for." Gentlemen who traveled widely had seen enough "sports of nature" at home and abroad to be more open-minded.[93] In a famous passage, John Locke related how the King of Siam had doubted the testimony of the Dutch ambassador that water could harden in the winter, since this contradicted the narrow experience of generations of Siamese.[94] Glanvill argued that the very absurdity of accounts of witchcraft made them credible, since they were too strange to be imagined: "For these circumstances being exceedingly unlikely, judging by the measures of common belief, 'tis the greater probability that they are not fictitious."[95] At least among some seventeenth-century English natural philosophers, to be incredulous was a sign of little learning and narrow experience, not of enlightened skepticism.

Balanced between the risks of credulity and incredulity, many natural philosophers had epistemological grounds for erring on the side of open-mindedness. Baconians, especially, argued that Aristotelian natural philosophy required correction by exceptions and singularities long ignored by its axioms. Recalling the exploits of Alexander, Bishop Thomas Sprat contended that true history could sometimes rival any romance for marvels.[96] The strange facts purveyed by the annals of the Royal Society of London and the Paris Académie Royale des Sciences in their first decades amply bore him out.[97]

By the mid-eighteenth century, all of this had become "vulgar credulity." The vulgar, once blamed by Browne for literal-mindedness,[98] now allegedly lacked the discipline and clarity to distinguish historical facts from poetical fancies. In the view of the learned, vulgar fear of the marvelous had given way to vulgar love of the marvelous. In the mid-seventeenth century, vulgar credulity had referred to an over-willingness to see wonders as prodigies; by the mid-eighteenth century, it had come to refer to an over-willingness to see wonders at all. When René Antoine Réaumur reported in 1712 on the regeneration of the limbs of crustaceans, he admitted that savants had here gone too far in distrusting reports by simple fishermen, but insisted that it was ordinarily prudent to "be on guard against the marvelous, which the vulgar always willingly credits."[99] Throughout the eighteenth century the Paris Académie treated reports of meteorite falls with skepticism verging on ridicule, for such phenomena reeked of the prodigious and the "love of the marvelous."[100] The *Encyclopédie* article on the "Marvels of the Dauphiné," a topographical marvel complex that dated at least back to Gervase of Tilbury, fairly sneered at those who had once admired the "burning fountain" or the "inaccessible mountain": "The ignorance of natural history and credulity have discovered marvels in an infinity of things which, when viewed with unprejudiced eyes, are found to be false or within the order of nature."[101]

Nature's Decorum

The learned rejection of wonder and wonders in the early eighteenth century partook of metaphysics and snobbery in almost equal measure. The "order of nature," like "enlightenment," was defined largely by what or who was excluded. Marvels and vulgarity played symmetrical and overlapping roles in this process of exclusion: the order of nature was the anti-marvelous; the enlightened were the anti-vulgar; and, by a kind of contrapositive analogy, marvels were vulgar. Enlightened natural philosophy, like enlightened drama, restricted itself not so much to the true, as to the verisimilar. Marvels were possible in both natural and civil history, but they were not probable, in the double sense of frequent and plausible. Nor were they seemly, for they deviated from the decorous order of things and customs. Stones might in fact fall from the sky, and kings might vent their passions with the same exclamations and gestures as peasants, but these facts violated decorum. If the Enlightenment had a physiognomy, it was the incredulous, ironic, and faintly patronizing smile of the savant or man of letters confronted with such a breach of *vraisemblance*.

The order of nature that banished marvels in the early eighteenth century was not identical to the order of nature laid down by natural laws in the seventeenth. Although the terminology of *leges naturae* had precedents in medieval optics and grammar, it attained wide currency in natural philosophy only in the seventeenth century, in close association with a theology of extreme divine voluntarism.[102] God imposed laws upon creation by legislative fiat and could revoke or alter them at will. This was, at least in principle, a natural order porous not only to marvels but to what Boyle called "those signal and manifest interpositions we call miracles."[103] But even the anti-voluntarist view of Bayle or Leibniz, that God's perfect and constant nature prevented him from suspending his own laws, was not incompatible with marvels.[104] Rare or secret concatenations of matter in motion might produce the most bizarre effects. Indeed, for ambitious naturalizers like Bacon and Descartes, the irregularities of nature were gauntlets thrown down to philosophical explanation; Descartes had claimed that his mechanical philosophy could explain "the most wonderful effects" to be found on earth.[105] Neither in principle nor in practice did the doctrine of natural laws, even as realized in the mechanical philosophy, exclude marvels from natural philosophy.

To accomplish this exclusion, a further premise was required: Nature was governed by immutable laws, *and* these laws insured that natural phenomena were always regular and uniform. This second premise flew in the face not only of daily experience, but also of much hard-won experimental and observational knowledge in the late seventeenth and early eighteenth centuries. Not only did learned journals devote reams of paper to reporting on singularities and anomalies; even more mundane phenomena displayed a disconcerting amount of variability under careful investigation. Christiaan Huygens in Amsterdam could not replicate the air pump experiments Robert Boyle had performed in London;[106] Leibniz could not achieve the static electrical effects that Otto Guericke had, even with a sulphur globe of Guericke's own manufacture;[107] Wilhelm Homberg followed Johann Bernoulli's recipe for glowing barometers to the letter, but still failed to produce luminescence.[108] Even if the experimenter and instrumentation were held constant, effects proved elusive and capricious. In 1670 the French natural philosopher Adrian Auzout concluded his account of the declination of the magnetic needle in something like despair: if the flux of magnetic matter meandered like a river, "there would be no hopes of finding a regular Hypothesis for that change, forasmuch as it would depend from causes that have no regularity at all in them, as most of the Mutations of Nature are."[109] A reporter to the Paris Académie on the effects of wind on thermometers sighed over the futility of repeating observations and experiments, because the results "destroy each other, and render the facts as difficult to establish, as the causes are to discover."[110] No wonder Hooke, the Royal Society's paid experimenter, did not think it worth the trouble to strive for "Mathematical Exactness" in natural philosophy.[111]

Philosophers taking the first steps toward a metaphysics of nature orderly in effects as well as in causes acknowledged and even admired the variety of nature, but insisted that it was achieved with minimal means. First among the "Rules of Reasoning" Newton appended to Book III of the *Principia* was an admonition to economy in natural philosophy: "We are to admit no more causes of natural things than such as are both true and sufficient to explain their appearances... for Nature is pleased with simplicity and affects not the pomp of superfluous causes."[112] In the late-seventeenth-century debate over the origins of figured stones, Hooke and John Ray objected to the "plastic virtues" explanation advanced by Robert Plot, Martin Lister, and others — not because it was anthropomorphic, but because it was the wrong kind of anthropomorphism. Commenting upon the shell-shaped stones of different colors and hardness, Ray could not bring himself to believe that they were sports of nature, because nature's "infinite prudence" required that all its forms serve some function beyond pure ornamentation.[113] Ray confessed he did not know "what Nature's designs are" in the case of the shell-stones, but he was certain that nature did not frivolously embellish them.

Fontenelle developed this theme in a lighter vein in his 1686 dialogue on the imaginative implications of the new astronomy. Nature was like the opera, in which the sets and machines hidden backstage created "a grand spectacle" (fig. 9.4). But nature carried off this "surprising magnificence" with a remarkable frugality: "There is nothing more beautiful than a great design that one executes at little expense."[114] Here, wed to the parsimony of the good bourgeois, was the princely prodigality of the spectacular fêtes of the young Louis XIV, in which mechanical dragons snorted fire and a crystalline palace materialized from nowhere. Nature still had her marvels, but she manufactured them on the cheap.[115]

By degrees, nature's simplicity and parsimony of means became joined to sobriety of ends. Once again, Fontenelle's writings trace the transformation. In his 1699 history of the Paris Académie, he extolled the pleasures of natural philosophy, in which nature "always following invariable laws, diversified to infinity her effects." But although the variety of nature offered pleasures comparable to those of studying "the astonishing difference of the manners and opinions of Peoples," natural philosophy was nonetheless more edifying than history, for it revealed "the traces of infinite Intelligence and Wisdom" rather than "the irregular effects of the passions and caprices" of men.[116] Still later writings emphasize that "Nature is not capricious"[117] and that "all her works are, so to speak, equally serious."[118] Marvels accorded ill with this sobriety, and serious nature became uniform nature. In 1699 Fontenelle had exclaimed that "the most curious treatises of History could hardly rival Phosphors,"[119] but when in 1730 the academician Charles Dufay succeeded in making many kinds of ordinary stones, from topaz to marble, glow in the dark, Fontenelle happily forecasted an end to other unique and therefore marvelous phenomena.[120] Rendered predictable and commonplace, erstwhile marvels would disappear. The transition from regularity of causes to regularity of effects was complete.

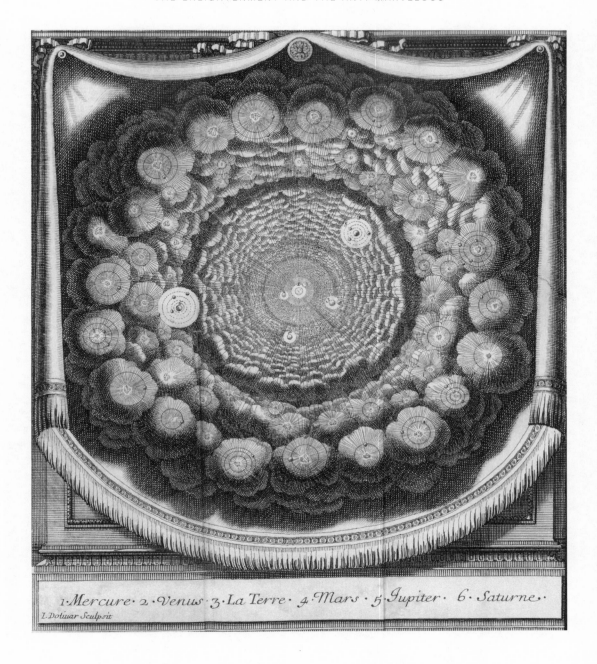

1·*Mercure*· 2·*Venus*· 3·*La Terre*· 4·*Mars*· 5·*Jupiter*· 6·*Saturne*·

I. *Dolivar Sculpsit*

Figure 9.4. The spectacle of the solar system
Bernard de Fontenelle, *Entretiens sur la pluralité des mondes* (Paris, 1686), frontispiece.

Like stage scenery, framed by curtains, the solar system presents an opera-like spectacle to the Marquise and her interlocutor in Fontenelle's *Entretiens* of 1686. In his preface, Fontenelle emphasized that he hoped above all to entertain his readers with astronomical novelties.

353

Fontenelle made a career out of popularizing natural philosophy for an elite urban audience that frequented the salons rather than the court, and it is possible that his metaphysical shift from marvelous to uniform nature paralleled a shift in cultural values from princely magnificence to bourgeois domesticity. Given the prominent role of women in the new intellectual sociability of the salons, it was perhaps strategic that Fontenelle's interlocutor in his dialogue on the plurality of worlds was a charming and quick-witted Marquise — especially since the women in his posthumous literary utopia were not even taught to read and write.[121] Whatever the relationship between the new metaphysics of uniform nature and the salon milieu, there is ample evidence that the metaphysics of uniform nature went hand in hand with a moral and aesthetic sensibility. Just as late-seventeenth-century feminists and other social critics had argued for a new nobility based on utility,[122] early-eighteenth-century savants called for a new natural philosophy to serve utility.[123] The proper attitude for the study of wondrous nature had been disinterested — and pointedly useless — curiosity; the proper attitude for the study of uniform nature was public-spirited utility. Even the frankly recreational public course on experimental physics offered by the Abbé Jean Antoine Nollet in Paris during the 1730s and 1740s was pitched to "the most reasonable curiosity" of Nollet's audience, and Nollet emphasized the utility above the entertainment value of his demonstrations (fig. 9.5).[124] In the Paris Académie Royale des Sciences after circa 1730, the new ethos of utility was linked to an increasing emphasis on the replication of experiments and the stabilization of effects.[125] In order to be made useful, nature had to be made uniform. Phenomena which persisted in rarity or variability or irregularity — meteorites, electrophosphorescence, figured stones — were no longer fit objects of scientific inquiry.

The emphasis on simplicity and uniformity penetrated to natural theology as well. In a 1746 memoir for the Berlin Akademie der Wissenschaften, the natural philosopher Pierre de Maupertuis developed his law of least action in the context of a criticism of natural theological arguments for the existence of God that appealed to "the marvels of nature." Maupertuis was severe with philosophers who had inferred God's existence and above all his perfection from the mere elaborateness of nature's construction: "Ability in execution is not enough; it is necessary that the motive be reasonable. One does not admire but rather blames the Worker; and he would be that much more to blame if he would have employed more skill to construct a machine that would have no utility, or whose effects would be dangerous." Ornament and ingenuity for their own sake were inefficient and therefore inelegant. Maupertuis declared an end to speculation about "the most marvelous objects" of the universe in favor of those "universal rules" of motion that illuminated rather than astonished the mind.[126] Neither God nor nature could any longer be admired for mere intricacy of workmanship without uniformity or utility.

Just as diligent curiosity replaced delighted curiosity during this period, discipline replaced pleasure in natural inquiry.[127] Boyle had delighted in the properties of an artificial phosphor; natural philosophical demonstrations of such substances often deliberately imitated the princely marvels of the court.[128] One of the phosphors that had enchanted Boyle had been distilled from human urine, a proof, Boyle thought, that God had meant to encourage the study of the lowliest things by hiding therein "so glorious and excellent thing, as a self-shining substance."[129] Perhaps thus encouraged, the chemist Homberg attempted to extract from human excrement a clear, odorless oil that would allegedly fix mercury in silver; he ended up with a highly flammable phosphor instead. But his account of his researches otherwise contrasts starkly with Boyle's on phosphors; whereas Boyle emphasized spectacle and eerie beauty, Homberg talks about hard work. In order to gather the raw materials for his distillation, Homberg shut himself up for three months with four strapping young men hired for the purpose, fed them on nothing but "the best Gonesse bread . . . [and] the best champagne," made sure they took long walks every day in the garden to promote regularity, and harvested their feces for his experiment. The only wonder he could muster was understandably feeble: "It is astonishing how the quantity of matter that a man makes at one time, which weighs about ten or twelve ounces, having been dried in a double-boiler, reduces to one ounce."[130]

Yet the metaphysics of uniformity did have its own standards of beauty. Fontenelle's ideal of an idyllic landscape was a fully domesticated countryside where "abundance reigned everywhere, [where] the order and symmetry were admirable."[131] Dissatisfied with fonts developed by "caprice" and "chance," the Paris Académie designed its own typeface, in which the letters were in strict proportion to one another and which "besides [displaying] a pleasant regularity will also be of an advantageous convenience for printers in the future," thus promoting both uniformity and utility.[132] The Scottish philosopher Francis Hutcheson simply turned the early modern aesthetic of variety on its head: "In every part of the world we call beautiful there is a vast uniformity amidst an almost infinite variety." Prototypical for Hutcheson were the pleasures afforded by the fixed periods of planets, seasons, and eclipses, which "charm the astronomer, and make his tedious calculations pleasant."[133] Whereas some forty years earlier the English naturalist Robert Plot had delighted in how nature had orna-

355

9.5.1

9.5.2

Figure 9.5. Wonder tamed
9.5.1–2. Jean-Antoine Nollet, *Leçons de physique expérimentale* (Paris, 1743–48), vol. 1, frontispiece; vol. 4, p. 319.

Like Fontenelle sixty years earlier (fig. 9.4), the Abbé Jean-Antoine Nollet aimed to entertain as
well as to instruct a lay audience with his popular physics lectures. But he drew the line at spectacle: "I have never claimed to make my lessons a spectacle of pure amusement, where one sees
repeated, without plan or choice, a large number of experiments capable only of busying the
eyes." Nollet's experiments were tamed wonders, surprising but controlled, with a didactic point
to make; here he shows an experimental cabinet (fig. 9.5.1) and the transformation of light into
heat (fig. 9.5.2).

357

mented the world with so many varieties of flowers,[134] Hutcheson was certain that the pleasures of botany lay in "what great uniformity and regularity of figure is found in each particular plant, leaf and flower!"

An anti-marvelous aesthetic of art mirrored the anti-marvelous aesthetic of nature, with "verisimilitude" in art corresponding to "order" in nature. Verisimilitude defined what was plausible in a work of literature or the fine arts with reference not so much to historical or natural fact as to the opinions of the audience as to what was possible or proper. Throughout the sixteenth and seventeenth centuries, critics pitted verisimilitude against two apparently opposed but sometimes coincident aesthetic ideals: historicity and the marvelous. The rediscovery and Latin translation of Aristotle's *Poetics* by Giorgio Valla in 1498 burdened poets and dramatists with two apparently incompatible aims: to delight their audiences with marvels, but at the same time to command their belief by lifelike portrayals of characters and events. Paralleling the sixteenth-century fascination with the wonders of art and nature, literary criticism of this period, particularly but not exclusively in Italy, forged an evaluative vocabulary centered on the marvelous.[135] By equating artistry with the depiction of marvels, critics effectively drove wedges between poetry and, on the one hand, the facts of civil and natural history and, on the other, the possibilities dictated by verisimilitude, defined in terms of common opinion.[136] Bacon was expressing a widely held view around 1600 when he described poetry as "feigned history" and "not tied to the laws of matter," therefore licensed to inject "more rareness, and more unexpected and alternative variations" than were strictly consistent with natural or historical fact.[137]

But seventeenth-century debates redrew the triangular opposition of history, verisimilitude, and the marvelous by sometimes allying the historical and the marvelous against the verisimilar. Given the intense interest in natural and artificial marvels displayed in cabinets, described in texts, and sung in ballads, it is not surprising that the discussion focused on the true marvelous. In the controversy unleashed by Pierre Corneille's *Le Cid* (1636), Jean Chapelain of the newly established Académie Française argued that poetry's first obligation was to verisimilitude rather than to truth, particularly extraordinary truths; Corneille countered that the heroic deeds described in dramas like his *Le Cid* might surpass common opinion founded on common experience, but nonetheless belonged in poetry because true.[138] In the course of the first half of the eighteenth century, both history and marvels lost out to verisimilitude. Short of contradicting well-known historical facts, the *Encyclopédie* directed poets and artists to respect the cardinal rule of verisimilitude: "A verisimilar fact is a fact possible in the circumstances where one lays the scene. Fictions without verisimilitude, and events prodigious to excess, disgust readers whose judgment is formed."[139]

But what was possible? Artistic possibility was ruled less by fact than by opinion. The *Encyclopédie* advised painters who strived for "poetic," as opposed to merely "mechanical," verisimilitude to observe the rules of decorum, "to give always to persons the passions fitting to them, following their age, their dignity.... The astonishment of the king must not be that of a man of the people."[140] Fiction was no servile imitation of nature, but its charge to create a more perfect nature did not completely unbridle the imagination. The marvelous, monstrous, and fantastic imaginations betrayed verisimilitude and proportion, risking the "debauchery of genius."[141] Much of the anti-marvelous criticism in art and literature likened styles that had broken with the constraints of verisimilitude to monsters, especially of hybrid species.[142]

Although the critiques of the marvelous in natural philosophy and aesthetics were not identical, they shared a visceral response to the marvelous that mingled disgust and even outrage with incredulity. Echoes of the aesthetic response to breaches of decorum can sometimes be heard in scientific responses to breaches of natural order. Writing in 1769 to the Paris Académie, the naturalist de la Faille, Perpetual Secretary of the Académie de La Rochelle, could hardly contain his indignation at the naturalists, ancient and modern, who had believed that *Concha anatifera adherens* hatched barnacle geese, for this "monstrous generation" was "contrary to the natural order."[143] De la Faille was not much concerned with the specific observations recorded by earlier naturalists like Sir Robert Moray almost a century earlier, who had opened "multitudes of little Shells; having within them little Birds perfectly shap'd."[144] Rather, he was exercised by their blatant disregard for "the natural order" — a disregard which, like the disregard for decorum in art and literature, menaced a well-regulated world with monsters.

The naturalist Georges Leclerc Buffon accorded monsters only three pages of the numerous volumes of his monumental *Histoire naturelle*, since, as he explained, they did not belong to "the ordinary facts of nature."[145] Although his classification of monsters came mostly from Aristotle, the specific examples were somewhat more up-to-date, culled alike from the popular press and the memoirs of learned societies. Buffon's cases, all duly dated and referenced, bear witness to the fact that marvels still circulated at many levels of Enlightenment culture. But Buffon's uncharacteristic brevity is still louder testimony that they had been banished to the margins of natural history and natural philosophy. No longer objects of wonder and desire on the far eastern and western rims of the European world, or objects of power encased in reliquaries or displayed at the courts of princes, marvels had been exiled to the hinterlands of vulgarity and learned indifference. The passions of wonder and curiosity that had singled them out as objects of knowledge had been decoupled and transformed; the culture of princely secrets and magnificence that had imbued them with power and prestige did not survive the turn of the eighteenth century. Nearly a hundred years later, Johann Wolfgang Goethe grasped that wonders belonged in the unfashionable provinces. In his long poem *Hermann und Dorothea*, a village apothecary nostalgically recalled once taking coffee in a garden grotto shimmering with shells and coral, "now of course covered with dust, and half decayed," and replaced by neat white slats and wooden benches. Goethe's apothecary utters what could have been the Enlightenment's epitaph for wonders: "for everything is supposed to be different and tasteful, ... Everything is simple and smooth."[146]

The Wistful Counter-Enlightenment

Who or what killed off wonders among the learned and the powerful in the Enlightenment? Goethe's contemporary and sometime-friend Friederich Schiller implied that the science that had stolen nature's soul and enslaved her to the law of gravitation was the guilty party:

> Feeling not the joy she bids me share,
> Ne'er entranc'd by her own majesty,
> Knowing her own guiding spirit ne'er,
> Ne'er made happy by *my* ecstasy,
> Senseless even to her Maker's praise,
> Like the pendule-clock's dead, hollow tone,
> Nature Gravitation's law obeys
> Servilely, — her Godhead flown.[147]

In its broad outlines, this accusation has been endlessly repeated and elaborated ever since. Charles Dickens added utilitarian ethics and statistics to science in his novel *Hard Times* (1854), in which "wonder, idleness, and folly" are Gradgrind's antonyms to facts and industry. Max Weber threw in secularization when he coined the oft-quoted phrase "disenchantment of the world" for the post-seventeenth-century rationalization of nature and society.[148] Mingling in almost equal measure elegy and condescension for an "enchanted" past abundant in wonder and wonders, this tradition might be called the wistful Counter-Enlightenment.

This persistent tradition has greatly distorted the historical understanding of wonder as both passion and object. Its nostalgia for an age of wonders, supposedly snuffed out by an age of reason, is rooted in an image of Enlightenment as the cultural and intellectual analogue of the transition from childhood to adulthood. To believe in wonders and to experience wonder are on this interpretation akin to the lost pleasures of childhood, which depend on an ignorance whose euphemism is innocence. Hence the miscellaneous quality of what can count as a wonder in this tradition — the legendary and the magical, the grotesque and the uncanny, folklore and fairy tales, alchemy and astrology: all are wondrous by the negative and anachronistic criterion that no educated adult now credits them.[149] To outgrow wonders is to mature into rationality, a process that is, for this tradition of cultural criticism, as sadly irreversible as adulthood. It need hardly be pointed out that the wistful Counter-Enlightenment is in this respect simply the photographic negative of the Enlightenment, accepting the familiar opposition of the rational (or the scientific) and the marvelous with the minor modification of a reversed valuation of its poles. Immanuel Kant was, after all, defending enlightenment when he defined it as "man's release from his self-incurred tutelage."[150]

But it was neither rationality nor science nor even secularization that buried the wondrous for European elites. Enlightenment savants did not embark on anything like a thorough program to test empirically the strange facts collected so assiduously by their seventeenth-century predecessors or to offer natural explanations for them. The philosophical project of naturalizing marvels arguably reached its zenith in the early seventeenth century, and the members of late-seventeenth- and early-eighteenth-century scientific societies were considerably less skeptical about marvelous reports of barnacle geese or carbuncles than Emperor Frederick II had been in the thirteenth century. Leading Enlightenment intellectuals did not so much debunk marvels as ignore them. On metaphysical, aesthetic, and political grounds, they excluded wonders from the realm of the possible, the seemly, and the safe. Protestant and Catholic clergymen aided and abetted philosophers and naturalists in banishing marvels and, in practice if not in precept, miracles as well. The quiet exit of demons from respectable theology coincides in time and corresponds in structure almost exactly with the disappearance of the preternatural in respectable natural philosophy. Intermediate between the natural and supernatural, both demons and marvels were not only unruly but usurping. Demons and enthusiasts counterfeited God's signs and wonders in order to misappropriate the wonder due to God; Boyle and others feared that too great an admiration for nature's wonders would lead to the same idolatrous outcome. There were thus strong reasons for intellectual elites of several stripes to suppress wonders and revile wonder, but majuscule Reason, scientific or otherwise, did not figure prominently among them.

Nor did some end of cultural childhood. The wonders wielded by medieval and Renaissance princes and feared by seventeenth-century civil and ecclesiastical authorities were not the charming fairy tales devised by Charles Perrault or the fabulous beasts of folklore and mythology. There was nothing innocent about the portents hawked in broadsides, nothing naive about the explanations advanced by preternatural philosophers, nothing childish about the strange facts reported in the *Philosophical Transactions of the Royal Society of London* or the *Histoire et Mémoires de l'Académie Royale des Sciences*. Wonders could and did sometimes entertain, but they could also awe and terrify. Even when they delighted, wonders were condensations of power — at their strongest, the power of God to turn the world upside down; at their mildest, the power of things to captivate the attention and send "animal spirits" rushing to the brain. Depending on context, wonders could command veneration, loyalty, fascination, or insatiable curiosity. Whatever its entail, the passion of wonder was a potent one, and also potentially dangerous. The unprecedented diffusion of marvels in texts and things in the early modern period broke the monopoly of princes and the church on the power of wonder; the response of intellectuals after 1660, especially those for whom the memory of civil and religious disturbances was still vivid, was to disarm a weapon that had fallen into the wrong hands. They did not view the neutralization of wonder as the putting aside of childish things.

The passion of wonder did not wholly disappear from the edicts of the Republic of Letters in the Enlightenment, but it shifted its objects and altered its texture almost beyond recognition. Natural theologians exclaimed over the exquisite workmanship of an insect's wings or the human eye, but not over that of monstrosities or kidney stones in the shape of crosses or stars. When Kant wrote that "[t]wo things fill the mind with ever new and increasing wonder and awe [*Bewunderung und Ehrfurcht*], the more often and the more seriously reflection concentrates upon them," he had in mind the objects of astronomy and of ethics.[151] For Kant these were the foremost examples of absolute regularity, poles apart from Descartes' "surprises of the soul." Kant had only contempt for the wonder of the *Wunderkammer*, "a taste for all that is *rare*, little though its inherent worth otherwise might be": this "spirit of minutiae" came close to being "the opposite of the sublime."[152] The true sublime, evoked by jagged mountain peaks and crashing thunder, was, like wonder, admixed with "astonishment [*Verwunderung*] amounting almost to terror, the awe and thrill of devout feeling." But in contrast to the wonder inspired by breaches in the natural order, and also in contrast to Kant's conception of the merely beautiful, the sublime stopped short of both terror and also the aesthetic freedom of play. The true sublime was, Kant claimed, superior to both in that it exercised "a law-ordained function, which is the genuine characteristic of human morality, where reason has to impose its dominion upon sensibility." The sublime turned the imagination, so suspiciously susceptible to wonder and fertile in wonders for early Enlightenment thinkers, into "an instrument of reason."[153]

Kant's sublime in effect made wonder an expression of the natural and moral order of laws. From the High Middle Ages to the Enlightenment, wonders had defined the order of nature by marking its limits; wonder had been the peculiarly cognitive passion that registered the breach of boundaries. Despite the vast changes in European culture witnessed by the six centuries spanned by this study, one strand remained unbroken, threading its way through the most diverse ways of acting, thinking, and feeling. For Voltaire as for Aquinas, for the Burgundian princes as for portent-struck peasants, the natural order was also a moral order in the broad and somewhat old-fashioned sense of moral as all that pertains to the human, from the political to the aesthetic. Hence the aberrations of nature were always charged with moral meaning, whether as warnings from an angry God, sports of a playful nature, or blemishes in the uniformity of the universe. Even in his transmutation of errant wonder into the law-abiding sublime, Kant paid tribute to the duality of the orders they bounded: "the starry heavens above me and the moral law within me."[154]

Epilogue

Do wonder and wonders still seize us? In some sense we live in a world saturated with wonders. The cultivated admire the wonders displayed in science museums and explained in books: volcanic eruptions, huge geodes, meteor showers, magnetism. The not-so-cultivated admire the wonders reported in the *Weekly World News*: monstrous births, UFOs, prodigious feasts and fasts. Depending on the company they keep, some wonders are respectable and others are disreputable; but none threatens the order of nature and society. Scientists have yet to explain many, perhaps most, wonders, but they subscribe to an ontology guaranteeing that all are in principle explicable. If the first criterion for distinguishing respectable from disreputable marvels is whether they are real, the second is whether there are explanations to reassure us that apparent exceptions only confirm nature's laws. In practice, the second criterion often decides the first. As for the social order, pomp and circumstance still accompany the powerful, but no collection of monsters, birds of paradise, and figured stones confirms their might. Even portents have been tamed: when hundreds of television viewers discerned the face of Jesus in a broadcast photograph of a nebula, no one anticipated riots.

Yet there are some striking continuities between earlier and contemporary responses to wonders. The tabloids sold in grocery stores, like sections of the *Guinness Book of Records*, contain many of the wonders in early modern broadsides. Indeed, some of these tabloid wonders so closely duplicate seventeenth-century oddities — a stockroom clerk who changes sex, a baby who sprouts a gold tooth — that one suspects their authors of pillaging the *Philosophical Transactions* and the *Journal des Sçavans*. If tone is a reliable guide to response, the tabloid wonders evoke neither the horror of portents nor the curiosity of strange facts, but rather the pleasure of natural sports. How to tell if your neighbors are extraterrestrials? Observe whether they mow their lawn on national holidays, advises one article. On closer inspection, the *Weekly World News* and its like are the descendants of the *Athenian Gazette*, rather than of Lycosthenes' *Chronicon* or the *Philosophical Transactions*.

Whether these wonders are real or not bears only obliquely on the pleasure they afford. Belief is a continuum, and although wonders must be real to evoke horror, they need only be well imagined to give pleasure. As in medieval romances and travelogues, some measure of verisimilitude is required for any intense response. In order to delight in them, we must be able to see in the mind's eye the juggling automata of the romance, the two-headed goat of the broadside, or the Martian of the tabloid. Illustrations, whether paintings, woodcuts, or photographs, do not so much prove the existence of marvels as help us to imagine them. Tabloid wonders exploit the paradox of highly implausible happenings reported with a show of journalistic rectitude. Some wonders of science fiction are of this sort: extraordinary settings described with much circumstantial detail. But most works of science fiction make no such pretense. Their pleasures spring from the strangeness of the counter-reality — an alternative nature, technology, or culture — and the richness of the literary depiction. The world of wonders created in these novels must be self-consistent and dense enough to sustain fictional life, but not real-life belief. As in the case of tabloid wonders, certain conventions live on. Just as medieval and early modern writers placed their wonders at the rim of the known world, so authors of science fiction often situate their stories outside known space and time; and just as Mandeville made islands the unit of experimentation with different natures and cultures, so science fiction writers often take planets as the sites of alternative worlds.

Such pleasurable wonders are frankly popular; indeed, they are so popular as to border on the undignified. In this they differ from earlier wonders hoarded by princes and studied by naturalists. "Serious" journalists scorn the tabloids, "serious" novelists shun science fiction, and their readers often follow suit. The same goes for the wonders of art: most objects that once adorned the *Wunderkammern* could now qualify for an exhibition of high kitsch. Once the cherished possessions of princes, the cherrystones carved in a hundred facets, the tableaux of seashells and coral, the ornamented ostrich eggs, and the delicate towers of turned ivory have sunk, at least in the view of art critics, to the cultural level of paintings on black velvet. Only if presented as historical artifacts — for example, in exhibitions on the evolution of collecting — do these objects command a place in large metropolitan museums. To display them without ironic or historical brackets is the mark of the provincial. To take an example from a country in which the center strongly dominates the periphery, consider the fate of the Habsburg collection in Austria. Its alabaster bowls and paintings are now displayed in the renowned Kunsthistorisches Museum of Vienna; its gilded rhinoceros horns and *Handsteine* have been relegated to the Schloss Ambras in provincial Innsbruck. The same is true of naturalia: the grand Naturhistorisches Museum in Vienna boasts geodes and skeletons of dinosaurs; the Haus der Natur in Salzburg still displays monsters pickled in brine.

Among the learned, wonders and wonder are often objects of mild condescension. They belong in the classroom or the museum. Indeed, wonder and wonders define the professional intellectual by contrast: seriousness of purpose, thorough training, habits of caution and exactitude are all opposed to a wonder-seeking sensibility. One may enter a scientific career through wonder, but one cannot persist in wonder, at least not in public before one's peers. William James thought science would be renewed by more attention to the "dust-cloud of exceptional observations," but his colleagues regarded his own investigations of spiritualism with skepticism and mockery. Investigators like William Corliss still compile enormous Baconian histories of strange phenomena, but they must resort to private publication.

Science and technology are fertile in their own wonders, but their spectacles are staged on television and in museums, planetaria, and Omnimax theaters, not at professional meetings. Even the museum wonders are regular irregularities — recurring meteor showers rather than unique monsters — and all are scrupulously explained. They are edifying wonders, intended to instruct as well as entertain. The wonder they evoke redounds to the greater glory of the science that explains them or the technology that creates them, translating into popular support for both science and technology: NASA, that fount of scientific and technological wonders, is perhaps the only governmental agency ever to receive voluntary contributions. But scientists and engineers do not wield the power of wonders in the manner of princes like Philip the Good, religious sects like the Jansenists, or even naturalists like Girolamo Cardano. However heaped with honors, only one or two scientists (Newton and Einstein) have commanded the full-strength wonder of awe. Like Cardano, scientists may become the public's guides to natural wonders, but professional taboos prevent them from claiming, with Cardano, to be the greatest wonders of their age.

Whether vulgar or edifying, an odor of the popular now clings to wonders. They still please and instruct, as they have since the twelfth century, but they no longer buttress regimes, subvert religions, or reform learning. The discontinuities of the tradition are as striking as the continuities. If the screaming headlines of the latest tabloid recall the

367

heavy black letter of an early modern broadside, their readerships none-theless diverge. One cannot imagine a diarist of the social and literary stature of Samuel Pepys — Leonard Woolf, say, or Edmund Wilson — faith-fully recording monsters he read about or saw. To be a member of a mod-ern elite is to regard wonder and wonders with studied indifference; enlightenment is still in part defined as the anti-marvelous. But deep inside, beneath tasteful and respectable exteriors, we still crave wonders. Sitting wide-eyed under a planetarium sky or furtively leafing through the *Weekly World News* in the checkout line, we wait for the rare and extraor-dinary to surprise our souls.

Photo Credits

1.1. Pierpont Morgan Library, New York.

1.2.1. By permission of the Dean and Chapter of Hereford Cathedral and the Hereford Mappa Mundi Trust, UK.

1.2.2. By permission of the Dean and Chapter of Hereford Cathedral and the Hereford Mappa Mundi Trust, UK.

1.3.1. British Library, London.

1.3.2. Bibliothèque Nationale, Paris.

1.4.1. Bodleian Library, Oxford.

1.4.2. Bibliothèque Nationale, Paris.

1.5.1–2. Bodleian Library, Oxford.

1.6.1. Fitzwilliam Museum, Cambridge.

1.6.2. Bibliothèque Nationale, Paris.

1.7. Centre Guillaume-le-Conquérant, Bayeux.

1.8.1. Courtesy Rosamond Purcell.

1.8.2. By permission of Ministero per i Beni Culturali e Ambientali, Florence.

1.9.1–2. National Library of Ireland, Dublin.

2.1.1. Bibliothèque Nationale, Paris.

2.1.2. Bibliothèque Nationale, Paris.

2.1.3. Bibliothèque Nationale, Paris.

2.2.1. Halberstadt Cathedral Treasure, Halberstadt.

2.2.2. Bodleian Library, Oxford.

2.3.1. General Research Division, New York Public Library, Astor, Lenox, and Tilden Foundations.

2.3.2. Bibliothèque Nationale, Paris.

2.3.3. Bibliothèque Nationale, Paris.

2.4.1–2. Art and Architecture Collection, Miriam and Ira D. Wallach Division of Art, Prints and Photography. The New York Public Library, Astor, Lenox, and Tilden Foundation.

2.5.1. Galleria dell'Accademia, Venice.

2.5.2. Marcello Bertoni Fotografo, Florence.

2.6.1. Art and Architecture Collection, Miriam and Ira D. Wallach Division of Art, Prints and Photography. The New York Public Library, Astor, Lenox, and Tilden Foundation.

2.6.2. Courtesy Rosamond Purcell.

2.7. Bibliothèque Nationale, Paris.

2.8.1–2. Herzog August Bibliothek, Wolfenbüttel.

2.9.1–2. Bibliothèque Nationale, Paris.

2.10. Cleveland Museum of Art, Cleveland.

2.11. Bibliothèque Nationale, Paris.

3.1. Bibliothèque Royale Albert I, Brussels.

4.1.1. Biblioteca General de la Universidad, Valencia.

4.2. Biblioteca Medicea Laurenziana, Florence.

4.3.1–3. By Permission of the Houghton Library, Harvard University, Cambridge, MA.

4.4.1. Rare Books and Manuscripts. The New York Public Library, Astor, Lenox, and Tilden Foundation.

4.4.2. By Permission of the Houghton Library, Harvard University, Cambridge, MA.

4.5.1–3. By Permission of the Houghton Library, Harvard University, Cambridge, MA.

4.6.1–2. By Permission of the Houghton Library, Harvard University, Cambridge, MA.

5.1. By Permission of the Houghton Library, Harvard University, Cambridge, MA.

5.2.1. Biblioteca Marciana, Venice.

5.2.2. Staatsbibliothek, Munich.

5.2.3. By Permission of the Houghton Library, Harvard University, Cambridge, MA.

5.3.1. Wellcome Institute for the History of Medicine, London.

5.3.2. By Permission of the Houghton Library, Harvard University, Cambridge, MA.

5.4. By Permission of the Philadelphia Museum of Art, Philadelphia, PA. Purchased: Smith Kline Beecham Corporation Fund.

5.5.1–2. By Permission of the Houghton Library, Harvard University, Cambridge, MA.

5.6.1–3. Kunsthistorisches Museum, Vienna.

5.7.1–2. British Library, London.

5.8.1–4. Niedersächsische Staats- und Universitätsbibliothek Göttingen.

5.9. Skoklosters Slott, Skokloster, Sweden.

6.1. Department of Special Collections, University of Chicago Library.

6.2.1–2. Niedersächsische Staats- und Universitätsbibliothek Göttingen.

6.3.1–2. Niedersächsische Staats- und Universitätsbibliothek Göttingen.

7.1.1–3. Uppsala University, Sweden.

7.2.1. Grünes Gewölbe, Dresden.

7.2.2. Research Libraries, New York Public Library.

7.3.1–2. By Permission of the Houghton Library, Harvard University, Cambridge, MA.

7.4.1. Museo del Prado, Madrid.

7.4.2. Historisches Museum, Frankfurt am Main.

7.5.1. Ulmer Museum, Ulm, Germany.

7.5.2. Museum of Comparative Zoology, Harvard University, Cambridge, MA.

7.6.1. Sammlungen Schloss Ambras/Kunsthistorisches Museum, Vienna.

7.6.2. Grünes Gewölbe, Dresden.

7.6.3. Sammlungen Schloss Ambras/Kunsthistorisches Museum, Vienna.

7.7.1. Sammlungen Schloss Ambras/Kunsthistorisches Museum, Vienna.

7.7.2. Staatliche Museen, Kassel, Germany.

7.7.3. Sammlungen Schloss Ambras/Kunsthistorisches Museum, Vienna.

7.8.1. Bibliothèque du CNAM, Paris.

7.8.2. Herzog Anton Ulrich-Museum, Braunschweig, Germany

7.9.1–2. Department of Special Collections, University of Chicago.

7.10. Kunsthistorisches Museum, Vienna.

8.1.1–6. Musée de Louvre, Paris.

9.1.1–2. Niedersächsische Staats- und Universitätsbibliothek Göttingen.

9.1.3. Department of Special Collections, University of Chicago .

9.2.1–2. Bibliothèque Nationale, Paris.

9.3. Bibliothèque Nationale, Paris.

9.4. By Permission of the Houghton Library, Harvard University, Cambridge, MA.

9.5.1–2. By Permission of the Houghton Library, Harvard University, Cambridge, MA.

Abbreviations

DSB	=	*Dictionary of Scientific Biography*
HARS	=	*Histoire de l'Académie Royale des Sciences*
JS	=	*Journal des Sçavans*
MARS	=	*Mémoires de l'Académie Royale des Sciences*
PT	=	*Philosophical Transactions of the Royal Society of London*

Notes

PREFACE

1. In Michel Foucault, *Foucault Live (Interviews, 1966–84)*, trans. John Johnston, ed. Sylvère Lotringer (New York: Semiotext[e], 1989), pp. 198–99.

2. Katharine Park and Lorraine Daston, "Unnatural Conceptions: The Study of Monsters in Sixteenth- and Seventeenth-Century France and England," *Past and Present* 92 (1981), pp. 20–54. As usual, the French were in the vanguard, especially Georges Canguilhem, *La Connaissance de la vie* (Paris: Librairie Vrin, 1965), Jean Céard's remarkable edition of Ambroise Paré, *Des Monstres et prodiges* [1573] (Geneva: Librairie Droz, 1971), and his *La Nature et les prodiges: L'Insolite au XVIe siècle en France* (Geneva: Librairie Droz, 1977).

3. Francis Bacon, *The Advancement of Learning* [1605], in James Spedding, Robert Leslie Ellis, and Douglas Denon Heath (eds.), *The Works of Francis Bacon*, 14 vols. (London: Longman, Green, Longman, & Roberts, 1857–74), vol. 3, p. 266. Unless otherwise noted, all references are to this edition.

INTRODUCTION

1. Robert Boyle, "Some Observations About Shining Flesh, Both of Veal and Pullet, and That without any Sensible Putrefaction in those Bodies" [1672], in *The Works of the Honourable Robert Boyle*, 6 vols., ed. Thomas Birch, facsimile reprint with an Introduction by Douglas McKie (1772; Hildesheim: Georg Olms, 1965–66), vol. 3, pp. 651–55, on p. 651.

2. René Descartes, *The Passions of the Soul* [1642], 70, trans. Stephen H. Voss (Indianapolis: Hackett, 1989), p. 56.

3. F. Bacon, *The Advancement of Learning* [1605], in his *Works*, vol. 3, p. 330.

4. Throughout this book we use the term "naturalists" in roughly the sense of Aristotle's "hoi physikoi," to refer to those who engage in the systematic study of nature, including natural history and natural philosophy, as well as disciplines like astronomy, optics, and, especially, medicine, which had been autonomous since antiquity. We hope thereby to avoid both cumbersome listings of different groups and the anachronism of the term "scientist," with its misleading associations of professionalization.

5. Aristotle, *Metaphysics*, 1.2, 982b10–18.

6. Jacques Le Goff, "The Marvelous in the Medieval West," in idem, *The Medieval Imagination*, trans. Arthur Goldhammer (Chicago: Chicago University Press, 1988), pp. 27–44, on p. 27 (our emphasis). The same assumptions define the essays in Michel Meslin (ed.), *Le Merveilleux: L'Imaginaire et les croyances en Occident* (Paris: Bordas, 1984), as well as, to some degree, the otherwise very useful studies of Claude Kappler, *Monstres, démons et merveilles à la fin du Moyen Age* (Paris: Payot, 1988), and Daniel Poirion, *Le Merveilleux dans la littérature française du Moyen Age* (Paris: Presses Universitaires de France, 1982). A signal exception to this tendency is Caroline Walker Bynum's fine essay, "Wonder," *American Historical Review* 102 (1997), pp. 1–26, which she allowed us to see in draft form. Le Goff's more recent article, "Le Merveilleux scientifique au Moyen Age," in Jean–François Bergier (ed.), *Zwischen Wahn, Glaube und Wissenschaft: Magie, Astrologie, Alchemie und Wissenschaftsgeschichte* (Zurich: Verlag der Fachvereine, 1988), pp. 87–113, despite the anachronism of its title, defines its topic more carefully.

7. Ibid.

8. *Dictionnaire historique de la langue française*, 2 vols., ed. Alain Rey (Paris: Dictionnaires Le Robert, 1992), s.v. "merveille"; and A. Walde and J.B. Hoffmann, *Lateinisches etymologisches Wörterbuch*, 3 vols., 3rd ed. (Heidelberg: Carl Winter Universitätsverlag, 1938–56), s.v. "mirus."

9. Hjalmar Frisk, *Etymologisches Wörterbuch der griechischen Sprache*, 2 vols. in 3 (Heidelberg: Carl Winter Universitätsverlag, 1960–66), s.v. "thauma."

10. Carlo Battisti and Giovanni Alessio, *Dizionario etimologico italiano*, 5 vols. (Florence: G. Barbèra, 1950–57), s.v. "meraviglia"; Walther von Wartburg, *Französisches etymologisches Wörterbuch*, 24 vols. (Basel: Helbing & Lichtenhahn, Zbinden Druck und Verlag AG/Tübingen: Paul Siebeck, 1940–83), s.v. "merveille"; Jacob and Wilhelm Grimm, *Deutsches Wörterbuch*, 33 vols. (Leipzig: S. Hirzel, 1854–1984; repr. Munich: Deutscher Taschenbuch Verlag, 1991), s.v. "Wunder"; and *The Oxford English Dictionary*, 20 vols., 2nd ed. (Oxford: Clarendon Press, 1989), s.v. "Wonder," "Marvel."

11. Frances A. Yates, "The Hermetic Tradition in Renaissance Science," in Charles S. Singleton (ed.), *Art, Science, and History in the Renaissance* (Baltimore: Johns Hopkins University Press, 1968), pp. 255–74. See also her *Giordano Bruno and the Hermetic Tradition* (Chicago: University of Chicago Press, 1964).

12. Cf. Le Goff, "The Marvelous," in idem, *The Medieval Imagination*.

CHAPTER ONE: THE TOPOGRAPHY OF WONDER

1. Gervase of Tilbury, *Otia imperialia ad Ottonem IV imperatorem,* in Gottfried Wilhelm Leibniz (ed.), *Scriptores rerum Brunsvicensium....* (Hanover: Nicolaus Foerster, 1707), p. 960. On the editions of this text and on Gervase himself, see Jacques Le Goff, "Une Collecte ethnographique en Dauphiné au début du XIIIe siècle," in his *L'Imaginaire médiéval* (Paris: Gallimard, 1985), pp. 40–56. A partial edition, with notes, appears in Felix Liebrecht (ed.), *Des Gervasius von Tilbury Otia imperialia in einer Auswahl neu herausgegeben und mit Anmerkungen begleitet* (Hanover: Carl Rümpler, 1856).

2. Gervase of Tilbury, *Otia imperialia*, p. 960.

3. On the relationship between miracles and marvels, see Chapter Three.

4. Gervase of Tilbury, *Otia imperialia*, p. 960; cf. Augustine, *City of God* [*De civitate Dei*], 21.4, trans. Henry Bettenson, with an introduction by John O'Meara (London: Penguin, 1984), pp. 969–70.

5. On this tradition see James S. Romm, *The Edges of the Earth in Ancient Thought: Geography, Exploration, and Fiction* (Princeton: Princeton University Press, 1992), pp. 92–93; and Elizabeth Rawson, *Intellectual Life in the Late Roman Republic* (Baltimore: Johns Hopkins University Press, 1985), Ch. 17. The principal texts are edited in A. Giannini, *Paradoxographorum graecorum reliquiae* (Milan: Istituto Editoriale Italiano, 1966). We are grateful to Glenn Most for help in understanding the complexities of this material.

6. Gervase of Tilbury, *Otia imperialia*, pp. 962 and 986.

7. In addition to the *dracs*, see, e.g., ibid., pp. 963 and 965, on a nut tree that fruited and flowered simultaneously and a local refectory that repelled flies.

8. Gerald of Wales, *Giraldus Cambrensis in Topographia Hibernie*, Preface, ed. John J. O'Meara, *Proceedings of the Royal Irish Academy* 52, sect. C, no. 4 (Dublin, 1949), p. 114.

9. Mandeville's work survives in over 300 manuscripts, in languages ranging from Czech to Irish: Christian Deluz, *Le Livre de Jehan de Mandeville: Une "géographie" au XIVe siècle* (Louvain–La–Neuve: Institut d'Etudes Médiévales de l'Université Catholique de Louvain, 1988), pp. 370–82. For vernacular versions of Gervase of Tilbury, see Le Goff, "Une Collecte," in idem, *L'Imaginaire médiéval*, p. 44, n. 2; on Gerald of Wales, Robert Bartlett, *Gerald of Wales, 1146–1223* (Oxford: Clarendon Press, 1982), p. 214. The fundamental study of medieval and early modern European writing on travel and topography is Mary B. Campbell, *The Witness and the Other World: Exotic European Travel Writing, 400–1600* (Ithaca, NY: Cornell University Press, 1988).

10. Ronald Latham, "Introduction," in *The Travels of Marco Polo*, trans. Ronald Latham (Harmondsworth: Penguin Books, 1958), p. 26.

11. Ranulph Higden, *Polychronicon*, 1.34, in *Polychronicon Ranulphi Higden monachi cestrensis, together with the English Translation of John Trevisa and an Unknown Writer of the Fifteenth Century*, 9 vols., ed. Churchill Babington, Rolls Series 41 (London, 1865–86), vol. 1, p. 361. See in general, Romm, *Edges*, and Claude Sutto, "L'Image du monde à la fin du Moyen Age," in Guy-H. Allard (ed.), *Aspects de la marginalité au Moyen Age* (Montreal: L'Aurore, 1975), pp. 58–69.

12. Gerald of Wales, *Topographia Hibernie*, Preface, p. 119; translation in Gerald of Wales, *History and Topography of Ireland*, trans. John J. O'Meara (Harmondsworth: Penguin, 1982), p. 31. In this and in following quotations, we have occasionally altered the translation slightly to correspond more closely to the Latin text.

13. See Rudolf Wittkower, "Marvels of the East: A Study in the History of Monsters," *Journal of the Warburg and Courtauld Institutes* 5 (1942), pp. 159–97; Romm, *Edges*, Ch. 3; John Block Friedman, *The Monstrous Races in Medieval Art and Thought* (Cambridge, MA: Harvard University Press, 1981); Claude Lecouteux, *Les Monstres dans la littérature alle-*

mande du Moyen Age: Contribution à l'étude du merveilleux médiéval, 3 vols. (Göppingen: Kümmerle Verlag, 1982), esp. vol. 1, pp. 237–54; Bruno Roy, "En marge du monde connu: Les races de monstres," in Allard, *Aspects,* pp. 71–81; Campbell, *Witness,* Ch. 2; and Kappler, *Monstres,* Ch. 1. On the Alexander tradition, see Lloyd L. Gunderson, "Introduction," in *Alexander's Letter to Aristotle about India,* trans. Gunderson (Meisenheim am Glan: Anton Hain, 1980), and in general George Cary, *The Medieval Alexander,* ed. D.J.A. Ross (Cambridge: Cambridge University Press, 1967), esp. pp. 14–16.

14. On the Arabic tradition, see M. Tawfiq Fahd, "Le Merveilleux dans la faune, la flore et les minéraux," in Mohamed Arkoun et al., *L'Étrange et le merveilleux dans l'Islam médiéval: Actes du colloque tenu au Collège de France à Paris, en mars 1974* (Paris: Editions J.A., 1978), pp. 117–35; and Aziz al-Azmeh, "Barbarians in Arab Eyes," *Past and Present* 134 (1992), pp. 3–18.

15. See *Alexander's Letter* and in general Campbell, *Witness,* pp. 68–69; Friedman, *Monstrous Races,* pp. 145–46.

16. *Liber monstrorum,* ed. and trans. Franco Porsia (Bari: Dedalo Libri, 1976), p. 126; see also Friedman, *Monstrous Races,* pp. 149–53; and Ann Knock, "The 'Liber monstrorum': An Unpublished Manuscript and Some Reconsiderations," *Scriptorium* 32 (1978), pp. 19–28.

17. *Liber monstrorum,* 1.40, p. 194.

18. *Marvels of the East: A Full Reproduction of the Three Known Copies,* ed. Montague Rhodes James (Oxford: Roxburghe Club, 1929). See in general, Campbell, *Witness,* Ch. 2. For a discussion of the work, its sources, and the relevant manuscripts and illustrations, see Patrick McGurk and Ann Knock, "The Marvels of the East," in Patrick McGurk et al., *An Eleventh-Century Anglo-Saxon Illustrated Miscellany: British Library Cotton Tiberius B.V Part I, Together with Leaves from British Library Cotton Nero D.II* (Copenhagen: Rosenkilde and Bagger, 1983), pp. 88–103. Note that the *Liber monstrorum* received the title "Marvels of the East" from its modern editors.

19. Gerald of Wales, *History and Topography,* 3.121, p. 124. On Gerald's later revisions of this work, see Bartlett, *Gerald,* pp. 212–13. On his participation in the English pacification of Ireland, see O'Meara, "Introduction," in Gerald of Wales, *History and Topography,* pp. 12–14, and Bartlett, *Gerald,* pp. 15–16 and in general Chs. 1 and 6.

20. Gerald of Wales, *History and Topography,* 3.109, p. 118.

21. Ibid., 1.11, p. 41. On marvels, miracles, and the natural order in Gerald, see Bartlett's excellent discussion in *Gerald of Wales,* Chs. 4–5; on hybrids and horror in late medieval and early modern thought, see Arnold I. Davidson, "The Horror of Monsters," in James J. Sheehan and Morton Sosna (eds.), *The Boundaries of Humanity: Humans, Animals, Machines* (Berkeley: University of California Press, 1991), pp. 36–67.

22. Gerald of Wales, *History and Topography,* 2. intro., p. 58. See Urban T. Holmes, "Gerald the Naturalist," *Speculum* 11 (1936), pp. 110–21; and Antonia Grandsen, "Realistic Observation in Twelfth-Century England," *Speculum* 47 (1973), pp. 29–51, esp. pp. 50–51.

23. See Michel Mollat, *Les Explorateurs du XIIIe au XVIe siècle: Premiers regards sur des mondes nouveaux* (Paris: J.-C. Lattès, 1984); Kappler, *Monstres*, Chs. 2–3; and especially Campbell, *Witness*, Ch. 3.

24. Cited in Kappler, *Monstres*, p. 51.

25. Polo, *Travels*, pp. 288–89; and idem, *Milione. Le divisament dou monde: Il Milione nelle redazioni toscana e franco–italiana*, ed. Gabriella Ronchi (Milan: Arnaldo Mondadori, 1982), 180, p. 579 (Franco–Italian version).

26. Latham, "Introduction," in ibid., pp. 16–19 and 24–25.

27. For an analysis and repertory of these themes, see Douglas Kelly, *The Art of Medieval French Romance* (Madison: University of Wisconsin Press, 1992), Ch. 5; Edmond Faral, "Le Merveilleux et ses sources dans les descriptions des romans français du XIIe siècle," in his *Recherches sur les sources latines des contes et romans courtois du Moyen Age* (Paris: Honoré Champier, 1913), pp. 307–88; and Poirion, *Le Merveilleux*, Chs. 2–6. On the specific relationship between romance and travel literature, see Campbell, *Witness*, pp. 106–12, and Poirion, *Le Merveilleux*, pp. 105–108.

28. Polo, *Milione*, 1, p. 305 (Franco–Italian version); and Adenet le Roi, *Li Roumans de Cléomadès*, ed. André van Hasselt, 2 vols. (Brussels: Victor Devaux, 1865), vol. 1, p. 1. See Katharine Park, "The Meanings of Natural Diversity: Marco Polo on the 'Division' of the World," in Michael McVaugh and Edith Sylla (eds.), *Text and Context in Ancient and Medieval Science* (Leiden: Brill, 1997), pp. 134–47.

29. The same shift also occurs in the work of Jewish travel writers; compare, for example, the work of the late twelfth-century Petachia of Ratisbon with that of the fifteenth-century Meshullam ben Menahem of Volterra, in Elkan Nathan Adler, *Jewish Travellers in the Middle Ages: Nineteen Firsthand Accounts* (New York: Dover Publications, 1987).

30. *Mandeville's Travels*, 4, ed. M.C. Seymour (London: Oxford University Press, 1968), p. 15. On mirabilia in *Mandeville's Travels*, see Deluz, *Jehan de Mandeville*, pp. 211–33.

31. On European representations of Eastern idolatry, see Michel Camille, *The Gothic Idol: Ideology and Image-Making in Medieval Art* (Cambridge: Cambridge University Press, 1989), pp. 151–64.

32. For example, Kappler, *Monstres*, pp. 20–21, 37 and Ch. 7 (in which the monstrous races are interpreted primarily as expressions of castration anxiety); Lecouteux, *Monstres*, esp. vol. 1, pp. 330–32; and Roy, "En marge." To a large degree, this misconception stems from the tendency to conflate the monstrous races with individual monstrous births, which had a very different and much more sinister cultural meaning; see above.

33. See Peter Stallybrass and Allon White, *The Politics and Poetics of Transgression*, intro. (Ithaca: Cornell University Press, 1986), esp. pp. 17–18.

34. Campbell, *Witness*, pp. 82–84; see also Stephen Greenblatt, *Marvelous Possessions: The Wonder of the New World* (Chicago: University of Chicago Press, 1991), Ch. 2.

35. On the medieval *fortuna* of Pliny, see Charles G. Nauert (Jr.), "Caius Plinius Secundus," in F. Edward Cranz and Paul Oskar Kristeller (eds.), *Catalogus translationum et*

commentariorum: Medieval and Renaissance Latin Translations and Commentaries, 6 vols. to date (Washington, DC: Catholic University of America Press, 1960–1986), vol. 4, pp. 297–422, esp. pp. 302–304; and Arno Borst, *Das Buch der Naturgeschichte: Plinius und seine Leser im Zeitalter des Pergaments* (Heidelberg: Winter, 1995).

36. On marvels in Pliny, see Jean Céard, *La Nature et les prodiges: L'insolite au XVIe siècle* (Geneva: E. Droz, 1977), pp. 12–20.

37. Pliny, *Natural History* [*Historia naturalis*], 7.2.32, 10 vols., Loeb Classical Library (Cambridge, MA: Harvard University Press, 1935–63), vol. 2 (trans. H. Rackham), p. 527.

38. Ibid., 18.1.1, vol. 5 (trans. H. Rackham), p. 189.

39. James of Vitry, *Libri duo, quorum prior Orientalis, sive Hierosolymitanae, alter Occidentalis historiae* (Douai: Balthazar Bellerus, 1597; repr. Westmead, England: Gregg International, 1971), *Historia Orientalis*, 92, pp. 215–16. Among other sources, James was commenting on the injunctions to tolerance in Romans 14.

40. Odoric of Pordenone, *Relatio*, 31, in *Sinica franciscana*, ed. Anastasius van den Wyngaert, vol. 1: *Itinera et relationes fratrum minorum saeculi XIII et XIV* (Quaracchi: Collegium S. Bonaventurae, 1929), pp. 482–83.

41. On the illustrations in this manuscript, see Millard Meiss, *French Painting in the Time of Jean de Berry: The Boucicaut Master* (London: Phaidon, 1968), pp. 42–46 and 116–22 (catalogue of images); and Rudolf Wittkower, "Marco Polo and the Pictorial Tradition of the Marvels of the East," in his *Allegory and the Migration of Symbols* (New York: Thames and Hudson, 1977), pp. 76–81. They are reproduced in full in Henri Omont (ed.), *Livre des merveilles . . . Réproduction des 265 miniatures du manuscrit français 2810 de la Bibliothèque Nationale* (Paris: Berthaud Frères-Catala Frères, [1908]).

42. Cited in Jacques Le Goff, "Le Merveilleux dans l'Occident médiéval," in Arkoun, *L'Étrange*, pp. 61–79, on p. 67.

43. Jordan of Sévérac, *Mirabilia descripta, or Wonders of the East* [c. 1330], trans. Henry Yule (New York/London: Burt Franklin, 1863), p. 55. See Roy, "En marge," pp. 77–79.

44. Augustine, *City of God*, 21.4, p. 970. For a comprehensive survey of Augustine's writing on marvels, see Céard, *Nature*, pp. 21–29.

45. Augustine, *City of God*, 21.7, p. 977.

46. Ibid., 21.8, p. 980.

47. Idem, *On Christian Doctrine* [*De doctrina christiana*], 2.16, trans. D.W. Robertson (Jr.) (Indianapolis: Bobbs-Merrill, 1958), pp. 50–51.

48. M.R. James, "Ovidius *De mirabilibus mundi*," in E.C. Quiggin (ed.), *Essays and Studies presented to William Ridgeway* (1913; repr. Freeport, NY: Books for Libraries Press, 1966), p. 297.

49. For an overview of this literature, see Jerry Stannard, "Natural History," in David C. Lindberg (ed.), *Science in the Middle Ages* (Chicago: University of Chicago Press, 1978), pp. 429–60.

50. Richard Barber (trans.), *Bestiary, Being an English Version of the Bodleian Library, Oxford MS. Bodley 764, with all the Original Miniatures Reproduced in Facsimile* (Wood-

bridge: Boydell Press, 1993), pp. 50, 62, 142, 120; cf. Gerald of Wales, *History and Topography*, 1.11 and 1.12–3, pp. 41–42 and 42–43. On bestiaries, see in general Wilma George and Brunsdon Yapp, *The Naming of the Beasts: Natural History in the Medieval Bestiary* (London: Duckworth, 1991); and Florence McCulloch, *Medieval Latin and French Bestiaries*, 2nd ed. (Chapel Hill: University of North Carolina Press, 1962).

51. Joan Evans, *Magical Jewels of the Middle Ages and the Renaissance, Particularly in England* (Oxford: Clarendon Press, 1922), pp. 64–67.

52. Augustine, *City of God*, 21.6, p. 976.

53. E.R. Smits, "Vincent of Beauvais: A Note on the Background of the *Speculum*," in W.J. Aerts, E.R. Smits, and J.B. Voorbij (eds.), *Vincent of Beauvais and Alexander the Great: Studies on the Speculum Maius and its Translation into Medieval Vernaculars* (Groningen: Egbert Forsten, 1986), pp. 1–9. On medieval encyclopedias, see in general Maria Teresa Beonio-Brocchieri Fumagalli, *Le enciclopedie dell'Occidente medioevale* (Turin: Loescher, 1981); Maurice de Gandillac, "Encyclopédies pré-médiévales et médiévales," in *Encyclopédies et Civilisations* = *Cahiers d'Histoire Mondiale* 9/3 (1966), pp. 483–518; and Pierre Michaud-Quantin, "Les petites encyclopédies du XIIIe siecle," in ibid., pp. 580–95. We exclude from this literature more specialized works of philosophy, such as Albertus Magnus's *De animalibus*, which will be discussed in Chapter Three.

54. Bartholomaeus Anglicus, *De proprietatibus rerum*, proemium (n.p., n.publ., 12 June 1488), sig. a1r; see in general M.C. Seymour, "Introduction," in Seymour et al., *Bartholomaeus Anglicus and his Encyclopedia* (Aldershot: Variorum, 1992), esp. pp. 1–17.

55. Thomas of Cantimpré, *De natura rerum: Editio princeps secundum codices manuscriptos*, 6.1, ed. H. Boese (Berlin: Walter de Gruyter, 1973), p. 232.

56. Vincent of Beauvais, *Speculum naturale*, prologue, 3 vols. (n.p., n.d. [Strassburg?: Adolf Rusch?, 1479?], vol. 1, fol. [1r]. See in general Smits, "Vincent of Beauvais"; C. Oursel, "Un exemplaire du 'Speculum Majus' de Vincent de Beauvais provenant de la bibliothèque de Saint Louis," *Bibliothèque de l'École des Chartes* 85 (1924), pp. 251–62; and Michel Lemoine, "L'Oeuvre encyclopédique de Vincent de Beauvais," *Encyclopédies et Civilisations* = *Cahiers d'Histoire Mondiale* 9/3 (1966), pp. 571–79, especially the outline of books of the *Speculum* on pp. 574–75.

57. Ibid., 30.10–11, vol. 3, fol. [120v].

58. E.g., ibid., 30.11; 31.3, vol. 3, fol. [154r].

59. Smits, "Vincent of Beauvais," p. 8; and J.C. Seymour, "Some Medieval French Readers of *De proprietatibus rerum*," *Scriptorium* 28 (1974), pp. 100–103. See also Lynn Thorndike, *A History of Magic and Experimental Science*, 8 vols. (New York: Columbia University Press, 1923–58), vol. 2, pp. 375–76; and idem, "The Properties of Things of Nature Adapted to Sermons," *Medievalia et Humanistica* 12 (1958), pp. 78–83.

60. *Gesta Romanorum*, 175, ed. Hermann Oesterley (Berlin: Weidmann, 1872), p. 575.

61. Bartholomaeus' book proved particularly popular with vernacular readers: by 1309 it had appeared in Italian, and translations followed into French, Provençal, English, and Spanish. All of these circulated in large numbers of manuscripts, making Bartholomaeus'

work one of the favorite books of private fifteenth-century libraries, even before the advent of printing: Donal Byrne, "Rex imago Dei: Charles V of France and the 'Livre des propriétés des choses'," *Journal of Medieval History* 7 (1981), pp. 97–113, on pp. 98–99. The French translation (1372) was the work of Jean Corbechon, who also translated the compilation of travel narratives known as the *Livre des merveilles du monde*, at the request of King Charles V.

62. Byrne, "Rex imago Dei," pp. 99 and 102.

63. Idem, "The Boucicaut Master and the Iconographical Tradition of the 'Livre des propriétés des choses,'" *Gazette des Beaux-Arts*, ser. 6, 92 (1978), pp. 149–64; see also Millard Meiss, *French Painting: The Boucicaut Master*, pp. 58–59 and 122–23. On princely owners of vernacular manuscripts of this work, see Byrne, "Boucicaut Master," pp. 152 and 156; and idem, "Two Hitherto Unidentified Copies of the 'Livre des propriétés des choses,' from the Royal Library of the Louvre and the Library of Jean de Berry," *Scriptorium* 31 (1977), pp. 90–98.

64. Vincent of Beauvais, *Speculum naturale*, 32.118–120, vol. 2, fols. 302v–302r.

65. Thomas of Cantimpré, *De natura rerum*, 8.3, p. 278; cf. Aristotle, *Generation of Animals*, 4.4, 770a24–8. On the literary tradition of the *ansibena* (more properly, *amphisbena*), see Lecouteux, *Monstres*, vol. 2, p. 167.

66. *Liber monstrorum*, 1.1, 1.8, 1.25, pp. 137, 153, 179; cf. Augustine, *City of God*, 16.8, p. 663.

67. See George Economou, *The Goddess Natura in the Middle Ages* (Cambridge: Harvard University Press, 1972), and, on this general transformation, Marie-Dominique Chenu, *Nature, Man and Society in the Twelfth Century: Essays on New Theological Perspectives in the Latin West*, ed. and trans. Jerome Taylor and Lester K. Little (Chicago: University of Chicago Press, 1968), Ch. 2. Additional references in F.J.E. Raby, "*Nuda Natura* and Twelfth-Century Cosmology," *Speculum* 43 (1968), pp. 72–77.

68. A number of modern commentators have failed to recognize this fundamental distinction; see, for example, Kappler, *Monstres*, and Lecouteux, *Monstres*. Confusion of the two kinds of monsters renders Porsia's introduction to the *Liber monstrorum* almost entirely beside the point.

69. Gerald of Wales, *History and Topography*, 1.11–2, pp. 42–43.

70. Ibid., 2.60, p. 77. Gerald was considerably less committed to the portentous meaning of such occurrences than many of his contemporaries and consistently left open the possibility that they were purely natural, with no higher meaning at all. See in general Bartlett, *Gerald*, Ch. 4.

71. On Cicero, Pliny, and Augustine, see Céard, *Nature*, pp. 7–31. On the classical prodigy tradition, see Raymond Bloch, *Les Prodiges dans l'antiquité classique (Grèce, Etrurie et Rome)* (Paris: Presses Universitaires de France, 1963). On prodigies and portents in medieval Europe, see Paul Rousset, "Le Sens du merveilleux à l'époque féodale," *Le Moyen Age*, ser. 4, 11 = 62 (1956), pp. 25–37; and Céard, *Nature*, pp. 31–40.

72. Isidore of Seville, *Etymologiarum sive originum libri XX*, 11.3.1–4, ed. W.M. Lindsay,

2 vols. (Oxford: Clarendon Press, 1911), vol. 2. Cf. Augustine, *City of God*, 21.8, pp. 980–83.

73. James of Vitry, *Historia occidentalis*, 1, pp. 260–61.

74. Rousset, "Sens," pp. 26 and 35–36; and Céard, *Nature*, pp. 33–35.

75. Guibert of Nogent, *Self and Society in Medieval France: The Memoirs of Abbot Guibert of Nogent*, 3.11, trans. John F. Benton (Toronto: University of Toronto Press, 1984), pp. 189–90; Latin in idem, *Autobiographie [De vita ipsius]*, ed. and trans. Edmond-René Labande (Paris: Les Belles Lettres, 1981), p. 377.

76. William of Malmesbury, *De gestis regum anglorum libri quinque*, 2 vols., ed. William Stubbs (Wiesbaden: Kraus Reprint, 1964), vol. 1, p. 276. For other contemporary references to this comet, see Rousset, "Sens," pp. 27–28 and 30; and Kelly, *Art*, pp. 156–57.

77. Gerald of Wales, *History and Topography*, 1.11, pp. 41–42. Gerald also allegorized other less remarkable species; see, for example, his discussion of the eagle (1.9, pp. 39–40), the crane (1.10, pp. 40–41), the osprey (1.12, pp. 42–43), and the kingfisher (1.14, pp. 44–45).

78. For one rare exception to this rule, see *Gesta Romanorum*, 176, p. 576, where a male child very much like that described by Guibert was interpreted as a figure of the interrelationship of body and soul.

79. Isidore of Seville, *Etymologiarum sive originum libri XX*, 11.3.5.

80. Davidson, "Horror," p. 48 (quotation) and throughout; cf. Mary Douglas, *Purity and Danger: An Analysis of the Concepts of Pollution and Taboo* (London: Ark Paperbacks, 1984), esp. Chs. 2–3. For a useful formal repertory of such human-animal hybrids in medieval texts, see Kappler, *Monstres*, pp. 149–57.

81. Gerald of Wales, *History and Topography*, 2.55, p. 74.

82. Ibid., 2.56, p. 75. Gerald made his disapproval of such conjunctions even more explicit in later recensions, adding to his poem four more lines in which the ox-man hybrid was described as "nature's revenge" (*naturae vindicta*): idem, *Topographia Hibernie*, 2.54, pp. 145–47.

83. Idem, *History and Topography*, 3.93, p. 103.

84. Giovanni Villani, *Cronica*, 9.79, ed. Francesco Gherardi Dragomanni, 4 vols. (Florence: Sansone Coen, 1845), vol. 2, pp. 197–98; Villani gives the date as 1316 (Florentine style).

85. Francesco Petrarca, *Rerum memorandarum libri*, 4.120, ed. Giuseppe Billanovich (Florence: G.C. Sansoni, 1943), pp. 270–71.

86. For further texts relating to this event, see Luigi Belloni, "L'ischiopago tripode trecentesco dello spedale fiorentino di Santa Maria della Scala," *Rivista di storia delle scienze mediche e naturali* 41 (1950), pp. 1–14, on pp. 2–4.

87. Friedman, *Monstrous Races*, Ch. 9.

88. Campbell, *Witness*, pp. 65–80 and 144–47. Campbell's discussions in Ch. 2 and pp. 138–48 deal with many of the themes discussed in this section from the point of view of a literary historian, emphasizing fictionality rather than belief. See also Suzanne Fleisch-

mann, "On the Representation of History and Fiction in the Middle Ages," *History and Theory* 22 (1983), pp. 278–310, on pp. 289–90.

89. Caroline Walker Bynum, "Wonder," *American Historical Review* 102 (1997), pp. 1–26, esp. p. 23. The same considerations apply to romance and miracle stories; concerning the former, see Kelly, *Art*, esp. pp. 13–31 and 146–50. On the latter, Klaus Schreiner, "'Discrimen veri ac falsi.' Ansätze und Formen der Kritik in der Heiligen- und Reliquienverehrung des Mittelalters," *Archiv für Kulturgeschichte* 48/1 (1966), pp. 1–53, esp. p. 24; and idem, "Zum Wahrheitsverständnis im Heiligen- und Reliquienwesen des Mittelalters," *Saeculum: Jahrbuch für Universalgeschichte* 17 (1966), pp. 131–69, on p. 162.

90. James of Vitry, *Historia orientalis*, 92, p. 215, citing Romans 14.

91. See Brian Stock, *The Implications of Literacy: Written Language and Models of Interpretation in the Eleventh and Twelfth Centuries* (Princeton: Princeton University Press, 1983), p. 31.

92. Paul Veyne, *Did the Greeks Believe in Their Myths? An Essay on the Constitutive Imagination*, trans. Paula Wissing (Chicago: Chicago University Press, 1988).

93. Ibid., p. 12; see in general Ch. 1 and pp. 43–46. The efforts of those ancient writers with a special interest in separating topographical fact from fiction, such as Strabo, were unstable and ultimately ineffective; see Romm, *Edges*, pp. 95–103.

94. On this criterion see Veyne, *Greeks*, p. 14.

95. Tertullian, *De carne Christi*, 5, in *Opera omnia*, in *Patrologia cursus completus*, 221 vols., ed. J.P. Migne (Paris: J.P. Migne, 1844–1904), ser. 1, vol. 2; see also Augustine, *City of God*, 21.4–8, pp. 968–83.

96. Schreiner, "'Discrimen,'" pp. 12–13.

97. Gervase of Tilbury, *Otia imperialia*, pp. 960–61; and Odoric of Pordenone, *Relatio*, 31, pp. 482–83.

98. Ibid., 38, p. 494.

99. Greenblatt, *Possessions*, p. 34.

100. Lucienne Carasso-Bulow, *The Merveilleux in Chrétien de Troyes' Romances* (Geneva: Librairie Droz, 1976), pp. 23–25. For an excellent discussion of rhetorical strategies used to establish credibility in travel narratives, see Kappler, *Monstres*, pp. 55–68.

101. Gerald of Wales, *History and Topography*, 2.33, p. 56. On the impressive quality of some of Gerald's own observations, see Holmes, "Gerald."

102. Gerald of Wales, *History and Topography*, 1.2, p. 35.

103. Augustine, *City of God*, 21.7, pp. 978–79.

104. Jordan of Sévérac, *Mirabilia*, trans. Yule, p. 17.

105. Odoric of Pordenone, *Relatio*, 38, p. 494.

106. Both cited in Kappler, *Monstres*, p. 55.

107. Gervase of Tilbury, *Otia imperialia*, p. 963. Albertus Magnus, *Book of Minerals* [*De mineralibus*], 1.1.7, trans. Dorothy Wyckoff (Oxford: Clarendon Press, 1967), p. 28.

108. Frederick II, *De arte venandi cum avibus*, ed. C.A. Willemsen (Leipzig: Insula, 1942), p. 55. On Frederick's natural studies and his special interest in mirabilia, see

Charles Homer Haskins, *Studies in the History of Mediaeval Science*, 2nd ed. (Cambridge, MA: Harvard University Press, 1927), pp. 263–64, 292–98, 321.

109. Erich Holdefleiss, *Der Augenscheinsbeweis im mittelalterlichen deutschen Strafverfahren* (Stuttgart: W. Kohlhammer, 1933), pp. 2–10.

110. Polo, *Travels*, 6, p. 253.

111. *Journal d'un bourgeois de Paris, 1405–1449*, ed. Alexandre Tuetey (Paris: H. Champion, 1881), no. 309, pp. 153–54.

112. Ibid., no. 508, pp. 238–39.

113. According to Alexandre Tuetey, the anonymous Parisian chronicler's account of the Aubervilliers twins differs on several key points from that of his contemporary Clément de Fauquembergue: ibid., p. 238, n. 4. There are similar discrepancies in contemporary representations of the Tuscan twins of 1317; see Belloni, "L'ischiopago."

CHAPTER TWO: THE PROPERTIES OF THINGS

1. Percy Ernest Schramm and Florentine Mütherich, *Denkmale der deutschen Könige und Kaiser: Ein Beitrag zur Herrschergeschichte von Karl dem Grossen bis Friedrich II, 768–1250* (Munich: Prestal Verlag, 1962), p. 72. On Frederick's menagerie, see Gustave Loisel, *Histoire des ménageries d'Antiquité à nos jours*, 2 vols. (Paris: Octave Doin/Henri Laurens, 1912), vol. 1, pp. 247–55.

2. Allen J. Grieco, "The Social Politics of Pre-Linnaean Botanical Classification," *I Tatti Studies: Essays in the Renaissance* 4 (1992), pp. 131–49.

3. *Le Trésor de Saint-Denis: Musée du Louvre, Paris, 12 mars–17 juin 1991* (Paris: Bibliothèque Nationale, Réunion des Musées Nationaux, 1991); further bibliography in ibid., p. 36, n. 2. Very little has been written on medieval collecting, with the recent exception of Krzysztof Pomian, "Collezionismo," in *Enciclopedia dell'arte medievale*, 6 vols. to date, gen. ed. Marina Righetti Tosti-Croce (Rome: Istituto della Enciclopedia Italiana, 1991–), vol. 5, pp. 156–60. Most discussions tend to present it principally as background to sixteenth- and seventeenth-century developments. The most useful of these are Krzysztof Pomian, *Collectors and Curiosities: Paris and Venice, 1500–1800* [1987], trans. Elizabeth Wiles-Portier (Cambridge: Polity Press, 1990), Ch. 1; and Adalgisa Lugli, *Naturalia et mirabilia: Il collezionismo enciclopedico nelle wunderkammern d'Europa* (Milan: Gabriele Mazzotta, 1983), Ch. 1. Lugli's suggestive and well-documented study tends nonetheless to distort the significance of the medieval material by projecting back onto it early modern concerns.

4. Details concerning these objects in *Trésor*, pp. 142–43, 223–25, 310–11.

5. Schramm and Mütherich, *Denkmale*, p. 67. See in general ibid., pp. 67–70; and D. Heinrich Otte, *Handbuch der kirchlichen Kunst-Archäologie des deutschen Mittelalters*, 2 vols., 5th ed., (Leipzig: T.O. Weigel, 1883), vol. 1, pp. 213–24.

6. James Raine, *Saint Cuthbert, with an Account of the State in Which His Remains Were Found upon the Opening of His Tomb in Durham Cathedral, in the Year MDCCCXXVII* (Durham: George Andrews, 1828), pp. 121–30. On eagle stones, see Antoine Schnapper, *Le Géant, la licorne et la tulipe: Collections et collectionneurs dans la France du XVIIe siècle, I:*

Histoire et histoire naturelle (Paris: Flammarion, 1988), p. 26; on beryls, Evans, *Magical Jewels*, pp. 63 and 89, and Albertus Magnus, *Book of Minerals*, p. 76; on griffins and griffin horns, Alcouffe, *Trésor*, p. 224. This particular griffin claw is now in the British Museum, Department of Medieval and Later Antiquities, no. OA24; see Mark Jones (ed.), *Fake? The Art of Deception* (London: British Museum Publications, 1990), p. 85.

7. Examples in Joseph Braun, *Die Reliquiare des christlichen Kultes und ihre Entwicklung* (Freiburg im Breisgau: Herder, 1940), pp. 129–31 and nos. 184–87; see also Otte, *Handbuch*, vol. 1, p. 213.

8. Braun, *Reliquiare*, pp. 113–37 and nos. 188–89, 207–209, 220, 310–11, 315.

9. Pomian, "Collezionismo," p. 158. See in general Denis Bethell, "The Making of a Twelfth-Century Relic Collection," in G.J. Cuming and Derek Baker (eds.), *Popular Belief and Practice: Papers Read at the Ninth Summer Meeting and the Tenth Winter Meeting of the Ecclesiastical History Society* (Cambridge: Cambridge University Press, 1972), pp. 64–72; Patrick J. Geary, *Furta Sacra: Thefts of Relics in the Central Middle Ages*, 2nd ed. (Princeton: Princeton University Press, 1990), esp. Chs. 3–5; and Jonathan Sumption, *Pilgrimage: An Image of Medieval Religion* (Totowa, NJ: Rowman and Littlefield, 1975), esp. Ch. 2.

10. Most inventories grouped possessions by the place they were stored, as in the case of the inventory of the shrine of St. Cuthbert, in Raine, *Saint Cuthbert*, pp. 121–30. Exceptionally, they might be recorded by form or function (e.g, crosses, reliquaries, silver and gold images, liturgical vessels, candlesticks, incense burners, "jewelry for the body," etc.) and/or mode of acquisition (gifts, purchases, etc.), as in the case of the collection of John, Duke of Berry; see Jules Guiffrey (ed.), *Inventaires de Jean Duc de Berry*, 2 vols. (Paris: Ernest Leroux, 1894–96).

11. On the very different associations of the *musaeum*, see Paula Findlen, "The Museum: Its Classical Etymology and Renaissance Genealogy," *Journal of the History of Collections* 1 (1989), pp. 59–78.

12. *Trésor*, pp. 24–25; Pierre Héliot and Marie-Laure Chastang, "Quêtes et voyages de reliques au profit des églises françaises du Moyen Age," *Revue d'Histoire Ecclésiastique* 59 (1964), pp. 789–823, and 60 (1965), pp. 5–53. Private collectors used their collections in the same way on occasion, as when John of Berry sold prized books and jewels and melted down gold objects to raise funds to pay his troops: Millard Meiss, *French Painting in the Time of Jean de Berry: The Late Fourteenth Century and the Patronage of the Duke*, 2 vols. (London and New York: Phaidon, 1967), vol. 1, p. 34.

13. Braun, *Reliquiare*, p. 140.

14. Cited in Odell Shepard, *The Lore of the Unicorn* (New York: Avenel Books, 1982), p. 106; see also Schnapper, *Géant*, pp. 88–89. From the later fifteenth century, unicorn horn was also thought to be an antidote to plague.

15. Shepard, *Lore*, pp. 107–108. Shepard identifies the emperor as John VI, who came to Venice to seek military aid in the first half of the fifteenth century.

16. Cited in Evans, *Magical Jewels*, p. 116. Similar examples in Guiffrey, *Inventaires*, vol. 1, pp. 93, 618, 619, 630, 631.

17. Ibid., p. 159. The bezoar was a concretion found in the stomachs of certain Eastern ruminants; see Schnapper, *Géant*, p. 32.

18. Guiffrey, *Inventaires*, vol. 1, p. 145.

19. Schnapper, *Géant*, p. 28. See Guiffrey, *Inventaires*, vol. 1, pp. 165–66. These presumably worked on sympathetic principles; the two serpent jaws, one decorated and one plain, in the Duc de Berry's collection (pp. 84 and 153) may have served the same function.

20. Evans, *Magical Jewels*, pp. 117–19. On Charles V's collection, see Pomian, "Collezionismo," p. 159. Eagle-stones, hollow stones with smaller stones inside, were thought to aid in pregnancy and childbirth; see Schnapper, *Géant*, pp. 26–27.

21. For the natural philosophical explanation of such properties, see Chapter Three.

22. William of Auvergne, *De universo*, 2.3.23, in his *Opera omnia*, ed. Giovanni Domenico Traiano (Venice: Damiano Zenaro, 1591), p. 1003, cols. 1–2.

23. Gerald of Wales, *History and Topography*, 1.32, p. 56.

24. W.S. Heckscher, "Relics of Pagan Antiquity in Medieval Settings," *Journal of the Warburg and Courtauld Institutes* 1 (1938), pp. 204–20, on p. 212.

25. See above Chapter One.

26. *Annales basilienses*, ad annum 1276, cited in Schramm and Mütherich, *Denkmale*, p. 73.

27. Suger of Paris, *De rebus in administratione sua gestis*, 33, trans. Erwin Panofsky, in Panofsky (ed. and trans.), *Abbot Suger on the Abbey Church of St.-Denis and Its Art Treasures*, ed. Gerda Panofsky-Soergel (Princeton: Princeton University Press, 1979), pp. 61–62 (translation emended for clarity); cf. Ezekiel 28.13.

28. On the multiple readings of precious stones, Meiss, *French Painting: The Late Fourteenth Century*, vol. 1, pp. 50–54 and 69–70; Panofsky, *Abbot Suger*, p. 188; and Evans, *Magical Jewels*, pp. 72–80. See also Gabrielle M. Spiegel, "History as Enlightenment: Suger and the *Mos Anagogicus*," in Paula Lieber Gerson (ed.), *Abbot Suger and Saint-Denis: A Symposium* (New York: Metropolitan Museum of Art, 1986), pp. 151–58, on p. 155.

29. Quoted in Panofsky, "Introduction," to *Abbot Suger*, p. 12.

30. Suger of Paris, *De consecratione ecclesiae Sancti Dionysii*, 2, trans. Erwin Panofsky, in Panofsky, *Abbot Suger*, pp. 88–89. Suger was relatively liberal about displaying his treasure, for which he was castigated for greed and worldliness by Bernard of Clairvaux; see Panofsky, "Introduction" to *Abbot Suger*, pp. 10–14. On the display of relics and relic collections, see in general Sumption, *Pilgrimage*, p. 35; Gustav Klemm, *Zur Geschichte der Sammlungen für Wissenschaft und Kunst in Deutschland* (Zerbst: G.A. Kummer, 1837), pp. 139–41; and Anton Legner (ed.), *Reliquien, Verehrung und Verklärung: Skizzen und Noten zur Thematik und Katalog zur Ausstellung der Kölner Sammlung Louis Peters im Schnütgen-Museum* (Cologne: Greven & Bechtold, 1989), pp. 25–29.

31. Richard Trexler, "Ritual Behavior in Renaissance Florence: The Setting," *Medievalia et Humanistica* 4 (1973), pp. 125–44, on pp. 131–32. See also Geary, *Furta Sacra*, p. 25.

32. Suger of Paris, *De consecratione*, in Panofsky, *Abbot Suger*, p. 115.

33. See Otte, *Handbuch*, pp. 213–24; Lugli, *Naturalia*, pp. 16–20; Julius von Schlosser, *Die Kunst- und Wunderkammern der Spätrenaissance: ein Beitrag zur Geschichte des Sammelwesens* (Leipzig: Klinkhardt & Biermann, 1908), pp. 14–15; and Klemm, *Geschichte*, p. 143.

34. Giovanni Boccaccio, *Genealogia deorum*, 4; and Mattäus Faber, *Kurtzgefasste historische Nachricht von der Schloss- und academischen Stifftskirche zu Aller-Heiligen in Wittenberg....* (Wittenberg: Gerdesische Witwe, 1717), pp. 140–42. On suspended ostrich eggs, see Isa Ragusa, "The Egg Reopened," *The Art Bulletin* 53 (1971), pp. 435–43; Creighton Gilbert, "'The Egg Reopened' Again," ibid., 56 (1974), pp. 252–58, and the references therein.

35. Guillaume Durand, *Rationale divinorum officiorum*, 1.3, ed. Niccolò Doard (Venice: Matteo Valentino, 1589), p. 12.

36. Ibid. Cf. Thomas of Cantimpré, *De natura rerum*, 5.110, p. 226.

37. See Karl Jordan, *Henry the Lion: A Biography*, trans. P.S. Falla (Oxford: Clarendon Press, 1986), pp. 153–54 and 201–208.

38. Pomian, "Collezionismo," p. 157.

39. See Richard Goldthwaite, "The Empire of Things: Consumer Demand in Renaissance Italy," in F.W. Kent and Patricia Simons, with J.C. Eade (eds.), *Patronage, Art and Society in Renaissance Italy* (Oxford: Clarendon Press, 1987), pp. 153–75; and idem, *Wealth and the Demand for Art in Italy, 1300–1600* (Baltimore: Johns Hopkins University Press, 1993).

40. Marco Spallanzani and Giovanna Gaeta Bertelà (eds.), *Libro d'inventario dei beni di Lorenzo il Magnifico* (Florence: Amici del Bargello, 1992); Pomian, "Collezionismo," pp. 158–59. See also Wolfgang Liebenwein, *Studiolo: Storia e tipologia di uno spazio culturale* (Ferrara: Panini, 1977), pp. 43–56.

41. See Otte, *Handbuch*, p. 214.

42. Guiffrey, *Inventaires*, vol. 1, pp. 76–78, 93, 152–54, 165, 196, 303. On John as collector, see in general Meiss, *French Painting: The Late Fourteenth Century*, Chs. 2–3.

43. Ibid., vol. 1, p. 151.

44. Jules Guiffrey, "La Ménagerie du Duc Jean de Berry, 1370–1403," *Mémoires de la Société des Antiquaires du Centre, Bourges* 22 (1899), pp. 63–84; and Meiss, *French Painting: The Late Fourteenth Century*, pp. 31–32.

45. Relics were manipulated in similar ways; see Trexler, "Ritual Behavior," pp. 128–32.

46. On the Duke as connoisseur, see Meiss, *French Painting: The Late Fourteenth Century*, pp. 69–70 and 303–305.

47. Suger of Paris, *De administratione*, 26, in Panofsky, *Abbot Suger*, p. 47.

48. Albertus Magnus, *Book of Minerals*, 2.3.3, pp. 130–33; he opted for the latter explanation.

49. See Gerald of Wales, *History and Topography*, 2.51, pp. 69; and Geoffrey of Monmouth, *The Historia regum Britanniae of Geoffrey of Monmouth*, 8.10–12, ed. and trans. Acton Griscom (London/New York: Longmans, Green and Co., 1929), pp. 409–14.

50. *Eneas, roman du XII siècle*, ll. 422–40, ed. J.-J. Salverda de Grave, 2 vols. (Paris: Librairie Honoré Champion, 1964), vol. 1, p. 13. On wonderful works of art in medieval romance, see in general Otto Söhring, "Werke bildender Kunst in altfranzösischen Epen," *Romanische Forschungen* 12 (1900), pp. 491–640; Gianfelice Peron, "Meraviglioso e verosimile nel romanzo francese medievale: Da Benoît de Sainte-Maure a Jean Renart," in Diego Lanza and Oddone Longo (eds.), *Il meraviglioso e il verosimile tra antichità e medioevo* (Florence: Olschki, 1989), pp. 293–323, esp. pp. 296–303; Faral, "Merveilleux," esp. pp. 321–50; and, on the sources of the marvels in *Eneas*, idem, "Ovide et quelques romans français du XIIe siècle," in ibid., pp. 73–157.

51. *Eneas*, ll. 7459–718, vol. 2, pp. 49–55. The caladrius was a wonderful bird that was supposed to be able to predict whether a sick person would live or die.

52. Benoît de Sainte Maure, *Le Roman de Troie*, ll. 13341–409, 6 vols., ed. Léopold Constans (Paris: Firmin Didot, 1904–12), vol. 2, pp. 293–98.

53. Ibid., ll. 14631–56; trans. Penny Sullivan in her "Medieval Automata: the 'Chambre de Beautés' in Benoît's *Roman de Troie*," *Romance Studies* 6 (1985), pp. 1–20, on p. 13. Sullivan's article includes a translation of the entire description.

54. Ibid., ll. 14657–918; trans. Sullivan, "Medieval Automata," pp. 13–16 (quotation on p. 14). On automata in medieval literature, see Merriam Sherwood, "Magic and Mechanics in Medieval Fiction," *Studies in Philology* 44 (1947), pp. 567–92; Söhring, "Werke," pp. 580–98; Sullivan, "Medieval Automata"; William Eamon, "Technology as Magic in the Late Middle Ages and Renaissance," *Janus* 70 (1983), pp. 171–212, on pp. 174–79; Faral, "Merveilleux," pp. 328–35; and Alfred Chapuis and Edouard Gélis, *Le Monde des automates: Étude historique et technique*, 2 vols. (Paris, 1928; repr. Geneva: Slatkine, 1984), Ch. 5.

55. Benoît, *Troie*, ll. 14668–69; trans. Sullivan, "Medieval Automata," p. 14. On the identification of engineer, magician, and natural philosopher, see Eamon, "Technology as Magic," p. 185; on the figure of the sorcerer in romance, see Robert-Léon Wagner, *"Sorcier" et "magicien": Contribution à l'histoire du vocabulaire de la magie* (Paris: E. Droz, 1939), esp. pp. 65–109.

56. On the courtly associations of magic, see Richard Kieckhefer, *Magic in the Middle Ages* (Cambridge: Cambridge University Press, 1989), pp. 95–115; and William Eamon, *Science and the Secrets of Nature: Books of Secrets in Medieval and Early Modern Culture* (Princeton: Princeton University Press, 1994), pp. 67–68.

57. *Aymeri de Narbonne*, ll. 3520–22, 2 vols., ed. Louis Demaison (Paris: Firmin Didot, 1887), pp. 148–49. For antecedents and analogues of this kind of automaton, see Gerard Brett, "The Automata of the Byzantine 'Throne of Solomon,'" *Speculum* 29 (1954), pp. 477–87, on pp. 484–85.

58. See Michael Nerlich, *Ideology of Adventure: Studies in Modern Consciousness, 1100–1750*, 2 vols., trans. Ruth Crowley (Minneapolis: University of Minnesota Press, 1987), vol. 1, pp. 6–14; and Erich Köhler, *Ideal und Wirklichkeit in der höfischen Epik: Studien zur Form der frühen Artus- und Gralsdichtung* (Tübingen: M. Niemeyer, 1970).

59. Benoît, *Troie*, ll. 14685–710; trans. Sullivan, "Medieval Automata," p. 14.

60. Ibid., ll. 14864–900; trans. Sullivan, "Medieval Automata," p. 16. For other romance automata with similar functions, see John Cohen, *Human Robots in Myth and Science* (London: Allen and Unwin, 1966), Ch. 4.

61. Ibid., ll. 14926–29; trans. Sullivan, "Medieval Automata," p. 16.

62. On this tradition, see Henri Omont, "Les Sept Merveilles du monde au Moyen Age," *Bibliothèque de l'École des Chartes* 43 (1882), pp. 40–59; J. Lanowski, "Weltwunder," in August Friedrich von Pauly, Georg Wissowa, Wilhelm Kroll et al. (eds.), *Realencyclopädie der classischen Altertumswissenschaft*, suppl. vol. 10 (Stuttgart: Alfred Druckenmüller, 1965), col. 1020–30; and T. Dombart, *Die sieben Weltwunder des Altertums*, 2nd ed., ed. Jörg Dietrich (Munich: Heimeran, 1967).

63. *De septem miraculis hujus mundi*, 7, ed. Henri Omont, in Omont, "Sept Merveilles," p. 52. The same author composed a companion treatise on the seven natural or divinely created wonders of the world, which appears in the same manuscript, transcribed in ibid., pp. 52–54.

64. Gervase of Tilbury, *Otia imperialia*, p. 963. On the tradition of Vergil as a magician, see John Webster Spargo, *Virgil the Necromancer: Studies in Virgilian Legends* (Cambridge, MA: Harvard University Press, 1934).

65. Alexander Neckam, *De naturis rerum libri duo. . . .* , 2.174, ed. Thomas Wright (London: Longman, Green, Longman, Roberts, and Green, 1863), p. 310. Vergil's structure, sometimes known as the "Salvatio Romae," had appeared in the widely circulated treatise on the seven wonders of the world attributed to Bede: *De septem miraculis mundi ab hominibus factis*, 1 (in Omont, "Sept Merveilles," pp. 47–48). Neckam was apparently the first to attribute it to the author of the *Aeneid*; see Spargo, *Virgil*, pp. 11 and 118–19.

66. Neckam, *De naturis rerum*, 2.172, pp. 281–82.

67. On problems of transmission, see Brett, "Automata," p. 480; and Derek J. de Solla Price,"Automata and the Origins of Mechanism and Mechanistic Philosophy," *Technology and Culture* 5 (1964), pp. 1–23, on pp. 15–17. On Islamic clocks and automata, see Chapuis and Gelis, *Monde des automates*, Ch. 3; and E. Wiedemann and F. Hauser, *Über die Uhren im Bereich der islamischen Kultur* (Halle: E. Karras, 1915).

68. Robert's description of the Hippodrome referred to "images of men and women and horses and oxen and camels and bears and lions and many kinds of animals cast in copper, which were so well fashioned and formed so naturally that there is not in Pagany or Christendom a master who could portray or fashion images so well as those images were fashioned, and they used in times gone by to perform by magic, but they no longer performed at all": translated in Brett, "Automata," pp. 484–85. See also Söhring, "Werke," pp. 585–86.

69. William of Rubruck, *The Journey of William of Rubruck to the Eastern Parts of the World, 1253–55 [Itinerarium ad partes orientales]*, trans. William Woodville Rockhill (London: Hakluyt Society, 1904), p. 208. Rubruck attributed this work to a French goldsmith, Guillaume Boucher, who had been taken captive by Mangu Khan.

70. Polo, *Travels*, p. 110; and Odoric of Pordenone, *Relatio*, 26, p. 473. Remains of a musical organ with dancing peacocks were found in the Mongol palace in Peking; see Emil Bretschneider, *Archaeological and Historical Researches on Peking and Its Environs* (Shanghai: American Presbyterian Mission Press, 1876), p. 28.

71. *Mandeville's Travels*, ed. Seymour, p. 166.

72. On the typology and vocabulary of magic in romance, see Wagner, *"Sorcier,"* esp. pp. 65–98; and Sherwood, "Magic and Mechanics." On the association of magic and the Orient, see Kappler, *Monstres*, pp. 65–68. On demons and automata, see Arthur Dickson, *Valentine and Orson: A Study in Late Medieval Romance* (New York: Columbia University Press, 1929), esp. pp. 195–99.

73. Bacon's edition of the Latin translation, together with his introduction and annotations, was edited by Robert Steele in *Opera hactenus inedita Rogerii Bacon*, 16 fascs. (Oxford: Clarendon Press, 1905–40), fasc. 5, pp. 1–175. See W.F. Ryan and Charles B. Schmitt (eds.), *Pseudo-Aristotle, The Secret of Secrets: Sources and Influence* (London: Warburg Institute, 1982) and, for the *Secret of Secrets*'s influence on Bacon, Eamon, *Science*, pp. 47–49. On Bacon's *scientia experimentalis* (with references and appropriate historiographical cautions), see David C. Lindberg, "Science as Handmaiden: Roger Bacon and the Patristic Tradition," *Isis* 78 (1987), pp. 518–36, on pp. 518–20 and 533–34.

74. Roger Bacon, *Epistola fratris Rogerii Baconis de secretis operibus artis et naturae, et de nullitate magiae*, 6, in R. Bacon, *Opera quaedam hactenus inedita*, 3 vols., ed. J.S. Brewer (London: Longman, Green, Longman and Roberts, 1859), vol. 1, p. 536; the translation by Tenney L. Davis, *Roger Bacon's Letter concerning the Marvelous Power of Art and of Nature and concerning the Nullity of Magic* (Easton, PA: Chemical, 1923), is unreliable. On Bacon's reputation for magic, see A.G. Molland, "Roger Bacon as Magician," *Traditio* 30 (1974), pp. 445–60. Asbestos was often called salamander skin.

75. R. Bacon, *Epistola*, 5, in *Opera*, ed. Brewer, vol. 1, p. 535; cf. *Secretum secretorum*, 3.19, in *Opera*, ed. Steele, fasc. 5, p. 153.

76. R. Bacon, *Epistola*, 4, in *Opera*, ed. Brewer, vol. 1, p. 533. Cf. Cary, *Medieval Alexander*, p. 238.

77. For Bacon's ties to the culture of the papal court, see Agostino Paravicini Bagliani, *Medicina e scienze della natura alla corte dei papi nel Duecento* (Spoleto: Centro Italiano di Studi sull'Alto Medioevo, 1991), pp. 327–61.

78. R. Bacon, *Epistola*, 1, in *Opera*, ed. Brewer, vol. 1, p. 523. Claims of this sort concerning the powers of art were partly responsible for a theological backlash against alchemy and magic in the later Middle Ages: see William Newman, "Technology and Alchemical Debate in the Late Middle Ages," *Isis* 80 (1989), pp. 423–45, on pp. 437–42; and Wagner, *"Sorcier,"* pp. 119–30.

79. On Bacon and demons, see Molland, "Bacon as Magician," pp. 458–59. On other thirteenth-century philosophical discussions of the power and operation of demons, see the commentary of Joseph Bernard McAllister in Thomas Aquinas, *The Letter of Saint Thomas Aquinas De occultis operibus naturae ad quemdam militem ultramontanum*, ed., trans.,

and comm. Joseph Bernard McAllister (Washington, DC: Catholic University of America Press, 1939), pp. 60–69.

80. See in general Carlo Cipolla, *Clocks and Culture 1300–1700* (New York: W.W. Norton, 1978), pp. 15–36; and Brian Stock, "Science, Technology, and Economic Progress in the Early Middle Ages," in Lindberg, *Science in the Middle Ages*, Ch. 1.

81. Villard of Honnecourt, *Album de Villard de Honnecourt, architecte du XIIIe siècle*, ed. Henri Omont (Paris: Berthaud Frères, 1906); partial translation of inscriptions in *The Sketchbook of Villard de Honnecourt*, ed. Theodore Bowie (Bloomington: Indiana University Press, 1959).

82. See Silvio A. Bedini, "The Role of Automata in the History of Technology," *Technology and Culture* 5 (1964), pp. 24–42, on p. 33.

83. Sherwood, "Magic and Mechanics," pp. 589–90. See also Anne Hagopian Van Buren, "Reality and Literary Romance in the Park of Hesdin," in Elisabeth Blair Mac-Dougall et al. (eds.), *Medieval Gardens* (Washington, DC: Dumbarton Oaks, 1986), pp. 117–34; Chapuis and Gélis, *Monde des automates*, pp. 72–74.

84. Quoted in Van Buren, "Reality," p. 121.

85. Translated in Sherwood, "Magic and Mechanics," pp. 588–89.

86. For an overview of the culture of these courts, see Françoise Piponnier, *Costume et vie sociale: La Cour d'Anjou, XIVe–XVe siècle* (Paris: Mouton, 1970); Richard Vaughan, *Philip the Good: The Apogee of Burgundy* (London: Longmans, 1970), esp. Ch. 5; idem, *Charles the Bold: The Last Valois Duke of Burgundy* (London: Longmans, 1973), esp. Ch. 5; and the synthetic account in idem, *Valois Burgundy* (London: Allen Lane, 1975), esp. Ch. 8.

87. *Le Livre des faits de Jacques de Lalaing*, cited in Piponnier, *Costume*, p. 67.

88. See in general ibid., esp. Ch. 3; Gigliola Soldi Rondinini, "Aspects de la vie des cours de France et de Bourgogne par les dépêches des ambassadeurs milanais (seconde moitié du XVe siècle)," in *Adelige Sachkultur des Spätmittelalters: Internationaler Kongress Krems an der Donau, 22. bis 25. September 1980* (Vienna: Verlag der österreichischen Akademie der Wissenschaften, 1982), pp. 194–214; A. van Nieuwenhuysen, *Les Finances du Duc de Bourgogne Philippe le Hardi (1384–1404): Economie et politique* (Brussels: Editions de l'Université de Bruxelles, 1984), Ch. 4; and Werner Paravicini, "The Court of the Dukes of Burgundy: A Model for Europe?" in Ronald G. Asch and Adolf M. Birke (eds.), *Princes, Patronage, and the Nobility: The Court at the Beginning of the Modern Age, ca. 1450–1650* (London: Oxford University Press, 1991), pp. 69–102, esp. pp. 75–76.

89. Marie-Thérèse Caron, *La Noblesse dans le duché de Bourgogne, 1315–1477* (Lille: Presses Universitaires de Lille, 1987), p. 293; and Daniel Poirion, *Le Merveilleux*, pp. 115–20.

90. Loisel, *Ménageries*, vol. 1, pp. 238–55; cf. Vaughan, *Philip the Good*, p. 145.

91. Piponnier, *Costume*, pp. 182–85 and 237–42.

92. Léon de Laborde, *Les Ducs de Bourgogne: Etudes sur les lettres, les arts et l'industrie pendant le XVe siècle*, 3 vols. (Paris: Plon, 1849–52), vol. 1, p. 252. See in general Agathe Lafortune-Martel, *Fête noble en Bourgogne au XVe siècle. Le Banquet du Faisan (1454):*

Aspects politiques, sociaux et culturels (Montreal: Bellarmin, 1984), pp. 157–58; and van Nieuwenhuysen, *Finances*, pp. 397–98.

93. Olivier de la Marche, *Mémoires d'Olivier de la Marche* [fifteenth century], 1.29, 4 vols., ed. Henri Beaune and J. D'Arbaumont (Paris: Renouard, 1883–88), vol. 2, p. 362. On the duke's fools and dwarfs, see Laborde, *Ducs*, vol. 3, p. 509.

94. Vaughan, *Philip the Good*, pp. 137–39.

95. Georges Doutrepont, *La Littérature française à la cour des ducs de Bourgogne* (Paris: Honoré Champion, 1909), p. 263; see in general Ch. 1 (on books of romances and the chivalric tradition), Ch. 2 (on antique romances and the Alexander tradition), and pp. 236–64 (on the literature of pilgrimage and Crusade).

96. Account of Gabriel Tetzel, in *The Travels of Leo of Rozmital through Germany, Flanders, England, France, Spain, Portugal and Italy, 1465–67*, ed. and trans. Malcolm Letts (Cambridge: Cambridge University Press, 1957), pp. 27–28.

97. Vaughan, *Philip the Good*, p. 151.

98. Jean Le Fèvre, *Chronique de Jean Le Fèvre, seigneur de Saint-Rémy*, 163, 2 vols., ed. François Morand (Paris: Renouard, 1876–81), vol. 2, pp. 158–72.

99. On the history of the *entremets* and the spectacle associated with fourteenth- and fifteenth-century banquets, see Lafortune-Martel, *Fête noble*, Chs. 2–3; and Laura Hibbard Loomis, "Secular Dramatics in the Royal Palace, Paris, 1378, 1389, and Chaucer's 'Tregetoures,'" *Speculum* 33 (1958), pp. 242–55.

100. On the cultural associations of the wild man, see Richard Bernheimer, *Wild Men in the Middle Ages* (New York: Octagon Books, 1970) and Roger Bartra, *Wild Men in the Looking Glass: The Mythic Origins of European Otherness*, trans. Carl T. Berrisford (Ann Arbor: University of Michigan Press, 1994), esp. chs. 4–5; on the myth of the Trojans and the figure of Jason at the court of Burgundy, see Georges Doutrepont, "Jason et Gédéon, patrons de la Toison d'or," in *Mélanges offerts à Godefroid Kurth* (Paris: Champion, 1908), pp. 191–208; and Yvon Lacaze, "Le rôle des traditions dans la genèse d'un sentiment national au XVe siècle: La Bourgogne de Philippe le Bon," *Bibliothèque de l'École des Chartes* 129 (1971), pp. 303–85, on pp. 303–305.

101. Le Fèvre, *Chronique*, 164, vol. 2, pp. 172–74; Le Fèvre gave a long account of its statutes in Ch. 176. See also Vaughan, *Philip the Good*, p. 161.

102. W. Paravicini, "Court," esp. pp. 78–82; C.A.J. Armstrong, "The Golden Age of Burgundy: Dukes that Outdid Kings," in A.D. Dickens (ed.), *The Courts of Europe: Politics, Patronage and Royalty, 1400–1800* (London: Thames and Hudson, 1977), pp. 55–75, on pp. 60–72; and, for literary manifestations of this program, Charity Cannon Willard, "The Concept of True Nobility at the Burgundian Court," *Studies in the Renaissance* 14 (1967), pp. 33–48.

103. On this project of Philip's, see Vaughan, *Philip the Good*, esp. pp. 216–18 and 268–74; and Doutrepont, *Littérature française*, pp. 245–63. On Philip's use of the Argonautic myth, see Marie Tanner, *The Last Descendant of Aeneas: The Hapsburgs and the Mythic Image of the Emperor* (New Haven: Yale University Press, 1993), pp. 57–59 and 150–53.

104. Their relations appear in Guillebert de Lannoy, *Oeuvres de Ghillebert de Lannoy, voyageur, diplomate et moraliste*, ed. Charles Potvin, with J.-C. Houzeau (Louvain: P. et J. Lefever, 1878), and Bertrandon de la Broquière, *The Voyage d'Outremer*, trans. Galen R. Kline (New York: Peter Lang, 1988).

105. For a detailed analysis of this event, see Lafortune-Martel, *Fête noble*. The principal published contemporary accounts are the two very closely related narratives of Olivier de la Marche (a participant) and Mathieu d'Escouchy: de la Marche, *Mémoires*, 1.29, vol. 2, pp. 340–80; and d'Escouchy, *Chronique de Mathieu d'Escouchy*, 108–109, 3 vols., ed. G. du Fresne de Beaucourt (Paris: Jules Renard, 1863–64), vol. 2, pp. 113–237.

106. De la Marche, *Mémoires*, 1.29, vol. 2, p. 354.

107. Ibid., vol. 2, pp. 352–53.

108. Ibid., vol. 2, pp. 356–60; quotation on p. 356.

109. Ibid., vol. 2, pp. 357–61; quotation on p. 357.

110. Ibid., vol. 2, pp. 362–68; quotations on pp. 368 and 365.

111. See the letters of Jean de Pleine, cited in Vaughan, *Philip the Good*, p. 145, and Jehan de Molesme, in Jacques-Joseph Champollion-Figeac (ed.), *Documents historiques inédits tirés des collections manuscrites de la Bibliothèque Nationale et des archives ou des bibliothèques des départments*, 4 vols. (Paris: Firmin Didot, 1841–46), vol. 4, p. 462.

112. Ibid., p. 461.

113. Lauro Martines, *Power and Imagination: City-States in Renaissance Italy* (New York: Knopf, 1979), Ch. 10.

114. These connections were made explicit in a contemporary poem by Philip's chancellor, Philippe Bouton. See Lafortune-Martel, *Fête noble*, pp. 123–24; the poem is edited in Jean de la Croix Bouton, "Un Poème à Philippe le Bon," *Annales de Bourgogne* 42 (1970), pp. 5–29.

115. See above note 35.

116. See the vows reproduced in de la Marche, *Mémoires*, 1.30, vol. 2, pp. 381–94; and, especially, d'Escouchy, *Chronique*, 109, vol. 2, pp. 165–237.

117. See Vaughan, *Philip the Good*, pp. 365–72; and Armstrong, "Golden Age," p. 62.

118. De la Marche, "Traictié des nopces de Monseigneur le duc de Bourgoingne et de Brabant," in his *Mémoires*, vol. 4, p. 107; see also his *Mémoires*, 2.4, vol. 3, pp. 101–201.

CHAPTER THREE: WONDER AMONG THE PHILOSOPHERS

1. Adelard of Bath, *Die Quaestiones naturales des Adelardus von Bath*, 64, ed. Martin Müller, Beiträge zur Geschichte und Philosophie des Mittelalters 31, Heft 2 (Münster i. W.: Aschendorff, 1934), pp. 58–59.

2. Ibid., 1, p. 6.

3. Ibid., 30, p. 35.

4. See Haskins, *Studies*, pp. 20–42; Charles Burnett (ed.), *Adelard of Bath: An English Scientist and Arabist of the Early Twelfth Century* (London: Warburg Institute, 1987); and in general Marie-Thérèse D'Alverny, "Translations and Translators," in Robert L. Benson and

Giles Constable (eds.), *Renaissance and Renewal in the Twelfth Century* (Cambridge, MA: Harvard University Press, 1982), pp. 421–62, and David C. Lindberg, "The Transmission of Greek and Arabic Learning to the West," in Lindberg, *Science in the Middle Ages*, pp. 52–90.

5. Aristotle, *Metaphysics*, 1.2, 982b10–18, trans. W.D. Ross, in Aristotle, *The Complete Works of Aristotle: Revised Oxford Translation*, 2 vols., ed. Jonathan Barnes (Princeton: Princeton University Press, 1984), vol. 2, p. 1554. See also Patrizia Pinotti, "Aristotele, Platone e la meraviglia del filosofo," in Lanza and Longo, *Il meraviglioso*, pp. 29–55.

6. Avicenna, *De viribus cordis*, 1.10 (Venice: per Paganinum de Paganinis, 1507), fol. 547r; English translation in Bert Hansen's introduction to his *Nicole Oresme and the Marvels of Nature: A Study of His De causis mirabilium with Critical Edition, Translation, and Commentary* (Toronto: Pontifical Institute of Mediaeval Studies, 1985), p. 66.

7. Avicenna, *Liber de anima seu Sextus de naturalibus*, 5.1, 2 vols., ed. S. Van Riet (Leiden: E.J. Brill, 1968–72), vol. 2, pp. 73–74.

8. See James A. Weisheipl, "Curriculum of the Faculty of Arts at Oxford in the Early Fourteenth Century," *Mediaeval Studies* 26 (1984), pp. 143–85; and John Emery Murdoch, "The Unitary Character of Medieval Learning," in John Emery Murdoch and Edith Dudley Sylla (eds.), *The Cultural Context of Medieval Learning* (Dordrecht: D. Reidel, 1973), pp. 271–348, esp. 271–72. For an introduction to the voluminous historical literature on the universities, see Pearl Kibre and Nancy G. Siraisi, "The Institutional Setting: The Universities," in Lindberg, *Science in the Middle Ages*, pp. 120–44.

9. Roger Bacon, *Questiones supra libros prime philosophie Aristotelis (Metaphysica I, II, V–X)*, in *Opera*, ed. Steele, fasc. 10, pp. 18–19.

10. Idem, *Questiones altere supra libros prime philosophie Aristotelis (Metaphysica I–IV)*, in *Opera*, ed. Steele, fasc. 11, p. 21. Bacon's condemnation of wonder in these academic works reflects the sensitivity of evaluations of wonder to social context: when writing for a courtly patron, Pope Clement IV, Bacon invoked the wonderfulness of certain natural phenomena frequently and with appreciation. See, for example, *The Opus Majus of Roger Bacon*, trans. Robert Belle Burke, 2 vols. (Philadelphia: University of Pennsylvania Press, 1928), vol. 1, p. 24 and vol. 2, pp. 630–31.

11. Albertus Magnus, *Metaphysica*, 1.2.6, in his *Opera omnia*, 40 vols. to date, gen. ed. Bernhard Geyer (Cologne: Aschendorff, 1951–), vol. 16/1, p. 23; for a fuller translation of this passage see J.V. Cunningham, *Woe or Wonder: The Passional Effect of Shakespearean Tragedy* (Denver: University of Denver Press, 1951), pp. 79–80.

12. Albertus Magnus, *De bono*, 3.2, ed. Henricus Kühle et al., in his *Opera omnia*, ed. Geyer, vol. 28, pp. 201–206 (quotation on p. 206). Cf. John Damascene, *De fide orthodoxa*, 2.13–15.

13. Thomas Aquinas, *Summa theologiae*, 1–2.41.4, ed. by the Institute of Medieval Studies in Ottawa, 4 vols. (Ottawa: University of Ottawa Press, 1940–44), vol. 2, cols. 935b–936b. See also idem, *In Metaphysicam Aristotelis commentaria*, 1.3.55, ed. M.-R. Cathala (Turin: Pietro Marietti, 1915), pp. 19–20. On Thomas's association of wonder

with pleasure, see *Summa theologiae*, 1–2.32.8, and Cunningham, *Woe or Wonder*, pp. 81–83.

14. Thomas Aquinas, *Summa theologiae*, 3.15.8, vol. 4, col. 2524a. Cf. Augustine, *Two Books on Genesis against the Manichees*, 1.8, in *Saint Augustine on Genesis*, trans. Roland J. Teske (Washington DC: Catholic University of America Press, 1991), p. 62.

15. See above Chapter One.

16. On the medieval definition of *ars* (Greek *techne*) and its relationship to philosophy and *scientia* (Greek *episteme*), see James A. Weisheipl, "The Classification of the Sciences in Medieval Thought," *Mediaeval Studies* 27 (1965), pp. 54–90; and idem, "The Nature, Scope, and Classification of the Sciences," in Lindberg, *Science in the Middle Ages*, pp. 461–82.

17. Aristotle's most extended discussion of *episteme* appears in his *Posterior Analytics*; for a brief survey of Aristotle's thought on this subject, see William A. Wallace, *Causality and Scientific Explanation*, 2 vols. (Ann Arbor: University of Michigan, 1972), vol. 1, pp. 11–18.

18. Although the ghost of this Aristotelian ideal persists in our own view of "science," an English word that in its present sense dates only to the nineteenth century, it is important to note that the two are in no way convertible: science in the modern sense includes vast amounts of material that medieval philosophers would have rejected as lowly and fallible guesswork, while it excludes theology, which they considered to be the model of *scientia*, since it dealt with the immutable realm of the divine. See in general Sydney Ross, "'Scientist': The Story of a Word," *Annals of Science* 18 (1962), pp. 65–86.

19. See Eileen Serene, "Demonstrative Science," in Norman Kretzmann, Anthony Kenny, and Jan Pinborg (eds.), *The Cambridge History of Later Medieval Philosophy* (Cambridge: Cambridge University Press, 1982), pp. 496–517; and William A. Wallace, "Albertus Magnus on Suppositional Necessity in the Natural Sciences," in James A. Weisheipl (ed.), *Albertus Magnus and the Sciences: Commemorative Essays 1980* (Toronto: Pontifical Institute of Mediaeval Studies, 1980), pp. 103–28.

20. Aristotle, *Parts of Animals*, 1.1, 639b7–640b16; 2.1, 646a8–13. On the role of "histories" in the study of nature, see Roger French, *Ancient Natural History* (London: Routledge, 1994), pp. 1–5; for Aristotle's own work in this area, see ibid., Ch. 1.

21. See Benedict M. Ashley, "St. Albert and the Nature of Natural Science," in Weisheipl, *Albertus Magnus*, pp. 73–102, on p. 94.

22. On Albertus' work in natural history, see Robin S. Oggins, "Albertus Magnus on Falcons and Hawks," in Weisheipl, *Albertus Magnus*, Ch. 17; John M. Riddle and James A. Mulholland, "Albert on Stones and Minerals," in ibid., Ch. 8; and Jerry Stannard, "Albertus Magnus and Medieval Herbalism," in ibid., Ch. 14.

23. Albertus Magnus, *De vegetabilibus libri VII*, 6.1.1, in *Opera omnia*, 38 vols., ed. Auguste Borgnet (Paris: Ludovicus Vives, 1890–99), vol. 10, pp. 159–60.

24. Ibid., 7.1.1, p. 590; see Nancy G. Siraisi, "The Medical Learning of Albertus Magnus," in Weisheipl, *Albertus Magnus*, pp. 379–414, esp. pp. 383–87.

25. Katharine Park and Eckhard Kessler, "The Concept of Psychology," in Charles B. Schmitt, Quentin Skinner, Eckhart Kessler, and Jill Kraye (eds.), *The Cambridge History of Renaissance Philosophy* (Cambridge: Cambridge University Press, 1988), p. 456, n. 5.

26. Aristotle, *De mirabilibus auscultationibus*, 830b5–11, 832b4–5, trans. L.D. Dowdall, in Aristotle, *Works*, vol. 2, pp. 1272 and 1275.

27. Albertus Magnus, *Physics*, 2.1.17, in *Opera omnia*, ed. Borgnet, vol. 3, pp. 151–52.

28. Aristotle, *Parts of Animals*, 1.5, 645a17–25, trans. W. Ogle, in *Works*, vol. 1, p. 1004.

29. Albertus Magnus, *De vegetabilibus*, 6.1.8, in *Opera omnia*, ed. Borgnet, vol. 10, p. 167; ibid., 6.1.19, p. 182.

30. Idem, *Book of Minerals*, 2.2.1, pp. 69 and 71; 2.2.10, p. 102.

31. Jole Agrimi and Chiara Crisciani, *Edocere medicos: Medicina scolastica nei secoli XIII–XV* (Naples: Guerini e Associati, 1988), esp. Ch. 2; see also their "Medici e 'vetulae' dal Duecento al Quattrocento: Problemi di una ricerca," in Paolo Rossi (ed.), *Cultura popolare et cultura dotta nel Seicento* (Milan: F. Angeli, 1983), pp. 144–59; idem, "Per una ricerca su *experimentum-experimenta*: Riflessione epistemologica et tradizione medica (secoli XIII–XV)," in Pietro Janni and Innocenzo Mazzini (eds.), *Presenza del lessico greco e latino nelle lingue contemporanee* (Macerata: Università degli Studi di Macerata, 1990), pp. 9–49; and Eamon, *Science*, pp. 53–58.

32. Chiara Crisciani and Claude Gagnon, *Alchimie et philosophie au Moyen Age: Perspectives et problèmes* (Quebec: L'Aurore/Univers, 1980), p. 65.

33. Agrimi and Crisciani, *Edocere medicos*, Ch. 2 et passim.

34. Albertus Magnus, *Metaphysica*, 1.2.10, in *Opera omnia*, ed. Geyer, vol. 16/1, p. 26.

35. Ibid., 1.1.11, p. 16. Note that Albertus indicated here that the *artifex* can also learn in this way, but that he is further differentiated from the philosopher by the fact that his work is manual and operative, rather than speculative.

36. [Pseudo-]Albertus Magnus, *De mirabilibus mundi*, in idem, *De secretis mulierum, . . . eiusdem de virtutibus herbarum, lapidum et animalium, . . . item de mirabilibus mundi* (Lyon: n.publ., 1560), sig. x7v: "opus sapientis est facere cessare mirabilia rerum quae apparent in conspectu hominum." For other formulations of the same idea, see Bert Hansen's introduction to his *Nicole Oresme*, pp. 64–69.

37. R. Bacon, *Questiones altere supra libros prime philosophie Aristotelis*, in *Opera*, ed. Steele, fasc. 11, p. 1; see in general A.G. Molland, "Medieval Ideas of Scientific Progress," *Journal of the History of Ideas* 39 (1978), pp. 561–77.

38. This translation of the Arabic text of Aristotle's *Metaphysics*, which was associated with the commentary of Averroes, lacked the relevant chapters of Book I; see Roger Steele's introduction to Roger Bacon, *Questiones supra libros prime philosophie Aristotelis*, in *Opera*, ed. Steele, fasc. 10, pp. xi–xii.

39. Aristotle, *Metaphysics*, 6.2, 1027a12, in *Works*, vol. 2, p. 1621.

40. Ibid., 1027a20–21, p. 1622.

41. Robert McQueen Grant, *Miracle and Natural Law in Graeco-Roman and Early*

Christian Thought (Amsterdam: North–Holland, 1952). On the evolution of the idea of laws of nature in the later medieval and early modern period, see John R. Milton, "The Origin and Development of the Concept of the 'Laws of Nature,'" *Archives of European Sociology* 22 (1981), pp. 173–95; Jane E. Ruby, "The Origins of Scientific 'Law,'" *Journal of the History of Ideas* 47 (1986), pp. 341–59; and Francis Oakley, "Christian Theology and Newtonian Science: The Rise of the Concept of the Laws of Nature," *Church History* 30 (1961), pp. 433–57.

42. For a lucid exposition of these issues, see Hansen, *Nicole Oresme*, pp. 62–64 (with references); and idem, "Science and Magic," in Lindberg, *Science in the Middle Ages*, pp. 483–506, on pp. 484–89. On the personification of nature as an artisan, a common trope in high medieval literature, see Economou, *Goddess Natura*, Chs. 3–4 et passim.

43. On this new view of nature, see Chenu, *Nature*, Ch. 2; Richard C. Dales, "A Twelfth-Century Concept of the Natural Order," *Viator* 9 (1978), pp. 179–92; and Benedicta Ward, *Miracles and the Medieval Mind: Theory, Record and Event, 1000–1215* (Philadelphia: University of Pennsylvania Press, 1982), pp. 3–7. Caroline Bynum has presented a subtle analysis of the relationship between marvels and miracles in her "Wonder," *American Historical Review* 102 (1997), esp. pp. 6–12.

44. Thomas Aquinas, *Summa contra gentiles*, 3.99.9, trans. Vernon J. Bourke, 3 vols. in 4 (Notre Dame: University of Notre Dame Press, 1975), vol. 3, pt. 2, pp. 78–79 (translation here modified for accuracy).

45. On the theory of miracles, see Ward, *Miracles*, Ch. 1. On the distinction between the marvelous and the miraculous, see Hansen, *Nicole Oresme*, pp. 62–66; and Agrimi and Crisciani, "Ricerca," esp. pp. 14–18.

46. See above Chapter One.

47. Thomas Aquinas, *Summa contra gentiles*, 3.101.2, vol. 3, pt. 2, p. 82.

48. Ibid., 3.102.3, p. 83.

49. Augustine, *City of God*, 21.6, p. 976. See above Chapter One.

50. Augustine, *Confessions*, 10.35, 2 vols., trans. William Watts [1631], (Cambridge MA: Harvard University Press, 1988), vol. 2, p. 175; cf. 1 John 2.16. We have in general relied on this edition for the Latin text and made our own translations. On curiosity in Augustine and other late antique and patristic writers, see Henri-Irénée Marrou, *Saint Augustin et la fin de la culture antique*, 4th edn. (Paris: E. de Boccard, 1958), pp. 149–57; and André Labhardt, "Curiositas: Notes sur l'histoire d'un mot et d'une notion," *Museum Helveticum* 17 (1960), pp. 206–24.

51. Augustine, *Confessions*, 10.35, vol. 2, p. 174. See also ibid., 5.3, vol. 1, pp. 210–16. On curiosity's associations with magic, see Hans Joachim Mette, "Curiositas," in *Festschrift Bruno Snell* (Munich: C.H. Beck, 1956), pp. 227–35.

52. Augustine, *Confessions*, 10.35, vol. 2, p. 176.

53. Ibid., 10.29, vol. 2, pp. 148–50.

54. Ibid., 10.35, vol. 2, p. 180.

55. Ibid., p. 176.

56. Ibid., 5.3, vol. 1, p. 212. See also Corrado Bologna, "Natura, miracolo, magia nel pensiero cristiano dell'alto Medioevo," in P. Xella (ed.), *Magia: Studi di storia delle religioni in memoria di Raffaela Garosi* (Rome: Bulzoni, 1976), pp. 253–72, esp. pp. 262–64.

57. On medieval discussions of curiosity, see Eamon, *Science*, pp. 59–66. The accounts in Hans Blumenberg, *The Legitimacy of the Modern Age*, trans. Robert M. Wallace (Cambridge, MA: MIT Press, 1983), pt. III, Chs. 4–6, and, especially, Christian K. Zacher, *Curiosity and Pilgrimage: The Literature of Discovery in Fourteenth-Century England* (Baltimore: Johns Hopkins University Press, 1976), Ch. 2, are misleading, as their authors take the modern meaning of "curiosity" as their point of departure and project it back onto earlier texts.

58. Bernard of Clairvaux, *The Twelve Degrees of Humility and Pride* [comp. c. 1127], 10, trans. Barton R.V. Mills (London: The Macmillan Co., 1929), p. 70; Innocent III cited in Zacher, *Curiosity and Pilgrimage*, p. 18.

59. Albertus Magnus, *De bono*, 4.1.4, in *Opera omnia*, ed. Geyer, vol. 28, p. 442. For earlier expressions of this idea, see Labhardt, "*Curiositas*."

60. Thomas Aquinas, *Summa theologiae*, 2–2.167.1, vol. 3, col. 2245b.

61. Ramon Lull, *Felix, or the Book of Wonders*, prologue, in *Selected Works of Ramon Llull (1232–1316)*, 2 vols., trans. Anthony Bonner (Princeton: Princeton University Press, 1985), vol. 2, p. 659.

62. Ibid., p. 660.

63. Ibid., conclusion, p. 1103.

64. Bonner, Introduction to Lull, *Book of Wonders*, in *Selected Works*, vol. 2, pp. 655–58; and Gret Schib, *La Traduction française du 'Libre de meravelles' de Ramon Lull* (Schaffhausen: Bolli und Böcherer, 1969), pp. 13–16.

65. For example, R. Bacon, *Epistola*, 1, in *Opera*, ed. Brewer, vol. 1, pp. 523–24; on conjuring as a form of entertainment, see Kieckhefer, *Magic*, pp. 90–94.

66. Thomas Aquinas, *Summa contra gentiles*, 3.92.13; vol. 3, pt. 2, p. 48. Cf. Aristotle, *Metaphysics*, 5.30, 1025a13–19, where the example involves a man digging a hole for a plant. For Albertus, see above, n. 27.

67. For example, Albertus Magnus, *Book of Minerals*, 1.1.6, pp. 24–26; and Thomas Aquinas, *De occultis operibus naturae ad quemdam militem*, esp. pp. 192–95. On this doctrine and its history, see Brian Copenhaver, "Natural Magic, Hermetism, and Occultism in Early Modern Science," in David C. Lindberg and Robert S. Westman (eds.), *Reappraisals of the Scientific Revolution* (Cambridge: Cambridge University Press, 1990), pp. 261–301, esp. pp. 272–77; and idem, "A Tale of Two Fishes: Magical Objects in Natural History from Antiquity through the Scientific Revolution," *Journal of the History of Ideas* 52 (1991), pp. 373–98, esp. pp. 380–83.

68. Albertus Magnus, *Book of Minerals*, e.g., 1.1.7, p. 28; 2.3.3, p. 131.

69. [Pseudo-]Albertus Magnus, *De mirabilibus mundi*, sig. x8r-v, in idem, *De secretis mulierum*; Loris Sturlese, "Saints et magiciens: Albert le Grand en face d'Hermès Trismégiste," *Archives de Philosophie* 43 (1980), pp. 615–34, esp. pp. 627–28; and Paola Zam-

belli, "Scholastic and Humanist Views of Hermeticism and Witchcraft," in Ingrid Merkel and Allen G. Debus (eds.), *Hermeticism and the Renaissance: Intellectual History and the Occult in Early Modern Europe* (Washington: Folger Shakespeare Library, 1988), pp. 125–35, on p. 127. Cf. Avicenna, *Liber de anima*, 4.2–4, vol. 2, pp. 12–67. For a lucid discussion of this general problem, see Armand Maurer, "Between Reason and Faith: Siger of Brabant and Pomponazzi on the Magic Arts," *Mediaeval Studies* 18 (1956), pp. 1–18, esp. pp. 2–11.

70. For example, R. Bacon, *Epistola*, 3, in *Opera*, ed. Brewer, vol. 1, p. 529; Thomas Aquinas, *Summa contra gentiles*, 3.103.1–4, vol. 3, pt. 2, pp. 86–87; and idem, *De occultis operibus naturae*, p. 26.

71. Ibid., 5, p. 22.

72. Siger of Brabant, *Questiones in Metaphysicam*, cited in Maurer, "Between Reason and Faith," pp. 7–8.

73. Kieckhefer, *Magic*, p. 12. In this they followed vernacular usage; see Wagner, *"Sorcier,"* esp. pp. 98–119; Dickson, *Valentine and Orson*, pp. 191–216. On belief in and practice of demonic magic, see Kieckhefer, *Magic*, pp. 150–75.

74. Thomas Aquinas, *De occultis operibus naturae*, pp. 20–30; R. Bacon, *Epistola*, 1, in *Opera*, ed. Brewer, vol. 1, p. 523.

75. Siger, *Questiones in Metaphysicam*, cited in Maurer, "Between Reason and Faith," p. 9. On learned attitudes toward "popular" superstition in this area, see Kieckhefer, *Magic*, pp. 181–87, and in general Dieter Harmening, *Superstitio: Überlieferungs- und theoriegeschichtliche Untersuchungen zur kirchlich-theologischen Aberglaubensliteratur des Mittelalters* (Berlin: Erich Schmidt, 1979), esp. pp. 296–317.

76. [Pseudo-]Albertus Magnus, *De mirabilibus mundi*, sig. z8v–aa1r, in idem, *De secretis mulierum*.

77. William Eamon, "Books of Secrets in Medieval and Early Modern Science," *Sudhoffs Archiv* 69 (1985), pp. 26–49; Eamon, *Science*, Ch. 2, esp. p. 58; and Agrimi and Crisciani, "Ricerca."

78. Peter of Maricourt, *Epistola de magnete*, 1.10, trans. Joseph Charles Mertens, in Edward Grant (ed.), *A Source Book in Medieval Science* (Cambridge, MA: Harvard University Press, 1974), pp. 372–73. See Edward Grant, "Peter Peregrinus," *DSB*, vol. 10, pp. 532–40.

79. [Pseudo-]Albertus Magnus, *De mirabilibus mundi*, sig. cc2r, cc4r, cc3r, in idem, *De secretis mulierum*.

80. For a survey of these efforts, including Buridan's work on comets, see Hansen, *Nicole Oresme*, pp. 54–61.

81. Oresme, *De causis mirabilium*, prologue, trans. in Hansen, *Nicole Oresme*, p. 137. For Oresme's biography, see ibid., pp. 6–7 and the references therein.

82. For Oresme's attack on astrology, see Hansen, *Nicole Oresme*, pp. 17–25; on Henry of Hesse, see Thorndike, *History*, vol. 3, pp. 480–501.

83. William R. Jones, "Political Uses of Sorcery in Medieval Europe," *The Historian*

34 (1972), 670–87, esp. pp. 685–86; Nicolas Jourdain, "Nicolas Oresme et les astrologues de la cour de Charles V," *Revue des Questions Historiques* 18 (1875), pp. 136–59; George W. Coopland, *Nicole Oresme and the Astrologers: A Study of His Livre de divinacions* (Cambridge, MA: Harvard University Press, 1952); and Hilary M. Carey, *Courting Disaster: Astrology at the English Court and University in the Later Middle Ages* (London: Macmillan, 1992), esp. pp. 106–11 and 15–20.

84. Nicole Oresme, *De causis mirabilium*, 3, in Hansen, *Nicole Oresme*, p. 225.

85. Ibid., pp. 223–25 (slight adjustments in translation).

86. Ibid., pp. 247 and 241 (quotation). See in general Stefano Caroti, "*Mirabilia e monstra* nei *Quodlibeta* di Nicole Oresme," *History and Philosophy of the Life Science* 6 (1984), pp. 133–50.

87. Oresme, *De causis mirabilium*, 3, in Hansen, *Nicole Oresme*, p. 271.

88. Ibid., prologue, p. 139 (punctuation added for clarity); see also pp. 343–45.

89. Aristotle, *Ethics*, 2.2, 1104a5–11. See Hansen, *Nicole Oresme*, pp. 74–76.

90. See John Emery Murdoch, "The Analytical Character of Late Medieval Learning: Natural Philosophy without Nature," in Lawrence D. Roberts (ed.), *Approaches to Nature in the Middle Ages* (Binghamton, NY: Center for Medieval and Early Renaissance Studies, 1982), pp. 171–213, esp. p. 174.

CHAPTER FOUR: MARVELOUS PARTICULARS

1. See Marco Ariani, "Il *fons vitae* nell'immaginario medievale," in F. Cardini and M. Gabriele (eds.), *Exaltatio essentiae. Essentia exaltata* (Pisa: Pacini, 1992), pp. 140–65.

2. Giovanni Dondi, *De fontibus calidis agri Patavini consideratio*, 2, in Tommaso Giunta (ed.), *De balneis omnia quae extant apud Graecos, Latinos, et Arabas....* (Venice: Giunta, 1553), fol. 95v. For a brief summary of this treatise, see Thorndike, *History*, vol. 3, pp. 392–96; translation of this passage adapted from translation on p. 396. Dondi is better known to historians of science for his famous mechanical clock.

3. Dondi, *De fontibus*, 2, fol. 95v; cf. Aristotle, *Metaphysics*, 1.2, 982b10–18.

4. Dondi, *De fontibus*, 2, fol. 95v (our emphasis). Cf. Aristotle, *Parts of Animals*, 1.5, 645a17–26; and Avicenna, *De viribus cordis*, 1.10, fol. 547r.

5. See above Chapter Three.

6. For a vivid portrait of this environment in the sixteenth century, see Paula Findlen, "Courting Nature," in N. Jardine, J.A. Secord, and E.C. Spary (eds.), *Cultures of Natural History* (Cambridge: Cambridge University Press, 1996), pp. 57–74.

7. On therapeutics in this period, see Nancy G. Siraisi, *Medieval and Early Renaissance Medicine: An Introduction to Knowledge and Practice* (Chicago: Chicago University Press, 1990), pp. 141–52. On *practica*, see Danielle Jacquart, "Theory, Everyday Practice, and Three Fifteenth-Century Physicians," in Michael McVaugh and Nancy G. Siraisi (eds.), *Renaissance Medical Learning: The Evolution of a Tradition = Osiris* 6 (1990), pp. 140–60, on pp. 140–41; Agrimi and Crisciani, *Edocere medicos*, esp. Ch. 6; Luke Demaitre, "Scholasticism in Compendia of Practical Medicine, 1250-1450," in Nancy Siraisi and Luke Demaitre

(eds.), *Science, Medicine, and the University, 1200–1550: Essays in Honor of Pearl Kibre =
Manuscripta* 20/2–3 (1976), pp. 81–95; and Andrew Wear, "Explorations in Renaissance
Writings on the Practice of Medicine," in Andrew Wear, R.K. French, and I.M. Lonie
(eds.), *The Medical Renaissance of the Sixteenth Century* (Cambridge: Cambridge University
Press, 1985), pp. 118–45.

8. On the relationship between *theorica* and *practica*, see Agrimi and Crisciani, *Edocere
medicos*, esp. Chs. 1 and 5; Luke Demaitre, "Theory and Practice in Medical Education at
the University of Montpellier in the Thirteenth and Fourteenth Centuries," *Journal of the
History of Medicine and Allied Sciences* 30 (1975), pp. 103–23; and Michael R. McVaugh,
"The Development of Medieval Pharmaceutical Theory," in Arnald of Villanova, *Opera
medica omnia*, vol. 2: *Aphorismi de gradibus*, ed. McVaugh (Granada: Seminarium Historiae
Medicae Granatensis, 1975), esp. pp. 31–48 and 89–120.

9. See, e.g., Katharine Park, *Doctors and Medicine in Early Renaissance Florence* (Princeton: Princeton University Press, 1985), Ch. 3; and Michael R. McVaugh, *Medicine Before
the Plague: Practitioners and Their Patients in the Crown of Aragon, 1285–1345* (Cambridge:
Cambridge University Press, 1993).

10. On the intellectual and social environment and the work it supported, see Park,
Doctors and Medicine, pp. 198–220, and Nancy G. Siraisi, *Taddeo Alderotti and His Pupils: Two
Generations of Italian Medical Learning* (Princeton: Princeton University Press, 1981), esp.
Chs. 1–3. On the relations between natural philosopical and medical training: Paul Oskar
Kristeller, "Philosophy and Medicine in Medieval and Renaissance Italy," in Stuart F. Spicker
(ed.), *Organism, Medicine and Metaphysics* (Dordrecht: D. Reidel, 1978), pp. 29–40.

11. Domenico Barduzzi, *Ugolino da Montecatini* (Florence: Istituto Micrografico Italiano,
1915), p. 71. See in general Katharine Park, "Natural Particulars: Medical Epistemology,
Practice, and the Literature of Healing Springs," in Anthony Griffin and Nancy G. Siraisi
(eds.), *Natural Particulars: Nature and the Disciplines in Renaissance Europe* (Cambridge, MA:
MIT Press, 1999), pp. 347–68; Richard Palmer, "'In this our lightye and learned tyme': Italian Baths in the Era of the Renaissance," in Roy Porter (ed.), *The Medical History of Waters
and Spas* (London: Wellcome Institute for the History of Medicine, 1990), pp. 14–22.

12. Albertus Magnus, *Book of Minerals*, 1.1.7–9, pp. 26–35. See Park, "Meanings."

13. Savonarola, *De balneis*, 2.1, fol. 11r.

14. Ugolino composed his *Tractatus de balneis* in 1417 and revised and expanded it in
1420; Michele Savonarola's *De balneis et thermis naturalibus omnibus Italiae* dates from
1448–49: Ugolino of Montecatini, *Tractatus de balneis*, ed. and trans. Michele Giuseppe
Nardi (Florence: Olschki, 1950); and Michele Savonarola, *De balneis et thermis naturalibus
omnibus Italiae*, in Giunta, *De balneis*, fols. 1r–36v. On the career and work of Ugolino, see
Barduzzi, *Ugolino*. On Savonarola, see Danielle Jacquart, "Médecine et alchimie chez
Michel Savonarole," in Jean-Claude Margolin and Sylvain Matton (eds.), *Alchimie et philosophie à la Renaissance* (Paris: J. Vrin, 1993), pp. 109–22, on pp. 111–17. By the middle of
the fifteenth century, baths had become such a staple of medical writing for aristocratic
patrons that Pier Candido Decembrio took Ugolino's *Tractatus* (composed, according to

the author, at the request of his "students and certain colleagues") and improved its Latin style and its structure for presentation to Savonarola's patron, Borso d'Este: Barduzzi, *Ugolino*, p. 74. This revised version appears in Giunta, *De balneis*, fol. 47v–57v.

15. Dondi, *De fontibus*, fols. 94–108r; on Dondi, Tiziana Pesenti, "Dondi dall'Orologio, Giovanni," *Dizionario biografico degli italiani*, 44 vols. to date (Rome: Istituto della Enciclopedia Italiana, 1960–), vol. 41, pp. 96–104.

16. Dondi, *De fontibus*, 2, fol. 95r–v.

17. Peter of Eboli, *De balneis puteolanis*, ed. and trans. Carlo Marcora and Jane Dolman (Milan: Il Mondo Positivo, 1987); on this work and its cultural context, see Luigia Melillo Corleto, "La medicina tra Napoli e Salerno nel Medioevo: Note sulla tradizione dell'idroterapia," in *Atti del Congresso Internazionale su Medicina Medievale e Scuola Medica Salernitana* (Salerno, 1994), pp. 26–35.

18. See Claus Michael Kauffmann, *The Baths of Pozzuoli: A Study of the Medieval Illuminations of Peter of Eboli's Poem* (Oxford: Bruno Cassirer, 1959), esp. pp. 68–86; Jonathan J.G. Alexander et al., *The Painted Page: Italian Renaissance Book Illumination, 1450–1550* (Munich/New York: Prestel, 1994), p. 65.

19. Savonarola, *De balneis*, 2.3, fol. 18v.

20. Ibid., 2.1, fol. 11v.

21. See Park, *Doctors and Medicine*, pp. 114–15 and 216–18.

22. Ugolino of Montecatini, *Tractatus*, p. 111.

23. Savonarola, *De balneis*, 2.3, fols. 20r and 27r.

24. Dondi, *De fontibus*, 2, fol. 25v. On this aspect of alchemical writing, see Chiara Crisciani and Glaude Gagnon, *Alchimie et philosophie au Moyen Age: Perspectives et problèmes* (Quebec: L'Aurore/Univers, 1980), pp. 73–74; and Michela Pereira, *L'Oro dei filosofi: Saggio sulle idee di un alchimista del Trecento* (Spoleto: Centro Italiano di Studi sull'Alto Medioevo, 1992), pp. 91–92.

25. See Jacquart, "Médecine et alchimie," and Park, "Natural Particulars."

26. Savonarola, *De balneis*, 2.7, fol. 36r.

27. Ugolino of Montecatini, *Tractatus*, pp. 49 and 123 (quotation). Francesco of Siena had composed his own treatise on baths, dedicated to Duke Galeazzo Visconti; see Thorndike, *History*, vol. 3, pp. 534–38.

28. Marsilio Ficino, *Three Books on Life [De triplici vita]* [1489], 2.15, ed. and trans. Carol V. Kaske and John R. Clark (Binghamton, NY: Medieval and Renaissance Texts and Studies, 1989), p. 213. General accounts of Ficino's magical medicine appear in Brian Copenhaver, "Astrology and Magic," in Schmitt et al., *Cambridge History*, pp. 264–300, on pp. 274–85; Giancarlo Zanier, *La medicina astrologica e sua teoria: Marsilio Ficino e i suoi critici contemporanei* (Rome: Edizioni dell'Ateneo e Bizarri, 1977); and Daniel Pickering Walker, *Spiritual and Demonic Magic from Ficino to Campanella* (London: The Warburg Institute, 1958), pp. 3–53 and 75–84.

29. Ficino, *Three Books on Life*, 3.1, p. 247.

30. Ibid., 3.16, p. 323.

31. Brian Copenhaver, "Scholastic Philosophy and Renaissance Magic in the *De vita* of Marsilio Ficino," *Renaissance Quarterly* 37 (1984), pp. 523–24; Giancarlo Zanier, "Ricerche sull'occultismo a Padova nel secolo XV," in Antonino Poppi (ed.), *Scienza e filosofia all'università di Padova nel Quattrocento* (Padua: Edizioni Lint, 1983), pp. 345–72; and idem, "Miracoli e magia in una *quaestio* di Giacomo da Forlì," *Giornale Critico della Filosofia Italiana*, ser. 4, 7 (1976), pp. 132–42.

32. On Ficino's Hermetic and Neoplatonic sources, see e.g. Eugenio Garin, *Astrology in the Renaissance: The Zodiac of Life*, trans. Carolyn Jackson and June Allen, with Clare Robertson (London: Routledge and Kegan Paul, 1983), pp. 61–79.

33. Antonio Benivieni, *De abditis nonnullis ac mirandis morborum et sanationum causis*, ed. Giorgio Weber (Florence: Leo S. Olschki, 1994); English translation of the partial posthumous edition by Charles Singer (Springfield, IL: Thomas, 1954). On the complicated textual history of this work, see Weber's introduction to his edition, pp. 7–33.

34. Benivieni, *De abditis*, p. 49.

35. Ibid., nos. 37 and 83, pp. 94 and 142. On the use of postmortems in private practice, see Katharine Park, "The Criminal and the Saintly Body: Autopsy and Dissection in Renaissance Italy," *Renaissance Studies* 47 (1994), pp. 1–33, on pp. 8–10. Benivieni referred to postmortems in about twenty of his more than two hundred cases (in manuscripts as well as in his published work).

36. Girolamo Cardano, *De admirandis curationibus et praedictionibus morborum*, in his *Opera omnia*, 10 vols. (Lyon: Jean-Antoine Huguetan and Marc-Antoine Ravaud, 1663), vol. 7, pp. 253–64. See Nancy G. Siraisi, "'Remarkable' Diseases, 'Remarkable' Cures, and Personal Experience in Renaissance Medical Texts," in idem, *Medicine and Italian Universities, 1250–1600* (Leiden: E.J. Brill, 2001); and idem, "Girolamo Cardano and the Art of Medical Narrative," *Journal of the History of Ideas* 52 (1991), pp. 581–602, esp. pp. 590–95.

37. Marcello Donati, *De medica historia mirabili libri sex* (Mantua: Francesco Osana, 1586); see Dario A. Franchini et al., *La scienza a corte: Collezionismo eclettico, natura e immagine a Mantova fra Rinascimento e Manierismo* (Rome: Bulzoni, 1979), pp. 58–59.

38. A. Paré, *Des monstres*.

39. Valerie I.J. Flint, *The Imaginative Landscape of Christopher Columbus* (Princeton: Princeton University Press, 1992), pp. 45–46 and 58–62.

40. Christopher Columbus, *The Diario of Christopher Columbus's First Voyage to America, 1492–1493*, ed. and trans. Oliver Dunn and James E. Kelley (Jr.) (Norman: University of Oklahoma Press, 1989), pp. 104–107. Cf. the passage from Marco Polo cited in Chapter One, at note 25. On Columbus's use of the rhetoric of wonder and romance, see Campbell, *Witness*, pp. 172–87; Greenblatt, *Possessions*, pp. 52–85; and Flint, *Imaginative Landscape*, pp. 115–46.

41. Sebastian Münster, *Cosmographey. Oder Beschreibung aller Länder Herrschaften und fürnemesten Stetten des gantzen Erdbodens* [1550] (Basel: Sebastianus Henricpetrus, 1588; repr. Grünwald bei München: Konrad Kölbl, 1977), p. 1367.

42. On the interplay of "new" and "old," see Michael T. Ryan, "Assimilating New Worlds in the Sixteenth and Seventeenth Centuries," *Comparative Studies in Society and History* 23 (1981), pp. 519–38; and Klaus A. Vogel, "'America': Begriff, geographische Konzeption und frühe Entdeckungsgeschichte in der Perspektive der deutschen Humanisten," in Karl Kohut (ed.), *Von der Weltkarte zum Kuriositätenkabinett: Amerika im deutschen Humanismus und Barock* (Frankfurt: Vervuert, 1995), pp. 11–43.

43. Greenblatt, *Possessions.* On Gerald, see above Chapter One.

44. Jean de Léry, *History of a Voyage to the Land of Brazil, otherwise Called America*, preface, trans. Janet Whatley (Berkeley: University of California Press, 1990), pp. lx–lxi; original French in idem, *Histoire d'un voyage fait en la terre du Brésil, autrement dite Amérique*, ed. Jean-Claude Morisot (Geneva: Droz, 1975), fol. 102v.

45. Girolamo Cardano, *De propria vita*, 41, in idem, *Ma vie*, ed. and trans. Jean Dayre (Paris: Honoré Champion, 1935), p. 121.

46. For an overview of these developments, see Harold J. Cook, "Physicians and Natural History," in Jardine et al., *Cultures of Natural History*, pp. 91–105; Karen Meier Reeds, *Botany in Medieval and Renaissance Universities* (New York: Garland, 1991), esp. pp. 14–23 and 47–54; and Paula Findlen, *Possessing Nature: Museums, Collecting, and Scientific Culture in Early Modern Italy* (Berkeley: University of California Press, 1994), esp. pp. 248–56. The most important studies of sixteenth-century natural history are Giuseppe Olmi, *L'inventario del mondo: Catalogazione della natura e luoghi del sapere nella prima età moderna* (Bologna: Il Mulino, 1992); Reeds, *Botany*; Karl H. Dannenfeldt, *Leonhard Rauwolf: Sixteenth-Century Physician, Botanist, and Traveller* (Cambridge MA: Harvard University Press, 1968); and, especially, Findlen, *Possessing Nature*, on which we have drawn heavily throughout this section.

47. See Findlen, *Possessing Nature*, esp. Chs. 4–5.

48. Garcia de Orta, *Aromatum et simplicium aliquot medicamentorum apud Indos nascentium historia* (Antwerp: C. Plantin, 1567); Nicolas Monardes, *Historia medicinal de las cosas que se traen de nuestras Indias Occidentales que sirven en medicina* [1565–74] (Seville: Padilla, 1988); and Gonzalo Fernandez de Oviedo y Valdes, *La historia general de las Indias* (Seville: Juan Cromberger, 1535). On Oviedo's influence, see Giuseppe Olmi, "'Magnus campus': I naturalisti italiani di fronte all'America nel secolo XVI," in idem, *L'inventario*, pp. 223–26; and Antonello Gerbi, *Nature in the New World: From Christopher Columbus to Gonzalo Fernandez de Oviedo*, trans. Jeremy Moyle (Pittsburgh: University of Pittsburgh Press, 1985), pp. 129–408.

49. Pierre Belon, *Les Observations de plusieurs singularitez et choses memorables trouvées en Grece, Asie, Iudée, Egypte, Arabie, et autres pays estranges* (Paris: G. Corrozet, 1553); Dannenfeldt, *Leonhardt Rauwolf*; and Findlen, *Possessing Nature*, pp. 163–64.

50. Cited in Olmi, "'Magnus campus'," p. 40.

51. For an introduction to this literature, see William B. Ashworth (Jr.), "Natural History and the Emblematic World View," in Lindberg and Westman, *Reappraisals*, pp. 303–32; and idem, "Remarkable Humans and Singular Beasts," in Joy Kenseth (ed.), *The*

Age of the Marvelous (Hanover, NH: Hood Museum of Art, Dartmouth College, 1991), pp. 113–32, esp. pp. 114–21.

52. Nicholas Monardes, *Joyfull Newes out of the Newe Founde Worlde*, trans. John Frampton (London: W. Norton, 1577).

53. A. Paré, *Des monstres*. The illustrations taken from Thevet's *Cosmographie universelle*, 2 vols. (Paris: P. L'Huilier, 1575), appeared only in the second edition of Paré's *Oeuvres* (1585).

54. The literature on collecting in this period is voluminous and of very high quality. Essential studies include Pomian, *Collectors*; Findlen, *Possessing Nature*; Oliver Impey and Arthur MacGregor (eds.), *The Origins of Museums: The Cabinet of Curiosities in Sixteenth- and Seventeenth-Century Europe* (Oxford: Clarendon Press, 1985); Lugli, *Naturalia*; Giuseppe Olmi, "Dal 'Teatro del mondo' ai mondi inventariati: Aspetti e forme del collezionismo nell'età moderna"; idem, "Ordine e fama: Il museo naturalistico in Italia nei secoli XVI–XVII," in idem, *L'inventario del mondo*, pp. 165–209 and 255–313; Schnapper, *Géant*; and Kenseth, *Age*. Also useful is Barbara Jeanne Balsiger, *The Kunst- und Wunderkammern: A Catalogue Raisonné of Collecting in Germany, France and England, 1565–1750*, 2 vols. (Ph.D. diss., University of Pittsburgh, 1971).

55. On medical men as collectors, see Findlen, *Possessing Nature*, passim; Pomian, *Collectors*, pp. 99–105; H.D. Schepelern, "Natural Philosophers and Princely Collectors: Worm, Paludanus and the Gottorp and Copenhagen Collections," in Impey and MacGregor, *Origins*, pp. 121–27; Schnapper, *Géant*, pp. 219–23; and William Schupbach, "Some Cabinets of Curiosities in European Academic Institutions," in Impey and MacGregor, *Origins*, pp. 169–78. On sixteenth- and seventeenth-century princely *Wunderkammern*, see Chapter Seven.

56. Quoted in Lucia Tongiorgi Tomasi, *Immagine e natura: L'immagine naturalistica nei codici e libri a stampa delle Biblioteche Estense e Universitaria, secoli XV–XVII* (Modena: Panini, 1984), p. 131. On Aldrovandi's collection, see Findlen, *Possessing Nature*, pp. 17–31 et passim; and Giuseppe Olmi, *Ulisse Aldrovandi: Scienza e natura nel secondo Cinquecento* (Trent: Libera Università degli Studi di Trento, 1976), pp. 78–89.

57. Cited in Findlen, *Possessing Nature*, p. 130.

58. Giovanni Battista Olivi, *De reconditis et praecipuis collectaneis ab honestissimo et solertissimo Francisco Calceolari Veronensi in Musaeo adservatis* (Venice: Paolo Zanfretto, 1584), n.p.

59. Ibid., p. 29.

60. Ibid., pp. 36, 27, 26, 33. On the idea of "total substance," closely related to the medieval idea of specific form as an explanation for occult causation, see Wear, "Explorations," pp. 141–44, and Linda Deer Richardson, "The Generation of Disease: Occult Causes and Diseases of the Total Substance," in Wear et al., *Medical Renaissance*, pp. 175–94, on pp. 181–87.

61. Olivi, *De reconditis*, p. 5.

62. See Paula Findlen, "The Economy of Scientific Exchange in Early Modern Italy,"

in Bruce Moran (ed.), *Patronage and Institutions* (Woodbridge: Boydell & Brewer, 1991), pp. 5–24.

63. Franchini, *Scienza*, p. 38.

64. Johann Kentmann, *Calculorum qui in corpore ac membris hominum innascuntur, genera XII depicta descriptaque, cum historiis singulis admirandis*, in Conrad Gesner (ed.), *De omni rerum fossilium genere, gemmis, lapidibus, metallis, et huiusmodi* (Zurich: Jakob Gesner, 1565), separate pag. See Hugh Torrens, "Early Collecting in the Field of Geology," in Impey and MacGregor, *Origins*, pp. 204–206.

65. A. Paré, *Des monstres*; Marcello Donati, *De medica historia mirabili libri sex*, fol. 301v. On Donati and his collection, see Franchini, *Scienza*, pp. 56–62.

66. Olivi, *De reconditis*, p. 47.

67. See Findlen, *Possessing Nature*, pp. 291–392.

68. Cited in Olmi, "'Magnus campus'," pp. 235–36. See in general ibid., pp. 231–44.

69. Olivi, *De reconditis*, p. 47. See also Olmi, "'Magnus campus'," pp. 244–45.

70. Olmi, "'Magnus campus'," pp. 231–45; Findlen, *Possessing Nature*, pp. 1–4 and 248–56 et passim; and Pomian, *Collectors*, pp. 101–103.

71. The straddling of classificatory boundaries was a continuing theme in both medieval and early modern writing on the marvelous, whether embodied in the vegetable lamb or the petrified tree; see Pomian, *Collectors*, p. 76.

72. Eamon, *Science*, pp. 281–85, developing ideas in Carlo Ginzburg, "Clues: Roots of an Evidential Paradigm," in his *Clues, Myths, and the Historical Method*, trans. John and Anne C. Tedeschi (Baltimore: Johns Hopkins University Press, 1989), pp. 96–125.

73. Oresme, *De causis mirabilium*, 4, in Hansen, *Nicole Oresme*, pp. 278–79; see above Chapter Three.

74. Copenhaver, "Science," pp. 398–400; idem, "Astrology and Magic," p. 287; Maurer, "Between Reason and Faith"; and Zambelli, "Scholastic and Humanistic Views."

75. Alfonso Ingegno, "The New Philosophy of Nature," in Schmitt et al., *Cambridge History*, p. 236–63, on. p. 238. The bibliography on Ficino's philosophy is enormous; for a general introduction to this aspect of his work, see Garin, *Astrology*, pp. 61–78, and Walker, *Magic*, pp. 3–53 and 75–84.

76. Ficino, *Three Books on Life*, 3.16, pp. 322–23 and 3.12, p. 301. See Copenhaver, "Scholastic Philosophy," pp. 523–24.

77. Ficino, *Theologia platonica de immortalitate animorum*, 13.4, in his *Opera omnia*, 2 vols. in 4 (Basel, 1576; repr. Turin: Bottega d'Erasmo, 1959), vol. 1, pt. 1, p. 328.

78. Ibid., p. 320.

79. See, for example, ibid., 4.4, p. 229.

80. See above Chapter Three, and in particular, Paola Zambelli, "L'Immaginazione e il suo potere: Da al-Kindi e Avicenna al Medioevo latino e al Rinascimento," in Albert Zimmermann and Ingrid Craemer-Ruegenberg (eds.), *Orientalische Kultur und Europäisches Mittelalter* (Berlin: Walter de Gruyter, 1985), pp. 188–206; Zambelli, "Scholastic and Humanist Views," esp. pp. 127–29; and Maurer, "Between Reason and Faith."

81. Andrea Cattani, *Opus de intellectu et de causis mirabilium effectuum* ([Florence]: [Bartolomeo de' Libri?], [c. 1504]), sigs. e2r–f8v; on this work, see Eugenio Garin, *Medioevo e Rinascimento* (Bari: Giuseppe Laterza, 1961), pp. 42–45. Cattani was a practicing physician and taught philosophy at the Florentine *studio*.

82. Pietro Pomponazzi, *De naturalium effectuum causis sive de incantationibus*, 10 and 12 (Basel, 1567; repr. Hildesheim/New York: Georg Olms, 1970), pp. 159–60 and 236–39. On Pomponazzi's use of *historiae*, see ibid., 7 and 12, pp. 100–101 and 296. This first edition was posthumous; reworked by the editor, it must be used with care. On Pomponazzi's influence, see Giancarlo Zanier, *Ricerche sulla diffusione e fortuna del "De incantationibus" di Pomponazzi* (Florence: La Nuova Italia, 1975).

83. Pomponazzi, *De naturalium effectuum causis*, esp. Ch. 1.

84. On early modern demonology and its treatment of marvelous phenomena, see Stuart Clark, "The Scientific Status of Demonology," in Brian Vickers (ed.), *Occult and Scientific Mentalities in the Renaissance* (Cambridge: Cambridge University Press, 1984), pp. 351–74.

85. Johannes Trithemius, *Octo quaestiones* [1508], cited in Noel L. Bran, "The Proto-Protestant Assault upon Church Magic: The 'Errores Bohemanorum' According to the Abbot Trithemius (1462–1516)," *Journal of Religious History* 12 (1982), pp. 9–22, on p. 19.

86. Ficino, *Theologia platonica*, 13.5, in *Opera omnia*, vol. 1, pt. 1, p. 335. See also Andrea Cattani, *De intellectu*, sig. f8r. On Agrippa, see Walker, *Magic*, pp. 85–86 and 90–93.

87. Cornelius Agrippa of Nettesheim, *De occulta philosophia libri tres*, ed. V. Perrone Compagni (Leiden: E.J. Brill, 1992). On Agrippa's sources, see Compagni, "Introduction," in ibid., pp. 23–24.

88. For Agrippa's checkered career at court, see Compagni, "Introduction," in ibid, pp. 4–8. On the courtly tenor of cultural life in Lorenzo's circle, see E.B. Fryde, "Lorenzo de' Medici: High Finance and the Patronage of Art and Learning," in Dickens, *Courts*, pp. 77–98.

89. See Ingegno, "New Philosophy of Nature."

90. Siraisi, "Girolamo Cardano," esp. p. 586. The most useful extended study of Cardano's philosophy is Alfonso Ingegno, *Saggio sulla filosofia di Cardano* (Florence: La Nuova Italia, 1980); see also Alessandro Simili, *Gerolamo Cardano, lettore e medico a Bologna* (Bologna: Azzoguidi, 1969).

91. These works were frequently revised and reprinted; for their publishing history, see Ingegno, *Saggio*, pp. 209–10, n. 1.

92. Cardano, *De rerum varietate*, 8.43, in his *Opera omnia*, vol. 3, p. 162. Jean Céard discusses these themes in the work of Cardano; see his *Nature*, pp. 230–39. On the theme of nature at play, Paula Findlen, "Jokes of Nature and Jokes of Knowledge: The Playfulness of Scientific Discourse in Early Modern Europe," *Renaissance Quarterly* 43 (1990), pp. 293–331.

93. Cardano, *De subtilitate*, 6, in *Opera omnia*, vol. 3, p. 458. See Ingegno, *Saggio*, p. 210, n. 1.

94. Cardano, *De rerum varietate*, 5.19, in *Opera omnia*, vol. 3, p. 53.

95. Ibid., 2.13, p. 32.

96. Ibid., 1.4, p. 15.

97. Ibid., 8.43, p. 161.

98. Ibid., 1.7, p. 23.

99. Ibid., 1.4, p. 14.

100. Ibid., 12.62, p. 239.

101. Idem, *De subtilitate*, 16, in *Opera omnia*, vol. 3, p. 605.

102. On the special nature of prodigies among marvels, see above Chapter One and below Chapter Five.

103. Cardano, *De rerum varietate*, 18.97, in *Opera omnia*, vol. 3, pp. 340–41 (quotation on p. 340).

104. See above Chapter Two.

105. In Agrippa, *De occulta philosophia*, p. 72.

106. Levinus Lemnius, *De miraculis occultis naturae libri III* (Antwerp: Christopher Plantin, 1574), n.p.

107. Cardano, *De subtilitate*, 1, in *Opera omnia*, vol. 3, p. 357. On subtlety, see Eamon, *Science*, pp. 279–80.

108. Eamon, *Science*, p. 225.

109. Giovanni Battista della Porta, *Magiae naturalis sive de miraculis rerum naturalium libri III*, 1.5 (Naples: Matthias Cancer, 1558), p. 7. On this work, see Eamon, *Science*, pp. 210–27.

110. Ibid. On the doctrine of signatures, see Massimo Luigi Bianchi, *Signatura rerum: Segni, magia e conoscenza da Paracelso a Leibniz* (Rome: Edizioni dell'Ateneo, 1987), as well as the review by Paula Findlen, "Empty Signs? Reading the Book of Nature in Renaissance Science," *Studies in the History and Philosophy of Science* 21 (1990), pp. 511–18. The doctrine of correspondence did not demand this aristocratic ontology, as the more populist version of it proposed by Paracelsus makes clear; see Walter Pagel, *Paracelsus: An Introduction to Philosophical Medicine in the Era of the Renaissance* (Basel/New York: S. Karger, 1958), pp. 62–71.

111. Della Porta, *Magiae naturalis*, dedicatory epistle to Philip II, sig. aiir.

112. Julius Caesar Scaliger, *Exotericarum exercitationum liber XV de subtilitate ad Hieronymum Cardanum* [1557] (Hanover: Wechel, 1584), esp. pp. 287 and 812. See in general Ian Maclean, "The Interpretation of Natural Signs: Cardano's *De subtilitate* versus Scaliger's *Exercitationes*," in Vickers, *Mentalities*, pp. 231–52.

113. Eamon, *Science*, pp. 221–27. These ideas also circulated and were developed among contemporary Jewish intellectuals; see David B. Ruderman, *Kabbalah, Magic, and Science: The Cultural Universe of a Sixteenth-Century Jewish Physician* (Cambridge, MA: Harvard University Press, 1988), pp. 69–73.

114. Cook, "Physicians," p. 91.

115. Céard surveys much of this literature in his *Nature*; see also Park and Daston, "Unnatural Conceptions," pp. 35–43.

116. F. Claudius Caelestinus, *De his que mundo mirabiliter eveniunt, ubi de sensuum erroribus et potentiis animae, ac de influentia caelorum*, ed. Oronce Finé (Paris: Simon de Colines, 1542). Pseudo-Albertus' *De mirabilibus* was reprinted countless times, including in English translation, mostly in conjunction with another treatise attributed to Albertus Magnus called variously *Liber aggregationis* and *Liber secretorum . . . de virtutibus herbarum, lapidum et animalium quorundam*.

117. See in general Kenseth, *Age*, esp. Kenseth, "The Age of the Marvelous. An Introduction," pp. 25–60.

Chapter Five: Monsters: A Case Study

1. Girolamo Cardano, *De rerum varietate*, 8.43, in his *Opera omnia*, vol. 3, p. 162.

2. Cited in Dieter Wuttke, "Sebastian Brants Verhältnis zu Wunderdeutung und Astrologie," in Walter Besch (ed.), *Studien zur deutschen Literatur und Sprache des Mittelalters: Festschrift für Hugo Moser zum 65. Geburtstag* (Berlin: Erich Schmidt Verlag, 1974), pp. 272–86, on p. 284, n. 25. See also Claude Kappler, "L'Interprétation politique du monstre chez Sébastien Brant," in M.T. Jones-Davies (ed.), *Monstres et prodiges au temps de la Renaissance* (Paris: Jean Touzot, 1980), pp. 100–10. Facsimiles of these and other broadsides in Sebastian Brant, *Flugblätter des Sebastian Brant*, ed. Paul Heitz (Strassburg: Heitz and Mündel, 1915).

3. Benvenuto Cellini, *The Autobiography of Benvenuto Cellini*, trans. George Bull (London: Penguin, 1956), p. 20; and Cardano, *De rerum varietate*, 1.7, in *Opera omnia*, vol. 3, p. 26; see also Laurence Breiner, "The Basilisk," in Malcolm South (ed.), *Mythical and Fabulous Creatures* (Westport, CT: Greenwood Press, 1986), pp. 113–22, on p. 115.

4. Cornelius Gemma, *De naturae divinis characterismis, seu raris et admirandis spectaculis, causis, indiciis, proprietatibus rerum in partibus singulis universi*, 1.6 (Antwerp: Christopher Plantin, 1575), pp. 75–76; see Céard, *Nature*, pp. 365–73.

5. See above Chapter One.

6. Park and Daston, "Unnatural Conceptions."

7. Georges Canguilhem, "Monstrosity and the Monstrous," *Diogène* 40 (1962), pp. 27–42; and Céard, *Nature*, esp. pp. 442–54.

8. Luca Landucci, *A Florentine Diary from 1450 to 1516 by Luca Landucci, Continued by an Anonymous Writer till 1542 with Notes by Iodoco Del Badia*, trans. Alice de Rosen Jervis (New York: E.P. Dutton and Co., 1927), pp. 249–50 (translation slightly altered for accuracy); Italian text in idem, *Diario fiorentino dal 1450 al 1516*, ed. Iodoco Del Badia (Florence: G.C. Sansoni, 1883), p. 314.

9. Idem, *Florentine Diary*, p. 250. On the construction of this monster and contemporary responses to it, see Ottavia Niccoli, *Prophecy and People in Renaissance Italy*, trans. Lydia G. Cochrane (Chicago: University of Chicago Press, 1990), pp. 35–51; and Rudolf

Schenda, "Das Monstrum von Ravenna: Eine Studie zur Prodigienliteratur," *Zeitschrift für Volkskunde* 56 (1960), pp. 209–25.

10. See above Chapter One.

11. Giovanni Villani, *Cronica*, 9.79, vol. 2, p. 198. See also Belloni, "L'ischiopago," pp. 1–14.

12. Petrarch, *Rerum memorandarum libri*, 4.120, p. 270.

13. Landucci, *Florentine Diary*, pp. 12, 47, 272.

14. Niccoli, *Prophecy and People*, esp. Chs. 1–2.

15. Wuttke, "Sebastian Brants Verhältnis;" idem, "Sebastian Brant und Maximilian I: Eine Studie zu Brants Donnerstein-Flugblatt des Jahres 1492," in *Die Humanisten in ihrer politischen und sozialen Umwelt*, ed. Otto Herding and Robert Stupperich (Boppard: Harald Boldt, 1976), pp. 141–76; and idem, "Wunderdeutung und Politik: Zu den Auslegungen der sogenannten Wormser Zwillinge des Jahres 1495," in *Landesgeschichte und Geistesgeschichte: Festschrift für Otto Herding zum 65. Geburtstag*, ed. Kaspar Elm, Eberhard Gönner, and Eugen Hillenbrand (Stuttgart: W. Kohlhammer, 1977), pp. 217–44. See also Irene Ewinkel, *De monstris: Deutung und Funktion von Wundergeburten auf Flugblättern im Deutschland des 16. Jahrhunderts* (Tübingen: Max Niemeyer, 1995), pp. 102–18 and 227–37.

16. Niccoli, *Prophecy and People*, pp. xii–xiv (quotation on p. xii); see also Céard, *Nature*, pp. 75–82. On early modern broadside literature, see *Bibliotheca Lindesiana: Catalogue of Early English Broadsides, 1505–1897* (Aberdeen: Aberdeen University Press, 1898); J.-P. Seguin, *L'Information en France avant le périodique* (Paris: G.-P. Maisonneuve et Larose, 1964); idem, *L'Information en France de Louis XII à Henri II* (Geneva: Travaux d'Humanisme et Renaissance, 1961); Bruno Weber, *Wunderzeichen und Winkeldrucker 1543–1586* (Zürich: Urs Graf, 1972); Eugen Holländer, *Wunder, Wundergeburt und Wundergestalt: Einblattdrucke des fünfzehnten bis achtzehnten Jahrhunderts* (Stuttgart: Ferdinand Enke, 1921), esp. pp. 61–126 and 271–350; Rudolf Schenda, "Die deutschen Prodigiensammlungen des 16. und 17. Jahrhunderts," *Archiv für Geschichte des Buchwesens* 4 (1963), pp. 638–710; and idem, *Die französischen Prodigiensammlungen in der zweiten Hälfte des 16. Jahrhunderts*, Münchener Romanistische Arbeiten, Heft 16 (Munich: Hueber, 1961).

17. Davidson, "Horror," esp. pp. 36–57; see above Chapter One.

18. John 9.1–3, on the man born blind, was often cited in support of this position.

19. Johannes Multivallis Tornacensis' continuation of the chronicle of Eusebius of Caesarea, in *Eusebii Caesariensis episcopi chronicon* (Paris: Henri Etienne, 1512), fol. 175r–v.

20. For examples of illustrations in chronicles, see Loren MacKinney, *Medical Illustrations in Medieval Manuscripts* (London: Wellcome Historical Medical Library, 1965), checklist entries 3.2, 24.2, 31.1, pp. 106, 113, 114; for diaries, *Journal d'un bourgeois de Paris, 1405–1449*, ed. Alexandre Tuetey (Paris: H. Champion, 1881), pp. 238–39.

21. Jurgis Baltrusaitis, *Réveils et prodiges: Le gothique fantastique* (Paris: Armand Colin, 1960), pp. 275–93 (demons) and 305–13 (memory images and visual allegories); Camille, *Gothic Idol*, esp. pp. 57–72 (idols); and David Freedberg, *The Power of Images: Studies in the*

History and Theory of Response (Chicago: University of Chicago Press, 1989), pp. 161–91 (images as objects of meditation).

22. For a definitive account of this literature, see Céard, *Nature*, pp. 93–271.

23. Konrad Lycosthenes, *Prodigiorum ac ostentorum chronicon* (Basel: H. Petri, 1557). On Lycosthenes' work and influence, see Céard, *Nature*, pp. 161–62 and 186–91; and Jean-Claude Margolin, "Sur quelques prodiges rapportés par Conrad Lycosthène," in Jones-Davies, *Monstres et prodiges*, pp. 42–54.

24. On Oresme, see above Chapter Three.

25. Lycosthenes, *Chronicon*, dedicatory epistle, n.p.

26. Fredericus Nausea, *Libri mirabilium septem* (Cologne: Peter Quentel, 1532). For some critical Catholic writers see Ewinkel, *De monstris*, pp. 32–34.

27. See, for example, Lycosthenes' dedicatory epistle to his edition of Julius Obsequens, *Prodigiorum liber* (Basel: Oporinus, 1552), p. 9.

28. See in general Robin Bruce Barnes, *Prophecy and Crisis: Apocalypticism in the Wake of the Lutheran Reformation* (Stanford: Stanford University Press, 1988).

29. 2 Esdras 5.4–8, *New Revised Standard Version* in *The Complete Parallel Bible* (New York/Oxford: Oxford University Press, 1989), p. 2593; on this text and its influence, see Ottavia Niccoli, "'Menstruum quasi Monstruum': Monstrous Births and Menstrual Taboo in the Sixteenth Century," trans. Mary M. Gallucci, in Edward Muir and Guido Ruggiero (eds.), *Sex and Gender in Historical Perspective* (Baltimore: Johns Hopkins University Press, 1990), pp. 1–25, esp. pp. 13–14. Another text frequently cited in this connection was Luke 21.25–26.

30. See Niccoli, *Prophecy and People*, pp. 189–90.

31. Martin Luther and Philipp Melanchthon, *Deuttung der czwo grewlichen Figuren, Bapstesels czu Rom und Munchkalbs zu Freijberg ijnn Meijsszen funden*, in *D. Martin Luthers Werke*, 102 vols. to date (Weimar: H. Bohlau, 1883–), vol. 11, pp. 370–85; see also Stefano Caroti, "Comete, portenti, causalità naturale e escatologia in Filippo Melantone," in *Scienze, credenze occulte, livelli di cultura*, Istituto Nazionale di Studi sul Rinascimento (Florence: Leo S. Olschki, 1982), pp. 393–426.

32. Barnes, *Prophecy and Crisis*, pp. 87–91 and 253.

33. John Knox, *The First Blast of the Trumpet against the Monstrous Regiment of Women*, in *The Works of John Knox*, 6 vols., ed. David Laing (Edinburgh: Woodrow Society, 1846–64), vol. 4, pp. 373–74. A selection of English broadsides appears in Simon McKeown (ed.), *Monstrous Births: An Illustrative Introduction to Teratology in Early Modern England* (London: Indelible Inc., 1991). For references to similar French material, see J.-P. Seguin, *L'Information en France avant le périodique*, pp. 121–23. Dudley Wilson, *Signs and Portents: Monstrous Births from the Middle Ages to the Enlightenment* (London: Routledge, 1993), is unconvincing in its conceptual outlines and treatment of the learned literature, but gives a good sense of the flavor of this material; see esp. pp. 33–56.

34. See Bernard Capp, *English Almanacs, 1500–1800: Astrology and the Popular Press* (Ithaca: Cornell University Press, 1979), pp. 164–79; William E. Burns, *An Age of Wonders:*

Prodigies, Providence and Politics in England, 1580–1727 (Ph.D. diss., University of California at Davis, 1994); and Jerome Friedman, *The Battle of the Frogs and Fairford's Flies: Miracles and the Pulp Press during the English Revolution* (New York: St. Martin's Press, 1993).

35. *Histoires prodigieuses*, 6 vols., ed. Pierre Boaistuau et al., 1. P. Boaistuau, 2. Claude de Tesserant, 3. François de Belle-Forest, 4. Rod. Hoyer, 5. trad. du latin de M. Arnauld Sorbin, par F. de Belle-Forest, 6. par I.D.M. [Jean de Marconville] de divers autheurs anciens & modernes (Lyon: Jean Pillehotte, 1598), vol. 1, pp. 29–30. On the various French editions of this work and their dates of publication, see A. Paré, *Des monstres*, pp. 205–206. Boaistuau's contribution appeared in English translation as Edward Fenton, *Certaine Secrete Wonders of Nature* (London: Henry Bynneman, 1569).

36. *Histoires prodigieuses*, vol. 1 (Pierre Boaistuau), pp. 314–15.

37. *Histoires prodigieuses*, vol. 2 (Claude de Tesserant), p. 404.

38. *Histoires prodigieuses*, vol. 4 (Rod. Hoyer), p. 958.

39. *Histoires prodigieuses*, vol. 6 (Jean de Marconville), pp. 110–14.

40. Matthew Paris, *The Illustrated Chronicles of Matthew Paris: Observations of Thirteenth-Century Life*, trans. Richard Vaughan (Cambridge: Corpus Christi College/Alan Sutton, 1993), p. 113.

41. Lycosthenes, *Chronicon*, p. 490; Benivieni, *De abditis*, pp. 169–71.

42. Landucci, *Florentine Diary*, p. 272 (translation emended); idem, *Diario fiorentino*, p. 343.

43. Cited in Niccoli, *Prophecy and People*, p. 33. For another reference to a license of this sort, see Hyder E. Rollins (ed.), *The Pack of Autolycus, or Strange and Terrible News . . . as Told in Broadside Ballads of the Years 1624–1693* (Cambridge, MA: Harvard University Press, 1927), p. 7.

44. Lucia Pioppi, *Diario (1541–1612)*, ed. Rolando Bussi (Modena: Panini, [1982]), p. 7.

45. See for example Jean Céard's analysis of Paré's many sources: A. Paré, *Des monstres*, pp. xix–xxv.

46. J. Paul Hunter, *Before Novels: The Cultural Contexts of Eighteenth-Century English Fiction* (New York/London: W.W. Norton, 1990), p. 197; Georges May, "L'Histoire a-t-elle engendré le roman? Aspects français de la question au seuil du siècle des lumières," *Revue d'Histoire Littéraire de la France* 55 (1955), pp. 155–76; and Lennard J. Davis, *Factual Fictions: The Origins of the English Novel* (New York: Columbia University Press, 1983), pp. 54–71.

47. Sections on monsters appear in, e.g., Lemnius, *De miraculis occultis naturae*; Jakob Rueff, *De conceptu et generatione hominis, et iis quae circa haec potissimum considerantur* (Zurich: Christophorus Froschoverus, 1559); and Johannes Schenck von Grafenberg, *Observationum medicarum rarum, novarum, admirabilium et monstrosarum*, 2 vols. (Freiburg i. B.: Martin Böckler, 1596). Treatises on monsters included A. Paré, *Des monstres* [1573]; Martin Weinrich, *De ortu monstrorum commentarius* (n.p., Heinricus Osthusius, 1595); Johann Georg Schenck, *Monstrorum historia memorabilis* (Frankfurt: M. Becker, 1609); and Fortunio Liceti, *De monstrorum caussis, natura, et differentiis* [1616], 2nd ed. (Padua: Paolo

Frambotto, 1634). Although Céard dates the emergence of a specialized medical discourse about monsters to c. 1600 (*Nature*, p. 443 and Ch. 8 passim), mid-sixteenth-century medical writers on monsters seem already to have taken it for granted.

48. See Rueff, *De conceptu*, 5.3, fol. 51v; and A. Paré, *Des monstres*, 19, pp. 62–68.

49. See also Caspar Peucer, *Commentarius de praecipuis generibus divinationum* (Wittemberg: Johannes Crato, 1560), fol. 442r–v and the discussion of this passage below in the next section. On Luther's effective distinction between biblical miracles and prodigies see Ewinkel, *De monstris*, pp. 25–34.

50. Illustration and text of broadside to be found in ibid., p. 375.

51. Examples in Wilson, *Signs and Portents*, p. 44 and 60; Seguin, *L'Information en France avant le périodique*, p. 120.

52. See Niccoli, "Menstruum quasi Monstruum," pp. 3–4.

53. Rueff, *De conceptu*, 5.3, fol. 48r–v.

54. On Melanchthon's treatment of this problem in his *Initia doctrinae physicae*, see Ewinkel, *De monstris*, pp. 137–39.

55. Peucer, *Commentarius*, fol. 442r–v.

56. Franchini, *Scienza*, p. 108.

57. Ibid., p. 137.

58. See Elisabeth Scheicher, *Die Kunstkammer. Kunsthistorisches Museum, Sammlungen Schloss Ambras* (Innsbruck: Kunsthistorisches Museum, 1977), pp. 149–53. As a child Gonsalvus was brought as a gift to King Henri II of France; his hairy son Arrigo was presented by the Duke of Parma to Cardinal Farnese as a gift: see Roberto Zapperi, "Ein Haarmensch auf einem Gemälde von Agostino Carracci," in Michael Hagner (ed.), *Der falsche Körper: Beiträge zu einer Geschichte der Monstrositäten* (Göttingen: Wallstein, 1995), pp. 45–55.

59. See above Chapter Four.

60. Samuel Pepys, *The Diary of Samuel Pepys* [1659–69], 8 vols. in 3, ed. Henry B. Wheatley (London: G. Bell and Sons, 1924), vol. 8, p. 174. See in general Thomas Frost, *The Old Showman, and the Old London Fairs* (London: Tinsley Brothers, 1874), pp. 22–80; R.W. Muncey, *Our Old English Fairs* (London: Sheldon Press, 1936), pp. 45–60; and Henry Morley, *Memoirs of Bartholomew Fair* (London: Chapman and Hall, 1859), esp. Ch. 16.

61. Pierre de l'Estoile, *Mémoires-Journaux*, 12 vols., ed. Gustave Brunet (Paris: Lemerre, 1875–96); James Paris Duplessis, *A Short History of Human Prodigious and Monstrous Births of Dwarfs, Sleepers, Giants, Strong Men, Hermaphrodites, Numerous Births, and Extreme Old Age &c.*, Sloane MS 5246, British Library, London; see also Duplessis's letter to Sloane describing how he gathered his information, Add. MS 4464, fol. 44r, British Library, London. On l'Estoile and Duplessis, see Wilson, *Signs and Portents*, pp. 78–84 and 90–100.

62. *Histoires prodigieuses*, vol. 3 (François de Belleforest), p. 698.

63. "A True and Certain Relation of a Strange Birth, which Was Born at Stonehouse, in the Parish of Plymouth, the 20th of October, 1635, in the Church of Plymouth, at the Interring of the Said Birth," in McKeown, *Monstrous Births*, p. 46. The preacher considered that it was easier to delight in such monsters when they were not encountered in the flesh.

64. Thomas Lupton, *A Thousand Notable Things* (London: John Charlewood, 1586), n.p.

65. Guillaume Bouchet, *Les Serées de Guillaume Bouchet, Sieur de Brocourt* [1584], 20, 6 vols., ed. C.E. Roybet (Paris: Alphonse Lemerre, 1873), vol. 3, pp. 250–66. Although this edition is based on the version published in 1608, the twentieth *serée* first appeared in 1598.

66. Théophraste Renaudot, *Recueil general des questions traitées és conferences du Bureau d'Adresse*, conf. 100 (4 February 1636), 5 vols. (Paris: Jean-Baptiste Loyson, 1660), vol. 2, pp. 825–68. On Renaudot's political career and the Bureau d'Adresse, see Howard M. Solomon, *Public Welfare, Science, and Propaganda in Seventeenth-Century France: The Innovations of Théophraste Renaudot* (Princeton: Princeton University Press, 1972).

67. Bouchet, *Les Serées*, "Discours de l'autheur sur son Livre des Serees," vol. 1, n.p.; and Renaudot, *Recueil*, "Avis au Lecteur," vol. 1, n.p.

68. In contrast, more conservative conversation manuals, aimed at gentry and would-be gentry, recommended sober history over "all the *Helpes to Discourse* which our weake Pamphletters can publish" and warned against conversation that "may seem either fabulous or impertinent;" see Richard Brathwait, *The English Gentleman: Containing Sundry Excellent Rules or Exquisite Observations Tending to Direction of every Gentleman, of Selected Ranke and Qualitie; How to Demeane or Accommodate Himselfe in the Manage of Publike and Private Affaires* (London: John Haviland for Robert Bostock, 1630), pp. 210–11. On the seventeenth-century gentlemanly etiquette of relating marvelous tales, see Steven Shapin, *A Social History of Truth: Civility and Science in Seventeenth-Century England* (Chicago/London: University of Chicago Press, 1994), pp. 80–81 et passim.

69. "A Certaine Relation of the Hog-faced Gentlewoman Called Mistris *Tannakin Skinker*, Who Was Borne at *Wirkham* a Neuter Towne betweene the Emperour and the Hollander, Scituate on the River *Rhyne*" (London: printed by J.O. for F. Grove, 1640), in Edward William Ashbee, *Occasional Facsimile Reprints of Rare and Curious Tracts of the 16th and 17th Centuries*, 2 vols. (London: privately printed, 1868–72), vol. 2, no. 16; also reprinted in McKeown, *Monstrous Births*, pp. 49–60.

70. *The Athenian Gazette: Or Casuistical Mercury* 1 (17–30 March 1691); see also *The History of the Athenian Society for the Resolving all Nice and Curious Questions, by a Gentleman who got Secret Intelligence of Their Whole Proceedings* [by L.R. = Charles Gildon] (London: n. publ., [1691]), a parody ("a *Second-Best History* of the *Second-Best Institution*") of Thomas Sprat's *History of the Royal Society* [1667], ed. Jackson I. Cope and Harold Whitmore Jones (St. Louis: Washington University Press, 1958). On the members of the Athenian Society, see Stephen Parks, "John Dunton and The Works of the Learned," *Library*, ser. 5, 23 (1968), pp. 13–24.

71. Realdo Colombo, *De re anatomica libri XV*, 15 (Venice: Nicholas Bevilacqua, 1559), pp. 268 and 263–64. Andreas Vesalius projected but never published a second volume of his *De humani corporis fabrica* (1543), to be devoted to illness and monstrosity: Nancy Siraisi, "Establishing the Subject: Vesalius and Human Diversity in *De humani corporis fabrica*, Books 1 and 2," *Journal of the Warburg and Courtauld Institutes* 57 (1994), pp. 60–88, on

p. 71. See also William L. Straus (Jr.) and Owsei Temkin, "Vesalius and the Problem of Variability," *Bulletin of the History of Medicine* 14 (1943), pp. 609–33.

72. Colombo, *De re anatomica*, 15, p. 262. On Cardano, see above Chapter Four.

73. Ibid.

74. Liceti, *De monstrorum caussis*, pp. 1 and 7.

75. Ibid., pp. 29–30.

76. Ibid., p. 43.

77. Thomas Browne, *Religio medici* (1643; facs. repr. Menston: Scolar Press, 1970), pp. 30 and 35; Gottfried Wilhelm Leibniz, *Nouveaux essais sur l'entendement humain* [comp. 1703–05, publ. 1765], 3.3.14, in *Philosophische Schriften*, ed. Leibniz-Forschungsstelle der Universität Münster, vol. 6 (Berlin: Akademie-Verlag, 1962), p. 293.

78. On the ludic overtones of the variety of nature with respect to monsters, see Céard, *Nature*, pp. 273–91; and Findlen, "Jokes," pp. 292–331.

79. Benedetto Varchi, *La prima parte delle lezzioni di M. Benedetto Varchi nella quale si tratta della natura della generazione del corpo humano, et de' mostri*, 1. proemio (Florence: Giunta, 1560), fol. [6v].

80. Ibid., 2. proemio, fol. 94v.

81. See above Chapter Four.

82. Varchi, *Prima parte*, 1. proemio, fol. [7r].

83. Canguilhem, "Monstrosity," p. 34.

84. Varchi, *Prima parte*, 2.2, fols. 104v–5r.

85. Céard, *Nature*, pp. 442–43; cf. Aristotle, *De generatione animalium*, 4.3, 767b6–7.

86. Quoted in Céard, *Nature*, p. 445.

87. Jean Riolan (Jr.), *Discours sur les hermaphrodits* (Paris: Pierre Ramier, 1614), pp. 26 and 67–68; cf. Colombo, *De re anatomica*, p. 268.

88. Jacques Duval, *Traité des hermaphrodits, parties génitales, accouchemens des femmes, etc.* (1612; repr. Paris: Isidore Liseux, 1880). For an account of the medical controversy surrounding the Le Marcis case, see Katharine Park, "The Rediscovery of the Clitoris: French Medicine and the *Tribade*," in Carla Mazzio and David Hillman (eds.), *The Body in Parts: Discourses and Anatomies in Early Modern Europe* (New York: Routledge, 1997), pp. 171–93, on pp. 179–84.

89. Riolan, *Discours*, pp. 46–52 and 29.

90. Ibid., p. 5.

91. Robert Boyle, "Observables upon a Monstrous Head," *Philosophical Transactions* (*PT*) 1 (1665/6), pp. 85–86; and "Extrait d'une lettre ecrite de Besançon le onzième de février par M. l'Abbé Doiset à M. l'Abbé Nicusse, & communiquée par ce dernier à l'auteur du Journal touchant un monstre né à deux lieues de cette ville," *Journal des Sçavans* (*JS*), Année 1682, pp. 71–72.

92. Bernard de Fontenelle, *Histoire du renouvellement de l'Académie Royale des Sciences en M.DC.XCIX. et les éloges historiques* (Amsterdam: Pierre de Coup, 1709), p. 11.

93. For accounts of the eighteenth-century debates over the causes of monsters in the

Académie des Sciences, see Jacques Roger, *Les Sciences de la vie dans la pensée française du XVIIIe siècle*, 2nd ed. (Paris: Armand Colin, 1971), pp. 186–89 and 397–403; Patrick Tort, *L'Ordre et les monstres: Le Débat sur l'origine des déviations anatomiques au XVIIIe siècle* (Paris: Le Sycomore, 1980); Javier Moscoso, "Vollkommene Monstren und unheilvolle Gestalten: Zur Naturalisierung der Monstrosität im 18. Jahrhundert," in Hagner, *Der falsche Körper*, pp. 56–72; and Michael Hagner, "Vom Naturalienkabinett zur Embryologie: Wandlungen der Monstrositäten und die Ordnung des Lebens," in ibid., pp. 73–107. Moscoso counts 130 articles on monsters in the *Histoire et Mémoires de l'Académie Royale des Sciences* between 1699 and 1770, and 96 in the *Philosophical Transactions* between 1665 and 1780: p. 60.

94. Alexis Littre, "Observations sur un foetus humain monstrueux," *Histoire de l'Académie Royale des Sciences (HARS)*, Année 1701 (Paris, 1704), pp. 88–94.

95. "Diverses observations anatomiques. Item IV," *HARS*, Année 1715 (Paris, 1718), pp. 13–14.

96. Moscoso, "Vollkommene Monstren," pp. 63–69; and Hagner, "Vom Naturalienkabinett zur Embryologie," pp. 96–97.

97. On the maternal imagination debate, for example, see Nicholas Malebranche, *De la Recherche de la vérité* [1674–75], 2/1.7–8, 6th ed. [1712] in Geneviève Rodis-Lewis (ed.), *Oeuvres de Malebranche*, 20 vols. (Paris: J. Vrin, 1962–67), vol. 1, pp. 234–45; and the detailed rebuttal of Eustache Marcot, "Memoire sur un enfant monstrueux," *MARS*, Année 1716 (Paris, 1718), pp. 329–47; also J. Blondel, *The Strength of Imagination in Pregnant Women Examin'd* (London: J. Peale, 1727). Marie-Hélène Huet, *Monstrous Imagination* (Cambridge, MA/London: Harvard University Press, 1993) discusses theories of the maternal imagination from the Renaissance through the Enlightenment. For general accounts of seventeenth- and eighteenth-century embryological theories, see Cesare Taruffi, *Storia della teratologia*, 8 vols. (Bologna: Regia Tipografia, 1881–1894), vol. 1, Ch. 5, pp. 176–332; Joseph Needham, *A History of Embryology*, 2nd ed. (Cambridge: Cambridge University Press, 1959); Charles W. Bodemer, "Embryological Thought in Seventeenth-Century England," in Charles W. Bodemer and Lester S. King (eds.), *Medical Investigation in Seventeenth Century England* (Los Angeles: University of California Press, 1968), pp. 3–23; and Shirley A. Roe, *Matter, Life, and Generation: Eighteenth-Century Embryology and the Haller-Wolff Debate* (Cambridge: Cambridge University Press, 1981).

98. "An Account Concerning a Woman Having a Double Matrix; as the Publisher Hath Englished It out of French, Lately Printed at Paris, where the Body Was Opened," *PT* 4 (1669), pp. 969–70 (quotation on p. 969).

99. Jacques-François-Marie du Verney, "Observations sur deux enfans joints ensemble," *MARS*, Année 1706 (Paris, 1707), pp. 418–521 (quotation on p. 431).

100. Jean Mery, "Sur un exomphale monstrueuse," *HARS*, Année 1716 (Paris, 1718), pp. 17–18 (quotation on p. 18); see also *MARS*, Année 1716 (Paris, 1718), pp. 136–45.

101. Canguilhem, "Monstrosity," p. 34.

102. See above Chapter Three.

103. "Sur un agneau foetus monstrueux," *HARS*, Année 1703 (Paris, 1705), pp. 28–32, on p. 28.

104. See above Chapter Four.

105. Robert Boyle, *A Free Inquiry into the Vulgarly Received Notion of Nature: Made in an Essay Addressed to a Friend* [1685/6], in Thomas Birch (ed.), *The Works of the Honourable Robert Boyle* [1772], 6 vols., (facs. repr. Hildesheim: Georg Olms, 1965–66), vol. 5, p. 201.

106. Ibid., pp. 197–98.

107. Ibid., p. 199.

108. Ibid., p. 220.

109. Samuel Clarke, *A Discourse Concerning the Unchangeable Obligations of Natural Religion, and the Truth and Certainty of the Christian Revelation* (London: James Knapton, 1706; repr. Stuttgart/Bad Cannstatt: Friedrich Frommann Verlag 1964), p. 355.

110. Lorraine Daston, "Marvelous Facts and Miraculous Evidence in Early Modern Europe," *Critical Inquiry* 18 (1991), pp. 93–124, on pp. 120–24 [repr. in James Chandler, Arnold I. Davidson, and Harry Harootunian (eds.), *Questions of Evidence: Proof, Practice, and Persuasion across the Disciplines* (Chicago/London: University of Chicago Press, 1994), pp. 243–74].

111. Ernst Kris and Otto Kurz, *Legend, Myth, and Magic in the Image of the Artist* [1934], trans. Alastair Lang and rev. Lottie M. Newman (New Haven/London: Yale University Press, 1979), pp. 47–59; Rensselaer W. Lee, "*Ut Pictura Poesis*: The Humanistic Theory of Painting," *The Art Bulletin* 22 (1940), pp. 197–269, on pp. 207–12; and Martin Kemp, "From 'Mimesis' to 'Fantasia': The Quattrocento Vocabulary of Creation, Inspiration, and Genius in the Visual Arts," *Viator* 8 (1977), pp. 347–98.

112. Baxter Hathaway, *Marvels and Commonplaces: Renaissance Literary Criticism* (New York: Random House, 1968), pp. 45–58 et passim; Bernard Weinberg, *A History of Literary Criticism in the Italian Renaissance*, 2 vols. (Chicago: University of Chicago Press, 1961), vol. 1, pp. 534–622 et passim; and James Paul Biester, *Strange and Admirable: Style and Wonder in the Seventeenth Century*, (Ph.D. diss., Columbia University, 1990). Giorgio Valla published the first Latin translation of the *Poetics* in 1498.

113. Gregorio Comanini, *Il Figino, overo della fine della pittura*, quoted in Piero Falchetta (ed.), *The Arcimboldo Effect: Transformations of the Face from the 16th to the 20th Century* (New York: Abbeville Press, 1987), pp. 189–90. See Thomas DaCosta Kaufmann, *The Mastery of Nature: Aspects of Art, Science, and Humanism in the Renaissance* (Princeton: Princeton University Press, 1993), pp. 126–35.

114. Quoted in A.J. Smith (ed.), *John Donne: The Critical Heritage* (London: Routledge, 1975), p. 100.

115. The *Kunstkammer* of the Archduke Ferdinand II also contained portraits of a dwarf (in formal court dress, with gold chains and sword), a cripple, and a giant: see Scheicher, *Die Kunstkammer*, pp. 149–53. Pedro's son Arrigo was painted by Agostino Carracci, along with a dwarf, a fool, and exotic animals: see Zapperi, "Ein Haarmensch," pp. 45–47.

116. Quoted in Edward W. Tayler (ed.), *Literary Criticism of Seventeenth-Century England* (New York: Knopf, 1967), p. 216.

117. Biester, *Strange and Admirable*, pp. 222–24; cf. Horace, *Ars poetica* [c. 15 B.C.], ll. 1–13, in idem, *Satires, Epistles and Ars Poetica*, trans. H. Rushton Fairclough (Cambridge, MA: Harvard University Press, 1970), pp. 451–53.

118. "Merveilleux, adj. (Littérat.)," in Denis Diderot and Jean d'Alembert (eds.), *Encyclopédie, ou Dictionnaire raisonné des sciences, des arts et des métiers*, 35 vols. (Paris: Briasson, David l'Aîné, Le Breton, Durand; and Paris/Neufchastel: Samuel Faulche, 1751–80), vol. 10 (1765), pp. 393–95, on p. 393.

119. Jean d'Alembert, *Preliminary Discourse to the Encyclopedia of Diderot* [1751], trans. Richard N. Schwab (Indianapolis: Bobbs-Merrill, 1963), p. 51, original in *Discours préliminaire*, in Diderot and d'Alembert, *Encyclopédie*, vol. 1, p. xvi.

120. [Jean-François Marmontel], "Fiction (Belles-Lettres)," in Diderot and d'Alembert, *Encyclopédie*, vol. 6, p. 680.

121. Ibid., p. 682.

122. Louis Chevalier de Jaucourt, "Vraisemblance (Poésie)," in ibid., vol. 17 (1765), p. 484. The images of bird-serpent and tiger-lamb couplings are borrowed from the opening lines of Horace's *Ars poetica*, p. 451: "sed non ut placidis coeant immitia, non ut serpentes avibus geminentur, tigribus agni."

123. Marmontel, "Fiction," in Diderot and d'Alembert, *Encyclopédie*, vol. 6, p. 682.

124. François M. A. de Voltaire, "Monstres," *Dictionnaire philosophique*, in *Oeuvres complètes de Voltaire*, 52 vols. (Paris: Garnier Frères, 1877–85), vol. 20, pp. 108–109.

125. Wolfgang Amadeus Mozart and Emanuel Schikaneder, *Die Zauberflöte* [1791] (Stuttgart: Reclam, 1991), 1.12, p. 23.

126. Liceti, *De monstrorum caussis*, p. 47.

CHAPTER SIX: STRANGE FACTS
1. Gottfried Wilhelm Leibniz, "An Odd Thought Concerning a New Sort of Exhibition ("Drôle de pensée, touchant une nouvelle sorte de Représentation")" [1675], in Philip Wiener (ed.), *Leibniz: Selections* (New York: Charles Scribner's Sons, 1951), pp. 585–94.

2. Findlen, *Possessing Nature*, pp. 17–24 and 357–58.

3. Mario Biagioli, "Galileo the Emblem Maker," *Isis* 81 (1990), pp. 230–58, on pp. 241–43.

4. Céard, *Nature*, pp. 229–30; and A. Paré, *Des monstres*, pp. xv, xxi, xxii, xxiv.

5. Bernard Palissy, *Discours admirables*, in *Les Oeuvres de Bernard Palissy*, ed. Anatole France (Paris: Charavay Frères, 1880), pp. 330–31.

6. [Renaudot], *Recueil*. The original conferences were held every Monday afternoon, 22 August 1633 – 1 September 1642: see Howard M. Solomon, *Public Welfare*, Ch. 3.

7. On the controversy provoked by the accessibility of Paré's work to women, see A. Paré, *Des monstres*, pp. xiv–xvi.

8. On exchanges among scholars see Findlen, *Possessing Nature*, pp. 293–45; and Anne Goldgar, *Impolite Learning: Conduct and Community in the Republic of Letters, 1680–1750* (New Haven/London: Yale University Press, 1995), pp. 12–53.

9. On the seventeenth-century meanings of these terms see: Walter E. Houghton, "The English Virtuoso in the Seventeenth Century," *Journal of the History of Ideas* 3 (1941), pp. 51–73 and 190–219; Robert A. Greene, "Whichcote, Wilkins, 'Ingenuity,' and the Reasonableness of Christianity," *Journal of the History of Ideas* 42 (1981), pp. 227–52; Ernest B. Gilman, *The Curious Perspective. Literary and Pictorial Wit in the Seventeenth Century* (New Haven/London: Yale University Press, 1978); and Jean Céard (ed.), *La Curiosité à la Renaissance* (Paris: Société d'Edition d'Enseignement Supérieur, 1986). The connotations of these terms (and of the related but distinct "virtuoso") coincided only in part, and "ingenious" was more narrowly English than "curious," which latter had a Europe-wide circulation.

10. See above Chapter Eight; also Katie Whitaker, "The Culture of Curiosity," in Jardine et al., *Cultures of Natural History*, pp. 75–90.

11. Charles G. Nauert (Jr.), "Humanists, Scientists, and Pliny: Changing Approaches to a Classical Author," *American Historical Review* 84 (1979), pp. 72–85.

12. Sir Walter Ralegh, *The Discoverie of the large and bewtiful Empire of Guiana* [1596], ed. V.T. Harlow (London: The Argonaut Press, 1928), pp. 56–57.

13. Eamon, *Science*, pp. 194–233.

14. Descartes, *Principia philosophiae* [1644], 187, in *Oeuvres de Descartes*, 12 vols., ed. Charles Adam and Paul Tannery (Paris: Léopold Cerf, 1897–1910), vol. 8, p. 314.

15. F. Bacon, *Novum organum*, 2.28, in *Works*, vol. 4, p. 169.

16. W.E. Knowles Middleton, *The Experimenters: A Study of the Accademia del Cimento* (Baltimore/London: Johns Hopkins University Press, 1971).

17. Harcourt Brown, *Scientific Organizations in Seventeenth-Century France (1620–1680)* (Baltimore: Williams & Wilkins, 1934), pp. 70–118.

18. Palissy, *Discours admirables*, in *Oeuvres*, p. 328.

19. See above Chapter Four.

20. Scipion Dupleix, *La Physique, ou Science des choses naturelles* [1640], 7.4, ed. Roger Ariew (Paris: Fayard, 1990), p. 426.

21. Liceti, *De monstrorum caussis*, pp. 254–57.

22. Meric Casaubon, *A Treatise Concerning Enthusiasm, As it Is an Effect of Nature: but Is Mistaken by Many for either Divine Inspiration, or Diabolical Possession* (London: R.D. for Tho. Johnson, 1655), p. 41.

23. The distinction between philosophy and history, including that between natural philosophy and natural history, originates in Aristotle (*Poetics*, 1451b1–7; *Parts of Animals*, 639a13–640a10) and continued to be standard well into the eighteenth century: see for example Thomas Hobbes, *Leviathan* [1651], ed. Colin B. Macpherson (London: Penguin, 1968), 1.9, pp. 147–48; and Jean d'Alembert, "Discours préliminaire des éditeurs," in Diderot and d'Alembert, *Encyclopédie*, vol. 1, especially the chart "Système détaillé des

connaissances humaines" and the accompanying explanation. See also Gianna Pomata, "'Observatio' ovvero 'Historia': Note su empirismo e storia in eta moderna," *Quaderni Storici* n.s. 91/1996, pp. 173–98.

24. F. Bacon, *Advancement*, in *Works*, vol. 3, p. 331; cf. idem, *Novum organum*, 2.29, in *Works*, vol. 4, pp. 168–69.

25. Idem, *Advancement*, in *Works*, vol. 3, p. 328.

26. Ibid., p. 330.

27. Ibid., pp. 330–31. Bacon's use of "superstition" here refers to the possible agency of demons, as he makes clear in his allusion to James I's own work on demonology (p. 301).

28. Idem, *De augmentis scientiarum* [1623], 2.2, in *Works*, vol. 1, p. 496.

29. Idem, "Phaenomena universi," in *Works*, vol. 3, p. 688, translated in "The Phenomena of the Universe; Or, Natural History for the Basis of Natural Philosophy," in Basil Montagu (ed.), *Lord Bacon's Works*, 16 vols. (London: William Pickering, 1825–34), vol. 15, p. 126.

30. Idem, *Advancement*, in *Works*, ed. Spedding et al., vol. 3, p. 331.

31. On the relationship between the wonders of art and nature in Bacon's natural philosophy, see below Chapter Seven.

32. Giovanni Battista della Porta, *Natural Magick* [*Magiae naturalis*, 1558] (London, 1658; repr. New York: Basic Books, 1957), pp. 8 and 14.

33. F. Bacon, *Sylva Sylvarum: Or, a Natural History in Ten Centuries* [1627], in *Works*, vol. 2, p. 641. This was at once a charge of immorality (dabbling in necromancy) and philosophical unprofessionalism (abandoning the realm of the purely natural).

34. Idem, *Advancement*, in *Works*, vol. 3, pp. 361–62.

35. Idem, "Description of the Intellectual Globe," in *Works*, vol. 5, pp. 507–508.

36. Idem, "The Great Instauration," in *Works*, vol. 4, p. 25.

37. Idem, *Novum organum*, 1.12–14, in *Works*, vol. 4, pp. 48–49.

38. Idem, "Description of the Intellectual Globe," in *Works*, vol. 5, p. 508.

39. Idem, *Novum organum*, 1.18, in *Works*, vol. 4, p. 50.

40. Idem, "Preparative towards a Natural and Experimental History," in *Works*, vol. 4, p. 254.

41. See Paula Findlen, "Courting Nature," in Jardine et al., *Cultures of Natural History*, pp. 57–74, for the rise of natural history as a discipline in early modern Europe.

42. F. Bacon, *Novum organum*, 1.17, 1.19–22, in *Works*, vol. 4, pp. 49–50.

43. Ibid., 1.104, p. 97; 1.103, p. 96.

44. Ibid., 1.61, pp. 62–63; 1.21, p. 50; 1.36, p. 53; 1.50, p. 58; 1.102, p. 96; 1.103 and 1.106, pp. 96 and 98.

45. Ibid., 1.26–28, pp. 51–52.

46. See above Chapter Four.

47. F. Bacon, *Novum organum*, 2.52, in *Works*, vol. 4, pp. 246–48; 2.2–3, p. 120.

48. Ibid., 2.27, pp. 164–67. On sympathies and antipathies, cf. 2.50, pp. 242–45;

on the axioms of the *Philosophia Prima*, see F. Bacon, *Advancement*, in *Works*, vol. 3, pp. 346–49.

49. Idem, *Novum organum*, 2.28, in *Works*, vol. 4, p. 168.

50. Ibid., 2.29, pp. 168–69.

51. Ibid., 2.30, pp. 169–70.

52. Ibid., 2.31, p. 170.

53. Ibid., 2.32, p. 173.

54. Ibid., 2.28, p. 168.

55. Ibid., 2.5, p. 123. On the relationship between nature's habits and laws in Bacon, see Friederich Steinle, "The Amalgamation of a Concept — Laws of Nature in the New Sciences," in Friedel Weinert (ed.), *Laws of Nature: Essays on the Philosophical, Scientific and Historical Dimensions* (Berlin/New York: De Gruyter, 1995), pp. 316–68, on p. 337.

56. Lemnius, *Secrets*, p. 7; cf. the dedicatory epistle by Richard Le Blanc, Cardano's French translator, to Princess Margaret of France, promising things "not only delightful when one knows them, but very useful for several purposes:" Cardano, *De la subtilité*, p. ii.

57. F. Bacon, "Preparative towards a Natural and Experimental History," in *Works*, vol. 4, p. 255.

58. Idem, *Novum organum*, 2.27, in *Works*, vol. 4, pp. 166–67.

59. Idem, "Phaenomena universi," in *Works*, vol. 3, p. 688, transl. in "The Phenomena of the Universe," in *Works,* ed. Montagu, vol. 15, p. 126.

60. Idem, *Advancement*, in *Works*, ed. Spedding et al., vol. 3, p. 267. On Bacon's ambivalent views on wonder, see above Chapter Eight.

61. Ibid., in *Works*, vol. 3, p. 294.

62. Idem, "Preparative," in *Works*, vol. 4, p. 255.

63. Idem, *Novum organum*, 2.29, in *Works*, vol. 4, pp. 168–69.

64. On the social context of Bacon's antipathy toward popular natural history, see Paula Findlen, "Disciplining the Discipline: Francis Bacon and the Reform of Natural History in the Seventeenth Century," in Donald Kelley (ed.), *History and the Disciplines in Early Modern Europe* (Rochester: University of Rochester Press, 1997), pp. 239–60.

65. F. Bacon, "Description of the Intellectual Globe," in *Works*, vol. 5, pp. 510–11.

66. Francis Bacon, *The Elements of the Common Lawes of England* [1630], Preface, sig. B3r (Amsterdam/New York: Theatrum Orbis Terrarum/De Capo Press, 1969), n.p.

67. Idem, "Plan," in *Works*, vol. 4, p. 29.

68. Each of these words has a somewhat different history and field of associations: for example, *Grimms Wörterbuch* gives the first usage of *Thatsache* as 1756 (as a translation of *res facti*); *That* was the earlier standard translation of the Latin *factum*, *gestum*, or *actum*.

69. F. Bacon, *Elements*, p. 65. Bacon's nineteenth-century English translators systematically and anachronistically rendered his *res* ("things") as "facts." For example, "temere a rebus abstractae" is rendered as "overhastily abstracted from the facts" (*Novum organum*,

"Proem," in *Works*, vol. 4, p. 7); "omnia a rebus ipsis petenda sunt" as "must go to facts themselves for everything" ("Plan," in *Works*, vol. 4, p. 28); "quantum de Naturae ordine re vel mente observaverit" as "so much and so much only as he has observed in fact or in thought of the course of nature" (*Novum organum*, 1.1, in *Works*, vol. 4, p. 47).

70. See the entry on "Fact" in the *Oxford English Dictionary*, which lists the earliest usage as 1632. The French *fait* seems to have retained its primary sense as a deed or relation of deeds (cf. Latin *res gestae*) well into the eighteenth century: see for example the entry for "Fait" in the *Dictionnaire de l'Académie Française*, 2 vols. (Lyon: Joseph Duplain, 1777).

71. F. Bacon, *Elements*, pp. 68–69.

72. Idem, "Plan," in *Works*, vol. 4, pp. 29–30.

73. "A Letter from Mr. St. *Georg Ash*, Sec. of the *Dublin Society*, to one of the *Secretaries* of the *Royal Society*; Concerning a *Girl* in *Ireland*, Who Has Several *Horns* Growing on Her Body," *PT* 15 (1685), pp. 1202–1204; Joseph Pitton de Tournefort, "Description d'un champignon extraordinaire," *MARS* 1666/99, 11 vols. in 14 (Paris: Martin/Coignard/Guerin, 1733), read 3 April 1692, vol. 10, pp. 69–70; Littre, "Observations," read 5 April 1701; "An Account of *Four Suns*, Which Very Lately Appear'd in *France*, and of *Two Rainebows*, Unusually Posited, Seen in the Same Kingdom, Somewhat Longer agoe," *PT* 1 (1665/6), pp. 219–22; and Gian Domenico Cassini, "Nouveau phenomene rare et singulier d'une Lumiere Celeste, qui a paru au commencement de Printemps de cette Année 1683," in "Découverte de la Lumiere Celeste qui paroist dans le Zodiaque," *MARS* 1666/99 (Paris, 1733), vol. 8, pp. 179–278, on pp. 182–84.

74. Robert Hooke, "A General Scheme of the Present State of Natural Philosophy, and How Its Defects May Be Remedied by a Methodical Proceeding in the Making Experiments and Collecting Observations, Whereby to Compile a Natural History, as the Solid Basis for the Superstructure of True Philosophy," in *The Posthumous Works of Robert Hooke* [1705], ed. Richard Waller with an Introduction by Richard S. Westfall (New York/London: Johnson Reprint Corporation, 1969), p. 44.

75. Letter of Huygens to Leibniz, 9 October 1690, in Christiaan Huygens, *Oeuvres complètes*, 22 vols., ed. Société Hollandaise des Sciences (Amsterdam: Swets & Zeitlinger N.V., 1888–1967), vol. 9, pp. 496–99, on p. 497.

76. Edmé Mariotte, "Expérience curieuse et nouvelle," *MARS* 1666/99 (Paris, 1733), 1682, vol. 10, p. 445.

77. Cassini, "Sur un nouveau phenomene," pp. 378–79.

78. On related aspects of the history of early modern facts, see Lorraine Daston, "Baconian Facts, Academic Civility, and the Prehistory of Objectivity," in Allan Megill (ed.), *Rethinking Objectivity* (Durham/London: Duke University Press, 1994), pp. 37–63; idem, "Marvelous Facts"; and idem, "The Cold Light of Facts and the Facts of Cold Light: Luminescence and the Transformation of the Scientific Fact, 1600–1750," *EMF: Studies in Early Modern France* 3 (1997), pp. 17–44.

79. Aristotle, *Parts of Animals*, 646a8–9. See also Gianna Pomata, "A New Way of

Saving the Phenomena: From 'Recipe' to 'Historian' in Early Modern Medicine," forthcoming.

80. Idem, *History of Animals*, 490a21 and 490b26.

81. Peter Dear, "Jesuit Mathematical Science and the Reconstitution of Experience in the Early Seventeenth Century," *Studies in the History and Philosophy of Science* 18 (1987), pp. 133–75; see also idem, "Totius in Verba: Rhetoric and Authority in the Early Royal Society," *Isis* 76 (1985), pp. 145–61; and Dear, *Discipline and Experience: The Mathematical Way in the Scientific Revolution* (Chicago/London: University of Chicago Press, 1995).

82. Aristotle, *Posterior Analytics*, 78a22–79a15.

83. Hobbes, *Leviathan*, 1.9, p. 148.

84. The seventeenth-century medical "observation" shared many features with the "fact" of natural and civil history, and compendia of medical oddities corresponded closely to the strange facts of coeval natural history and natural philosophy. On these early modern medical compendia see Gianna Pomata, "Uomini mestruati. Somiglianza e differenza fra i sessi in Europa in età moderna," *Quaderni Storici*, n.s. 79/27 (1992), pp. 51–103.

85. Jean Domat, *Les Loix civiles dans leur ordre naturel* [1691], 3 vols., 2nd ed. (Paris: La Veuve de Jean-Baptiste Coignard, 1691–97), vol. 2, p. 347.

86. On skepticism, see Richard Popkin, *The History of Scepticism from Erasmus to Descartes*, rev. ed. (Assen: Van Gorcum, 1964); on degrees of certainty, Barbara J. Shapiro, *Probability and Certainty in Seventeenth-Century England* (Princeton: Princeton University Press, 1983); also Lorraine Daston, *Classical Probability in the Enlightenment* (Princeton: Princeton University Press, 1988), pp. 58–67.

87. Boyle, *Some Considerations about the Reconcileableness of Reason and Religion* (originally published as *Possibility of the Resurrection* [1675]), in *Works*, vol. 4, p. 182.

88. Sprat, *History*, p. 99.

89. Boyle, "General Heads for the Natural History of a Country, Great or Small; Drawn out for the Use of Travellers and Navigators," in *Works*, vol. 5, pp. 733–43, on p. 734.

90. John Locke, *An Essay Concerning Human Understanding*, 3.6. esp. 16, 2 vols., ed. Alexander C. Fraser (New York: Dover, 1959), vol. 2, pp. 56–70, esp. p. 70; Gottfried Wilhelm Leibniz, *Nouveaux essais*, 3.3.14, 3.6.14, 3.6.17, 4.4.13.

91. Cardano, *De subtilitate*, in *Opera omnia*, vol. 3, p. 1.

92. F. Bacon, "The Great Instauration," [1620], Preface, in *Works*, vol. 4, p. 18.

93. See above Chapter Four.

94. Descartes, *Principia philosophiae* [1644], 187, in *Oeuvres*, vol. 8, p. 314; *Principes de la philosophie* [1647], 187, ibid., vol. 9, pp. 308–309.

95. F. Bacon, *Sylva Sylvarum*, in *Works*, vol. 2, p. 648.

96. The Royal Society was granted letters patent by Charles II, but only the Académie Royale des Sciences was state-financed.

97. Harcourt Brown, *Scientific Organizations*, pp. 1–118.

98. On the rhetorical models for early modern scientific dialogues, see Jean Dietz

Moss, "The Rhetoric of Proof in Galileo's Writings on the Copernican System," in William A. Wallace (ed.), *Reinterpreting Galileo* (Washington, DC: Catholic University Press, 1986), pp. 179–204; on Galileo's relationship to courtly culture, see Mario Biagioli, *Galileo Courtier* (Chicago/London: University of Chicago Press, 1993), esp. pp. 111, 123, 126 concerning Castiglione's influence.

99. Fontenelle, *Histoire du renouvellement*, Reglement 26, pp. 36–37.

100. Sprat, *History*, p. 33. On the values of civility in the early Royal Society, see Steven Shapin and Simon Schaffer, *Leviathan and the Air-Pump: Hobbes, Boyle and The Experimental Life* (Princeton: Princeton University Press, 1985), pp. 72–76.

101. Académie Royale des Sciences, *Mémoires pour servir à l'histoire naturelle des animaux et des plantes* [1669] (Amsterdam: Pierre Mortier, 1736), Preface, p. 3r. On the botanical part of this project, see Alice Stroup, *A Company of Scientists: Botany, Patronage, and Community at the Seventeenth-Century Parisian Academy of Sciences* (Berkeley/Los Angeles/Oxford: University of California Press, 1990), pp. 65–88.

102. Claude Perrault, *Mémoires pour servir à l'histoire naturelle des animaux* [1671] (Paris: La Compagnie des Libraires, 1733), pp. viii–ix.

103. Sprat, *History*, p. 91.

104. Shapin and Schaffer, *Leviathan*, p. 321.

105. See, for example, Gian Domenico Cassini, "Sur un nouveau phenomene ou sur une lumiere celeste," *HARS*, in *MARS* 1666/99 (Paris, 1733), 1683, vol. 1, pp. 378–81, concluding section "De la Nature de cette Lumiere."

106. Edmond Halley, "An Account of the Late Surprizing Appearance of the *Lights* Seen in the *Air*, on the Sixth of *March* last; with an Attempt to Explain the Principal *Phaenomena* thereof," *PT* 29 (1714–16), pp. 406–28, on p. 406.

107. Boyle, "A Short Account of some Observations Made by Mr. Boyle, about a Diamond, That Shines in the Dark," in *Works*, vol. 4, p. 799.

108. Idem, "The Aerial Noctiluca," in *Works*, vol. 4, pp. 393–94.

109. Shapiro, *Probability*, pp. 44–54; Steven Shapin, "Robert Boyle and Mathematics: Reality, Representation, and Scientific Practice," *Science in Context* 2 (1988), pp. 373–404; and Shapin and Schaffer, *Leviathan*, pp. 22–26 and 165–68.

110. Descartes, *Discours de la méthode* [1637], 6, in *Oeuvres*, vol. 6, pp. 64–65 and 73.

111. Richard S. Westfall, *Never at Rest: A Biography of Isaac Newton* (Cambridge: Cambridge University Press, 1980), pp. 238–52.

112. F. Bacon, *Novum organum*, 1.46, in *Works*, vol. 4, p. 56.

113. Idem, *New Atlantis* [1627], in *Works*, vol. 3, pp. 164–65.

114. Descartes, *Discours de la méthode*, 6, in *Oeuvres*, vol. 6, pp. 72–73. On the actual role of servants in performing experiments, see Shapin, *Social History*, pp. 355–408.

115. Roger Hahn, *The Anatomy of a Scientific Institution. The Paris Academy of Sciences, 1666–1803* (Berkeley/Los Angeles/London: University of California Press, 1971), pp. 7–8; and Brown, *Scientific Organizations*, pp. 78–79.

116. Fontenelle, *Histoire du renouvellement*, pp. 25–26.

117. "Unless you see signs and wonders you will not believe." John 4.48, *New Revised Standard Version*, p. 2842.

118. The doctrine of the cessation of miracles had Augustinian precedents, but was revived forcefully by Jean Calvin in the "Epistre" to his *Institutions de la religion chrestienne* (n.p. [Geneva: Jean Gérard? or Michel du Bois?], 1541).

119. The sixteenth- and seventeenth-century reforms were codified in Pope Benedict XIV, *De beatificatione servorum Dei et beatorum canonizatione* (Bologna: Longhi, 1734–38); on medieval and early modern procedures, see Damian Joseph Blaher, *The Ordinary Processes in Causes of Beatification and Canonization* (Washington, D.C.: Catholic University of America Press, 1949), pp. 1–52.

120. Jean de Viguerie, "Le Miracle dans la France du XVIIe siècle," *XVIIe Siècle* 35 (1983), pp. 313–31; Barbara Schuh, *"Von vilen und mancherley seltzamen Wunderzeichen": Die Analyse von Mirakelbüchern und Wallfahrtsquellen*, Halbgraue Reihe, ed. Max-Planck-Institut für Geschichte Göttingen (St. Katharinen: Scripta Mercaturae, 1989), pp. 19–21; and Rebekka Habermas, "Wunder, Wunderliches, Wunderbares. Zur Profanisierung eines Deutungsmusters in der Frühen Neuzeit," in Richard van Dülmen (ed.), *Armut, Liebe, Ehre. Studien zur historischen Kulturforschung* (Frankfurt a.M.: Fischer Taschenbuch Verlag, 1988), pp. 38–66.

121. "Nouveautez du commencement de l'année. La Comete," *JS*, Année 1681, pp. 9–12, on p. 12.

122. "A Narrative of Divers Odd Effects of a Dreadful Thunder-clap, at Stralsund in Pomerania 19/20 June 1670; Taken out of a Relation, there Printed by Authority in High Dutch," *PT* 5 (1670), pp. 2084–87, on p. 2084.

123. "Diverses observations de Physique generale. Item V," *HARS*, Année 1703 (Paris, 1705), pp. 18–19.

124. Sprat, *History*, pp. 358–59.

125. Sprat, *History*, p. 55–56.

126. See above Chapter Nine.

127. Edmond Halley, "An Account of the Cause of the Late Remarkable Appearance of the Planet Venus, Seen this Summer, for Many Days together, in the Day Time," *PT* 29 (1714–16), pp. 466–68, on p. 468.

128. Etienne Binet [René François], *Essay des merveilles de nature et des plus nobles artifices* (Rouen: Romain de Beauvais et Iean Osmont, 1621), p. 72.

129. Thomas Browne, *Pseudodoxia Epidemica: Or, Enquiries into Very Many Received Tenents and Commonly Presumed Truths* [1646], 3.6, 2 vols., ed. Robin Robbins (Oxford: Clarendon Press, 1981), vol. 1, pp. 178–80.

130. "Histoire de l'enfant de Vilne en Lithuanie, à la dent d'or," *JS*, Année 1681, pp. 401–402, on p. 402.

131. Oldenburg to Boyle, 3 November 1664, in Boyle, *Works*, vol. 6, pp. 170–72. Boyle himself was transmitting the report of a Dr. Turbervill.

132. Boyle, "Short Account," in *Works*, vol. 1, p. 793; Benvenuto Cellini, *Abhandlun-*

gen über die Goldschmiedekunst und die Bildhauerei [1568], trans. Ruth and Max Fröhlich (Basel: Gewerbemuseum Basel, 1974), pp. 40–41.

133. Samuel Butler, "The Elephant in the Moon" [comp. c. 1676, publ. 1759], quoted in Michael McKeon, *The Origins of the English Novel 1600–1740* (Baltimore/London: The Johns Hopkins University Press, 1987), p. 72.

134. *JS*, Année 1684, p. 47.

135. "Diverses observations de Physique generale. Item X," *HARS*, Année 1703 (Paris, 1705), p. 22.

136. See below Chapter Nine concerning the shift to skepticism thereafter.

137. Shapin, *Social History*, pp. 221 and 287.

138. On relationships to foreign members and correspondents in the Paris Academy, see Hahn, *Anatomy*, pp. 77–78; on the Royal Society, see Michael Hunter, *The Royal Society and Its Fellows 1660–1700: The Morphology of an Early Scientific Institution* (Chalfont St. Giles: British Society for the History of Science, 1982), pp. 107–10.

139. Cassini, "Sur un nouveau phenomene," p. 380.

140. See Shapin, *Social History*, pp. 3–41, for an extensive discussion of the role of trust in science.

141. Domat, *Loix*, vol. 2, p. 348.

142. Boyle, "Short Account," in *Works*, vol. 1, pp. 796 and 791.

143. Steinle, "Amalgamation," pp. 335 and 337–41.

144. "Sur le Phosphore du Barometre," *HARS*, Année 1700 (Paris, 1703), pp. 5–8, on p. 8.

145. Hooke, "General Scheme," in *Posthumous Works*, p. 38.

146. Shapin, *Social History*, p. 344.

147. Quoted in Christian Licoppe, *La Formation de la pratique scientifique. Le Discours de l'expérience en France et en Angleterre (1630–1820)* (Paris: Éditions la Découverte, 1996), p. 199.

148. See below Chapter Nine.

149. Antoine Arnauld and Pierre Nicole, *La Logique, ou L'Art de penser* [1662], ed. Pierre Clair and François Girbal (Paris: Presses Universitaires de France, 1965), p. 340.

150. Locke, *Essay* [1689], 4.15.4–6, vol. 2, pp. 365–68.

151. Ibid., 4.15.5, pp. 366–67.

152. Pierre Bayle, *Pensées diverses sur la comète* [1681], 2 vols., ed. A. Prat (Paris: Société Nouvelle de Librairie et Edition, 1939), vol. 1, p. 38.

153. Eustache Marcot, "Memoire." Marcot's letter to the Academy of Science is preserved in the Pochette de Séance 29 mai 1715, Archives de l'Académie des Sciences, Paris.

CHAPTER SEVEN: WONDERS OF ART, WONDERS OF NATURE

1. On the *Kunstschrank*, which has been in the possession of the University of Uppsala since 1695, see Thomas Heinemann, *The Uppsala Art Cabinet* (Uppsala: Almqvist &

Wiksell, 1982); also Hans-Olof Boström, "Philipp Hainhofer and Gustavus Adolphus' *Kunstschrank* in Uppsala," in Impey and MacGregor, *Origins*, pp. 90–101.

2. Quoted in Boström, "Philipp Hainhofer," p. 99.

3. Quoted in Jurgis Baltrusaitis, *An Essay on the Legend of Forms* [1983], trans. Richard Miller (Cambridge, MA/London: MIT Press, 1989), p. 63.

4. A. Paré, *Des monstres*, p. 139.

5. Aristotle, *Physics*, 2.1, 192b9–19, in *Works*, vol. 1, p. 329.

6. Ibid., 2.1, 193a28–29 and 193b9–11.

7. Gérard Paré, *Les Idées et les lettres au XIIIe siècle: Le Roman de la Rose* (Montreal: Centre de Psychologie et de Pédagogie, 1947), pp. 67–68; and Economou, *Goddess Natura*, pp. 26–27. See above Chapter Two.

8. Guillaume de Lorris and Jean de Meun, *The Romance of the Rose*, trans. Charles Dahlberg (Hanover, NH/London: University Press of New England, 1983), p. 272.

9. George Puttenham, *The Art of English Poesie* [1589], ed. Gladys D. Willcock and Alice Walker [1936] (Folcroft: Folcroft Press, 1969), pp. 303–304.

10. John Case, *Lapis philosophicus* [1599], quoted in Charles B. Schmitt, "John Case on Art and Nature," *Annals of Science* 33 (1976), pp. 543–59, on p. 547; see also Elisabeth Blair MacDougall, "A Paradise of Plants: Exotica, Rarities, and Botanical Fantasies," in Kenseth, *Age*, pp. 145–57, on p. 153, for early modern grafting techniques and their reception.

11. Ferrante Imperato, *Dell'Historia naturale di Ferrante Imperato Napolitano libri XXVIII* (Naples: Constantio Vitale, 1599), Preface, n.p.

12. Palissy, *Discours admirables*, in *Oeuvres*, p. 212.

13. A.J. Close, "Commonplace Theories of Art and Nature in Classical Antiquity and in the Renaissance," *Journal of the History of Ideas* 30 (1969), pp. 467–86.

14. Thomas DaCosta Kaufmann, *The School of Prague: Painting at the Court of Rudolf II* (Chicago/London: University of Chicago Press, 1988), p. 94.

15. Edward William Tayler, *Nature and Art in Renaissance Literature* (New York/London: Columbia University Press, 1964), p. 21.

16. On the rich early modern terminology for such collections, see Findlen, "Museum," pp. 60–63; also C.F. Neickelius, *Museographia, oder Einleitung zum rechten Begriff und nützlicher Anlegung der Museorum oder Raritäten-Kammern* (Leipzig/Breslau: Michael Hubert, 1727), pp. 2–3; and Johann Daniel Major, *Unvorgreiffliches Bedencken von Kunst- und Naturalien-Kammern insgemein* [1674], repr. in D. Michael Bernhard Valentini, *Museum Museorum, oder Vollständige Schau-Bühne aller Materialien und Specereÿen nebst deren natür- lichen Beschreibung, Selection, Nutzen und Gebrauch* (Frankfurt am Main: Johann D. Zunner, 1704), pp. 4–11.

17. In addition to classics such as von Schlosser, *Die Kunst- und Wunderkammern der Spätrenaissance* [1908], and David Murray, *Museums: Their History and Their Use*, 2 vols. (Glasgow: James MacLehose and Sons, 1904), there is a large recent literature on early modern collections: Olmi, *Ulisse Aldrovandi*; Lugli, *Naturalia*; Findlen, *Possessing Nature*;

Impey and MacGregor, *Origins*; Pomian, *Collectors*; Schnapper, *Géant*; Kaufmann, *School*; Kenseth, *Age*; Elisabeth Scheicher, *Die Kunst- und Wunderkammern der Habsburger* (Vienna/Munich/Zurich: Molden Edition, 1979); Wolfgang Liebenwein, *Studiolo: Die Entstehung eines Raumtyps und seine Entwicklung bis um 1600* (Berlin: Gebr. Mann, 1977); and Horst Bredekamp, *Antikensehnsucht und Maschinenglauben: Die Geschichte der Kunstkammer und die Zukunft der Kunstgeschichte* (Berlin: Klaus Wagenbach, 1993).

18. Boström, "Philipp Hainhofer," p. 92; Baltrusaitis, *Essay*, p. 63; and Schnapper, *Géant*, pp. 220–21.

19. On the exchange networks of Italian collectors, see Findlen, *Possessing Nature*, pp. 348–60. On learned travel in the early modern period see Justin Stagl, *A History of Curiosity: The Theory of Travel 1550–1800* (Chur: Harwood, 1995); evidence on individual itineraries of naturalists can be found in Harold J. Cook, "The New Philosophy in the Low Countries," in Roy Porter and Mikulás Teich (eds.), *The Scientific Revolution in National Context* (Cambridge: Cambridge University Press, 1992), pp. 115–49, as well as in contemporary diaries and journals.

20. Michel de Montaigne, *Journal de voyage en Italie par la Suisse et l'Allemagne en 1580 et 1581*, ed. Maurice Rat (Paris: Editions Garnier Frères, 1955), p. 54. Ambras was already a tourist attraction by the late sixteenth century and remained one in the late eighteenth, an obligatory stop for Goethe en route to Italy: Scheicher, *Kunst- und Wunderkammern*, pp. 73 and 76; John Lough (ed.), *John Locke's Travels in France 1675–1679* (Cambridge: Cambridge University Press, 1953), pp. 5–6.

21. H.J. Witkam (ed.), *Catalogues of All the Chiefest Rarities in the Publick Anatomie Hall of the University of Leyden* (Leiden: University of Leiden Library, 1980).

22. Many catalogues are reproduced in part or in full in Balsiger, *The Kunst- und Wunderkammern*.

23. See, for example, Pierre Borel, *Les Antiquitez, raretez, plantes, mineraux, & autres choses considerables de la ville, & Comté de Castres d'Albigeois* (Castres: Arnaud Colomiez, 1649; repr. Geneve: Minkoff reprint, 1973), pp. 124–31; Jacob Spon, *Recherche des antiquités et curiosités de la ville de Lyon* [1673], ed. J.-B. Montfalcon (Lyon: Louis Perrin, 1857), pp. 248–58; Neickelius, *Museographia*, pp. 19–137; and Major, *Bedencken*, Appendix.

24. Neickelius, *Museographia*, pp. 455–57.

25. Major, *Bedencken*, p. 14; Erwin Neumann, "Das Inventar der rudolfinischen Kunstkammer von 1607/1611," in Magnus von Platen (ed.), *Queen Christina of Sweden. Documents and Studies* (Stockholm: P.A. Norstedt & Söner, 1966), pp. 262–65, on p. 263; Schnapper, *Géant*, pp. 105, 116, 238; Scheicher, *Kunst- und Wunderkammern*, pp. 81 and 171; and Joy Kenseth, "'A World of Wonders in One Closet Shut,'" in Kenseth, *Age*, pp. 80–101, on p. 85. The common inspiration for classification appears to have been Pliny, *Natural History*, 33–37.

26. Menzhausen, "Kunstkammer," p. 71; Kaufmann, *School*; H.D. Schepelern, "The Museum Wormianum Reconstructed. A Note on the Illustration of 1655," *Journal of the History of Collections* 1 (1990), pp. 81–85, on p. 83; and Oleg Neverov, "'His Majesty's

Cabinet' and Peter I's Kunstkammer," in Impey and MacGregor, *Origins*, pp. 54–61, on p. 60.

27. William Eamon, "Court, Academy, and Printing House: Patronage and Scientific Careers in Late-Renaissance Italy," in Moran, *Patronage*, pp. 25–50, on pp. 31 and 35; and Kaufmann, *Mastery*, pp. 178–79.

28. Mark S. Weil, "Love, Monsters, Movement, and Machines: The Marvelous in Theaters, Festivals, and Gardens," in Kenseth, *Age*, pp. 158–78; and John Shearman, *Mannerism* (Harmondsworth: Penguin, 1967), pp. 106–11.

29. Paolo Morigi, *Historia dell'antichità di Milano* [1592], quoted in Piero Falchetta, "Anthology of Sixteenth-Century Texts," in Falchetta, *The Arcimboldo Effect*, pp. 143–202, on p. 172.

30. Pomian, *Collectors*, pp. 78–99 and 121–38.

31. Schnapper, *Géant*, pp. 219–20. On the connections between medical simples and natural history collections in Italy, see Findlen, *Possessing Nature*, pp. 248–56.

32. Schnapper, *Géant*, p. 223.

33. See above Chapter Four. Other medical collectors included Girolamo Cardano in Milan, Francesco Calzolari of Verona, Ulisse Aldrovandi in Bologna, Felix Platter in Basel, Jan Swammerdam in Amsterdam, Bernard Paludanus in Enkhuizen, Basil Besler in Nuremburg, Michael Valentini in Giessen, Pierre Borel and Laurent Joubert in southern France, and Michele Mercati in Rome. The representation of medical men in this list ranges from apothecaries to professors: Cardano was a physician and held chairs of medicine at the Universities of Pavia and Bologna; Imperato was an apothecary, as was Calzolari; Mercati was papal physician; Aldrovandi had studied medicine as well as law and mathematics and taught materia medica at the University of Bologna; Worm held chairs in both medicine and Greek at the University of Copenhagen; Borel was a physician in Castres; Joubert was also a physician and chancellor of the University of Montpellier; Platter was chief physician of the city of Basel, where he also held the chair of applied medicine; Swammerdam took a medical degree in Leiden, although he did not practice. On the links between natural history and medical pursuits, see Cook, "Physicians," pp. 98–105.

34. Schupbach, "Cabinets of Curiosities," pp. 170–73; see also Arthur MacGregor, "'A Magazin of All Manner of Inventions': Museums in the Quest for 'Solomon's House' in Seventeenth-Century England," *Journal of the History of Collections* 1 (1989), pp. 207–12,- on p. 209, concerning the museum established by the Royal College of Physicians in 1654.

35. Philippe Tamizey de Larroque (ed.), *Boniface de Borilly. Lettres inédits écrites à Peiresc (1618–1631)* (Aix-en-Provence: Garcin et Didier, 1891), p. 48; and Schnapper, *Géant*, p. 95.

36. Olmi, *Ulisse Aldrovandi*, pp. 57–60.

37. Laura Laurencich-Minelli, "Museography and Ethnographical Collections in Bologna during the Sixteenth and Seventeenth Centuries," in Impey and MacGregor, *Origins*, pp. 17–23, on p. 21; cf. Schnapper, *Géant*, p. 11 concerning parallel evidence from guidebooks and vistors' reports. A recent attempt to reconstruct the museum of Olaus

Worm in Copenhagen indicates that at least some of the engravings were tolerably exact representations of actual physical arrangements: Schepelern, "Museum," pp. 81–82.

38. For this view of early modern collections, see *inter alia* Lugli, *Naturalia*, pp. 73–78 and 93–99; also Kaufmann, *Mastery*, p. 181.

39. Samuel Quicchelberg, *Inscriptiones vel tituli theatri amplissimi, complectentis rerum universitatis singulas materias et imagines exemias....* (Munich: Adam Berg, 1565). For details concerning Quicchelberg's career, especially his association with the court and *Kunstkammer* of Duke Albrecht V of Bavaria, see Eva Schulz, "Notes on the History of Collecting and of Museums in the Light of Selected Literature of the Sixteenth to Eighteenth Century," *Journal of the History of Collections* 2 (1990), pp. 205–18, on p. 206.

40. Ibid.

41. Cf. Schnapper, *Géant*, p. 10, concerning the lack of influence of the "theater of the universe" ideal on actual collections, which were highly selective. But for an opposing view, see Patricia Falguières, "Fondation du théâtre ou methode de l'exposition universelle: les inscriptions de Samuel Quicchelberg (1565)," *Les Cahiers du Musée National d'Art Moderne* 40 (1992), pp. 91–109.

42. In early modern usage, "encyclopaedia" (or "cyclopedia") meant either a basic course of studies (literally, the "circle of learning") or a comprehensive account of all knowledge: see the *Oxford English Dictionary* entry.

43. Analogous contrasts might be drawn between the alleged encyclopedism of the early modern collections and medieval encyclopedism: see above Chapter One.

44. Borel, *Catalogue de choses rares qui sont dans le Cabinet de Maistre Pierre Borel, Médecin de Castres....* [1645], quoted in Balsiger, *The Kunst- und Wunderkammern*, p. 99.

45. Pomian, *Collectors*, Ch. 1.

46. Nehemiah Grew, *Musaeum Regalis Societatis. Or, A Catalogue & Description of the Natural and Artificial Rarities Belonging to the Royal Society and Preserved at Gresham College* (London: W. Rawlins, 1681), Preface, n.p.

47. Schnapper, *Géant*, pp. 22–33.

48. On the fusion of wonder and curiosity in seventeenth-century natural philosophy, see below Chapter Eight.

49. Robert Hooke, *Micrographia: Or Some Physical Descriptions of Minute Bodies Made by Magnifying Glasses, with Observations and Inquiries Thereupon* (London: John Martyn and James Allestry, 1665), p. 184.

50. Fontenelle, *Histoire du renouvellement*, p. 16.

51. "Account of *L'Architecture navale*, avec le *routier* des Indes Orientales & Occidentales: Par le Sieur Dassié," *PT* 12 (1677), pp. 879–83, on p. 880.

52. Balthasar Monconys, *Voyages de M. de Monconys*, 4 vols. (Paris: Pierre Delaulne, 1695), vol. 3, pp. 103–108. Monconys here describes items from the collection of the Saxon Elector in Dresden, which he visited in November, 1663.

53. Zacharias Konrad von Uffenbach, *Merkwürdige Reisen durch Niedersachsen, Holland und Engelland*, 3 vols. (Ulm: Gaum, 1753–54), vol. 2, pp. 545–50, quoted in Michael Hunter,

"The Cabinet Institutionalized: The Royal Society's 'Repository' and Its Background," in Impey and MacGregor, *Origins*, pp. 159–68, on p. 167; and John Evelyn, *The Diary of John Evelyn*, ed. E.S. de Beer, 6 vols. (Oxford: Oxford University Press, 1955), vol. 3, p. 108.

54. See, for example, Etienne Binet, *Essay*, p. 496, on automata.

55. A. Paré, *Des Monstres*, p. 117.

56. H.W. Janson, "The Image Made by Chance," in Millard Meiss (ed.), *De Artibus Opuscula LX: Essays in Honor of Erwin Panofsky* (New York: New York University Press, 1961), pp. 254–66, on pp. 260–61.

57. Scheicher, *Kunst-und Wunderkammern*, p. 171.

58. Cf. how very similar objects were depicted in Dutch still lifes: Norman Bryson, *Looking at the Overlooked: Four Essays on Still Life Painting* (Cambridge, MA: Harvard University Press, 1990), p. 126.

59. Kenseth, "World," p. 95.

60. L.C. Schmalling, "Beschreibung des Naturalienkabinetts des Herrn Pastor Göze, in Quedlinburg, und seiner microscopischen Experimente," *Hannoverisches Magazin* 20 (1782), pp. 966–75, on p. 973.

61. Sullivan, "Medieval Automata." See above Chapter Two.

62. Musée National des Techniques (ed.), *Jacques Vaucanson* (Ivry: Salles & Grange, 1983); see also Read Benhamou, "From *Curiosité* to *Utilité*: The Automaton in Eighteenth-Century France," *Studies in Eighteenth-Century Culture* 17 (1987), pp. 91–105; A.J. Turner, "Further Documents Concerning Jacques Vaucanson and the Collections of the Hôtel de Mortagne," *Journal of the History of Collections* 2 (1990), pp. 41–45; Barbara Maria Stafford, *Artful Science: Enlightenment Entertainment and the Eclipse of Visual Education* (Cambridge, MA/London: MIT Press, 1994), pp. 191–95.

63. Bredekamp, *Antikensehnsucht*, pp. 49–51.

64. Pliny, *Natural History*, 35.36.65, vol. 9, p. 309–11. On the currency of the story from antiquity through the seventeenth century, see Kris and Kurz, *Legend*, pp. 62–64.

65. Cornelius de Bie, *Het Gulden Cabinet van edel vry Schilder Const* (Antwerp: Ian Meysens, 1661), pp. 217–19.

66. Janson, "Image," p. 258; Kris and Kurz, *Legend*, pp. 47–59; and Svetlana Alpers, *The Art of Describing: Dutch Art in the Seventeenth Century* (Chicago: University of Chicago Press, 1983), pp. xvii–xxv.

67. Bryson, *Looking*, p. 134.

68. Warnke, *Artist*, pp. 148–55.

69. Ernst Kris, "Der Stil 'Rustique': Die Verwendung des Naturabgusses bei Wenzel Jamnitzer und Bernard de Palissy," *Jahrbuch der Kunsthistorischen Sammlungen* N.F. 1 (1926), pp. 137–208, on p. 195.

70. Walter Melion, "Prodigies of Nature, Wonders of the Hand: Karel van Mander on the Depiction of Wondrous Sights," Paper delivered to Pacific Northwest Renaissance Conference, March, 1994.

71. Kaufmann, *Mastery*, pp. 41–45.

72. Ibid., pp. 45–46.

73. Giuseppe Olmi, "Science-Honour-Metaphor: Italian Cabinets of the Sixteenth and Seventeenth Centuries," in Impey and MacGregor, *Origins*, pp. 5–16, on pp. 7–8; cf. Henrietta McBurney, "Cassiano dal Pozzo as Scientific Commentator: Ornithological Texts and Images of Birds from the Museo Cartaceo," in Elizabeth Cropper, Giovanna Perini, and Francesco Solinas (eds.), *Documentary Culture: Florence and Rome from Grand-Duke Ferdinand I to Pope Alexander VII* (Bologna/Baltimore: Nuova Alfa Editoriale/Johns Hopkins University Press, 1992), pp. 349–62.

74. Karen Meier Reeds, "Renaissance Humanism and Botany," *Annals of Science* 33 (1976), pp. 519–42; on the theme of idealization in Renaissance art theory, see Lee, "*Ut Pictura Poesis*," pp. 197–269.

75. Peter Parshall, "Imago Contrafacta: Images and Facts in the Northern Renaissance," *Art History* 16 (1993), pp. 554–79.

76. David Freedberg, "Ferrari on the Classification of Oranges and Lemons," in Cropper et al., *Documentary*, pp. 287–306, on p. 288. Cf. Alix Cooper, "The Museum and the Book: The *Metallotheca* and the History of an Encyclopaedic Natural History in Early Modern Italy," *Journal of the History of Collections* 7 (1995), pp. 1–23. esp. pp. 5–6.

77. Kris, "Stil," pp. 137–56.

78. Quoted in ibid., p. 158.

79. Palissy, "Devis d'une grotte pour la royne mère du roy," in *Discours admirables*, in *Oeuvres*, pp. 465–71, on p. 470. On the resemblances among the *Wunderkammern*, grottoes, and the gardens at Bomarzo, Tivoli, and Pratolino, see Weil, "Love," pp. 169–75.

80. On Palissy's theory of fossils, see his *Discours admirables*, in *Oeuvres*, pp. 330–37.

81. Aristotle's four causes were the material, efficient, formal, and final. So, for example, the material cause of a house would be the wood, brick, mortar, etc. of which it is made; the efficient cause, the actual operations of the builder; the formal cause, the blueprints of the architect; and the final cause, the purpose that the house serves (to provide shelter): Aristotle, *Physics*, 2.3, 194b16–195b30.

82. Ibid., 2.8, 199a12–18.

83. Scheicher, *Kunst- und Wunderkammern*, p. 23.

84. Gesner, *De omni rerum fossilium genere*, pp. 86–96; Ulisse Aldrovandi, *Musaeum metallicum* (Bologna: J.B. Ferronius, 1648), p. 762; Athanasius Kircher, *Mundus subterraneus* [1664], 2 vols. (Amsterdam: Janssonio-Waesbergiana, 1678), vol. 2, pp. 22–48; and Robert Plot, *The Natural History of Oxfordshire* (Oxford: Theater, 1677), pp. 80–130. For other descriptions of figured stones in collections, see Cardano, *De la subtilité*, in *Les Livres de Hierome Cardanus*, p. 318v; Imperato, *Historia*, p. 663; and Grew, *Musaeum*, pp. 262, 265, 268, 284, 307, 311.

85. On these theories and the debate over fossils, see Frank Dawson Adams, *The Birth and Development of the Geological Sciences* (Baltimore: Williams & Wilkins, 1938), pp. 81–132 and 254–59; Martin J.S. Rudwick, *The Meaning of Fossils: Episodes in the History of Palaeontology*, 2nd ed. (Chicago/London: University of Chicago Press, 1985), pp. 1–84;

Paolo Rossi, *The Dark Abyss of Time: The History of the Earth and the History of Nations from Hooke to Vico* [1979], trans. Lydia G. Cochrane (Chicago/London: University of Chicago Press, 1984), pp. 7–23; and Norma E. Emerton, *The Scientific Reinterpretation of Form* (Ithaca/London: Cornell University Press, 1984), pp. 19–75.

86. On the notion of the *lusus naturae* in Renaissance natural philosophy and collecting, see Céard, *Nature*; and Findlen, "Jokes," pp. 292–331.

87. Plot, *Natural History*, p. 121.

88. Lough, *Locke's Travels*, p. 177.

89. Claude du Molinet, *Le Cabinet de la Bibliothèque de Sainte Geneviève* (Paris: Antoine Dezallier, 1692), p. 218.

90. F. Bacon, *Gesta Grayorum* [1594], in *Works*, vol. 8, pp. 334–35.

91. Idem, *New Atlantis*, in *Works*, vol. 3, pp. 156–64.

92. Idem, *Advancement*, in *Works*, vol. 3, p. 267. On Bacon's ambivalent views on wonder in natural history and natural philosophy, see above Chapter Eight.

93. Idem, *Gesta Grayorum*, in *Works*, vol. 8, p. 335.

94. Idem, *New Atlantis*, in *Works*, vol. 3, p. 164.

95. Idem, *Advancement*, in *Works*, vol. 3, p. 331; cf. idem, *Novum organum*, 2.29, in *Works*, vol. 4, p. 169.

96. On Bacon's natural history, particularly the history of "nature erring or varying," see below Chapter Six.

97. F. Bacon, "Phaenomena universi" in *Works*, vol. 3, p. 688, translated in "The Phenomena of the Universe," in *Works*, ed. Montagu, vol. 15, p. 126.

98. Idem, "Description of the Intellectual Globe," in *Works*, ed. Spedding et al., vol. 5, p. 506; cf. the very similar passage in the enlarged Latin version of *The Advancement of Learning*, *De augmentis scientiarum*, 2.2, in *Works*, vol. 1, p. 496.

99. Idem, *Advancement*, in *Works*, vol. 3, p. 333.

100. See above Chapter Eight for Descartes' views on the cognitive uses and abuses of the passions of wonder (*admiration*) and astonishment (*étonnement*).

101. René Descartes, *Discours de la méthode*, 5, in *Oeuvres*, vol. 6, pp. 55–56.

102. Idem, *Principes de la philosophie* [Latin 1644, French 1647], 203, in *Oeuvres*, vol. 9, pp. 321–22.

103. Idem, *Discours de la méthode* [1637], 5, in *Oeuvres*, vol. 6, p. 50.

104. Ibid., pp. 56–57.

105. Idem, *Meditationes de prima philosophia* [1641], 2, in *Oeuvres*, vol. 7, p. 32.

106. On this aspect of the Leibniz-Clarke correspondence, see Steven Shapin, "Of Gods and Kings: Natural Philosophy and Politics in the Leibniz-Clarke Disputes," *Isis* 72 (1981), pp. 187–215, and Francis Oakley, *Omnipotence, Covenant, and Order. An Excursion into the History of Ideas from Abelard to Leibniz* (Ithaca/London: Cornell University Press, 1984), pp. 92–113.

107. Walter Charleton, *The Darknes of Atheism Dispelled by the Light of Nature. A Physico-Theological Treatise* (London: William Lee, 1652), pp. 115, 121, 113, 133.

108. The *Free Inquiry* was first published in 1686, followed by an abridged Latin translation, *Tractatus de ipsa natura* (1687). On the possible relationships between the treatises of Boyle and More, see John Henry, "Henry More versus Robert Boyle: The Spirit of Nature and the Nature of Providence," in Sarah Hutton (ed.), *Henry More (1614–1687). Tercentenary Studies* (Dordrecht/Boston/London: Kluwer, 1990), pp. 55–76.

109. On the close connection between the theology of divine voluntarism and seventeenth-century conceptions of natural law, see Milton, "Origin," pp. 173–95.

110. Boyle, *Free Inquiry*, in *Works*, vol. 5, p. 161.

111. Ibid., p. 164. On Boyle's arguments concerning anomalies and the autonomy of nature, see above Chapter Five.

112. Boyle, *Free Inquiry*, in *Works*, vol. 5, p. 163. The Strasbourg clock, built (1570–74) by a team of artisans under the direction of the Swiss mathematician Conrad Dasypodius, was celebrated as an object of wonder throughout Europe: Cahn, *Masterpieces*, p. 91.

113. On Leibniz's critique of Boyle, see Catherine Wilson, "De Ipsa Natura: Sources of Leibniz's Doctrine of Force, Activity and Natural Law," *Studia Leibnitiana* 19 (1987), pp. 148–72.

114. Leibniz, Letter to Thomasius (20/30 April 1709), quoted in Wilson, "De Ipsa Natura," p. 163.

115. Cudworth, Ralph, *The True Intellectual System of the Universe* [1678] (Hildesheim/New York: Georg Olms, 1977), p. 162.

116. Ibid., p. 156.

117. Henry More, *The Immortality of the Soul* (London: James Flesher, 1662), p. 199.

118. Ibid., pp. 175 and 203.

119. John Beaumont, "Two Letters Written by Mr. *John Beaumont* Junior of *Stony-Easton* in *Somerset-shire*, Concerning Rock-Plants and Their Growth," *PT* 11 (1676), pp. 724–42.

120. Hooke, *Micrographia*, pp. 110–11. Compare the very similar passages in John Ray, *Observations: Topographical, Moral, & Physiological* (London: John Martyn, 1673), p. 124.

121. Browne, *Religio medici*, 1.16, p. 35.

122. Hooke, *Micrographia*, pp. 2 and 8.

123. Schnapper, *Géant*, pp. 246–48; and Major, *Bedencken*, pp. 14–15.

124. David Hume, *Dialogues Concerning Natural Religion* [1779], ed. Norman Kemp Smith (New York: Macmillan, 1947), pp. 178–79 et passim.

125. Kenseth, "Introduction," p. 54.

126. Boyle, *Free Inquiry*, in *Works*, vol. 5, p. 165.

127. Cudworth, *True Intellectual System*, p. 172.

CHAPTER EIGHT: THE PASSIONS OF INQUIRY

1. Isaac Newton, "New Theory about Light and Colours," *PT* 7 (1672), pp. 3075–87, repr. in H.S. Thayer (ed.), *Newton's Philosophy of Nature: Selections from His Writings* (New York: Hafner, 1953), pp. 68–81, on p. 68.

2. Simon Schaffer, "Glass Works: Newton's Prisms and the Uses of Experiment," in David Gooding, Trevor Pinch, and Simon Schaffer (eds.), *The Uses of Experiment: Studies in the Natural Sciences* (Cambridge: Cambridge University Press, 1989), pp. 67–104, esp. pp. 80–85.

3. Hooke, *Micrographia*, Preface, n.p.

4. René Descartes, *Les Passions de l'âme* [1649], 70 and 75–76, in *Oeuvres*, vol. 11, pp. 380–81 and 384–85.

5. Hobbes, *Leviathan*, 1.6, p. 124.

6. See above Chapter Three on wonder and curiosity in medieval natural philosophy. On the shifting meanings of curiosity in particular from antiquity through the Enlightenment, see Krzysztof Pomian, "Curiosity and Modern Science," Sylvia Lemmie England Memorial Lecture, Victoria and Albert Museum, January, 1993.

7. Aristotle, *Metaphysics*, 1.2, 982b10–20. On the etymology of "wonder," see above, Introduction. In English, French, and Italian the words for "curiosity" stem from the Latin *curiositas*, itself deriving from *cura*: care, solicitude, or pains bestowed on something. The early modern vernaculars (except German) preserve this sense of care, adding the nuance of excessive fastidiousness or delicacy, as well as the meaning of a desire to know: in addition to the entries in the *Oxford English Dictionary* and the *Dictionnaire historique de la langue française*, see Françoise Charpentier, Jean Céard, and Gisèle Mathieu-Castellani, "Préliminaires," in Céard, *Curiosité*, pp. 7–23. Early modern German had two sets of words with which to translate the Latin *curiositas*: *Fürwitz* and *Neugierde* (cf. *Neubegierde*, *Neugierigkeit*). *Fürwitz* is the older word and connotes an unseemly haste to know, especially to know what does not concern one. *Neugierde* ("greed for the new") appears to be of early modern coinage and emphasizes the desire for novelties, as well as the general desire to know: see entries in *Grimms Wörterbuch*.

8. On associations with magic, see Mette, "Curiositas," pp. 227–35.

9. Plutarch, *La curiosità* [*Peri Polypragmosynis*], Italian trans. and ed. Emidio Pettine (Salerno: Edizioni Kibotion, 1977), 517c–e, pp. 32–35.

10. Labhardt, "*Curiositas*," pp. 206–24, esp. pp. 212–14.

11. Augustine of Hippo, *Confessions*, 5.3, vol. 1, pp. 210–13; cf. Psalm 8.8. See above Chapter Three on this passage.

12. For a brief survey of the main Christian authorities, see André Cabassut, "Curiosité," *Dictionnaire de Spiritualité* (Paris: Beauchesne, 1953), vol. 2, col. 2654–62. As late as the early eighteenth century conservative moralists still invoked Augustinian language of the "concupiscence of the eyes" in discussing curiosity: see for example Jacques Bénigne Bossuet, *Traité de la concupiscence* [1731], ed. Charles Urbain and E. Levesque (Paris: Editions Fernand Roches, 1930), Ch. 8.

13. André Godin, "Erasme: 'Pia/Impia Curiositas,'" in Céard, *Curiosité*, pp. 25–36; and Françoise Charpentier, "Les *Essais* de Montaigne: Curiosité/incuriosité," in ibid., pp. 111–21.

14. Hans Blumenberg, *Der Prozeß der theoretischen Neugierde* (Frankfurt am Main:

Suhrkamp, 1988); and Carlo Ginzburg, "High and Low: The Theme of Forbidden Knowledge in the Sixteenth and Seventeenth Centuries," *Past and Present* 73 (1976), pp. 28–41.

15. F. Bacon, "The Great Instauration," in *Works*, vol. 4, p. 20.

16. See Blumenberg, *Prozeß*; Ginzburg, "High and Low;" Céard, *Curiosité*, pp. 14–15; and Patrick Brantlinger, "To See New Worlds: Curiosity in *Paradise Lost*," *Modern Language Quarterly* 33 (1972), pp. 355–69.

17. Hobbes, *Leviathan*, 1.6, p. 124. See also Jeffrey Barnouw, "Hobbes' Psychology of Thought: Endeavours, Purpose and Curiosity," *History of European Ideas* 10 (1989), pp. 519–45.

18. Hobbes, *Leviathan*, 1.3, p. 96.

19. Ibid., 1.6, p. 124.

20. On the psychology of greed and avarice in connection with the development of a money economy, see Georg Simmel, *The Philosophy of Money* [1907], trans. Tom Bottomore and David Frisby (London: Routledge and Kegan Paul, 1978), pp. 238–46.

21. F. Bacon, *Advancement*, in *Works*, vol. 3, p. 317.

22. Marin Mersenne, *Les Questions theologiques, physiques, morales et mathematiques* [1634], repr. in Mersenne, *Questions inouyes* (Paris: Fayard, 1985), Qu. 23, p. 302.

23. Ibid., Qu. 46, p. 397.

24. René Descartes, *La Recherche de la vérité par la lumière naturelle* [post. 1701], in *Oeuvres*, vol. 10, p. 500.

25. Blumenberg, *Prozeß*, pp. 129–34. See above Chapter Three.

26. Mersenne, *Questions*, Qu. 1, p. 212; Qu. 46, p. 399.

27. François Maximilien Misson, *A New Voyage to Italy. With Curious Observations on Several Other Countries, as, Germany, Switzerland, Savoy, Geneva, Flanders, and Holland* [1691] (London: T. Goodwin, 1699), pp. 130–31. The cabinet here described is that of the Count Mascardo in Verona. On the objects and emotions associated with early modern collections, see above Chapter Seven.

28. André Morize, *L'Apologie du luxe au XVIIIe siècle* [1909] (Geneva: Slatkine Reprints, 1970); and Albert Hirschman, *The Passions and the Interests: Political Arguments for Capitalism before its Triumph* (Princeton: Princeton University Press, 1977).

29. Belon, *Observations*, p. 18.

30. F. Bacon, *Advancement*, in *Works*, vol. 3, p. 294.

31. For example, "a very curious Person, studying Physick at *Leyden*" who reported a case of unusually contracted skin pores, or "the curious Dr. Beale" who donated an extraordinary kidney stone to the Royal Society Repository, to take two examples at random: "Some Observations of Odde Constitutions of Bodies," *PT* 1 (1665/6), pp. 138–39, on p. 138; "Observables: Touching Petrification," *PT* 1 (1665/6), pp. 320–21, on p. 320.

32. Sprat, *History*, pp. 67–68. On "disinterestedness" in early modern science, see Peter Dear, "From Truth to Disinterestedness in the Seventeenth Century," *Social Studies of Science* 22 (1992), pp. 619–31.

33. Fontenelle, *Histoire du renouvellement*, p. 12.

34. Christian Licoppe, *Formation*, pp. 116–17.

35. Pomian, *Collectors*, pp. 53–60.

36. Nicholas Barbon, *Discourse of Trade* (1690), quoted in Christopher J. Berry, *The Idea of Luxury: A Conceptual and Historical Investigation* (Cambridge: Cambridge University Press, 1994), p. 112.

37. François Poulain de la Barre, *De l'education des dames pour la conduite de l'esprit*, 2 vols. (Paris: Antoine Dezalier, 1679), vol. 1, p. 183.

38. The January 1681 number, for example, includes under this rubric a report on the recently observed comet, "la plus considerable nouveauté du commencement de cette Année," *JS*, Année 1681, pp. 9–12.

39. Cassini, "Sur un nouveau phenomene," pp. 378–79.

40. See above Chapter One.

41. Aristotle, *Metaphysics*, 1.2, 982b10–20 and 982a8–25. See above Chapter Three on medieval reactions to Aristotelian wonder in natural philosophy.

42. Hobbes, *Leviathan*, 1.6, p. 124.

43. Descartes, *Les Passions*, 75, in *Oeuvres*, vol. 11, p. 119.

44. Hooke, "General Scheme," in *Posthumous Works*, p. 6.

45. F. Bacon, *Novum organum*, 1.19–20, in *Works*, vol. 4, p. 50.

46. Sprat, *History*, p. 8.

47. F. Bacon, *Novum organum*, 2.31, in *Works*, vol. 4, p. 170.

48. Descartes, *Les Passions*, 75, in *Oeuvres*, vol. 11, p. 119.

49. Boyle, "New Experiments," in *Works*, vol. 3, pp. 157–69.

50. Idem, "Some Observations About Shining Flesh," in *Works*, vol. 3, pp. 651–55.

51. Idem, "Short Account," [1663] in *Works*, vol. 1, pp. 789–99.

52. On Boyle's use of servant-assistants, see Steven Shapin, *Social History*, pp. 355–408.

53. Boyle, "Short Account," in *Works*, vol. 1, p. 799. Boyle attempted to "lessen the wonder of all the formerly mentioned observations" of an apparently unique diamond by briskly rubbing other diamonds until they glowed.

54. Joseph Hall, *Occasional Meditations* [1633] in Frank Livingstone Huntley, *Bishop Hall and Protestant Meditation in Seventeenth-Century England: A Study with Texts of the Art of Divine Meditation (1606) and Occasional Meditations (1633)*, 101 and 120 (Binghamton, NY: Center for Medieval and Early Renaissance Studies, 1981), pp. 179 and 193.

55. Arthur Warwick, *Spare-Minutes; or, Resolved Meditations and Premeditated Resolutions* [1634] in John L. Leivsay (ed.), *The Seventeenth-Century Resolve* (Lexington: University Press of Kentucky, 1980), no. 37, pp. 132–43, on p. 141.

56. Boyle, "Some Observations About Shining Flesh," in *Works*, vol. 3, p. 655; cf. idem, "New Experiments," in *Works*, vol. 3, p. 164.

57. Idem, "The Aerial Noctiluca," [1680] in *Works*, vol. 4, p. 384.

58. Thomas Shadwell, *The Virtuoso* [1676], Act III, in *The Complete Works of Thomas Shadwell*, 5 vols., ed. Montague Summers (London: The Fortune Press, 1927), vol. 3, p.

142. On Hooke's furious response to the play, see R.S. Westfall's Introduction to Hooke, *Posthumous Works*.

59. Hooke, *Micrographia*, p. 184.

60. Ibid., Preface, n.p.

61. See David Summers, *The Judgment of Sense: Renaissance Naturalism and the Rise of Aesthetics* (Cambridge: Cambridge University Press, 1987), pp. 27, 37, 253, 311–12; John Shearman, *Mannerism* (Harmondsworth: Penguin, 1967), pp. 111–12; and Baxter Hathaway, *Marvels and Commonplaces*, pp. 57–59 et passim.

62. Thomas Hobbes, *"De Mundo" Examined*, 38.5; quoted in Barnouw, "Hobbes' Psychology," p. 539.

63. John Ray, *The Wisdom of God Manifested in the Works of the Creation*, 2nd ed. (London: John Smith, 1692), pp. 158–59.

64. Labhardt, *"Curiositas,"* p. 219.

65. Jean Dupèbe, "Curiosité et magie chez Johannes Trithemius," in Céard, *Curiosité*, pp. 71–97; see also della Porta, *Natural Magick*, pp. 1–2, for attempts to drive a wedge between the "wicked curiosity" of sorcery and blameless natural magic.

66. [Chevalier de Jaucourt], "Curiosité," in Diderot and d'Alembert, *Encyclopédie*, vol. 4, pp. 577–78.

67. Abraham Cowley, "To the Royal Society," in Sprat, *History*, n.p. (dedicatory ode).

68. Hooke, "General Scheme," in *Posthumous Works*, p. 6.

69. On the role of secrets in medieval and early modern natural philosophy, see Eamon, *Science*, esp. pp. 59–65 and 314–18, concerning curiosity.

70. On the attractions of strange facts for early scientific societies, see above Chapter Six.

71. Thomas Wright, "A Curious and Exact Relation of a *Sand-floud*, Which Hath Lately Overwhelmed a Great Tract of Land in the County of *Suffolk*," *PT* 3 (1668), pp 634–39.

72. John Winthrop, "An Extract of a Letter, Written by *John Winthrop Esq*., Governour of *Connecticut* in *New England*, to the Publisher, Concerning Some Natural Curiosities of Those Parts, especially a Very Strange and Very Curiously Contrived Fish, Sent for the Repository of the *R. Society*," *PT* 5 (1670), pp. 1151–53, on p. 1151.

73. Hooke, "General Scheme," in *Posthumous Works*, pp. 61–62.

74. Malebranche, *De la Recherche de la vérité* [1674–75], 5.8, 6th ed. [1712], in *Oeuvres*, vol. 2, pp. 204–207.

75. Descartes, *Les Passions*, 75, in *Oeuvres*, vol. 11, p. 119.

76. Ibid., 73, p. 118.

77. Ibid., 78, p. 121.

78. Hall, "Occasional Meditations," 122, pp. 188–89.

79. F. Bacon, "Valerius Terminus or The Interpretation or Nature," in *Works*, vol. 3, p. 246.

80. Ibid., pp. 218 and 246.

81. Idem, *Advancement*, in *Works*, vol. 3, p. 314.

82. Idem, *Novum organum*, 2.28, in *Works*, vol. 4, p. 168. See above Chapter Six on Bacon's use of wonders to reform natural philosophy.

83. Joseph Glanvill, *Essays on Several Important Subjects in Philosophy and Religion* [1676], quoted in Richard S. Westfall, *Science and Religion in Seventeenth-Century England* (New Haven: Yale University Press, 1958), p. 47.

84. Meric Casaubon, *On Credulity and Incredulity in Things Divine and Spiritual* (London: Brabazon Aylmer/Robert Dawlett, 1672), p. 27.

85. John Spencer, *A Discourse Concerning Prodigies* [1663], 2nd ed. (London: J. Field, 1665), sig. a5r.–v. On the links between wonder and enthusiasm, see below Chapter Nine.

86. Boyle, *Free Inquiry*, in *Works*, vol. 5, pp. 192 and 165. On Boyle's attack on nature's agency and autonomy, see above Chapter Seven.

87. Idem, *Of the High Veneration Man's Intellect Owes to God* [1685], in *Works*, vol. 5, p. 153.

88. Ibid., p. 152.

89. Ibid., pp. 152 and 153.

90. The first Boyle Lecture was delivered in London in 1691. Among the most prominent lecturers on natural theological themes were Richard Bentley, Samuel Clarke, William Whiston, and William Derham.

91. Jan Swammerdam, *Histoire générale des insectes* [1669] (Utrecht: Guillaume de Walcheren, 1682), pp. 3–4.

92. René Antoine Ferchault de Réaumur, *Mémoires pour servir à l'histoire des insectes*, 6 vols. (Paris: Imprimérie Royale, 1734–42), vol. 1, p. 10.

93. Ibid., p. 17.

94. John Arbuthnot, "An Argument for Divine Providence, Taken from the Regularity Observ'd in the Birth of Both Sexes," *PT* 27 (1710–12), pp. 186–90.

95. Bernard de Fontenelle, *La République des philosophes, ou Histoire des Ajaoiens* (Geneva: n.publ., 1768), p. 63.

96. Bernard de Fontenelle, *Entretiens sur la pluralité des mondes* [1686], ed. François Bott (Paris: Editions de l'Aube, 1990), p. 24.

97. Ibid., p. 67.

98. Fontenelle, *Histoire du renouvellement*, p. 21.

99. Ibid., pp. 17 and 20.

100. [Edmé François Gersaint], *Catalogue raisonné de coquilles, insectes, plantes marines, et autres curiosités naturelles* (Paris: Flahaut Fils, 1736), pp. 21–22 and 7.

101. David Hume, "Of Miracles," *An Enquiry Concerning Human Understanding* [1748], ed. Eric Steinberg (Indianapolis: Hackett, 1977), p. 78.

102. Idem, *A Treatise of Human Nature* [1739–40], 2nd ed., ed. P.H. Nidditch (Oxford: Oxford University Press, 1978), 2.3.10, p. 453.

103. Ibid., 2.1.8, p. 301.

104. Adam Smith, *The Principles Which Lead and Direct Philosophical Enquiries: Illus-*

trated by the History of Astronomy, ed. Joseph Black and James Hutton, in *The Works of Adam Smith*, 5 vols. (London: T. Cadell and W. Davies, 1811), vol. 5, pp. 75–76, 69–70, 79–81. For a roughly contemporary view linking admiration and astonishment to fear, albeit in the context of aesthetic theory rather than of natural philosophy, see Edmund Burke, *A Philosophical Inquiry into the Origin of Our Ideas of the Sublime and the Beautiful* [1757], ed. Adam Phillips (Oxford/New York: Oxford University Press, 1990), 2.2, p. 54.

105. Boyle, "A Short Memorial of Some Observations Made upon an Artificial Substance, That Shines without Any Precedent Illustration," [1677], in *Works*, vol. 4, pp. 366–70, on p. 368.

106. Malebranche, *De la Recherche de la vérité*, 5.8, vol. 2, p. 207.

107. [Chevalier de Jaucourt], "Curiosité," in Diderot and d'Alembert, *Encyclopédie*, vol. 4, pp. 577–78, on p. 578.

CHAPTER NINE: THE ENLIGHTENMENT AND THE ANTI-MARVELOUS

1. "Merveilleux, adj. (Littérat.)," in Diderot and d'Alembert, *Encyclopédie*, vol. 10, p. 395. Unsigned articles were often but not always by Diderot.

2. On fairs and coffeehouses, see John Altick, *The Shows of London* (Cambridge, MA/London: Harvard University Press, 1978), pp. 34–49, and Robert M. Isherwood, *Farce and Fantasy: Popular Entertainment in Eighteenth-Century Paris* (Oxford: Oxford University Press, 1986); concerning cabinets see, for example, the description of the Bibliothèque de Sainte–Geneviève in Germain Brice, *Description de la ville de Paris*, 9th ed. [1752], ed. Pierre Codet (Geneva: Librairie Droz/Paris: Librairie Minard, 1971), pp. 299–302; also C.F. Neickelius, *Museographia*, pp. 19–137, for a list of the important collections in early eighteenth–century Europe; Capp, *English Almanacs*, pp. 270–86; on marvels in the popular press, see Robert Darnton, *Mesmerism and the End of the Enlightenment in France* (Cambridge, MA/London: Harvard University Press, 1968), pp. 20–36; on electrical demonstrations, see John L. Heilbron, *Electricity in the 17th and 18th Centuries: A Study in Early Modern Physics* (Berkeley: University of California Press, 1979), pp. 229–39 and 264–75; on academies, see Daniel Roche, *Le Siècle des lumières en province. Académies et académiciens provinciaux 1680–1789*, 2 vols. (Paris/The Hague: Mouton, 1978), vol. 1, pp. 372–79; and on the Bibliothèque Bleue, see Robert Mandrou, *De la culture populaire aux XVIIe et XVIIIe siècles: La Bibliothèque Bleue de Troyes* (Paris: Editions Stock, 1964), pp. 37–38, 57–59, 70–75.

3. Spencer, *Discourse*, sig. a4.v.

4. Joy Kenseth, "Introduction," p. 54. Kenseth here stands in for a host of historians, including historians of science, who propound this interpretation.

5. See, for example, Pierre Bayle, *Pensées diverses, écrites à un Docteur de Sorbonne, à l'occasion de la comète qui parut au mois de Decembre 1680* (Rotterdam: Reinier Leers, 1683); and Edmond Halley, *Astronomiae cometicae synopsis* (Oxford: John Senex, 1705).

6. John Henry, "Occult Qualities and the Experimental Philosophy: Active Principles in Pre–Newtonian Matter Theory," *History of Science* 24 (1986), pp. 335–81; B.J.T. Dobbs,

The Foundations of Newton's Alchemy, or "The Hunting of the Greene Lyon" (Cambridge: Cambridge University Press, 1975); and H.G. Alexander (ed.), *The Leibniz–Clarke Correspondence* [1717] (Manchester: Manchester University Press, 1976), p. 30.

7. Isaac Newton, *Opticks or a Treatise of the Reflections, Refractions, Inflections & Colours of Light* [1704], Queries 28–31 (1730, 4th ed.; repr. London: G. Bell & Sons, 1931), pp. 362–406.

8. [François M. A. de Voltaire], "Imagination, Imaginer," in Diderot and d'Alembert, *Encyclopédie*, vol. 8, pp. 560–63, on p. 561.

9. John L. Heilbron, "Experimental Natural Philosophy," in G.S. Rousseau and Roy Porter (eds.), *The Ferment of Knowledge. Studies in the Historiography of Eighteenth–Century Science* (Cambridge: Cambridge University Press, 1980), pp. 357–87, on pp. 379–82.

10. Paolo Rossi, *Francesco Bacone: Della magia alla scienza* [1957] (Torino: Einaudi, 1974), pp. 39–40; and Schnapper, *Géant*, pp. 246–48.

11. Kay S. Wilkins, "Some Aspects of the Irrational in Eighteenth–Century France," *Studies on Voltaire and the Eighteenth Century* 140 (1975), pp. 107–201, on pp. 121–25; Capp, *English Almanacs*, p. 280; Académie Royale des Sciences, "Description anatomique de trois caméléons," *Mémoires pour servir à l'histoire naturelle des animaux, dressés par M. Perrault* (Paris: La Compagnie des Libraires, 1733), pp. 35–68; and Schnapper, *Géant*, pp. 91–93.

12. "Observations de Physique Générale," *HARS*, Année 1715 (Paris, 1718), pp. 3–4; Isaac Newton, "Vegetation of Metals" [c. 1669], quoted in John Henry, "The Scientific Revolution in England," in Porter and Teich, *Scientific Revolution*, pp. 178–209, on p. 185; and Robert Moray, "A Relation Concerning Barnacles," *PT* 12 (1677), pp. 925–27.

13. See above Chapter Six.

14. On the delicate economy of trust and status involved in such reports, see Steven Shapin, *Social History*, pp. 65–125 et passim.

15. On the late–seventeenth–century restoration of political and ecclesiastical authority, see Theodore Rabb, *The Struggle for Stability in Early Modern Europe* (Oxford/New York: Oxford University Press, 1975).

16. *JS*, Année 1681, pp. 9–12, 93–94, 145–56.

17. Bayle, *Pensées*, pp. 26–27.

18. Casaubon, *Treatise*, pp. 3–4.

19. Michael Heyd, *"Be Sober and Reasonable": The Critique of Enthusiasm in the Seventeenth and Early Eighteenth Centuries* (Leiden/New York: E.J. Brill, 1995), pp. 159–60.

20. Sprat, *History*, pp. 359–60.

21. On the history and historiography of enthusiasm in European perspective, see Michael Heyd, "The Reaction to Enthusiasm in the Seventeenth Century: Towards an Integrative Approach," *Journal of Modern History* 53 (1981), pp. 258–80; also Lotte Mulligan, "'Reason,' 'Right Reason,' and 'Revelation' in Mid–Seventeenth–Century England," in Vickers, *Mentalities*, pp. 375–401; Heyd, *"Be Sober and Reasonable"*.

22. Henry More, *Enthusiasmus Triumphatus, or, A Discourse of the Nature, Causes, Kinds, and Cure of Enthusiasme* (London: James Flesher, 1656), p. 1; on the context of

More's treatise, see Robert Crocker, "Mysticism and Enthusiasm in Henry More," in Hutton, *Henry More*, pp. 137–55.

23. Edward Stillingfleet, *Origines Sacrae, or A Rational Account of the Grounds of the Christian Faith*, 3rd ed. (London: Henry Mortlock, 1666). p. 109.

24. William Warburton, *A Critical and Philosophical Enquiry into the Causes of Prodigies and Miracles as Related by Historians* [1727], quoted in R.M. Burns, *The Great Debate on Miracles: From Joseph Glanvill to David Hume* (Lewisburg: Bucknell University Press, 1981), p. 76.

25. Clarke, *Discourse*, p. 347; and Thomas Sherlock, *The Tryal of the Resurrection of Jesus* (London: J. Roberts, 1729), p. 31.

26. Spencer, *Discourse*, sig. a3r.

27. Casaubon, *Treatise*, pp. 28–42; and More, *Enthusiasmus*, pp. 1–15. Casaubon, fearful of keeping company with suspected atheists such as Pomponazzi, allowed that there might sometimes be a "concurrence" of natural and diabolical causes in some cases of enthusiasm, but added that when the bodily distemper is "cured by physical drugs and potions, the Devil is driven away, and hath no more power over the same bodies": p. 31.

28. Joseph Glanvill, *Philosophia Pia, or, A Discourse of the Religious Temper and Tendencies of the Experimental Philosophy* [1671], in Bernhard Fabian (ed.), *Collected Works of Joseph Glanvill* (Hildesheim/New York: Georg Olms, 1970), vol. 5, pp. 48–50. Cf. Halley, "Account of the Cause of the Late Remarkable Appearance," p. 466.

29. Bert Hansen, *Nicole Oresme*, p. 137; and Pomponazzi, *De naturalium effectuum causis*, pp. 6–20. See above Chapters Three and Four.

30. Casaubon, *On Credulity*; More, *Immortality*; Joseph Glanvill, *Saducismus Triumphatus, or, Full and Plain Evidence Concerning Witches and Apparitions* [1666], 3rd ed. (London: Roger Tuckyr, 1700).

31. Spencer, *Discourse*, pp. 128–30.

32. "An Account of *Hevelius*, His *Prodromus Cometicus*, together with Some Animadversions Made upon It by a French Philosopher," *PT* 1 (1665/6), pp. 104–108, on p. 108.

33. See above Chapter Five.

34. Spencer, *Discourse*, sig. a3.v.

35. Halley, "Account of the Late Surprizing Appearance," pp. 411 and 416.

36. Christian Wolff, *Gedancken über das ungewöhnliche Phaenomen* (Halle: n.publ., 1716), pp. 4–8 and 31–32.

37. Descartes, *Les Passions*, 155–56, in *Oeuvres*, vol. 11, pp. 462–63.

38. See above Chapter Three.

39. On the politics of these two clashes between the Jansenists and the Jesuits, see Tetsuya Shiokawa, *Pascal et les miracles* (Paris: Librairie A.G. Nizet, 1977); and B. Robert Kreiser, *Miracles, Convulsions, and Ecclesiastical Politics in Early Eighteenth–Century Paris* (Princeton: Princeton University Press, 1978).

40. "Enthousiasme," in Diderot and d'Alembert, *Encyclopédie*, vol. 5, pp. 719–22, on p. 720.

41. P. Séjourné, "Superstition," *Dictionnaire de théologie catholique*, 16 vols. (Paris: Librairie Latouzey et Ané, 1941), vol. 14, Part 2, pp. 2763–824; Peter Burke, *Popular Culture in Early Modern Europe* (London: Temple Smith, 1978), p. 241; David Gentilcore, *From Bishop to Witch: The System of the Sacred in Early Modern Terra d'Otranto* (Manchester/New York: Manchester University Press, 1992), pp. 95–101.

42. Chevalier de Jaucourt, "Superstition," in Diderot and d'Alembert, *Encyclopédie*, vol. 15, pp. 669–70, on p. 670.

43. William Monter, *Ritual, Myth and Magic in Early Modern Europe* (Athens, OH: Ohio University Press, 1984), pp. 114–29; and Jean–Marie Goulemot, "Démons, merveilles et philosophie à l'âge classique," *Annales. Economies, Sociétés, Civilisations* 35 (1980), pp. 1223–50.

44. Jean–Baptiste Thiers, *Traité des superstitions qui regardent les sacremens* [1679], 4 vols., 4th ed. (Paris: Compagnie des Librairies, 1741), vol. 1, pp. 6–7, 109, 294–95.

45. Pierre Lebrun, *Histoire critique des pratiques superstitieuses* (Paris: n.publ., 1702), pp. 183–98.

46. Daston, "Marvelous Facts," pp. 120–24.

47. More, *Enthusiasmus*, p. 40.

48. Lebrun, *Histoire*, pp. 287–88.

49. William R. Newman, *Gehennical Fire: The Lives of George Starkey, an American Alchemist in the Scientific Revolution* (Cambridge, MA/London: Harvard University Press, 1994).

50. "Two Instances of Something Remarkable in *Shining Flesh*, from Dr. *J. Beal* of *Yeavel* in Somersetshire, in a Letter to the Publisher," *PT* 11 (1676), pp. 599–603, on p. 601.

51. Robert Hooke, "A Contrivance to Make the Picture of Anything Appear on a Wall, Cub–Board, or within a Picture–Frame, &c. in the Midst of a Light–Room in the Day–Time; or in the Night–Time in Any Room That Is Enlightened with a Considerable Number of Candles," *PT* 3 (1668), pp. 741–43, on p. 742.

52. More, *Immortality*, p. 169.

53. Kay S. Wilkins, "The Treatment of the Supernatural in the *Encyclopédie*," *Studies on Voltaire and the Eighteenth Century* 90 (1972), pp. 1757–71; and J.J.V. M. de Vet, "Miracles in the *Journal Litéraire*: A Case–Study on a Delicate Subject for Journalists of the Enlightenment," *Nederlands Archief voor Kerkgeschiednis* 67 (1987), pp. 186–98.

54. John Ray to Edward Lhwyd 1690, quoted in Charles Earle Raven, *John Ray: Naturalist* [1942], 2nd ed. (Cambridge: Cambridge University Press, 1986), p. 429; and Halley, "Account of the Late Surprizing Appearance," p. 411.

55. Murray Wright Bundy, "The Theory of the Imagination in Classical and Mediaeval Thought," *University of Illinois Studies in Language and Literature* 12 (1927), pp. 1–289, on pp. 260–74.

56. Benedict Spinoza, *Tractatus Theologico–Politicus* [1670], in *The Chief Works of Benedict de Spinoza*, ed. and trans. R.H.M. Elwes (New York: Dover, 1951), pp. 25–28.

57. Locke, *Essay*, 4.19.7, vol. 2, p. 432.

58. Anthony Ashley Cooper, third Earl of Shaftesbury, *A Letter Concerning Enthusiasm, to My Lord ***** (London: J. Morphew, 1708), p. 69.

59. Etienne de Condillac, *Traité des systèmes* [1749], in Georges Le Roy (ed.), *Oeuvres philosophiques de Condillac*, 3 vols. (Paris: Presses Universitaires de France, 1947–51), vol. 1, pp. 134–40.

60. Voltaire, "Imagination," in Diderot and d'Alembert, *Encyclopédie*, vol. 8, p. 561; and Condillac, *Traité*, p. 205.

61. Voltaire, "Imagination," in Diderot and d'Alembert, *Encyclopédie*, vol. 8, p. 562.

62. See, for example, [J.S. Bailly], *Rapport des commissaires chargés par le roi, de l'examen du magnétisme animal* (Paris: Imprimerie Royale, 1785), pp. 27–61.

63. Marcot, "Mémoire."

64. Bailly, *Rapport*, p. 18.

65. Natalie Zemon Davis, "Proverbial Wisdom and Popular Errors," in idem, *Society and Culture in Early Modern France* (London: Duckworth, 1975), pp. 227–67.

66. Laurent Joubert, *Erreurs populaires et propos vulgaires, touchant la medecine et le regime de sante* [1578] (Bordeaux: S. Millanges, 1579), pp. 11, 108, 574–575, 222.

67. Browne, *Pseudodoxia Epidemica*, 1.3, pp. 15, 17, 21.

68. Ibid., 1.8, p. 46.

69. Ibid., 1.11, p. 67.

70. "Nouveautez du commencement de l'année: La Comete," *JS*, Année 1681, pp. 9–12, on p. 10.

71. Bayle, *Pensées*, p. 173.

72. See above Chapter Eight.

73. Charles Blount, *Miracles, No Violations of the Laws of Nature* (London: Robert Sollers, 1683), p. 3. Blount's tract was a free translation of parts of Spinoza's *Tractatus Theologico–Politicus*.

74. Glanvill, *Philosophia Pia*, in *Works*, vol. 5, pp. 12–13.

75. Fontenelle, *Histoire du renouvellement*, p. 18.

76. Bernard de Fontenelle, *De l'origine des Fables* [1724], ed. J.-R. Carré (Paris: Librairie Félix Alcan, 1932), pp. 13–14, 33, 37.

77. Jacques Revel, "Forms of Expertise: Intellectuals and 'Popular' Culture in France (1650–1800)," in Steven L. Kaplan (ed.), *Understanding Popular Culture: Europe from the Middle Ages to the Nineteenth Century* (Berlin/New York: Mouton, 1984), pp. 255–73, on p. 262: "to be denounced for error or false belief now meant to be socially discredited."

78. Hume, "Of Miracles," pp. 79–80.

79. Ann Blair, "Tradition and Innovation in Early Modern Natural Philosophy: Jean Bodin and Jean–Cecile Frey," *Perspectives on Science* 2 (1994), pp. 428–54, on pp. 445–46.

80. See above Chapter Six.

81. "Diverses observations de physique generale. Item IX," *HARS*, Année 1703 (Paris, 1705), pp. 21–22.

82. Ibid., Item X, p. 22.

83. Bouquin, [Letter to the Royal Society of Science], Pochette de Séance 23 juillet 1721, Archives de l'Académie des Sciences, Paris.

84. Shapin, *Social History*, p. 237 et passim.

85. Hume, "Of Miracles," pp. 72–77; cf. Locke, *Essay*, 4.16.6, vol. 2, pp. 375–76.

86. Michel de Certeau, "Une Mutation culturelle et religieuse: Les Magistrats devant les sorciers au XVIIe siècle, " *Revue d'Histoire de l'Eglise en France* 55 (1969), pp. 300–19; Albert Soman, "Les Procès de sorcellerie au Parlement de Paris (1565–1640)," *Annales. Economies, Sociétés, Civilisations* 32 (1977), pp. 790–814, on pp. 794–96; and Keith Thomas, *Religion and the Decline of Magic* (New York: Scribner's, 1971), pp. 570–83.

87. Thomas, *Religion*, pp. 535–39 et passim; Certeau, "Mutation," p. 309.

88. Jacques Autun [Jacques de Chavanes], *L'Incrédulité sçavante et la crédulité ignorante au sujet des magiciens et des sorciers* (Lyon: Jean Molin, 1671), Préface, pp. 11–16. Cf. F.N. Taillepied, *Psichologie, ou Traité de l'apparition des esprits* (Paris: Guillaume Bichon, 1588), pp. 20–21 and 59–67; and Casaubon, *On Credulity*, pp. 28–39. On the arguments of French demonologists, see Jonathan L. Pearl, "French Catholic Demonologists and Their Enemies in the Late Sixteenth and Early Seventeenth Centuries," *Church History* 52 (1983), pp. 457–67; on the English, Thomas, *Religion*, pp. 570–75.

89. F. Bacon, *Advancement*, in *Works*, vol. 3, p. 331.

90. Ibid., p. 288.

91. Joubert, *Erreurs*, pp. 225 and 246–57.

92. Browne, *Pseudodoxia Epidemica*, 3.1, vol. 1, pp. 160–64. On the difficulties facing the naturalists, see Madeline Doran, "On Elizabethan 'Credulity' with Some Questions Concerning the Use of the Marvelous in Literature," *Journal of the History of Ideas* 1 (1940), pp. 151–76.

93. Joshua Childrey, *Britannica Baconica: Or, The Natural Rarities of England, Scotland, & Wales*, Preface (London: Printed for the Author, 1661), n.p.

94. Locke, *Essay*, 4.15.5, vol. 2, p. 367; cf. Hume, *Enquiry*, p. 76.

95. Glanvill, *Saducismus Triumphatus* (1681), quoted in Michael McKeon, *Origins*, p. 86.

96. Sprat, *History*, pp. 90–91 and 99.

97. See above Chapter Six.

98. Browne, *Pseudodoxia Epidemica*, 1.3, p. 16.

99. René Antoine Ferchault de Réaumur, "Sur les diverses reproductions qui se font dans les écrevisses, les omars, les crabes, & c. Et entr'autres sur celles de leurs jambes & de leurs écailles," *MARS*, Année 1712 (Paris, 1714), pp. 226–45, on p. 223.

100. Ron Westrum, "Science and Social Intelligence about Anomalies: The Case of Meteorites," *Social Studies of Science* 8 (1978), pp. 461–93, on pp. 466–76.

101. "Merveilles du Dauphiné, (*Hist. nat.*)," in Diderot and d'Alembert, *Encyclopédie*, vol. 10, pp. 392–93, on p. 392.

102. Ruby, "Origins," pp. 341–59; Oakley, *Omnipotence*, pp. 85–92; and Milton, "Origin," pp. 173–95.

103. Boyle, *Free Inquiry*, in *Works*, vol. 5, p. 216.

104. Bayle, *Pensées*, pp. 171–72; and *Leibniz–Clarke Correspondence*, p. 12.

105. René Descartes, *Principia philosophiae* [1644], 187, in *Oeuvres*, vol. 8, p. 314.

106. Shapin and Schaffer, *Leviathan*, pp. 265–76.

107. Heilbron, *Electricity*, p. 219.

108. Daston, "Cold Light," and Licoppe, *Formation*, pp. 100–02.

109. "An Observation of M. Adrian Auzout, a French Philosopher, Made in Rome (where He now Is) about the Beginning of this Year 1670, Concerning the Declination of the Magnet," *PT* 5 (1670), pp. 1184–87, on p. 1186.

110. "Sur l'effet du vent à l'egard du thermometre," *HARS*, Année 1710 (Paris, 1712), pp. 13–15, 544–46, on p. 13.

111. Hooke, "General Scheme," in *Posthumous Works*, pp. 37–38.

112. Isaac Newton, "Rules of Reasoning," in idem, *Mathematical Principles of Natural Philosophy* [1687], 2 vols., trans. Andrew Motte [1729], rev. Florian Cajori (Berkeley/Los Angeles/London: University of California Press, 1971), vol. 2, p. 398. Cf. Tillotson's claim, apropos of the cessation of miracles, that "wise God, who is never wanting in what is necessary, does not use to be lavish in that which is superfluous": John Tillotson, "The General and Effectual Publication of the Gospel by the Apostles" [1688], in G.W. Weldon (ed.), *Tillotson's Sermons* (London: Ward and Downey, 1886), pp. 223–38, on p. 236.

113. Ray, *Observations*, p. 124.

114. Fontenelle, *Entretiens*, pp. 22–23 and 28.

115. Orest Ranum, "Islands and the Self in a Ludovician Fête," in David Lee Rubin (ed.), *Sun King: The Ascendancy of French Culture during the Reign of Louis XIV* (Washington, D.C.: The Folger Shakespeare Library, 1992), pp. 17–34, on pp. 30–31.

116. Fontenelle, *Histoire du renouvellement*, pp. 16–18.

117. Idem, *République*, p. 44.

118. "Sur un agneau foetus monstrueux," *HARS*, Année 1703 (Paris, 1705), pp. 28–32, on p. 28.

119. Idem, *Histoire du renouvellement*, p. 17.

120. "Sur un grand nombre de Phosphores nouveaux," *HARS*, Année 1730 (Paris, 1732), pp. 48–52, on p. 52.

121. Fontenelle, *République*, p. 65. On the rôle of the Marquise in the *Entretiens*, see Erica Harth, *Cartesian Women: Versions and Subversions of Rational Discourse in the Old Regime* (Ithaca/London: Cornell University Press, 1992), pp. 123–67.

122. Carolyn C. Lougee, *Le Paradis des femmes: Women, Salons, and Social Stratification in Seventeenth-Century France* (Princeton: Princeton University Press, 1976), pp. 44–45.

123. On the role of the Paris Académie Royale des Sciences as technical consultant to the French regime, see Hahn, *Anatomy*, pp. 118–19.

124. Jean Antoine Nollet, *Leçons de physique experimentale*, 4 vols. (Paris: Chez les Freres Guerin, 1743–48), vol. 1, pp. xxix, xxxvii–xli.

125. Licoppe, *Formation*, p. 206.

126. Pierre Louis de Maupertuis, "Les Loix du mouvement et du repos déduites d'un Principe Metaphysique," *Mémoires de l'Académie Royale des Sciences et des Belles Lettres* [Berlin], Classe de Philosophie Speculative, 2 (1746), pp. 268–94, on pp. 276–79.

127. See above Chapter Eight.

128. Boyle, "A Short Memorial," in *Works*, vol. 4, pp. 366–70, on p. 368; and Jan Golinski, "A Noble Spectacle: Phosphorus and the Public Cultures of Science in the Early Royal Society," *Isis* 80 (1989), pp. 11–30.

129. Boyle, "Aerial Noctiluca," in *Works*, vol. 4, pp. 379–404, on p. 403.

130. Wilhelm Homberg, "Observations sur la matiere fecale," *MARS*, Année 1711 (Paris, 1714), pp. 39–47, on p. 40.

131. Fontenelle, *République*, pp. 18–19.

132. "Samedy 22me d'aout 1693," *Procès–Verbaux de l'Académie Royale des Sciences*, Archives de l'Académie des Sciences, Paris.

133. Francis Hutcheson, *An Inquiry Concerning Beauty, Order, Harmony, Design* [1725], ed. Peter Kivy (The Hague: Martinus Nijhoff, 1973), pp. 41–43.

134. Plot, *Natural History*, p. 121.

135. Hathaway, *Marvels and Commonplaces*, pp. 9–20 and 45–58.

136. Weinberg, *History*, vol. 1, p. 41.

137. F. Bacon, *Advancement*, in *Works*, vol. 3, p. 343.

138. Douglas Lane Patey, *Probability and Literary Form: Philosophic Theory and Literary Practice in the Augustan Age* (Cambridge: Cambridge University Press, 1984), pp. 79–83.

139. Chevalier de Jaucourt, "Vraisemblance (Poésie)," in Diderot and d'Alembert, *Encyclopédie*, vol. 17, p. 484.

140. Idem, "Vraisemblance *pittoresque* (Peinture)," in Diderot and d'Alembert, *Encyclopédie*, vol. 17, pp. 484–86, on pp. 484 and 485.

141. Marmontel, "Fiction," in Diderot and d'Alembert, *Encyclopédie*, vol. 6, pp. 679–83.

142. Chevalier de Jaucourt, "Vraisemblance (Poésie)," in Diderot and d'Alembert, *Encyclopédie*, vol. 17, p. 484, borrows a comparison from Horace's *Ars poetica* mocking the chimerical combinations of artists: "to couple birds with serpents, tigers with lambs." "Fiction," pp. 681–82, relates each form of aberrant imagination to a kind of monster: the marvelous to giants, the monstrous to mythical hybrids like Pegasus and the Sphinx, and the fantastic to "the mixture of neighboring species." See above Chapter Five for further aesthetic critiques of the monstrous.

143. M. de la Faille, "Mémoire dans lequel on examine le sentiment des Anciens & des Modernes sur l'origine des MACREUSES," [1769], *Mémoires de Mathématiques et de Physique, Présentés à l'Académie Royale des Sciences, par divers Savans & lus dans ses Assemblées* (Paris: Moutard/Panckoucke, 1780), pp. 331–44, on pp. 331 and 335.

144. Moray, "Relation," p. 926.

145. Georges–Louis Leclerc Comte de Buffon, *Oeuvres complètes de Buffon*, 12 vols., ed. Pierre Flourens (Paris: Garnier Frères, 1853), vol. 2, pp. 307–10, on p. 307.

146. Johann Wolfang Goethe, "III. Thalia: Die Bürger," *Hermann und Dorothea*, in *Goethes Werke. Hamburger Ausgabe*, 14 vols., ed. Erich Trunz (Hamburg: Christian Wegner Verlag, 1949–67), vol. 2, p. 459.

147. "The Gods of Greece" ["Die Götter Griechenlands," 1798/1800], in *The Poems of Schiller*, attempted in English by Edgar Alfred Bowring (London: John W. Parker & Son, 1851), p. 75.

148. Max Weber, "Wissenschaft als Beruf," [1919], translated as "Science as a Vocation," in H.H. Gerth and C. Wright Mills (eds.), *From Max Weber: Essays in Sociology* [1946] (New York: Oxford University Press, 1958), pp. 129–56, on p. 155.

149. For a recent example of the assimilation of this miscellany of genres (and more) to the early modern canon of marvels, and the whole lot to the opposite of Weber's "disenchantment," see Mark A. Schneider, *Culture and Enchantment* (Chicago/London: University of Chicago Press, 1993).

150. Immanuel Kant, *Was ist Aufklärung?* [1784], in *Critique of Practical Reason and Other Writings in Moral Philosophy*, ed. and trans. Lewis White Beck (Chicago, 1949; repr. New York/London: Garland Publishing, 1976), pp. 286–92, on p. 286.

151. Idem, *Kritik der praktischen Vernunft* [1788], in ibid., p. 258.

152. Idem, *Beobachtungen über das Gefühl des Schönen und Erhabenen* [1764], in *Observations on the Feeling of the Beautiful and Sublime*, trans. John T. Goldthwait (Berkeley/Los Angeles/Oxford: University of California Press, 1991), p. 71.

153. Idem, *Kritik der Urteilskraft* [1790], in *The Critique of Judgment*, trans. James Creed Meredith (Oxford: Clarendon Press, 1952), pp. 120–21. Kant admitted an "aesthetic" affinity between the sublime and enthusiasm, but condemned the latter because it was incompatible with "the exercise of free deliberation upon fundamental principles" and therefore did "not merit any delight on the part of reason" (ibid., pp. 124–25).

154. Idem, *Kritik der praktischen Vernunft* [1788], in *Critique of Practical Reason*, p. 258.

CAPTIONS

1. Codicological information in Kenseth, *The Age of the Marvelous*, pp. 221–22.

2. See McGurk and Knock, "Marvels of the East," pp. 88–103.

3. Barber, *Bestiary*, p. 121.

4. Ibid., pp. 30 and 63–64.

5. Gerald of Wales, *History and Topography*, p. 76.

6. Barber, *Bestiary*, pp. 137–38.

7. Hildegard Erlemann and Thomas Stangier, "Festum Reliquiarum," in *Reliquien. Verehrung und Verklärung*, p. 28; Schlosser, *Die Kunst- und Wunderkammern der Spätrenaissance*, pp. 18–20.

8. See Gilbert, "'The Egg Reopened' Again," esp. p. 256.

9. José Gestoso y Perez, *Sevilla monumental y artistica*, 3 vols. (Seville: La Andalucía Moderna, 1889–92), vol. 2, p. 99.

10. D.J.A. Ross, *Illustrated Medieval Alexander-Books in Germany and the Nether-

lands: A Study in Comparative Iconography (Cambridge: The Modern Humanities Research Association, 1971), pp. 66–67, 70–71.

11. Villard of Honnecourt, *The Sketchbook of Villard de Honnecourt*, pp. 128 and 66.

12. Bedini, "Role of Automata," p. 33.

13. Loomis, "Secular Dramatics in the Royal Palace," pp. 243–46.

14. Claire Richter Sherman, *Imaging Aristotle: Verbal and Visual Representation in Fourteenth-Century France* (Berkeley: University of California Press, 1995), esp. pp. 53–54.

15. Kauffmann, *The Baths of Pozzuoli*, pp. 82–83.

16. Codicological information from S. Gentile, S. Niccoli and P. Viti, *Marsilio Ficino e il ritorno di Platone: Mostra di manoscritti, stampe e documenti, 17 maggio – 16 giugno 1984* (Florence: Le Lettere, 1984), pp. 133–36.

17. On the iconography of the monsters of Florence and Ravenna, see Niccoli, *Prophecy and People*, pp. 35–51; Schenda, "Monstrum von Ravenna."

18. See Jurgis Baltrusaitis, *Le Moyen Age fantastique: Antiquités et exotismes dans l'art gothique* (Paris: Flammarion, 1981), p. 162.

19. Martin Luther and Philipp Melanchthon, *Deuttung der czwo grewlichen Figuren, Bapstesels czu Rom und Munchkalbs zu Freijberg ijnn Meijsszen funden* (Wittenberg, 1523), in Luther, *D. Martin Luther's Werke*, p. 381.

20. Scheicher, *Die Kunstkammer*, pp. 149, 153.

21. Du Verney, "Observations," p. 419.

22. Thomas DaCosta Kaufmann, *Variations on the Imperial Theme in the Age of Maximilian II and Rudolf II* (New York: Garland, 1978), pp. 99–102.

23. Gisela Luther, "Stilleben als Bilder der Sammelleidenschaft," in Gerhard Langemeyer and Hans-Albert Peters, eds., *Stilleben in Europa*, Westfalisches Landesmuseum für Kunst und Kunstgeschichte, Münster (Münster: Aschendorff, 1979), pp. 88–128.

24. Bodo Guthmüller, *Ovido Metamorphoseos Vulgare: Formen und Funktionen der volkssprachlichen Wiedergabe klassischer Dichtung in der italienischen Renaissance* (Boppard am Rhein: Harold Boldt, 1981); Ann Moss, *Ovid in Renaissance France: A Survey of the Latin Editions of Ovid and Commentaries Printed in France before 1600* (London: The Warburg Institute, 1982).

25. Anselm Boetius de Boodt, *Le parfaict ioallier, ou Histoire des pierreries* [1609], trans. Jean Bachou (Lyon: Jean-Antoine Huguetan, 1644), p. 279.

26. Jennifer Montagu, *The Expression of the Passions: The Original Influence of Charles le Brun's "Conférence sur l'expression générale et particulière"* (New Haven/London: Yale University Press, 1994).

27. David Hume, "Of Miracles," pp. 84-85.

28. Darnton, *Mesmerism and the End of the Enlightenment in France*, p. 63.

29. [Bailly], *Rapport*, p. 59.

Abbreviations

DSB	=	*Dictionary of Scientific Biography*
HARS	=	*Histoire de l'Académie Royale des Sciences*
JS	=	*Journal des Sçavans*
MARS	=	*Mémoires de l'Académie Royale des Sciences*
PT	=	*Philosophical Transactions of the Royal Society of London*

Bibliography

Manuscripts

Boaistuau, Pierre, *Histoires prodigieuses*, Wellcome MS 136, Wellcome Institute for the History of Medicine, London.

Bouquin, [Letter to the Académie Royale des Sciences], Pochette de Séance 23 juillet 1721, Archives de l'Académie des Sciences, Paris.

Duplessis, James Paris, Letter to Hans Sloane, Add. MS 4464, fol. 44r, British Library, London.

———, *A Short History of Human Prodigious and Monstrous Births of Dwarfs, Sleepers, Giants, Strong Men, Hermaphrodites, Numerous Births, and Extreme Old Age &c.*, Sloane MS 5246, British Library, London.

Marcot, Eustache, [Letter to the Académie Royale des Sciences], Pochette de Séance 29 mai 1715, Archives de l'Académie des Sciences, Paris.

"Samedy 22me d'aout 1693," *Procès-Verbaux de l'Académie Royale des Sciences*, Archives de l'Académie des Sciences, Paris.

Primary Literature

Académie Royale des Sciences, "Description anatomique de trois caméleons," in Perrault, *Mémoires*, pp. 35–68.

Académie Royale des Sciences, *Mémoires pour servir à l'histoire naturelle des animaux et des plantes* [1669] (Amsterdam: Pierre Mortier, 1736).

"An Account Concerning a Woman Having a Double Matrix; as the Publisher Hath Englished It out of French, Lately Printed at Paris, where the Body Was Opened," *PT* 4 (1669), pp. 969–70.

"An Account of *Four Suns*, Which very Lately Appear'd in *France*, and of *Two Raine-bows*, Unusually Posited, Seen in the Same Kingdom, Somewhat Longer agoe," *PT* 1 (1665/6), pp. 219–22.

"An Account of *Hevelius*, his *Prodromus Cometicus*, together with some Animadversions Made upon It by a French Philosopher," *PT* 1 (1665/6), pp. 104–108.

"Account of *L'Architecture navale*, avec le *routier* des Indes Orientales & Occidentales: Par le Sieur Dassié," *PT* 12 (1677), pp. 879–83.

Adelard of Bath, *Die Quaestiones naturales des Adelardus von Bath* [1111–16], ed. Martin Müller, Beiträge zur Geschichte und Philosophie des Mittelalters 31, Heft 2 (Münster i. W.: Aschendorff, 1934).

Adenet le Roi, *Li Roumans de Cléomadès*, 2 vols., ed. André van Hasselt (Brussels: Victor Devaux, 1865).

Adler, Elkan Nathan, *Jewish Travellers in the Middle Ages: Nineteen Firsthand Accounts* (New York: Dover Publications, 1987).

Agrippa of Nettesheim, Cornelius, *De occulta philosophia libri tres*, ed. V. Perrone Compagni (Leiden: E.J. Brill, 1992).

Albertus Magnus, *Book of Minerals* [comp. c. 1256], trans. Dorothy Wyckoff (Oxford: Clarendon Press, 1967).

————, *Opera omnia*, 38 vols., ed. Auguste Borgnet (Paris: Ludovicus Vives, 1890–95).

————, *Opera omnia*, 40 vols. to date, gen. ed. Bernhard Geyer (Cologne: Aschendorff, 1951–).

Albertus Magnus [Pseudo-], *De secretis mulierum, ... eiusdem de virtutibus herbarum, lapidum et animalium, ... item de mirabilibus mundi* [late 13th century] (Lyon: n.publ., 1560).

Aldrovandi, Ulisse, *Musaeum metallicum* (Bologna: J.B. Ferronius, 1648).

Alembert, Jean d', *Preliminary Discourse to the Encyclopedia of Diderot* [1751], trans. Richard N. Schwab (Indianapolis: Bobbs-Merrill, 1963).

Alexander's Letter to Aristotle about India, trans. Lloyd L. Gunderson (Meisenheim am Glan: Anton Hain, 1980).

Aquinas, Thomas, *In Metaphysicam Aristotelis commentaria* [?–1272?], ed. M.-R. Cathala (Turin: Pietro Marietti, 1915).

————, *The Letter of Saint Thomas Aquinas De occultis operibus naturae ad quemdam militem ultramontanum* [c. 1270], ed., trans., and comm. Joseph Bernard McAllister (Washinton, DC: Catholic University of America Press, 1939).

————, *Summa contra gentiles* [c. 1258–64], 3 vols. in 4, trans. Vernon J. Bourke (Notre Dame: University of Notre Dame Press, 1975).

————, *Summa theologiae*, ed. by the Institute of Medieval Studies in Ottawa, 4 vols. (Ottawa: University of Ottawa Press, 1940–44).

Arbuthnot, John, "An Argument for Divine Providence, Taken from the Regularity Observ'd in the Birth of Both Sexes," *PT* 27 (1710–12), pp. 186–90.

Aristotle, *The Complete Works of Aristotle: Revised Oxford Translation*, 2 vols., ed. Jonathan Barnes (Princeton: Princeton University Press, 1984).

Arnald of Villanova, *Arnaldi de Villanova Opera medica omnia*, ed. Michael R. McVaugh (Granada: Seminarium Historiae Medicae Granatensis, 1975).

Arnauld, Antoine, and Pierre Nicole, *La Logique, ou L'Art de penser* [1662], ed. Pierre Clair and François Girbal (Paris: Presses Universitaires de France, 1965).

Ars memorandi (Pforzheim: Thomas Anshelm, 1502).

Ashbee, Edward William, *Occasional Facsimile Reprints of Rare and Curious Tracts of the 16th and 17th Centuries*, 2 vols. (London: privately printed, 1868–72).

Augustine, *City of God*, trans. Henry Bettenson, with an introduction by John O'Meara (London: Penguin, 1984).

————, *Confessions* [397–401], 2 vols., trans. William Watts [1631], Loeb ed. (Cambridge, MA/London: Harvard University Press/William Heinemann, 1988).

————, *On Christian Doctrine* [396, 426], trans. D.W. Robertson (Jr.) (Indianapolis: Bobbs-Merrill, 1958).

————, *Saint Augustine on Genesis* [388–423], trans. Roland J. Teske (Washington DC: Catholic University of America Press, 1991).

Autun, Jacques [Jacques de Chavanes], *L'Incrédulité sçavante et la crédulité ignorante au sujet des magiciens et des sorciers* (Lyons: Jean Molin, 1671).

Auzout, Adrian, "An Observation of M. Adrian Auzout, a French Philosopher, Made in Rome (where He now Is) about the Beginning of this Year 1670, Concerning the Declination of the Magnet," *PT* 5 (1670), pp. 1184–87.

Avicenna, *De viribus cordis* (Venice: per Paganinum de Paganinis, 1507).

————, *Liber de anima seu Sextus de naturalibus*, 2 vols., ed. S. Van Riet (Leiden: E.J. Brill, 1968–72).

Aymeri de Narbonne, 2 vols., ed. Louis Demaison (Paris: Firmin Didot, 1887).

Bacon, Francis, *The Elements of the Common Lawes of England* [1630] (Amsterdam/New York: Theatrum Orbis Terrarum/Da Capo Press, 1969).

————, *Lord Bacon's Works*, 16 vols., ed. Basil Montagu (London: William Pickering, 1825–34).

————, *The Works of Francis Bacon*, ed. James Spedding, Robert Leslie Ellis, and Douglas Denon Heath, 14 vols. (London: Longman, Green, Longman, & Roberts, 1857–74).

Bacon, Roger, *Opera hactenus inedita Rogerii Bacon*, 16 fascs., ed. Robert Steele (Oxford: Clarendon Press, 1905–40).

————, *Opera quaedam hactenus inedita*, ed. J.S. Brewer, 3 vols. (London: Longman, Green, Longman & Roberts, 1859).

————, *The Opus Majus of Roger Bacon*, trans. Robert Belle Burke, 2 vols. (Philadelphia: University of Pennsylvania Press, 1928).

————, *Roger Bacon's Letter Concerning the Marvelous Power of Art and of Nature and Concerning the Nullity of Magic* [c. 1260], trans. Tenney L. Davis (Easton, PA: Chemical, 1923).

Barber, Richard (trans.), *Bestiary, Being an English Version of the Bodleian Library, Oxford MS. Bodley 764, with all the Original Miniatures Reproduced in Facsimile* (Woodbridge: Boydell Press, 1993).

Bartholomaeus Anglicus, *De proprietatibus rerum* (n.p., n.publ., 12 June 1488).

[Bailly, J.S.], *Rapport des commissaires chargés par le roi, de l'examen du magnétisme animal* (Paris: Imprimerie Royale, 1785).

Bayle, Pierre, *Pensées diverses sur la comète* [1681], 2 vols., ed. A. Prat (Paris: Société Nouvelle de Librairie et Edition, 1939).

Beal, J., "Two Instances of Something Remarkable in *Shining Flesh*, from Dr. *J. Beal* of

Yeavel in Somersetshire, in a Letter to the Publisher," *PT* 11 (1676), pp. 599–603.

Beaumont, John, "Two Letters Written by Mr. *John Beaumont* Junior of *Stony-Easton* in *Somerset-Shire*, Concerning Rock-Plants and Their Growth," *PT* 11 (1676), pp. 724–42.

Belon, Pierre, *Les Observations de plusieurs singularitez et choses memorables trouvées en Grece, Asie, Iudée, Egypte, Arabie, et autres pays estranges* (Paris: G. Corrozet, 1553).

———, *L'Histoire naturelle des estranges poissons marins* (Paris: Reynaud Chaudière, 1551).

Benedict XIV, Pope, *De servorum Dei beatificatione et beatorum canonizatione* (Bologna: Longhi, 1734–38).

Benivieni, Antonio, *De abditis nonnullis ac mirandis morborum et sanationum causis*, ed. Giorgio Weber (Florence: Leo S. Olschki, 1994).

———, *The Hidden Causes of Diseases* [1507], trans. Charles Singer (Springfield, IL: Thomas, 1954).

Benoît de Sainte Maure, *Le Roman de Troie* [1165], 6 vols., ed. Léopold Constans (Paris: Firmin Didot, 1904–12).

Bernard of Clairvaux, *The Twelve Degrees of Humility and Pride* [comp. c. 1127], trans. Barton R.V. Mills (London: The Macmillan Co., 1929).

Bertrandon de la Broquière, *The Voyage d'Outremer* [travel 1432–33, printed not before 1807], trans. Galen R. Kline (New York: Peter Lang, 1988).

Bie, Cornelius de, *Het Gulden Cabinet van edel vry Schilder Const* (Antwerp: Ian Meysens, 1661).

Binet, Etienne, [René François], *Essay des merveilles de nature et des plus nobles artifices* (Rouen: Romain de Beauvais et Jean Osmont, 1621).

Blondel, J., *The Strength of Imagination in Pregnant Women Examin'd* (London: J. Peale, 1727).

Blount, Charles, *Miracles, No Violations of the Laws of Nature* (London: Robert Sollers, 1683).

Boniface de Borilly, *Boniface de Borilly. Lettres inédites écrites à Peiresc (1618–1631)*, ed. Philippe Tamizey de Larroque (Aix-en-Provence: Garcin et Didier, 1891).

Boaistuau, Pierre, see *Histoires prodigieuses*.

Boodt, Anselm Boetius de, *Le parfaict ioallier, ou Histoire des pierreries* [1609], trans. Jean Bachou (Lyon: Jean-Antoine Huguetan, 1644).

Borel, Pierre, *Les Antiquitez, raretez, plantes, mineraux, & autres choses considerables de la ville, & Comté de Castres d'Albigeois* (Castres: Arnaud Colomiez, 1649; repr. Geneva: Minkoff reprint, 1973).

Bossuet, Jacques Bénigne, *Traité de la concupiscence* [1731], ed. Charles Urbain and E. Levesque (Paris: Editions Fernand Roches, 1930).

Bouchet, Guillaume, *Les Serées de Guillaume Bouchet, Sieur de Brocourt* [1584], 6 vols, ed. C.E. Roybet (Paris: Alphonse Lemerre, 1873).

Boyle, Robert, "Extrait d'une lettre ecrite de Besançon le onzième de février par M. l'Abbé Doiset à M. l'Abbé Nicusse, & communiquée par ce dernier à l'auteur du Journal touchant un monstre né à deux lieues de cette ville," *JS*, Année 1682, pp. 71–72.

———, "Observables upon a Monstrous Head," *PT* 1 (1665/6), pp. 85–86.

————, *The Works of the Honourable Robert Boyle*, 6 vols, ed. Thomas Birch, facsimile reprint with an Introduction by Douglas McKie (1772; Hildesheim: Georg Olms, 1965–66).

Brant, Sebastian, *Flugblätter des Sebastian Brant*, ed. Paul Heitz (Strassburg: Heitz and Mündel, 1915).

Brathwait, Richard, *The English Gentleman: Containing Sundry Excellent Rules or Exquisite Observations Tending to Direction of every Gentleman, of Selected Ranke and Qualitie; How to Demeane or Accommodate Himselfe in the Manage of Publike and Private Affaires* (London: John Haviland for Robert Bostock, 1630).

Brice, Germain, *Description de la ville de Paris*, 9th ed. [1752], ed. Pierre Codet (Geneva/Paris: Librairie Droz/Librairie Minard, 1971).

Browne, Thomas, *Pseudodoxia Epidemica: Or, Enquiries into Very Many Received Tenents and Commonly Presumed Truths* [1646], 2 vols., ed. Robin Robbins (Oxford: Clarendon Press, 1981).

————, *Religio medici* (1643; facs. repr. Menston: Scolar Press, 1970).

Buffon, Georges-Louis Leclerc, Comte de, *Oeuvres complètes de Buffon*, 12 vols., ed. Pierre Flourens (Paris: Garnier Frères, 1853).

Burke, Edmund, *A Philosophical Inquiry into the Origin of Our Ideas of the Sublime and the Beautiful* [1757], ed. Adam Phillips (Oxford/New York: Oxford University Press, 1990).

Caelestinus, Claudius, *De his que mundo mirabiliter eveniunt, ubi de sensuum erroribus et potentiis animae, ac de influentia caelorum*, ed. Oronce Finé (Paris: Simon de Colines, 1542).

Calvin, Jean, *Institution de la religion chrestienne* [1536] (n.p., n.publ. [Geneva: Jean Gérard? or Michel du Bois?], 1541).

Cardano, Girolamo, *De subtilitate libri XXI* (Nuremberg: Johannes Petrus, 1550).

————, *Les Livres de Hierome Cardanus medecin milanois, intitules de la subtilité, & subtiles inventions, ensemble les causes occultes, & les raisons d'icelles* [1550], trans. Richard Le Blanc (Paris: Charles l'Angelier, 1556).

————, *Ma vie* [1575/76], ed. and trans. Jean Dayre (Paris: Honoré Champion, 1935).

————, *Opera omnia*, 10 vols. (Lyons: Jean-Antoine Huguetan and Marc-Antoine Ravaud, 1663).

Casaubon, Meric, *On Credulity and Incredulity in Things Divine and Spiritual* (London: Brabazon Aylmer/Robert Dawlet, 1672).

————, *A Treatise Concerning Enthusiasme, As It Is an Effect of Nature: But Is Mistaken by Many for either Divine Inspiration, or Diabolical Possession* (London: R.D. for Tho. Johnson, 1655).

Cassini, Gian Domenico, "Description de l'apparence de trois soleils vûs en même temps sur l'horizon," *MARS* 1666/99, Année 1693 (read 31 January 1693), 11 vols. in 14 (Paris: Martin/Coignard/Guerin, 1733), vol. 11, pp. 234–40.

————, "Nouveau phenomene rare et singulier d'une Lumiere Celeste, qui a paru au commencement de Printemps de cette Année 1683," in "Découverte de la Lumiere Celeste qui paroist dans le Zodiaque," *MARS* 1666/99, 11 vols. in 14 (Paris: Martin/Coignard/Guerin, 1733), vol. 8, pp. 179–278, on pp. 182–84.

————, "Sur un nouveau phenomene ou sur une lumiere celeste," *HARS*, in *MARS* 1666/99, 11 vols. in 14 (Paris: Martin/Coignard/Guerin, 1733), 1683, vol. 1, pp. 378–81.

Catalogues of All the Chiefest Rarities in the Publick Anatomie Hall of the University of Leyden, ed. H.J. Witkam (Leiden: University of Leiden Library, 1980).

Cattani, Andrea, *Opus de intellectu et de causis mirabilium effectuum* ([Florence]: [Bartolomeo de Libri?], [c. 1504]).

Cellini, Benvenuto, *Abhandlungen über die Goldschmiedekunst und die Bildhauerei* [1568], trans. Ruth and Max Fröhlich (Basel: Gewerbemuseum Basel, 1974).

————, *The Autobiography of Benvenuto Cellini*, trans. George Bull (London: Penguin, 1956).

"A Certaine Relation of the Hog-faced Gentlewoman Called Mistris *Tannakin Skinker*, Who Was Borne at *Wirkham* a Neuter Towne betweene the Emperour and the Hollander, Scituate on the River *Rhyne*" (London: printed by J.O. for F. Grove, 1640).

Ceruti, Benedetto, and Andrea Chiocco, *Musaeum Francisci Calceolari Iunioris Veronensis* (Verona: Angelo Tamo, 1622).

Champollion-Figeac, Jacques-Joseph (ed.), *Documents historiques inédits tirés des collections manuscrites de la Bibliothèque Nationale et des archives ou des bibliothèques des départments*, 4 vols. (Paris: Firmin Didot, 1841–46).

Charleton, Walter, *The Darknes of Atheism Dispelled by the Light of Nature. A Physico-Theological Treatise* (London: William Lee, 1652).

Chavanes, Jacques de, see Jacques Autun.

Childrey, Joshua, *Britannica Baconica: Or, The Natural Rarities of England, Scotland, & Wales* (London: Printed for the Author, 1661).

Clarke, Samuel, *A Discourse Concerning the Unchangeable Obligations of Natural Religion, and the Truth and Certainty of the Christian Revelation* (London: James Knapton, 1706; repr. Stuttgart/Bad Cannstatt: Friedrich Frommann Verlag 1964).

Colombo, Realdo, *De re anatomica libri XV* (Venice: Nicholas Bevilacqua, 1559).

Columbus, Christopher, *The Diario of Christopher Columbus's First Voyage to America, 1492–1493*, ed. and trans. Oliver Dunn and James E. Kelley (Jr.) (Norman: University of Oklahoma Press, 1989).

Condillac, Etienne de, *Oeuvres philosophiques de Condillac*, 3 vols., ed. Georges Le Roy (Paris: Presses Universitaires de France, 1947–51).

Cudworth, Ralph, *The True Intellectual System of the Universe* [1678] (Hildesheim/New York: Georg Olms, 1977).

Descartes, René, *Oeuvres de Descartes*, 12 vols., ed. Charles Adam and Paul Tannery (Paris: Léopold Cerf, 1897–1910).

————, *The Passions of the Soul* [1649], trans. Stephen H. Voss (Indianapolis: Hackett, 1989).

Dictionnaire de l'Académie Française, 2 vols. (Lyon: Joseph Duplain, 1777).

Diderot, Denis, and Jean d'Alembert (eds.), *Encyclopédie, ou Dictionnaire raisonné des sciences, des arts et des métiers*, 35 vols. (Paris: Briasson, David l'Aîné, Le Breton, Durand, and Paris/Neufchastel: Samuel Faulche, 1751–80).

Dodoens, Rembert, *Historie of Plantes* [1554], trans. H. Lyte (Antwerp: Henry Loë, 1578).

Domat, Jean, *Les Loix civiles dans leur ordre naturel* [1691], 3 vols., 2nd ed. (Paris: La Veuve de Jean-Baptiste Coignard, 1691–97).

Donati, Marcello, *De medica historia mirabili libri sex* (Venice: Felix Valgrisius, 1588).

Dondi, Giovanni, *De fontibus calidis agri Patavini consideratio* [c. 1372–74], in Giunta, *De balneis*, fol. 94r-108r.

Dupleix, Scipion, *La Physique, ou Science des choses naturelles* [1640], ed. Roger Ariew (Paris: Fayard, 1990).

Durand, Guillaume, *Rationale divinorum officiorum*, ed. Niccolò Doard (Venice: Matteo Valentino, 1589).

Duval, Jacques, *Traité des hermaphrodits, parties génitales, accouchemens des femmes, etc.* (1612; repr. Paris: Isidore Liseux, 1880).

Eneas, roman du XIIe siècle, 2 vols., ed. J.-J. Salverda de Grave (Paris: Honoré Champion, 1964).

Escouchy, Mathieu d', *Chronique de Mathieu d'Escouchy* [fifteenth century], 3 vols., ed. G. du Fresne de Beaucourt (Paris: Jules Renard, 1863–64).

Eusebius of Caesarea, *Eusebii Caesariensis episcopi chronicon* [after 303] (Paris: Henri Etienne, 1512).

Evelyn, John, *The Diary of John Evelyn* [1680–84], 6 vols., ed. E.S. de Beer (Oxford: Oxford University Press, 1955).

Faber, Mattäus, *Kurtzgefasste historische Nachricht von der Schloss- und academischen Stifft-skirche zu Aller-Heiligen in Wittenberg....* (Wittenberg: Gerdesische Witwe, 1717).

Faille, de la, "Mémoire dans lequel on examine le sentiment des Anciens & des Modernes sur l'origine des MACREUSES" [1769], *Mémoires de mathématiques et de physique, présentés à l'Académie Royale des Sciences, par divers savans & lus dans ses assemblées* (Paris: Moutard/Panckoucke, 1780), pp. 331–44.

Falchetta, Piero, "Anthology of Sixteenth-Century Texts," in Falchetta, *The Arcimboldo Effect*, pp. 143–202.

Fenton, Edward, *Certaine Secrete Wonders of Nature* (London: Henry Bynneman, 1569).

Ficino, Marsilio, *Opera omnia*, 2 vols. in 4 (Basel, 1576; repr. Turin: Bottega d'Erasmo, 1959).

———, *Three Books on Life* [1489], ed. and trans. Carol V. Kaske and John R. Clark (Binghamton, NY: Medieval and Renaissance Texts and Studies, 1989).

Fontenelle, Bernard de, *De l'origine des fables* [1724], ed. J.-R. Carré (Paris: Librairie Félix Alcan, 1932).

———, *Entretiens sur la pluralité des mondes* [1686], ed. François Bott (Paris: Editions de l'Aube, 1990).

————, *Histoire du renouvellement de l'Académie Royale des Sciences en M.DC.XCIX. et les éloges historiques* (Amsterdam: Pierre de Coup, 1709).

————, *La République des philosophes, ou Histoire des Ajaoiens* (Geneva, n.publ., 1768).

François, René, see Etienne Binet.

Frederick II, *De arte venandi cum avibus* [c. 1245–50], ed. C.A. Willemsen (Leipzig: Insula, 1942).

Gemma, Cornelius, *De naturae divinis characterismis, seu raris et admirandis spectaculis, causis, indiciis, proprietatibus rerum in partibus singulis universi* (Antwerp: Christopher Plantin, 1575).

Geoffrey of Monmouth, *The Historia regum Britanniae of Geoffrey of Monmouth* [?–1139], ed. and trans. Acton Griscom (London/New York: Longmans, Green and Co., 1929).

Gerald of Wales, *Giraldi Cambrensis opera*, 8 vols., ed. J. S. Brewer. (Wiesbaden: Kraus Reprint, 1964–66).

————, *Giraldus Cambrensis in Topographia Hibernie*, ed. John J. O'Meara, Proceedings of the Royal Irish Academy 52, sect. C, no. 4 (Dublin, 1949).

————, *History and Topography of Ireland* [after 1185], trans. John J. O'Meara (Harmondsworth: Penguin, 1982).

[Gersaint, Edmé François], *Catalogue raisonné de coquilles, insectes, plantes marines, et autres curiosités naturelles* (Paris: Flahaut Fils, 1736).

Gervase of Tilbury, *Des Gervasius von Tilbury Otia imperialia in einer Auswahl neu herausgegeben und mit Anmerkungen begleitet* [c. 1211], ed. Felix Liebrecht (Hanover: Carl Rümpler, 1856).

————, *Otia imperialia ad Ottonem IV imperatorem* [c. 1211], in Leibniz, *Scriptores*.

Gesner, Conrad (ed.), *De omni rerum fossilium genere, gemmis, lapidibus, metallis, et huiusmodi* (Zurich: Jakob Gesner, 1565).

Gesta Romanorum [early 14th century], ed. Hermann Oesterley (Berlin: Weidmann, 1872).

Giannini, A., *Paradoxographorum graecorum reliquiae* (Milan: Istituto Editoriale Italiano, 1966).

Giunta, Tommaso (ed.), *De balneis omnia quae extant apud Graecos, Latinos, et Arabas....* (Venice: Giunta, 1553).

Glanvill, Joseph, *Collected Works of Joseph Glanvill*, 8 vols., ed. Bernhard Fabian (Hildesheim/New York: Georg Olms, 1970).

————, *Saducismus Triumphatus, or, Full and Plain Evidence Concerning Witches and Apparitions* [1666], 3rd ed. (London: Roger Tuckyr, 1700).

Goethe, Johann Wolfgang, *Goethes Werke. Hamburger Ausgabe*, 14. vols., ed. Erich Trunz (Hamburg: Christian Wegner Verlag, 1949–67).

Grant, Edward (ed.), *A Source Book in Medieval Science* (Cambridge, MA: Harvard University Press, 1974).

Grew, Nehemiah, *Musaeum Regalis Societatis. Or, A Catalogue & Description of the Natural and Artificial Rarities Belonging to the Royal Society and Preserved at Gresham College* (London: W. Rawlins, 1681).

Guibert of Nogent, *Autobiographie* [*De vita ipsius*], ed. and trans. Edmond-René Labande (Paris: Les Belles Lettres, 1981).

———, *Self and Society in Medieval France: The Memoirs of Abbot Guibert of Nogent*, trans. John F. Benton (Toronto: University of Toronto Press, 1984).

Halley, Edmond, "An Account of the Cause of the Late Remarkable Appearance of the Planet Venus, Seen this Summer, for Many Days together, in the Day Time," *PT* 29 (1714–16), pp. 466–68.

———, "An Account of the Late Surprizing Appearance of the *Lights* Seen in *Air*, on the Sixth of *March* Last; with an Attempt to Explain the Principal *Phaenomena* thereof," *PT* 29 (1714–16), pp. 406–28.

———, *Astronomiae cometicae synopsis* (Oxford: John Senex, 1705).

Higden, Ranulph, *Polychronicon Ranulphi Higden monachi cestrensis, together with the English Translation of John Trevisa and an Unknown Writer of the Fifteenth Century*, 9 vols., ed. Churchill Babington, Rolls Series 41 (London, 1865–86).

"Histoire de l'enfant de Vilne en Lithuanie, à la dent d'or," *JS*, Année 1681, pp. 401–402.

Histoires prodigieuses, 6 vols., ed. Pierre Boaistuau et al., 1. P. Boaistuau, 2. Claude de Tesserant, 3. François de Belle-Forest, 4. Rod. Hoyer, 5. trad. du latin de M. Arnauld Sorbin, par F. de Belle-Forest, 6. par I.D.M. [Jean de Marconville] de divers autheurs anciens & modernes (Lyon: Jean Pillehotte, 1598).

The History of the Athenian Society for the Resolving all Nice and Curious Questions, by a Gentleman Who Got Secret Intelligence of Their Whole Proceedings, by L.R. = Charles Gildon (London: n. publ., [1691]).

Hobbes, Thomas, *Leviathan* [1651], ed. Colin B. Macpherson (London: Penguin, 1968).

Homberg, Wilhelm, "Observations sur la matiere fecale," *MARS*, Année 1711 (Paris, 1714), pp. 39–47.

Hooke, Robert, "A Contrivance to Make the Picture of Anything Appear on a Wall, Cubboard, or within a Picture-Frame, &c. in the Midst of a Light-Room in the Day-Time; or in the Night-Time in Any Room that Is Enlightened with a Considerable Number of Candles," *PT* 3 (1668), pp. 741–43.

———, *Micrographia: Or Some Physical Descriptions of Minute Bodies Made by Magnifying Glasses, with Observations and Inquiries Thereupon* (London: John Martyn and James Allestry, 1665).

———, *The Posthumous Works of Robert Hooke* [1705], ed. Richard Waller with an Introduction by Richard S. Westfall (New York/London: Johnson Reprint Corporation, 1969).

Horace, *Satires, Epistles and Ars Poetica*, trans. H. Rushton Fairclough (Cambridge, MA: Harvard University Press, 1970).

Hume, David, *Dialogues Concerning Natural Religion* [1779], ed. Norman Kemp Smith (New York: Macmillan, 1947).

———, "Of Miracles," *An Enquiry Concerning Human Understanding* [1748], ed. Eric Steinberg (Indianapolis: Hackett, 1977).

459

————, *A Treatise of Human Nature* [1739–40], 2nd ed., ed. P.H. Nidditch (Oxford: Oxford University Press, 1978).

Hutcheson, Francis, *An Inquiry Concerning Beauty, Order, Harmony, Design* [1725], ed. Peter Kivy (The Hague: Martinus Nijhoff, 1973).

Huygens, Christiaan, *Oeuvres complètes*, 22 vols., ed. Société Hollandaise des Sciences (Amsterdam: Swets & Zeitlinger N.V., 1888–1967).

Imperato, Ferrante, *Dell'historia naturale di Ferrante Imperato Napolitano libri XXVIII* (Naples: Constantio Vitale, 1599).

Isidore of Seville, *Etymologiarum sive originum libri XX* [?-636], 2 vols., ed. W.M. Lindsay (Oxford: Clarendon Press, 1911).

James of Vitry, *Libri duo, quorum prior Orientalis, sive Hierosolymitanae, alter Occidentalis historiae* (Douai: Balthazar Bellerus, 1597; repr. Westmead: Gregg International Publishers, 1971).

Jean de Léry, *Histoire d'un voyage fait en la terre du Brésil, autrement dite Amérique* [1578], ed. Jean-Claude Morisot (Geneva: Droz, 1975).

————, *History of a Voyage to the Land of Brazil, otherwise Called America*, trans. Janet Whatley (Berkeley: University of California Press, 1990).

Jordan of Sévérac, *Mirabilia descripta: The Wonders of the East* [c. 1330], trans. Henry Yule (London: Hakluyt Society, 1863).

Joubert, Laurent, *Erreurs populaires et propos vulgaires, touchant la medecine et le regime de sante* [1578] (Bordeaux: S. Millanges, 1579).

Journal d'un bourgeois de Paris, 1405–1449, ed. Alexandre Tuetey (Paris: H. Champion, 1881).

Julius Obsequens, *Prodigiorum liber*, ed. Konrad Lycosthenes (Basel: Oporinus, 1552).

Kant, Immanuel, *The Critique of Judgment*, trans. James Creed Meredith (Oxford: Clarendon Press, 1952).

————, *Critique of Practical Reason and Other Writings in Moral Philosophy*, ed. and trans. Lewis White Beck (Chicago, 1949; repr. New York/London: Garland Publishing, 1976).

————, *Observations on the Feeling of the Beautiful and Sublime*, trans. John T. Goldthwait (Berkeley/Los Angeles/Oxford: University of California Press, 1991).

Kentmann, Johann, *Calculorum qui in corpore ac membris hominum innascuntur, genera XII depicta descriptaque, cum historiis singulis admirandis*, in Gesner, *De omni rerum fossilium genere*, separate foliation.

Kircher, Athanasius, *Mundus subterraneus* [1664], 2 vols. (Amsterdam: Janssonio-Waesbergiana, 1678).

Knox, John, *The Works of John Knox*, 6 vols., ed. David Laing (Edinburgh: Woodrow Society, 1846–64).

Landucci, Luca, *Diario fiorentino dal 1450 al 1516*, ed. Iodoco Del Badia (Florence: G.C. Sansoni, 1883).

————, *A Florentine Diary from 1450 to 1516 by Luca Landucci, Continued by an Anonymous*

Writer till 1542 with Notes by Iodoco Del Badia, trans. Alice de Rosen Jervis (New York: E.P. Dutton and Co., 1927).

Lannoy, Guillebert de, *Oeuvres de Ghillebert de Lannoy, voyageur, diplomate et moraliste*, ed. Charles Potvin, with J.-C. Houzeau (Louvain: P. et J. Lefever, 1878).

Le Fèvre, Jean, *Chronique de Jean Le Fèvre, seigneur de Saint-Rémy* [15th century], 2 vols., ed. François Morand (Paris: Renouard, 1876–81).

Le Miracle de la Sainte Hostie conservée dans les flammes à Faverney, en 1608. Notes et documents publiés à l'occasion du IIIe centenaire du miracle (Besançon: Imprimerie Jacquin, 1908).

Lebrun, Pierre, *Histoire critique des pratiques superstitieuses* (Paris: n.publ., 1702).

Leibniz, Gottfried Wilhelm, *The Leibniz-Clarke Correspondence* [1717], ed. H.G. Alexander (Manchester: Manchester University Press, 1976).

———, *Leibniz: Selections*, ed. Philip Wiener (New York: Charles Scribner's Sons, 1951).

———, *Nouveaux essais*, *Philosophische Schriften*, ed. Leibniz-Forschungsstelle der Universität Münster, vol. 6 (Berlin: Akademie-Verlag, 1962).

———, *Sämtliche Schriften*, ed. Preussische Akademie der Wissenschaften (Darmstadt: Otto Reichl Verlag, 1927).

———, (ed.), *Scriptores rerum Brunsvicensium. . . .* (Hanover: Nicolaus Foerster, 1707).

Leivsay, John L. (ed.), *The Seventeenth-Century Resolve* (Lexington: University Press of Kentucky, 1980).

Lemnius, Levinus, *De miraculis occultis naturae libri tres* (Antwerp: Christopher Plantin, 1574).

———, *Les Secrets miracles de nature* [1559] (Lyon: Jean Frellon, 1566).

L'Estoile, Pierre de, *Mémoires-Journaux*, 12 vols., ed. Gustave Brunet (Paris: Lemerre, 1875–96).

"A Letter from Mr. St. *Georg Ash*, Sec. of the *Dublin Society*, to one of the *Secretaries* of the *Royal Society*; Concerning a *Girl* in *Ireland*, Who Has Several *Horns* Growing on Her Body," *PT* 1 (1685), pp. 1202–204.

Liber monstrorum, ed. and trans. Franco Porsia (Bari: Dedalo Libri, 1976).

Liber monstrorum de diversis generibus [8th century?], ed. Corrado Bologna (Milan: Bompiani, 1977).

Liceti, Fortunio, *De monstrorum natura, caussis et differentiis libri duo* [1616], 2nd ed. (Padua: Paolo Frambotto, 1634).

Littre, Alexis, "Observations sur un foetus humain monstrueux," *HARS*, Année 1701 (Paris, 1704), pp. 88–94.

Livre des merveilles . . . Réproduction des 265 miniatures du manuscrit français 2810 de la Bibliothèque Nationale, ed. Henri Omont (Paris: Berthaud Frères-Catala Frères, [1908]).

Locke, John, *An Essay Concerning Human Understanding* [1689], 2 vols., ed. Alexander C. Fraser (New York: Dover, 1959).

Lorris, Guillaume de, and Jean de Meun, *The Romance of the Rose*, trans. Charles Dahlberg (Hanover, NH/London: University Press of New England, 1983).

Lull, Ramon, *Selected Works of Ramon Llull (1232–1316)*, 2 vols., trans. Anthony Bonner, (Princeton: Princeton University Press, 1985).

———, *La Traduction française du 'Libre de meravelles' de Ramon Lull*, ed. Gret Schib (Schaffhausen: Bolli und Böcherer, 1969).

Lupton, Thomas, *A Thousand Notable Things* (London: John Charlewood, 1586).

Luther, Martin, *D. Martin Luthers Werke*, 102 vols. to date (Weimar: H. Bohlau, 1883–).

Lycosthenes, Conrad, *Prodigiorum ac ostentorum chronicon* (Basel: H. Petri, 1557).

Major, Johann Daniel, *Unvorgreiffliches Bedencken von Kunst- und Naturalien-Kammern insgemein* [1674], repr. in Valentini, *Museum Museorum*.

Malebranche, Nicholas, *Oeuvres de Malebranche*, 20 vols., ed. Geneviève Rodis-Lewis (Paris: J. Vrin, 1962–67).

Mandeville's Travels [1480, comp. c. 1357], ed. M.C. Seymour (London: Oxford University Press, 1968).

Marche, Olivier de la, *Mémoires d'Olivier de la Marche* [15th century], 4 vols., ed. Henri Beaune and J. d'Arbaumont (Paris: Renouard, 1883–88).

Marcot, Eustache, "Memoire sur un enfant monstrueux," *MARS*, Année 1716 (Paris, 1718), pp. 329–47.

Mariotte, Edmé, "Expérience curieuse et nouvelle," *HARS*, in *MARS* 1666/99, 11 vols. in 14 (Paris: Martin/Coignard/Guerin, 1733), 1682, vol. 10, pp. 445.

Marvels of the East: A Full Reproduction of the Three Known Copies, ed. Montague Rhodes James (Oxford: Roxburghe Club, 1929).

The Marvels of Rome: Mirabilia urbis Romae [c. 1143], 2nd ed., ed. and trans. Francis Morgan Nichols and Eileen Gardiner (New York: Italica Press, 1986).

Maupertuis, Pierre Louis de, "Les Loix du mouvement et du repos déduites d'un principe metaphysique," *Mémoires de l'Académie Royale des Sciences et des Belles Lettres* [Berlin], Classe de Philosophie Speculative, 2 (1746), pp. 268–94.

McKeown, Simon (ed.), *Monstrous Births: An Illustrative Introduction to Teratology in Early Modern England* (London: Indelible Inc., 1991).

Mersenne, Marin, *Questions inouyes* [1634] (Paris: Fayard, 1985).

Mery, Jean, "Sur un exomphale monstrueuse," *HARS*, Année 1716 (Paris, 1718), pp. 17–18.

Misson, François Maximilien, *A New Voyage to Italy. With Curious Observations on several other Countries, as, Germany, Switzerland, Savoy, Geneva, Flanders, and Holland* [1691] (London: T. Goodwin, 1699).

Molinet, Claude du, *Le Cabinet de la Bibliothèque de Sainte Geneviève* (Paris: Antoine Dezallier, 1692).

Monardes, Nicolas, *Historia medicinal de las cosas que se traen de nuestras Indias Occidentales que sirven en medicina* [1565–74] (Seville: Padilla, 1988).

———, *Ioyfull Newes out of the Newe Founde Worlde*, trans. John Frampton (London: W. Norton, 1577).

Monconys, Balthasar, *Voyages de M. de Monconys*, 4 vols. (Paris: Pierre Delaulne, 1695).

Montaigne, Michel de, *Journal de voyage en Italie par la Suisse et l'Allemagne en 1580 et 1581*, ed. Maurice Rat (Paris: Editions Garnier Frères, 1955).

Montgeron, Louis-Basile Carré de, *La Verité des miracles operés à l'intercession de M. de Pâris et autres appelans*, 2 vols. (Utrecht: Les Libraires de la Compagnie, 1737–41).

Moray, Robert, "A Relation Concerning Barnacles," *PT* 12 (1677), pp. 925–27.

More, Henry, *Enthusiasmus Triumphatus, or, A Discourse of the Nature, Causes, Kinds, and Cure of Enthusiasme* (London: James Flesher, 1656).

————, *The Immortality of the Soul* (London: James Flesher, 1662).

Mozart, Wolfgang Amadeus, and Emanuel Schikaneder, *Die Zauberflöte* [1791] (Stuttgart: Reclam, 1991).

Münster, Sebastian, *Cosmographey. Oder Beschreibung aller Länder Herrschaften und fürnemesten Stetten des gantzen Erdbodens* [1550] (Basel: Sebastianus Henricpetrus, 1588; repr. Grünwald bei München: Konrad Kölbl, 1977).

"A Narrative of Divers Odd Effects of a Dreadful Thunder-Clap, at Stralsund in Pomerania, 19/20 June; Taken out of a Relation, there Printed by Authority in High Dutch," *PT* 5 (1670), pp. 2084–87.

"A Narrative of a Monstrous Birth in Plymouth, Octob. 22. 1670; together with the Anatomical Observations, Taken thereupon by William Durston Doctor in Physick, and Communicated to Dr. Tim Clerk," *PT* 5 (1670), pp. 2096–98.

Nausea, Fredericus, *Libri mirabilium septem* (Cologne: Peter Quentel, 1532).

Neckam, Alexander, *De naturis rerum libri duo...*, ed. Thomas Wright (London: Longman, Green, Longman, Roberts & Green, 1863).

Neickelius, C.F., *Museographia, oder Einleitung zum rechten Begriff und nützlicher Anlegung der Museorum oder Raritäten-Kammern* (Leipzig/Breslau: Michael Hubert, 1727).

New Revised Standard Version, in *The Complete Parallel Bible* (New York/Oxford: Oxford University Press, 1989).

Newton, Isaac, *Mathematical Principles of Natural Philosophy* [1687], 2 vols., trans. Andrew Motte [1729], rev. Florian Cajori (Berkeley/Los Angeles/London: University of California Press, 1971).

————, "New Theory about Light and Colours," *PT* 7 (1672), pp. 3075–87, repr. in *Newton's Philosophy of Nature*, pp. 68–81.

————, *Newton's Philosophy of Nature: Selections from His Writings*, ed. H.S. Thayer (New York: Hafner, 1953).

————, *Opticks, or A Treatise of the Reflections, Refractions, Inflections of Light* [1704] (1730, 4th ed.; repr. London: G. Bell & Sons, 1931).

Nollet, Jean Antoine, *Leçons de physique experimentale*, 4 vols. (Paris: Chez les Frères Guerin, 1743–48).

"Nouveautez du commencement de l'année: La Comete," *JS*, Année 1681, pp. 9–12.

"Observables: Touching Petrification," *PT* 1 (1665/6), pp. 320–21.

Odoric of Pordenone, *Relatio*, in *Sinica franciscana*, vol. 1, pp. 413–95.

Olivi, Giovanni Battista, *De reconditis et praecipuis collectaneis ab honestissimo et solertissimo*

Francisco Calceolario Veronensi in Musaeo adservatis (Venice: Paolo Zanfretto, 1584).

Oresme, Nicole, *De causis mirabilium* [comp. c. 1370], in Hansen, *Nicole Oresme*, pp. 135–363.

Orta, Garcia de, *Aromatum et simplicium aliquot medicamentorum apud Indos nascentium historia* (Antwerp: C. Plantin, 1567).

Ortus sanitatis (Mainz: Jacob Meydenbach, 1491).

Oviedo, Gonzalo Fernandez de, *La historia general de las Indias* (Seville: Juan Cromberger, 1535).

Palladino, Jacopo da Teramo, *Belial* (Augsburg: Günther Zainer, 1472).

Palissy, Bernard, *Les Oeuvres de Bernard Palissy*, ed. Anatole France (Paris: Charavay Frères, 1880).

Paré, Ambroise, *Des monstres et prodiges* [1573], ed. Jean Céard (Geneva: Librairie Droz, 1971).

Paris, Matthew, *The Illustrated Chronicles of Matthew Paris: Observations of Thirteenth-Century Life*, trans. Richard Vaughan (Cambridge: Corpus Christi College/Alan Sutton, 1993).

Pepys, Samuel, *The Diary of Samuel Pepys* [1659–69], ed. Henry B. Wheatley, 8 vols. in 3 (London: G. Bell and Sons, 1924).

Perrault, Claude, *Mémoires pour servir à l'histoire naturelle des animaux* [1671] (Paris: La Compagnie des Libraires, 1733).

Peter of Eboli, *De balneis puteolanis* [13th century], ed. and trans. Carlo Marcora and Jane Dolman (Milan: Il Mondo Positivo, 1987).

Peter of Maricourt, *Epistola de magnete*, trans. Joseph Charles Mertens, in Grant, *Source Book*, pp. 368–76.

Petrarch, Francesco, *Rerum memorandarum libri* [1343– ?], ed. Giuseppe Billanovich (Florence: G.C. Sansoni, 1943).

Peucer, Caspar, *Commentarius de praecipuis generibus divinationum* (Wittemberg: Johannes Crato, 1560).

Pioppi, Lucia, *Diario (1541–1612)*, ed. Rolando Bussi (Modena: Panini, 1982).

Platt, Hugh, *The Garden of Eden: Or, An Accurate Description of All Flowers and Fruits Now Growing in England* (London: William and John Leake, 1675).

Pliny, *Natural History*, 10 vols., Loeb Classical Library (Cambridge, MA: Harvard University Press, 1935–63).

Plot, Robert, *The Natural History of Oxfordshire* (Oxford: Theater, 1677).

Plutarch, *La curiosità*, ed. and trans. Emidio Pettine (Salerno: Edizioni Kibotion, 1977).

Polo, Marco, *Milione. Le divisament dou monde: Il Milione nelle redazioni toscana e franco-italiana*, ed. Gabriella Ronchi (Milan: Arnaldo Mondadori, 1982).

——, *The Travels of Marco Polo* [1298/99], trans. Ronald Latham (Harmondsworth: Penguin Books, 1958).

Pomponazzi, Pietro, *De naturalium effectuum causis sive de incantationibus* [comp. c. 1520] (Basel, 1567; repr. Hildesheim/New York: Georg Olms, 1970).

Porta, Giovanni Battista della, *Magiae naturalis, sive de miraculis rerum naturalium libri IIII*

(Naples: Matthias Cancer, 1558).

———, *Natural Magick* (London, 1658; repr. New York: Basic Books, 1957).

———, *Phytognomonica* (Naples: Orazio Salviani, 1588).

Poulain de la Barre, François, *De l'education des dames pour la conduite de l'esprit*, 2 vols. (Paris: Antoine Dezalier, 1679).

Puttenham, George, *The Art of English Poesie* [1589], ed. Gladys D. Willcock and Alice Walker [1936] (Folcroft: Folcroft Press, 1969).

Quicchelberg, Samuel, *Inscriptiones vel tituli theatri amplissimi, complectentis rerum universitatis singulas materias et imagines exemias. . . .* (Munich: Adam Berg, 1565).

Ralegh, Sir Walter, *The Discoverie of the Large and Bewtiful Empire of Guiana* [1596], ed. V.T. Harlow (London: The Argonaut Press, 1928).

Ray, John, *Observations: Topographical, Moral, & Physiological* (London: John Martyn, 1673).

———, *The Wisdom of God Manifested in the Works of the Creation* [1691], 2nd ed. (London: John Smith, 1692).

Réaumur, René Antoine Ferchault de, *Mémoires pour servir à l'histoire des insectes*, 6 vols. (Paris: Imprimérie Royale, 1734–42).

———, "Sur les diverses reproductions qui se font dans les écrevisses, les omars, les crabes, & c. et entr'autres sur celles de leurs jambes & de leurs écailles," *MARS,* Année 1712 (Paris, 1714), pp. 226–45.

Renaudot, Théophraste, *Recueil general des questions traitées és conferences du Bureau d'Adresse*, 5 vols. (Paris: Jean Baptiste Loyson, 1660).

Riolan (Jr.), Jean, *Discours sur les hermaphrodits* (Paris: Pierre Ramier, 1614).

Rollins, Hyder E. (ed.), *The Pack of Autolycus, or Strange and Terrible News . . . as Told in Broadside Ballads of the Years 1624–1693* (Cambridge, MA: Harvard University Press, 1927).

Rueff, Jakob, *De conceptu et generatione hominis, et iis quae circa haec potissimum considerantur* (Zurich: Christophorus Froschoverus, 1559).

Savonarola, Michele, *De balneis et thermis naturalibus omnibus Italiae*, in Giunta, *De balneis*, fol. 1r – 36v.

Scaliger, Julius Caesar, *Exotericarum exercitationum liber XV de subtilitate ad Hieronymum Cardanum* [1557] (Hanover: Wechel, 1584).

Schenck, Johann Georg, *Observationum medicarum rarum, novarum, admirabilium et monstrosarum*, 2 vols. (Freiburg i. B.: Martin Böckler, 1596).

———, *Monstrorum historia memorabilis* (Frankfurt: M. Becker, 1609).

Schiller, Friedrich, *The Poems of Schiller*, trans. Edgar Alfred Bowring (London: John W. Parker & Son, 1851).

Schmalling, L.C., "Beschreibung des Naturalienkabinetts des Herrn Pastor Göze, in Quedlinburg, und seiner microscopischen Experimente," *Hannoverisches Magazin* 20 (1782), pp. 966–75.

Shadwell, Thomas, *The Complete Works of Thomas Shadwell*, 5 vols., ed. Montague Summers

(London: The Fortune Press, 1927).

Shaftesbury, Anthony Ashley Cooper, *A Letter Concerning Enthusiasm, to My Lord* ***** (London: J. Morphew, 1708).

Sherlock, Thomas, *The Tryal of the Resurrection of Jesus* (London: J. Roberts, 1729).

Sinica franciscana, ed. Anastasius van den Wyngaert, vol. 1: *Itinera et relationes fratrum minorum saeculi XIII et XIV* (Quaracchi: Collegium S. Bonaventurae, 1929).

Smith, Adam, *The Works of Adam Smith*, 5 vols. (London: T. Cadell and W. Davies, 1811).

"Some Observations of Odde Constitutions of Bodies," *PT* 1 (1665/6), pp. 138–39.

Spencer, John, *A Discourse Concerning Prodigies* [1663], 2nd ed. (London: J. Field, 1665).

Spinoza, Benedict, *The Chief Works of Benedict de Spinoza*, ed. and trans. R.H.M. Elwes (New York: Dover, 1951).

Spon, Jacob, *Recherche des antiquités et curiosités de la ville de Lyon* [1673], ed. J.-B. Montfalcon (Lyon: Louis Perrin, 1857).

Sprat, Thomas, *History of the Royal Society* [1667], ed. Jackson I. Cope and Harold Whitmore Jones (St. Louis: Washington University Press, 1958).

Stillingfleet, Edward, *Origines Sacrae, or A Rational Account of the Grounds of the Christian Faith*, 3rd ed. (London: Henry Mortlock, 1666).

Suger of Paris, *De consecratione ecclesiae Santi Dionysii,* in Panofsky, *Abbot Suger,* pp. 82–121.

———, *De rebus in administratione sua gestis*, in Panofsky, *Abbot Suger*, pp. 40-81.

"Sur un agneau foetus monstrueux," *HARS*, Année 1703 (Paris, 1705), pp. 28–32.

"Sur l'effet du vent à l'egard du thermometre," *HARS*, Année 1710 (Paris, 1712), pp. 13–15 and 544–46.

"Sur un grand nombre de phosphores nouveaux," *HARS*, Année 1730 (Paris, 1732), pp. 48–52.

"Sur le phosphore du barometre," *HARS*, Année 1700 (Paris, 1703), pp. 5–8.

Swammerdam, Jan, *Histoire générale des insectes* [1669] (Utrecht: Guillaume de Walcheren, 1682).

Taillepied, F.N., *Psichologie, ou Traité de l'apparition des esprits* (Paris: Guillaume Bichon, 1588).

Tayler, Edward W. (ed.), *Literary Criticism of Seventeenth-Century England* (New York: Knopf, 1967).

Tertullian, *Opera omnia*, in *Patrologia cursus completus*, 221 vols., ed. J.P. Migne (Paris: J.P. Migne, 1844–1904), ser. 1, vol. 2.

Thevet, André, *La Cosmographie universelle*, 2 vols. (Paris: P. L'Huilier, 1575).

Thiers, Jean-Baptiste, *Traité des superstitions qui regardent les sacremens* [1679], 4 vols., 4th ed. (Paris: Compagnie des Librairies, 1741).

Thomas of Cantimpré, *De natura rerum: Editio princeps secundum codices manuscriptos* [c. 1240], ed. H. Boese (Berlin: Walter de Gruyter, 1973).

Tillotson, John, *Tillotson's Sermons*, ed. G.W. Weldon (London: Ward and Downey, 1886).

Tournefort, Joseph Pitton de, "Description d'un champignon extraordinaire," *HARS*, in

MARS 1666/99, 11 vols. in 14 (Paris: Martin/Coignard/Guerin, 1733), read 3 April 1692, vol. 10, pp. 69–70.

The Travels of Leo of Rozmital through Germany, Flanders, England, France, Spain, Portugal and Italy, 1465–67, ed. and trans. Malcolm Letts (Cambridge: Cambridge University Press, 1957).

Uffenbach, Zacharias Konrad von, *Merkwürdige Reisen durch Niedersachsen, Holland und Engelland*, 3 vols. (Ulm: Gaum, 1753–54).

Ugolino of Montecatini, *Tractatus de balneis* [comp. 1417–20], ed. and trans. Michele Giuseppe Nardi (Florence: Olschki, 1950).

Valentini, D. Michael Bernhard, *Museum Museorum, oder Vollständige Schau-Bühne aller Materialien und Specereÿen nebst deren natürlichen Beschreibung, Selection, Nutzen und Gebrauch* (Frankfurt am Main: Johann D. Zunner, 1704; 2nd ed. 1714).

Varchi, Benedetto, *La prima parte delle lezzioni di M. Benedetto Varchi nella quale si tratta della natura della generazione del corpo humano, et de' mostri* (Florence: Giunta, 1560).

Vaucanson, Jacques, *Le Mécanisme du fluteur* (Paris: J. Guerin, 1738).

Verney, Jacques-François-Marie du, "Observations sur deux enfans joints ensemble," *MARS*, Année 1706 (Paris, 1707), pp. 418–521.

Villani, Giovanni, *Cronica di Giovanni Villani*, 4 vols. in 2, ed. Francesco Gherardi Dragomanni (Florence: Sansone Coen, 1845).

Villard of Honnecourt, *Album de Villard de Honnecourt, architecte du XIIIe siècle* [c. 1235], ed. Henri Omont (Paris: Berthaud Frères, 1906).

————, *The Sketchbook of Villard de Honnecourt*, ed. Theodore Bowie (Bloomington: Indiana University Press, 1959).

Vincent of Beauvais, *Speculum naturale*, 3 vols. (n.p., n.d. [Strasbourg: Adolf Rusch?, 1479?]); a copy of this edition is on deposit at the Houghton Library of Harvard University.

Voltaire, François M. A. de, *Oeuvres complètes de Voltaire*, 52 vols., ed. Louis Moland (Paris: Garnier Frères, 1877–85).

Warwick, Arthur, *Spare-Minutes; or, Resolved Meditations and Premeditated Resolutions* [1634] in Leivsay, *The Seventeenth-Century Resolve*, no. 37, pp. 132–43.

Weinrich, Martin, *De ortu monstrorum commentarius* (n.p., Heinricus Osthusius, 1595).

William of Rubruck, *The Journey of William of Rubruck to the Eastern Parts of the World, 1253–55*, trans. William Woodville Rockhill (London: Hakluyt Society, 1900).

William of Auvergne, *Opera omnia*, ed. Giovanni Domenico Traiano (Venice: Damiano Zenaro, 1591).

William of Malmesbury, *De gestis regum anglorum libri quinque* [1120–40], ed. William Stubbs, 2 vols. (Wiesbaden: Kraus Reprint, 1964).

Winthrop, John, "An Extract of a Letter, Written by *John Winthrop Esq.*, Governour of *Connecticut* in *New England*, to the Publisher, Concerning Some Natural Curiosities of Those Parts, especially a Very Strange and Very Curiously Contrived Fish, Sent for the

Repository of the *R. Society*," *PT* 5 (1670), pp. 1151–53.

Wolff, Christian, *Gedancken über das ungewöhnliche Phaenomen* (Halle: n.publ., 1716).

Worm, Olaus, *Musaeum Wormianum seu Historia rerum rariorum* (Leyden: Elsevier, 1655).

Wright, Thomas, "A Curious and Exact Relation of a *Sand-floud*, Which Hath Lately Overwhelmed a Great Tract of Land in the County of *Suffolk*," *PT* 3 (1668), pp. 634–39.

Secondary Literature

Adams, Frank Dawson, *The Birth and Development of the Geological Sciences* (Baltimore: Williams & Wilkins, 1938).

Adelige Sachkultur des Spätmittelalters: Internationaler Kongress Krems an der Donau, 22. bis 25. September 1980 (Vienna: Verlag der österreichischen Akademie der Wissenschaften, 1982).

Aerts, W.J., E.R. Smits, and J.B. Voorbij (eds.), *Vincent of Beauvais and Alexander the Great: Studies on the Speculum Maius and Its Translation into Medieval Vernaculars* (Groningen: Egbert Forsten, 1986).

Agrimi, Jole, and Chiara Crisciani, *Edocere medicos: Medicina scolastica nei secoli XIII–XV* (Naples: Guerini e Associati, 1988).

———, "Medici e 'vetulae' dal Duecento al Quattrocento: Problemi di una ricerca," in Rossi, *Cultura popolare*, pp. 144–59.

———, "Per una ricerca su *experimentum-experimenta*: Riflessione epistemologica e tradizione medica (secoli XIII–XV)," in Janni and Mazzini, *Presenza*, pp. 9–49.

Al-Azmeh, Aziz, "Barbarians in Arab Eyes," *Past and Present* 134 (1992), pp. 3–18.

Alexander, Jonathan J.G. (ed.), *The Painted Page: Italian Renaissance Book Illumination, 1450–1550* (Munich/New York: Prestel, 1994).

Allard, Guy-H. (ed.), *Aspects de la marginalité au Moyen Age* (Montreal: L'Aurore, 1975).

Alpers, Svetlana, *The Art of Describing: Dutch Art in the Seventeenth Century* (Chicago: University of Chicago Press, 1983).

Altick, John, *The Shows of London* (Cambridge, MA/London: Harvard University Press, 1978).

Ariani, Marco, "Il *fons vitae* nell'immaginario medievale," in Cardini and Gabriele, *Exaltatio*, pp. 140–65.

Arkoun, Mohamed, et al. (eds.), *L'Étrange et le merveilleux dans l'Islam médiéval: Actes du colloque tenu au Collège de France à Paris, en mars 1974* (Paris: Editions J.A., 1978).

Armstrong, C.A.J., "The Golden Age of Burgundy: Dukes that Outdid Kings," in Dickens, *Courts*, pp. 55–75.

Arnold, Ulli, Joachim Menzhausen, and Gerd Spitzer, *Grünes Gewölbe Dresden* (Leipzig: Edition Leipzig, 1993).

Asch, Ronald G., and Adolf M. Birke (eds.), *Princes, Patronage, and the Nobility: The Court at the Beginning of the Modern Age, ca. 1450–1650* (London: Oxford University Press, 1991).

Ashley, Benedict M., "St. Albert and the Nature of Natural Science," in Weisheipl, *Albertus*

Magnus, pp. 73–102.

Ashworth (Jr.), William B., "Natural History and the Emblematic World View," in Lindberg and Westman, *Reappraisals*, pp. 303–32.

———, "Remarkable Humans and Singular Beasts," in Kenseth, *Marvelous*, pp. 113–32.

Atti del Congresso Internazionale su Medicina Medievale Salernitana e Scuola Medica (Salerno, 1994).

Balsiger, Barbara Jeanne, *The Kunst- und Wunderkammern: A Catalogue Raisonné of Collecting in Germany, France and England, 1565–1750*, 2 vols. (Ph.D. diss., University of Pittsburgh, 1971).

Baltrusaitis, Jurgis, *An Essay on the Legend of Forms* [1983], trans. Richard Miller (Cambridge, MA/London: MIT Press, 1989).

———, *Le Moyen Age fantastique: Antiquité et exotismes dans l'art gothique* (Paris: Flammarion, 1981).

———, *Réveils et prodiges: Le Gothique fantastique* (Paris: Armand Colin, 1960).

Barduzzi, Domenico, *Ugolino da Montecatini* (Florence: Istituto Micrografico Italiano, 1915).

Barnes, Robin Bruce, *Prophecy and Crisis: Apocalypticism in the Wake of the Lutheran Reformation* (Stanford: Stanford University Press, 1988).

Barnouw, Jeffrey, "Hobbes' Psychology of Thought: Endeavours, Purpose and Curiosity," *History of European Ideas* 10 (1989), pp. 519–45.

Bartlett, Robert, *Gerald of Wales, 1146–1223* (Oxford: Clarendon Press, 1982).

Bartra, Roger, *Wild Men in the Looking Glass: The Mythic Origins of European Otherness*, trans. Carl T. Berrisford (Ann Arbor: University of Michigan Press, 1994).

Battisti, Carlo, and Giovanni Alessio, *Dizionario etimologico italiano*, 5 vols. (Florence: G. Barbèra, 1950–57).

Bedini, Silvio A., "The Role of Automata in the History of Technology," *Technology and Culture* 5 (1964), pp. 24–42.

Belloni, Luigi, "L'ischiopago tripode trecentesco dello spedale fiorentino di Santa Maria della Scala," *Rivista di storia delle scienze mediche e naturali* 41 (1950), pp. 1–14.

Benhamou, Reed, "From *Curiosité* to *Utilité*: The Automaton in Eighteenth-Century France," *Studies in Eighteenth-Century Culture* 17 (1987), pp. 91–105.

Benson, Robert L., and Giles Constable (eds.), *Renaissance and Renewal in the Twelfth Century* (Cambridge, MA: Harvard University Press, 1982).

Beonio-Brocchieri Fumagalli, Maria Teresa, *Le enciclopedie dell'Occidente medioevale* (Turin: Loescher, 1981).

Bergier, Jean-François, (ed.), *Zwischen Wahn, Glaube und Wissenschaft: Magie, Astrologie, Alchemie und Wissenschaftsgeschichte* (Zurich: Verlag der Fachvereine, 1988).

Berry, Christopher J., *The Idea of Luxury: A Conceptual and Historical Investigation* (Cambridge: Cambridge University Press, 1994).

Besch, Walter (ed.), *Studien zur deutschen Literatur und Sprache des Mittelalters: Festschrift für Hugo Moser zum 65. Geburtstag* (Berlin: Erich Schmidt Verlag, 1974).

Bethell, Denis, "The Making of a Twelfth-Century Relic Collection," in Cuming and

Baker, *Popular Belief*, pp. 64–72.

Biagioli, Mario, *Galileo Courtier* (Chicago/London: University of Chicago Press, 1993).

———, "Galileo the Emblem Maker," *Isis* 81 (1990), pp. 230–58.

Bianchi, Massimo Luigi, *Signatura rerum: Segni, magia e conoscenza da Paracelso a Leibniz* (Rome: Edizioni dell'Ateneo, 1987).

Bibliotheca Lindesiana: Catalogue of Early English Broadsides, 1505–1897 (Aberdeen: Aberdeen University Press, 1898).

Biester, James Paul, *Strange and Admirable: Style and Wonder in the Seventeenth Century* (Ph.D. diss., Columbia University, 1990).

Blaher, Damian Joseph, *The Ordinary Processes in Causes of Beatification and Canonization* (Washington, D.C.: Catholic University of America Press, 1949).

Blair, Ann, "Tradition and Innovation in Early Modern Natural Philosophy: Jean Bodin and Jean-Cecile Frey," *Perspectives on Science* 2 (1994), pp. 428–54.

Bloch, Raymond, *Les Prodiges dans l'Antiquité classique (Grèce, Etrurie et Rome)* (Paris: Presses Universitaires de France, 1963).

Blumenberg, Hans, *The Legitimacy of the Modern Age* [1966], trans. Robert M. Wallace (Cambridge, MA: MIT Press, 1983).

———, *Der Prozeß der theoretischen Neugierde* (Frankfurt am Main: Suhrkamp, 1988).

Bodemer, Charles, "Embryological Thought in Seventeenth-Century England," in Bodemer and King, *Medical Investigation*, pp. 3–23.

Bodemer, Charles, and Lester S. King, *Medical Investigation in Seventeenth Century England* (Los Angeles: University of California Press, 1968).

Bologna, Corrado, "Natura, miracolo, magia nel pensiero cristiano dell'alto Medioevo," in Xella, *Magia*, pp. 253–72.

Boström, Hans-Olof, "Philipp Hainhofer and Gustavus Adolphus' *Kunstschrank* in Uppsala," in Impey and MacGregor, *Origins*, pp. 90–101.

Bouton, Jean de la Croix, "Un Poème à Philippe le Bon sur la Toison d'or," *Annales de Bourgogne* 42 (1970), pp. 5–29.

Bran, Noel L., "The Proto-Protestant Assault upon Church Magic: The 'Errores Bohemanorum' According to the Abbott Trithemius (1462–1516)," *Journal of Religious History* 12 (1982), pp. 9–22.

Brantlinger, Patrick, "To See New Worlds: Curiosity in *Paradise Lost*," *Modern Language Quarterly* 33 (1972), pp. 355–69.

Braun, Joseph, *Die Reliquiare des christlichen Kultes und ihre Entwicklung* (Freiburg im Breisgau: Herder, 1940).

Bredekamp, Horst, *Antikensehnsucht und Maschinenglauben: Die Geschichte der Kunstkammer und die Zukunft der Kunstgeschichte* (Berlin: Klaus Wagenbach, 1993).

Breiner, Laurence, "The Basilisk," in South, *Creatures*, pp. 113–22.

Bretschneider, Emil, *Archaeological and Historical Researches on Peking and Its Environs* (Shanghai: American Presbyterian Mission Press, 1876).

Brett, Gerard, "The Automata of the Byzantine 'Throne of Solomon,'" *Speculum* 29 (1954), pp. 477–87.

Brown, Harcourt, *Scientific Organizations in Seventeenth-Century France (1620–1680)* (Baltimore: Williams & Wilkins, 1934).

Bryson, Norman, *Looking at the Overlooked: Four Essays on Still Life Painting* (Cambridge, MA: Harvard University Press, 1990).

Bundy, Murray Wright, "The Theory of the Imagination in Classical and Mediaeval Thought," *University of Illinois Studies in Language and Literature* 12 (1927), pp. 1–289.

Burke, Peter, *Popular Culture in Early Modern Europe* (London: Temple Smith, 1978).

Burnett, Charles (ed.), *Adelard of Bath: An English Scientist and Arabist of the Early Twelfth Century* (London: Warburg Institute, 1987).

Burns, R.M., *The Great Debate on Miracles: From Joseph Glanvill to David Hume* (Lewisburg: Bucknell University Press, 1981).

Burns, William E., *An Age of Wonders: Prodigies, Providence and Politics in England, 1580–1727* (Ph.D. diss., University of California at Davis, 1994).

Bynum, Caroline Walker, "Wonder," *American Historical Review* 102 (1997), pp. 1–26.

Byrne, Donal, "The Boucicaut Master and the Iconographical Tradition of the 'Livre des propriétés des choses,'" *Gazette des Beaux-Arts*, ser. 6, 92 (1978), pp. 149–64.

———, "Rex imago Dei: Charles V of France and the *Livre des propriétés des choses*," *Journal of Medieval History* 7 (1981), pp. 97–113.

———, "Two Hitherto Unidentified Copies of the 'Livre des propriétés des choses,' from the Royal Library of the Louvre and the Library of Jean de Berry," *Scriptorium* 31 (1977), pp. 90–98.

Cabassut, André, "Curiosité," *Dictionnaire de Spiritualité*, col. 2654–61.

Camille, Michael, *The Gothic Idol: Ideology and Image-Making in Medieval Art* (Cambridge: Cambridge University Press, 1989).

Campbell, Mary B., *The Witness and the Other World: Exotic European Travel Writing, 400–1600* (Ithaca, NY: Cornell University Press, 1988).

Canguilhem, Georges, "Monstrosity and the Monstrous," *Diogène* 40 (1962), pp. 27–42.

Capp, Bernard, *English Almanacs, 1500–1800: Astrology and the Popular Press* (Ithaca, NY: Cornell University Press, 1979).

Carasso-Bulow, Lucienne, *The Merveilleux in Chrétien de Troyes' Romances* (Geneva: Librairie Droz, 1976).

Cardini, F., and M. Gabriele (eds.), *Exaltatio essentiae. Essentia exaltata* (Pisa: Pacini, 1992).

Carey, Hilary M., *Courting Disaster: Astrology at the English Court and University in the Later Middle Ages* (London: Macmillan, 1992).

Caron, Marie-Thérèse, *La Noblesse dans le duché de Bourgogne, 1315–1477* (Lille: Presses Universitaires de Lille, 1987).

Caroti, Stefano, "Comete, portenti, causalità naturale e escatologia in Filippo Melantone," in *Scienze*, pp. 393–426.

———, "*Mirabilia* e *monstra* nei *Quodlibeta* di Nicole Oresme," *History and Philosophy of the Life Sciences* 6 (1984), pp. 133–50.

Cary, George, *The Medieval Alexander*, ed. D.J.A. Ross (Cambridge: Cambridge University Press, 1967).

Céard, Jean (ed.), *La Curiosité à la Renaissance* (Paris: Société Française d'Edition d'Enseignement Supérieur, 1986).

——, *La Nature et les prodiges: L'Insolite au XVIe siècle* (Geneva: E. Droz, 1977).

Certeau, Michel de, "Une Mutation culturelle et religieuse: Les Magistrats devant les sorciers au XVIIe siècle," *Revue d'Histoire de l'Eglise en France* 55 (1969), pp. 300–19.

Chandler, James, Arnold I. Davidson, and Harry Harootunian (eds.), *Questions of Evidence: Proof, Practice, and Persuasion across the Disciplines* (Chicago/London: University of Chicago Press, 1994).

Chapuis, Alfred, and Edouard Gélis, *Le Monde des automates: Étude historique et technique*, 2 vols. (Paris, 1928; repr. Geneva: Slatkine, 1984).

Charpentier, Françoise, "Les *Essais* de Montaigne: Curiosité/incuriosité," in Céard, *Curiosité*, pp. 111–21.

Charpentier, Françoise, Jean Céard, and Gisèle Mathieu-Castellani, "Préliminaires," in Céard, *Curiosité*, pp. 7–23.

Chenu, Marie-Dominique, *Nature, Man, and Society in the Twelfth Century: Essays on New Theological Perspectives in the Latin West*, trans. J.R. Taylor and L.K. Little (Chicago: University of Chicago Press, 1968).

Cipolla, Carlo, *Clocks and Culture 1300–1700* (New York: W.W. Norton, 1978).

Clark, Stuart, "On the Scientific Status of Demonology," in Vickers, *Mentalities*, pp. 351–74.

Close, A.J., "Commonplace Theories of Art and Nature in Classical Antiquity and in the Renaissance," *Journal of the History of Ideas* 30 (1969), pp. 467–86.

Cohen, John, *Human Robots in Myth and Science* (London: Allen and Unwin, 1966).

Cook, Harold J., "The New Philosophy in the Low Countries," in Porter and Teich, *Scientific Revolution*, pp. 115–49.

——, "Physicians and Natural History," in Jardine, *Cultures of Natural History*, pp. 91–105.

Cooper, Alix, "The Museum and the Book: The *Metallotheca* and the History of an Encylopaedic Natural History Collection in Early Modern Italy," *Journal of the History of Collections* 7 (1995), pp. 1–23.

Coopland, George W., *Nicole Oresme and the Astrologers: A Study of His Livre de divinacions* (Cambridge, MA: Harvard University Press, 1952).

Copenhaver, Brian, "Astrology and Magic," in Schmitt, *Cambridge History*, pp. 264–300.

——, "Natural Magic, Hermetism, and Occultism in Early Modern Science," in Lindberg and Westman, *Reappraisals*, pp. 261–301.

——, "Scholastic Philosophy and Renaissance Magic in the *De vita* of Marsilio Ficino," *Renaissance Quarterly* 37 (1984), pp. 523–54.

——, "A Tale of Two Fishes: Magical Objects in Natural History from Antiquity through the Scientific Revolution," *Journal of the History of Ideas* 52 (1991), pp. 373–98.

Corleto, Luigia Melillo, "La medicina tra Napoli e Salerno nel Medioevo: Note sulla tradizione dell'idroterapia," in *Atti*, pp. 26–35.

Cranz, F. Edward, and Paul Oskar Kristeller (eds.), *Catalogus translationum et commentariorum: Medieval and Renaissance Latin Translations and Commentaries*, 6 vols. to date (Washington DC: Catholic University of America Press, 1960–86).

Crisciani, Chiara, and Claude Gagnon, *Alchimie et philosophie au Moyen Age: Perspectives et problèmes* (Quebec: L'Aurore/Univers, 1980).

Crocker, Robert, "Mysticism and Enthusiasm in Henry More," in Hutton, *Henry More*, pp. 137–55.

Cropper, Elizabeth, Giovanna Perini, and Francesco Solinas (eds.), *Documentary Culture: Florence and Rome from Grand-Duke Ferdinand I to Pope Alexander VII* (Bologna: Nuova Alfa Editoriale/Baltimore: Johns Hopkins University Press, 1992).

Cuming, G.J., and Derek Baker (eds.), *Popular Belief and Practice: Papers Read at the Ninth Summer Meeting and the Tenth Winter Meeting of the Ecclesiastical History Society* (Cambridge: Cambridge University Press, 1972).

Cunningham, J.V., *Woe or Wonder: The Emotional Effect of Shakespearean Tragedy* (Denver: University of Denver Press, 1951).

Dales, Richard C., "A Twelfth-Century Concept of the Natural Order," *Viator* 9 (1978), pp. 179–92.

D'Alverny, Marie-Thérèse, "Translations and Translators," in Benson and Constable, *Renaissance*, pp. 421–62.

Dannenfeldt, Karl H., *Leonhard Rauwolf: Sixteenth-Century Physician, Botanist, and Traveller* (Cambridge MA: Harvard University Press, 1968).

Darnton, Robert, *Mesmerism and the End of the Enlightenment in France* (Cambridge, MA/ London: Harvard University Press, 1968).

Daston, Lorraine, "Baconian Facts, Academic Civility, and the Prehistory of Objectivity," in Megill, *Rethinking Objectivity*, pp. 37–63.

———, *Classical Probability in the Enlightenment* (Princeton: Princeton University Press, 1988).

———, "The Cold Light of Facts and the Facts of Cold Light: Luminescence and the Transformation of the Scientific Fact, 1600–1750," *EMF: Studies in Early Modern France* 3 (1997), pp. 17–44.

———, "Marvelous Facts and Miraculous Evidence in Early Modern Europe," *Critical Inquiry* 18 (1991), pp. 93–124, repr. in Chandler et al., *Questions*, pp. 243–74.

Daston, Lorraine, and Katharine Park, "Hermaphrodites in Renaissance France," *Critical Matrix* 1 (1985), pp. 1–19.

Davidson, Arnold I., "The Horror of Monsters," in Sheehan and Sosna, *Boundaries*, pp. 36–67.

Davis, Lennard J., *Factual Fictions: The Origins of the English Novel* (New York: Columbia

University Press, 1983).

Davis, Natalie Zemon, *Society and Culture in Early Modern France* (London: Duckworth, 1975).

Dear, Peter, *Discipline and Experience: The Mathematical Way in the Scientific Revolution* (Chicago/London: University of Chicago Press, 1995).

———, "From Truth to Disinterestedness in the Seventeenth Century," *Social Studies of Science* 22 (1992), pp. 619–31.

———, "Jesuit Mathematical Science and the Reconstitution of Experience in the Early Seventeenth Century," *Studies in the History and Philosophy of Science* 18 (1987), pp. 133–75.

———, "*Totius in verba*: Rhetoric and Authority in the Early Royal Society," *Isis* 76 (1985), pp. 145–61.

Deluz, Christian, *Le Livre de Jehan de Mandeville: Une "géographie" au XIVe siècle* (Louvain-La-Neuve: Institut d'Etudes Médiévales de l'Université Catholique de Louvain, 1988).

Demaitre, Luke, "Scholasticism in Compendia of Practical Medicine, 1250–1450," in Siraisi and Demaitre, *Science*, pp. 81–95.

———, "Theory and Practice in Medical Education at the University of Montpellier in the Thirteenth and Fourteenth Centuries," *Journal of the History of Medicine and Allied Sciences* 30 (1975), pp. 103–23.

Dickens, A.D. (ed.), *The Courts of Europe: Politics, Patronage and Royalty, 1400–1800* (London: Thames and Hudson, 1977).

Dickson, Arthur, *Valentine and Orson: A Study in Late Medieval Romance* (New York: Columbia University Press, 1929).

Dictionary of Scientific Biography, ed. Charles Coulston Gillispie (New York: Charles Scribner's Sons, 1970–80).

Dictionnaire de Spiritualité, 17 vols., ed. Marcel Viller et al. (Paris: Beauchesne, 1937–94).

Dictionnaire de théologie catholique, 16 vols. (Paris: Librairie Latouzey et Ané, 1941).

Dictionnaire historique de la langue française, 2 vols., ed. Alain Rey (Paris: Dictionnaires Le Robert, 1992).

Dizionario biografico degli italiani, 44 vols. to date (Rome: Istituto della Enciclopedia Italiana, 1960–   ).

Dobbs, B.J.T., *The Foundations of Newton's Alchemy, or "The Hunting of the Greene Lyon"* (Cambridge: Cambridge University Press, 1975).

Dombart, T., *Die sieben Weltwunder des Altertums*, 2nd ed., ed. Jörg Dietrich (Munich: Heimeran, 1967).

Doran, Madeline, "On Elizabethan 'Credulity' with Some Questions Concerning the Use of the Marvelous in Literature," *Journal of the History of Ideas* 1 (1940), pp. 151–76.

Douglas, Mary, *Purity and Danger: An Analysis of the Concepts of Pollution and Taboo* [1966] (London: Ark Paperbacks, 1984).

Doutrepont, Georges, "Jason et Gédéon, patrons de la Toison d'or," *Mélanges offerts à*

Godefroid Kurth (Paris: Champion, 1908), pp. 191–208.

———, *La Littérature française à la cour des ducs de Bourgogne* (Paris: Honoré Champion, 1909).

Dupèbe, Jean, "Curiosité et magie chez Johannes Trithemus," in Céard, *Curiosité*, pp. 71–97.

Eamon, William, "Books of Secrets in Medieval and Early Modern Science," *Sudhoffs Archiv* 69 (1985), pp. 26–49.

———, "Court, Academy, and Printing House: Patronage and Scientific Careers in Late-Renaissance Italy," in Moran, *Patronage*, pp. 25–50.

———, *Science and the Secrets of Nature: Books of Secrets in Medieval and Early Modern Culture* (Princeton: Princeton University Press, 1994).

———, "Technology as Magic in the Late Middle Ages and Renaissance," *Janus* 70 (1983), pp. 171–212.

Economou, George D., *The Goddess Natura in Medieval Literature* (Cambridge, MA: Harvard University Press, 1972).

Elm, Kaspar, Eberhard Gönner, and Eugen Hillenbrand (eds.), *Landesgeschichte und Geistesgeschichte: Festschrift für Otto Herding zum 65. Geburtstag* (Stuttgart: W. Kohlhammer, 1977).

Emerton, Norma E., *The Scientific Reinterpretation of Form* (Ithaca/London: Cornell University Press, 1984).

Enciclopedia dell'arte medievale, 6 vols. to date, gen. ed. Marina Righetti Tosti-Croce (Rome: Istituto della Enciclopedia Italiana, 1991–).

Erlemann, Hildegard, and Thomas Stangier, "Festum Reliquiarum," in *Reliquien. Verehrung und Verklärung*, pp. 21–35.

Evans, Joan, *Magical Jewels of the Middle Ages and the Renaissance, Particularly in England* (Oxford: Clarendon Press, 1922).

Ewinkel, Irene, *De monstris: Deutung und Funktion von Wundergeburten auf Flugblättern im Deutschland des 16. Jahrhunderts* (Tübingen: Max Niemeyer, 1995).

Fahd, M. Tawfiq, "Le Merveilleux dans la faune, la flore et les minéraux," in Arkoun et al., *L'Étrange*, pp. 117–35.

Falchetta, Piero (ed.), *The Arcimboldo Effect: Transformations of the Face from the 16th to the 20th Century* (New York: Abbeville Press, 1987).

Falguières, Patricia, "Fondations du théâtre ou Méthode de l'exposition universelle: les Inscriptions de Samuel Quicchelberg (1565)," *Les Cahiers du Musée National d'Art Moderne* 40 (1992), pp. 91–109.

Faral, Edmond, "Le Merveilleux et ses sources dans les descriptions des romans français du XIIe siècle," in Faral, *Recherches*, pp. 307–88.

———, "Ovide et quelques romans français du XIIe siècle," in Faral, *Recherches*, pp. 1–157.

———, *Recherches sur les sources latines des contes et romans courtois du Moyen Age* (Paris: Honoré Champion, 1913).

Félibien, Michel, *Histoire de l'abbaye de Saint-Denis* (Paris: F. Léonard, 1706).

475

Festschrift Bruno Snell (Munich: C.H. Beck, 1956).

Findlen, Paula, "Courting Nature," in Jardine et al., *Cultures*, pp. 57–74.

———, "Disciplining the Disciplines: Francis Bacon and the Reform of Natural History in the Seventeenth Century," in Kelley, *History*, pp. 239–60.

———, "The Economy of Scientific Exchange in Early Modern Italy," in Moran, *Patronage*, pp. 5–24.

———, "Empty Signs? Reading the Book of Nature in Renaissance Science," *Studies in the History and Philosophy of Science* 21 (1990), pp. 511–18.

———, "Jokes of Nature and Jokes of Knowledge: The Playfulness of Scientific Discourse in Early Modern Europe," *Renaissance Quarterly* 43 (1990), pp. 292–331.

———, "The Museum: Its Classical Etymology and Renaissance Genealogy," *Journal of the History of Collections* 1 (1989), pp. 59–78.

———, *Possessing Nature: Museums, Collecting, and Scientific Culture in Early Modern Italy* (Berkeley: University of California Press, 1994).

Fleischmann, Suzanne, "On the Representation of History and Fiction in the Middle Ages," *History and Theory* 22 (1983), pp. 278–310.

Flint, Valerie I.J., *The Imaginative Landscape of Christopher Columbus* (Princeton: Princeton University Press, 1992).

Foucault, Michel, *Foucault Live (Interviews, 1966–84)*, trans. John Johnston, ed. Sylvère Lotringer (New York: Semiotext(e), 1989).

Franchini, Dario, et al., *La scienza a corte: Collezionismo eclettico, natura e immagine a Mantova fra Rinascimento e Manierismo* (Rome: Bulzoni, 1979).

Freedberg, David, "Ferrari on the Classification of Oranges and Lemons," in Cropper et al., *Documentary Culture*, pp. 287–306.

———, *The Power of Images: Studies in the History and Theory of Response* (Chicago: University of Chicago Press, 1989).

French, Roger, *Ancient Natural History* (London: Routledge, 1994).

Friedman, Jerome, *The Battle of the Frogs and Fairford's Flies: Miracles and the Pulp Press during the English Revolution* (New York: St. Martin's Press, 1993).

Friedman, John Block, *The Monstrous Races in Medieval Art and Thought* (Cambridge, MA: Harvard University Press, 1981).

Frisk, Hjalmar, *Etymologisches Wörterbuch der griechischen Sprache*, 2 vols. in 3 (Heidelberg: Carl Winter Universitätsverlag, 1960–66).

Frost, Thomas, *The Old Showman, and the Old London Fairs* (London: Tinsley Brothers, 1874).

Fryde, E.B., "Lorenzo de' Medici: High Finance and the Patronage of Art and Learning," in Dickens, *Courts*, pp. 77–98.

Gandillac, Maurice de, "Encyclopédies pré-médiévales et médiévales," in *Cahiers d'Histoire Mondiale* 9/3 (1966) = *Encyclopédies et Civilisations*, pp. 483–518.

Garin, Eugenio, *Astrology in the Renaissance: The Zodiac of Life*, trans. Carolyn Jackson and June Allen, with Clare Robertson (London: Routledge and Kegan Paul, 1983).

———, *Medioevo e Rinascimento* (Bari: Giuseppe Laterza, 1961).

Geary, Patrick J., *Furta Sacra: Thefts of Relics in the Central Middle Ages*, 2nd ed. (Princeton: Princeton University Press, 1990).

Gentilcore, David, *From Bishop to Witch: The System of the Sacred in Early Modern Terra d'Otranto* (Manchester/New York: Manchester University Press, 1992).

Gentile, S., S. Niccoli, and P. Viti, *Marsilio Ficino e il ritorno di Platone: Mostra di manoscritti, stampe e documenti, 17 maggio – 16 giugno 1984* (Florence: Le Lettere, 1984).

George, Wilma, and Brunsdon Yapp, *The Naming of the Beasts: Natural History in the Medieval Bestiary* (London: Duckworth, 1991).

Gerbi, Antonello, *Nature in the New World: From Christopher Columbus to Gonzalo Fernandez de Oviedo*, trans. Jeremy Moyle (Pittsburgh: University of Pittsburgh Press, 1985).

Gerson, Paula Lieber (ed.), *Abbot Suger and Saint-Denis: A Symposium* (New York: Metropolitan Museum of Art, 1986).

Gestoso y Perez, José, *Sevilla monumental y artistica*, 3 vols. (Seville: La Andalucía Moderna, 1889–92).

Gilbert, Creighton, "'The Egg Reopened' Again," *Art Bulletin* 56 (1974), pp. 252–58.

Gilman, Ernest B., *The Curious Perspective: Literary and Pictorial Wit in the Seventeenth Century* (New Haven/London: Yale University Press, 1978).

Ginzburg, Carlo, *Clues, Myths, and the Historical Method*, trans. John and Anne C. Tedeschi (Baltimore: Johns Hopkins University Press, 1989).

———, "Clues: Roots of an Evidential Paradigm," in Ginzburg, *Clues, Myths*, pp. 96–125.

———, "High and Low: The Theme of Forbidden Knowledge in the Sixteenth and Seventeenth Centuries," *Past and Present* 73 (1976), pp. 28–41.

Godin, André, "Erasme: 'Pia/Impia Curiositas,'" in Céard, *Curiosité*, pp. 25–36.

Goldgar, Anne, *Impolite Learning: Conduct and Community in the Republic of Letters, 1680–1750* (New Haven/London: Yale University Press, 1995).

Goldthwaite, Richard, "The Empire of Things: Consumer Demand in Renaissance Italy," in Kent et al., *Patronage*, pp. 153–75.

———, *Wealth and the Demand for Art in Italy, 1300–1600* (Baltimore: Johns Hopkins University Press, 1993).

Golinski, Jan, "A Noble Spectacle: Phosphorus and the Public Cultures of Science in the Early Royal Society," *Isis* 80 (1989), pp. 11–30.

Gooding, David, Trevor Pinch, and Simon Schaffer (eds.), *The Uses of Experiment: Studies in the Natural Sciences* (Cambridge: Cambridge University Press, 1989).

Goulemot, Jean-Marie, "Démons, merveilles et philosophie à l'âge classique," *Annales. Economies, Sociétés, Civilisations* 35 (1980), pp. 1223–50.

Grafton, Anthony, and Nancy Siraisi, *Renaissance Natural Philosophy and the Disciplines* (Chicago: University of Chicago Press, 1998).

———, *Natural Particulars: Nature and the Disciplines in Renaissance Europe* (Cambridge, MA: MIT Press, 1999).

Grandsen, Antonia, "Realistic Observation in Twelfth-Century England," *Speculum* 47 (1973), pp. 29–51.

Grant, Edward, "Peter Peregrinus," *DSB*, vol. 10, pp. 532–40.

Grant, Robert McQueen, *Miracle and Natural Law in Graeco-Roman and Early Christian Thought* (Amsterdam: North-Holland, 1952).

Greenblatt, Stephen, *Marvelous Possessions: The Wonder of the New World* (Chicago: University of Chicago Press, 1991).

Greene, Robert A., "Whichcote, Wilkins, 'Ingenuity,' and the Reasonableness of Christianity," *Journal of the History of Ideas* 42 (1981), pp. 227–52.

Grieco, Allen J., "The Social Politics of Pre-Linnaean Botanical Classification," *I Tatti Studies: Essays in the Renaissance* 4 (1992), pp. 131–49.

Grimm, Jacob and Wilhelm, *Deutsches Wörterbuch*, 33 vols. (Leipzig: S. Hirzel, 1854–1984; repr. Munich: Deutscher Taschenbuch Verlag, 1991).

Guiffrey, Jules (ed.), *Inventaires de Jean Duc de Berry*, 2 vols. (Paris: Ernest Leroux, 1894–96).

———, "La Ménagerie du Duc Jean de Berry, 1370–1403," *Mémoires de la Société des Antiquaires du Centre, Bourges* 22 (1899), pp. 63–84.

Guthmüller, Bodo, *Ovido Metamorphoseos Vulgare: Formen und Funktionen der volkssprachlichen Wiedergabe klassischer Dichtung in der italienischen Renaissance* (Boppard an Rhein: Harold Boldt, 1981).

Habermas, Rebekka, "Wunder, Wunderliches, Wunderbares. Zur Profanisierung eines Deutungsmusters in der Frühen Neuzeit," in Van Dülmen, *Armut,* pp. 38–66.

Hagner, Michael (ed.), *Der falsche Körper: Beiträge zu einer Geschichte der Monstrositäten* (Göttingen: Wallstein, 1995).

———, "Vom Naturalienkabinett zur Embryologie: Wandlungen der Monstrositäten und die Ordnung des Lebens," in Hagner, *Der falsche Körper*, pp. 73–107.

Hahn, Roger, *The Anatomy of a Scientific Institution: The Paris Academy of Sciences, 1666–1803* (Berkeley/Los Angeles/London: University of California Press, 1971).

Hansen, Bert, *Nicole Oresme and the Marvels of Nature: A Study of his De causis mirabilium with Critical Edition, Translation, and Commentary* (Toronto: Pontifical Institute of Mediaeval Studies, 1985).

———, "Science and Magic," in Lindberg, *Science*, pp. 483–506.

Harmening, Dieter, *Superstitio: Überlieferungs- und theoriegeschichtliche Untersuchungen zur kirchlich-theologischen Aberglaubensliteratur des Mittelalters* (Berlin: Erich Schmidt, 1979).

Harth, Erica, *Cartesian Women: Versions and Subversions of Rational Discourse in the Old Regime* (Ithaca/London: Cornell University Press, 1992).

Haskins, Charles Homer, *Studies in the History of Mediaeval Science*, 2nd ed. (Cambridge, MA: Harvard University Press, 1927).

Hathaway, Baxter, *Marvels and Commonplaces: Renaissance Literary Criticism* (New York: Random House, 1968).

Heckscher, W.S., "Relics of Pagan Antiquity in Medieval Settings," *Journal of the Warburg and Courtauld Institutes* 1 (1938), pp. 204–20.

Heilbron, John L., *Electricity in the 17th and 18th Centuries: A Study in Early Modern Physics* (Berkeley: University of California Press, 1979).

————, "Experimental Natural Philosophy," in Rousseau and Porter, *Ferment of Knowledge*, pp. 357–87.

Heinemann, Thomas, *The Uppsala Art Cabinet* (Uppsala: Almqvist & Wiksell, 1982).

Héliot, Pierre, and Marie-Laure Chastang, "Quêtes et voyages de reliques au profit des églises françaises du Moyen Age," *Revue d'Histoire Ecclésiastique* 59 (1964), pp. 789–823, and 60 (1965), pp. 5–53.

Henry, John, "Henry More versus Robert Boyle: The Spirit of Nature and the Nature of Providence," in Hutton, *Henry More*, pp. 55–76.

————, "Occult Qualities and the Experimental Philosophy: Active Principles in Pre-Newtonian Matter Theory," *History of Science* 24 (1986), pp. 335–81.

————, "The Scientific Revolution in England," in Porter and Teich, *Scientific Revolution*, pp. 178–209.

Herding, Otto, and Robert Stupperich (eds.), *Die Humanisten in ihrer politischen und sozialen Umwelt* (Boppard: Harald Boldt, 1976).

Heyd, Michael, *"Be Sober and Reasonable": The Critique of Enthusiasm in the Seventeenth and Early Eighteenth Centuries* (Leiden/New York: E.J. Brill, 1995).

————, "The Reaction to Enthusiasm in the Seventeenth Century: Towards an Integrative Approach," *Journal of Modern History* 53 (1981), pp. 258–80.

Hirschman, Albert, *The Passions and the Interests: Political Arguments for Capitalism before its Triumph* (Princeton: Princeton University Press, 1977).

Holdefleiss, Erich, *Der Augenscheinsbeweis im mittelalterlichen deutschen Strafverfahren* (Stuttgart: W. Kohlhammer, 1933).

Holländer, Eugen, *Wunder, Wundergeburt und Wundergestalt: Einblattdrucke des fünfzehnten bis achtzehnten Jahrhunderts* (Stuttgart: Ferdinand Enke, 1921).

Holmes, Urban T., "Gerald the Naturalist," *Speculum* 11 (1936), pp. 110–21.

Honour, Hugh, *The European Vision of America* (Cleveland: Cleveland Museum of Art, 1975).

Houghton, Walter E., "The English Virtuoso in the Seventeenth Century," *Journal of the History of Ideas* 3 (1941), pp. 51–73 and 190–219.

Huet, Marie-Hélène, *Monstrous Imagination* (Cambridge, MA/London: Harvard University Press, 1993).

Hunter, J. Paul, *Before Novels: The Cultural Contexts of Eighteenth-Century English Fiction* (New York/London: W.W. Norton, 1990).

Hunter, Michael, "The Cabinet Institutionalized: The Royal Society's 'Repository' and Its Background," in Impey and MacGregor, *Origins*, pp. 159–68.

————, *The Royal Society and Its Fellows 1660–1700: The Morphology of an Early Scientific Institution* (Chalfont St. Giles: British Society for the History of Science, 1982).

Huntley, Frank Livingstone, *Bishop Hall and Protestant Meditation in Seventeenth-Century England: A Study with Texts of the Art of Divine Meditation (1606) and Occasional Medita-*

tions (1633) (Binghamton, NY: Center for Medieval and Early Renaissance Studies, 1981).

Hutton, Sarah (ed.), *Henry More (1614–1687): Tercentenary Studies* (Dordrecht/Boston/London: Kluwer, 1990).

Impey, Oliver, and Arthur MacGregor (eds.), *The Origins of Museums: The Cabinets of Curiosities in Sixteenth- and Seventeenth-Century Europe* (Oxford: Clarendon Press, 1985).

Ingegno, Alfonso, "The New Philosophy of Nature," in Schmitt et al., *Cambridge History*, pp. 236–63.

———, *Saggio sulla filosofia di Cardano* (Florence: La Nuova Italia, 1980).

Isherwood, Robert M., *Farce and Fantasy: Popular Entertainment in Eighteenth-Century Paris* (Oxford: Oxford University Press, 1986).

Jacquart, Danielle, "Médecine et alchimie chez Michel Savonarole," in Margolin and Matton, *Alchimie* (Paris: J. Vrin, 1993), pp. 109–22.

———, "Theory, Everyday Practice, and Three Fifteenth-Century Physicians," in McVaugh and Siraisi, *Medical Learning*, pp. 140–60.

James, M.R., "Ovidius *De mirabilibus mundi*," in Quiggin, *Essays*, pp. 286–98.

Janni, Pietro, and Innocenzo Mazzini (eds.), *Presenza del lessico greco e latino nelle lingue contemporanee* (Macerata: Università degli Studi di Macerata, 1990).

Janson, H.W., "The Image Made by Chance," in Meiss, *De Artibus*, pp. 254–66.

Jardine, N., J.A. Secord, and E.C. Spary (eds.), *Cultures of Natural History* (Cambridge: Cambridge University Press, 1996).

Jones, Mark (ed.), *Fake? The Art of Deception* (London: British Museum Publications, 1990).

Jones, William R., "Political Uses of Sorcery in Medieval Europe," *The Historian* 34 (1972), pp. 670–87.

Jones-Davies, M.T. (ed.), *Monstres et prodiges au temps de la Renaissance* (Paris: Jean Touzot, 1980).

Jordan, Karl, *Henry the Lion: A Biography*, trans. P.S. Falla (Oxford: Clarendon Press, 1986).

Jourdain, Nicolas, "Nicolas Oresme et les astrologues de la cour de Charles V," *Revue des Questions Historiques* 18 (1875), pp. 136–59.

Kaplan, Steven L. (ed.), *Understanding Popular Culture: Europe from the Middle Ages to the Nineteenth Century* (Berlin/New York: Mouton, 1984).

Kappler, Claude, "L'Interprétation politique du monstre chez Sébastien Brant," in Jones-Davies, *Monstres*, 100–10.

———, *Monstres, démons et merveilles à la fin du Moyen Age* (Paris: Payot, 1988).

Kauffmann, Claus Michael, *The Baths of Pozzuoli: A Study of the Medieval Illuminations of Peter of Eboli's Poem* (Oxford: Bruno Cassirer, 1959).

Kaufmann, Thomas DaCosta, *The Mastery of Nature: Aspects of Art, Science, and Humanism in the Renaissance* (Princeton: Princeton University Press, 1993).

————, *The School of Prague: Painting at the Court of Rudolf II* (Chicago/London: University of Chicago Press, 1988).

————, *Variations on the Imperial Theme in the Age of Maximilian II and Rudolf II* (New York: Garland, 1978).

Kelley, Donald (ed.), *History and the Disciplines in Early Modern Europe* (Rochester: University of Rochester Press, 1997).

Kelly, Douglas, *The Art of Medieval French Romance* (Madison: University of Wisconsin Press, 1992).

Kemp, Martin, "From 'Mimesis' to 'Fantasia': The Quattrocento Vocabulary of Creation, Inspiration, and Genius in the Visual Arts," *Viator* 8 (1977), pp. 347–98.

Kenseth, Joy, "The Age of the Marvelous: An Introduction," in Kenseth, *Marvelous*, pp. 25–60.

———— (ed.), *The Age of the Marvelous* (Hanover, NH: Hood Museum of Art, Dartmouth College, 1991).

————, "'A World of Wonders in One Closet Shut,'" in Kenseth, *Marvelous*, pp. 80–101.

Kent, F.W. and Patricia Simons, with J.C. Eade (eds.), *Patronage, Art and Society in Renaissance Italy* (Oxford: Clarendon Press, 1987).

Kibre, Pearl, and Nancy G. Siraisi, "The Institutional Setting: The Universities," in Lindberg, *Science*, pp. 120–44.

Kieckhefer, Richard, *Magic in the Middle Ages* (Cambridge: Cambridge University Press, 1989).

Klemm, Gustav, *Zur Geschichte der Sammlungen für Wissenschaft und Kunst in Deutschland* (Zerbst: G.A. Kummer, 1837).

Knock, Ann, "The 'Liber monstrorum': An Unpublished Manuscript and Some Reconsiderations," *Scriptorium* 32 (1978), pp. 19–28.

Köhler, Erich, *Ideal und Wirklichkeit in der höfischen Epik: Studien zur Form der frühen Artus- und Gralsdichtung* (Tübingen: M. Niemeyer, 1970).

Kohut, Karl (ed.), *Von der Weltkarte zum Kuriositätenkabinett: Amerika im deutschen Humanismus und Barock* (Frankfurt: Vervuert, 1995).

Kreiser, B. Robert, *Miracles, Convulsions, and Ecclesiastical Politics in Early Eighteenth-Century Paris* (Princeton: Princeton University Press, 1978).

Kretzmann, Norman, Anthony Kenny, and Jan Pinborg (eds.), *The Cambridge History of Later Medieval Philosophy* (Cambridge: Cambridge University Press, 1982).

Kris, Ernst, "Der Stil 'Rustique': Die Verwendung des Naturabgusses bei Wenzel Jamnitzer und Bernard de Palissy," *Jahrbuch der Kunsthistorischen Sammlungen* N.F. 1 (1926), pp. 137–208.

Kris, Ernst, and Otto Kurz, *Legend, Myth, and Magic in the Image of the Artist* [1934], trans. Alastair Lang and rev. Lottie M. Newman (New Haven/London: Yale University Press, 1979).

Kristeller, Paul Oskar, "Philosophy and Medicine in Medieval and Renaissance Italy," in Spicker, *Organism*, pp. 29–40.

Labhardt, André, "*Curiositas*: Notes sur l'histoire d'un mot et d'une notion," *Museum Helveticum* 17 (1960), pp. 206–24.

Laborde, Léon de, *Les Ducs de Bourgogne: Etudes sur les lettres, les arts et l'industrie pendant le XVe siècle*, 3 vols. (Paris: Plon, 1849–52).

Lacaze, Yvon, "Le Rôle des traditions dans la genèse d'un sentiment national au XVe siècle: la Bourgogne de Philippe le Bon," *Bibliothèque de l'Ecole des Chartes* 129 (1971), pp. 303–85.

Lafortune-Martel, Agathe, *Fête noble en Bourgogne au XVe siècle. Le banquet du Faisan (1454): Aspects politiques, sociaux et culturels* (Montreal: Bellarmin, 1984).

Langemeyer, Gerhard, and Hans-Albert Peters (eds.), *Stilleben in Europa*, Westfälisches Landesmuseum für Kunst und Kunstgeschichte (Münster: Aschendorff, 1979).

Lanowski, J., "Weltwunder," in Pauly et al., *Realencyclopädie*, suppl. vol. 10, pp. 1020–30.

Lanza, Diego, and Oddone Longo (eds.), *Il meraviglioso e il verosimile tra Antichità e Medioevo* (Florence: Olschki, 1989).

Laurencich-Minelli, Laura, "Museography and Ethnographical Collections in Bologna during the Sixteenth and Seventeenth Centuries," in Impey and MacGregor, *Origins*, pp. 17–23.

Le Goff, Jacques, *L'Imaginaire médiéval* (Paris: Gallimard, 1985).

———, "Le Merveilleux dans l'Occident medieval," in Arkoun, *L'Étrange*, pp. 61–79.

———, *The Medieval Imagination*, trans. Arthur Goldhammer (Chicago: Chicago University Press, 1988).

———, "Le Merveilleux scientifique au Moyen Age," in Bergier, *Zwischen Wahn, Glaube und Wissenschaft*, pp. 87–113.

Lecouteux, Claude, *Les Monstres dans la littérature allemande du Moyen Age: Contribution à l'étude du merveilleux médiéval*, 3 vols. (Göppingen: Kümmerle Verlag, 1982).

Lee, Rensselaer W., "*Ut Pictura Poesis*: The Humanistic Theory of Painting," *The Art Bulletin* 22 (1940), pp. 197–269.

Legner, Anton (ed.), *Reliquien, Verehrung und Verklärung: Skizzen und Noten zur Thematik und Katalog zur Ausstellung der Kölner Sammlung Louis Peters im Schnütgen-Museum* (Cologne: Greven & Bechthold, 1989).

Lemoine, Michel, "L'Oeuvre encyclopédique de Vincent de Beauvais," in *Cahiers d'Histoire Mondiale* 9/3 (1966) = *Encyclopédies et Civilisations*, pp. 571–79.

Licoppe, Christian, *La Formation de la pratique scientifique: Le Discours de l'expérience en France et en Angleterre (1630–1820)* (Paris: Editions la Découverte, 1996).

Liebenwein, Wolfgang, *Studiolo: Die Entstehung eines Raumtyps und seine Entwicklung bis um 1600* (Berlin: Gebr. Mann, 1977).

———, *Studiolo: Storia e tipologia di uno spazio culturale* (Ferrara: Panini, 1977).

Lindberg, David C., "Science as Handmaiden: Roger Bacon and the Patristic Tradition," *Isis* 78 (1987), pp. 518–36.

——— (ed.), *Science in the Middle Ages* (Chicago: Chicago University Press, 1978).

————, "The Transmission of Greek and Arabic Learning to the West," in Lindberg, *Science*, pp. 52–90.

Lindberg, David C., and Robert S. Westman (eds.), *Reappraisals of the Scientific Revolution* (Cambridge: Cambridge University Press, 1990).

Livorno e Pisa: Due città e un territorio nella politica dei Medici (Pisa: Nistri-Lisci e Pacini, 1980).

Loisel, Gustave, *Histoire des ménageries d'Antiquité à nos jours*, 2 vols. (Paris: Octave Doin/Henri Laurens, 1912).

Loomis, Laura Hibbard, "Secular Dramatics in the Royal Palace, Paris, 1378, 1389, and Chaucer's 'Tregetoures,'" *Speculum* 33 (1958), pp. 242–55.

Lougee, Carolyn C., *Le Paradis des Femmes: Women, Salons, and Social Stratification in Seventeenth-Century France* (Princeton: Princeton University Press, 1976).

Lugli, Adalgisa, *Naturalia et mirabilia: Il Collezionismo enciclopedico nelle wunderkammern d'Europa* (Milan: Mazzota, 1983).

Luther, Gisela, "Stilleben als Bilder der Sammelleidenschaft," in Langemeyer, *Stilleben*, pp. 88–128.

MacDougall, Elisabeth Blair, "A Paradise of Plants: Exotica, Rarities, and Botanical Fantasies," in Kenseth, *Marvelous*, pp. 145–57.

MacDougall, Elisabeth Blair (ed.), *Medieval Gardens* (Washington, DC: Dumbarton Oaks, 1986).

MacGregor, Arthur, "'A Magazin of All Manner of Inventions': Museums in the Quest for 'Solomon's House' in Seventeenth-Century England," *Journal of the History of Collections* 1 (1989), pp. 207–12.

MacKinney, Loren, *Medical Illustrations in Medieval Manuscripts* (London: Wellcome Historical Medical Library, 1965).

Maclean, Ian, "The Interpretation of Natural Signs: Cardano's *De subtilitate* versus Scaliger's *Exercitationes*," in Vickers, *Mentalities*, pp. 231–52.

Mandrou, Robert, *De la culture populaire aux XVIIe et XVIIIe siècles: La Bibliothèque Bleue de Troyes* (Paris: Editions Stock, 1964).

Margolin, Jean-Claude, "Sur quelques prodiges rapportés par Conrad Lycosthène," in Jones-Davies, *Monstres*, pp. 42–54.

Margolin, Jean-Claude, and Sylvain Matton (eds.), *Alchimie et philosophie à la Renaissance* (Paris: J. Vrin, 1993).

Marrou, Henri-Irénée, *Saint Augustin et la fin de la culture antique*, 4th ed. (Paris: E. de Boccard, 1958).

Martines, Lauro, *Power and Imagination: City-States in Renaissance Italy* (New York: Knopf, 1979).

Maurer, Armand, "Between Reason and Faith: Siger of Brabant and Pomponazzi on the Magic Arts," *Mediaeval Studies* 18 (1956), pp. 1–18.

May, Georges, "L'Histoire a-t-elle engendré le roman? Aspects français de la question au seuil du siècle des lumières," *Revue d'Histoire Littéraire de la France* 55 (1955), pp. 155–76.

Mazzio, Carla, and David Hillman (eds.), *The Body in Parts: Discourses and Anatomies in Early Modern Europe* (New York: Routledge, 1997).

McBurney, Henrietta, "Cassiano dal Pozzo as Scientific Commentator: Ornithological Texts and Images of Birds from the Museo Cartaceo," in Cropper et al., *Documentary Culture*, pp. 349–62.

McCulloch, Florence, *Medieval Latin and French Bestiaries*, 2nd ed. (Chapel Hill: University of North Carolina Press, 1962).

McGurk, Patrick, et al., *An Eleventh-Century Anglo-Saxon Illustrated Miscellany: British Library Cotton Tiberius B.V Part I, together with Leaves from British Library Cotton Nero D.II* (Copenhagen: Rosenkilde and Bagger, 1983).

McGurk, Patrick, and Ann Knock, "The Marvels of the East," in McGurk et al., *Miscellany*, pp. 88–103.

McKeon, Michael, *The Origins of the English Novel 1600–1740* (Baltimore/London: The Johns Hopkins University Press, 1987).

McVaugh, Michael R., "The Development of Medieval Pharmaceutical Theory," in Arnald of Villanova, *Opera medica*, vol. 2: *Aphorismi de gradibus*, pp. 3–136.

———, *Medicine before the Plague: Practitioners and Their Patients in the Crown of Aragon, 1285–1345* (Cambridge: Cambridge University Press, 1993).

McVaugh, Michael R., and Nancy G. Siraisi (eds.), *Renaissance Medical Learning: The Evolution of a Tradition = Osiris* 6 (1990).

Megill, Allan (ed.), *Rethinking Objectivity* (Durham/London: Duke University Press, 1994).

Meiss, Millard (ed.), *De Artibus Opuscula LX: Essays in Honor of Erwin Panofsky* (New York: New York University Press, 1961).

———, *French Painting in the Time of Jean de Berry: The Boucicaut Master* (London: Phaidon, 1968).

———, *French Painting in the Time of Jean de Berry: The Late Fourteenth Century and the Patronage of the Duke*, 2 vols. (London and New York: Phaidon, 1967).

Melion, Walter, "Prodigies of Nature, Wonders of the Hand: Karel van Mander on the Depiction of Wondrous Sights," Paper delivered to Pacific Northwest Renaissance Conference, March 1994.

Menzhausen, Joachim, "Elector Augustus's *Kunstkammer*: An Analysis of the Inventory of 1587," in Impey and MacGregor, *Origins*, pp. 69–75.

Merkel, Ingrid, and Allen G. Debus (eds.), *Hermeticism and the Renaissance: Intellectual History and the Occult in Early Modern Europe* (Washington: Folger Shakespeare Library, 1988).

Meslin, Michel (ed.), *Le Merveilleux: L'imaginaire et les croyances en Occident* (Paris: Bordas, 1984).

Mette, Hans Joachim, "Curiositas," in *Festschrift Bruno Snell*, pp. 227–35.

Michaud-Quantin, Pierre, "Les Petites encyclopédies du XIIIe siècle," in *Cahiers d'Histoire Mondiale* 9/3 (1966) = *Encyclopédies et Civilisations*, pp. 580–95.

Middleton, W.E. Knowles, *The Experimenters: A Study of the Accademia del Cimento* (Baltimore/London: Johns Hopkins University Press, 1971).

Milton, John R., "The Origin and Development of the Concept of the 'Laws of Nature,'" *Archives of European Sociology* 22 (1981), pp. 173–95.

Molland, A.G., "Medieval Ideas of Scientific Progress," *Journal of the History of Ideas* 39 (1978), pp. 561–77.

——, "Roger Bacon as Magician," *Traditio* 30 (1974), pp. 445–60.

Mollat, Michel, *Les Explorateurs du XIIIe au XVIe siècle: Premiers regards sur des mondes nouveaux* (Paris: J.-C. Lattès, 1984).

Montagu, Jennifer, *The Expression of the Passions: The Origin and Influence of Charles le Brun's "Conférence sur l'expression générale et particulière"* (New Haven/London: Yale University Press, 1994).

Monter, William, *Ritual, Myth and Magic in Early Modern Europe* (Athens, OH: Ohio University Press, 1984).

Moran, Bruce (ed.), *Patronage and Institutions* (Woodbridge: Boydell & Brewer, 1991).

Morize, André, *L'Apologie du luxe au XVIIIe siècle* [1909] (Geneva: Slatkine Reprints, 1970).

Morley, Henry, *Memoirs of Bartholomew Fair* (London: Chapman and Hall, 1859).

Moscoso, Javier, "Vollkommene Monstren und unheilvolle Gestalten: Zur Naturalisierung der Monstrosität im 18. Jahrhundert," in Hagner, *Der falsche Körper*, pp. 56–72.

Moss, Ann, *Ovid in Renaissance France: A Survey of the Latin Editions of Ovid and Commentaries Printed in France before 1600* (London: The Warburg Institute, 1982).

Moss, Jean Dietz, "The Rhetoric of Proof in Galileo's Writings on the Copernican System," in Wallace, *Reinterpreting Galileo*, pp. 179–204.

Muir, Edward, and Guido Ruggiero (eds.), *Sex and Gender in Historical Perspective* (Baltimore: Johns Hopkins University Press, 1990).

Mulligan, Lotte, "'Reason,' 'Right Reason,' and 'Revelation' in Mid-Seventeenth-Century England," in Vickers, *Mentalities*, pp. 375–401.

Muncey, R.W., *Our Old English Fairs* (London: Sheldon Press, 1936).

Murdoch, John Emery, "The Analytical Character of Late Medieval Learning: Natural Philosophy without Nature," in Roberts, *Approaches*, pp. 171–213.

——, "The Unitary Character of Medieval Learning," in Murdoch and Sylla, *Cultural Context*, pp. 271–348.

Murdoch, John Emery, and Edith Dudley Sylla (eds.), *The Cultural Context of Medieval Learning* (Dordrecht: D. Reidel, 1973).

Murray, David, *Museums: Their History and Their Use*, 2 vols. (Glasgow: James MacLehose and Sons, 1904).

Musée National des Techniques, *Jacques Vaucanson* (Ivry: Salles & Grange, 1983).

Nauert (Jr.), Charles G., "Caius Plinius Secundus," in Cranz and Kristeller, *Catalogus translationum et commentariorum*, vol. 4, pp. 297–422.

——, "Humanists, Scientists, and Pliny: Changing Approaches to a Classical Author," *American Historical Review* 84 (1979), pp. 72–85.

Needham, Joseph, *A History of Embryology*, 2nd ed. (Cambridge: Cambridge University Press, 1959).

Nerlich, Michael, *Ideology of Adventure: Studies in Modern Consciousness, 1100–1750*, 2 vols., trans. Ruth Crowley (Minneapolis: University of Minnesota Press, 1987).

Neumann, Erwin, "Das Inventar der rudolfinischen Kunstkammer von 1607/1611," in Platen, *Queen Christina*, pp. 262–65.

Neverov, Oleg, "'His Majesty's Cabinet' and Peter I's Kunstkammer," in Impey and MacGregor, *Origins*, pp. 54–61.

Newman, William R., *Gehennical Fire: The Lives of George Starkey, an American Alchemist in the Scientific Revolution* (Cambridge, MA/London: Harvard University Press, 1994).

———, "Technology and Alchemical Debate in the Late Middle Ages," *Isis* 80 (1989), pp. 423–45.

Niccoli, Ottavia, "'Menstruum quasi Monstruum': Monstrous Births and Menstrual Taboo in the Sixteenth Century," trans. Mary M. Gallucci, in Muir and Ruggiero, *Sex and Gender*, pp. 1–25.

———, *Prophecy and People in Renaissance Italy* [1987], trans. Lydia G. Cochrane (Princeton: Princeton University Press, 1990).

Oakley, Francis, "Christian Theology and the Newtonian Science: The Rise of the Concept of the Laws of Nature," *Church History* 30 (1961), pp. 433–57.

———, *Omnipotence, Covenant, and Order: An Excursion into the History of Ideas from Abelard to Leibniz* (Ithaca/London: Cornell University Press, 1984).

Oggins, Robin S., "Albertus Magnus on Falcons and Hawks," in Weisheipl, *Albertus Magnus*, pp. 441–62.

Oldenburg, Henry, *The Correspondence of Henry Oldenburg*, ed. Marie Boas Hall and Rupert Hall, 13 vols. (Madison/Milwaukee: The University of Wisconsin Press, 1965–73).

Olmi, Giuseppe, "Dal 'Teatro del mondo' ai mondi inventariati: Aspetti e forme del collezionismo nell'età moderna" in Olmi, *L'Inventario*, pp. 165–209.

———, *L'inventario del mondo: Catalogazione della natura e luoghi del sapere nell prima età moderna* (Bologna: Il Mulino, 1992).

———, "'Magnus campus': I naturalisti italiani di fronte all'America nel secolo XVI," in Olmi, *L'inventario*, pp. 211–52.

———, "Ordine e fama: Il museo naturalistico in Italia nei secoli XVI-XVII," in Olmi, *L'inventario*, pp. 255–313.

———, "Science-Honour-Metaphor: Italian Cabinets of the Sixteenth and Seventeenth Centuries," in Impey and MacGregor, *Origins*, pp. 5–16.

———, *Ulisse Aldrovandi: Scienza e natura nel secondo Cinquecento* (Trent: Libera Università degli Studi di Trento, 1976).

Olschki, Leonardo, *Guillaume Boucher: A French Artist at the Court of the Khans* (Baltimore: Johns Hopkins University Press, 1946).

Omont, Henri, "Les Sept Merveilles du monde au Moyen Age," *Bibliothèque de l'École des Chartes* 43 (1882), pp. 40–59.

Otte, D. Heinrich, *Handbuch der kirchlichen Kunst-Archäologie des deutschen Mittelalters*, 2 vols., 5th ed. (Leipzig: T.O. Weigel, 1883).

Oursel, C., "Un Exemplaire du 'Speculum Majus' de Vincent de Beauvais provenant de la bibliothèque de Saint Louis," *Bibliothèque de l'École des Chartes* 85 (1924), pp. 251–62.

The Oxford English Dictionary, 20 vols., 2nd ed. (Oxford: Clarendon Press, 1989).

Pagel, Walter, *Paracelsus: An Introduction to Philosophical Medicine in the Era of the Renaissance* (Basel/New York: S. Karger, 1958).

Palmer, Richard, "'In this our lightye and learned tyme': Italian Baths in the Era of the Renaissance," in Porter, *Medical History*, pp. 14–22.

Panofsky, Erwin (ed. and trans.), *Abbot Suger on the Abbey Church of Saint-Denis and Its Art Treasure*, ed. Gerda Panofsky-Soergel (Princeton: Princeton University Press, 1979).

Paravicini Bagliani, Agostino, *Medicina e scienze della natura alla corte dei papi nel Duecento* (Spoleto: Centro Italiano di Studi sull'Alto Medioevo, 1991).

Paravicini, Werner, "The Court of the Dukes of Burgundy: A Model for Europe?" in Asch and Birke, *Princes*, pp. 69–102.

Paré, Gérard, *Les Idées et les lettres au XIIIe siècle: Le Roman de la Rose* (Montreal: Centre de Psychologie et de Pédagogie, 1947).

Park, Katharine, "The Criminal and the Saintly Body: Autopsy and Dissection in Renaissance Italy," *Renaissance Studies* 47 (1994), pp. 1–33.

———, *Doctors and Medicine in Early Renaissance Florence* (Princeton: Princeton University Press, 1985).

———, "The Meanings of Natural Diversity: Marco Polo on the 'Division' of the World," Sylla and McVaugh, *Text and Context*, pp. 134–47.

———, "Natural Particulars: Medical Epistemology, Practice, and the Literature of Healing Springs," in Grafton and Siraisi, *Natural Particulars*, pp. 347–68.

———, "The Rediscovery of the Clitoris: French Medicine and the *Tribade*," in Mazzio and Hillman, *Body in Parts*, pp. 171–93.

Park, Katharine, and Lorraine Daston, "Unnatural Conceptions: The Study of Monsters in Sixteenth- and Seventeenth-Century France and England," *Past and Present* 92 (1981), pp. 20–54.

Park, Katharine, and Eckhard Kessler, "The Concept of Psychology," in Schmitt et al., *Cambridge History*, pp. 455–63.

Parks, Stephen, "John Dunton and The Works of the Learned," *Library*, ser. 5, 23 (1968), pp. 13–24.

Parshall, Peter, "*Imago Contrafacta*: Images and Facts in the Northern Renaissance," *Art History* 16 (1993), pp. 554–79.

Patey, Douglas Lane, *Probability and Literary Form: Philosophic Theory and Literary Practice in the Augustan Age* (Cambridge: Cambridge University Press, 1984).

Pauly, August Friedrich von, Georg Wissowa, Wilhelm Kroll et al. (eds.), *Realencyclopädie der classischen Altertumswissenschaft* (Stuttgart: Metzler/Alfred Druckenmüller, 1895–).

Pearl, Jonathan L., "French Catholic Demonologists and Their Enemies in the Late Sixteenth and Early Seventeenth Centuries," *Church History* 52 (1983), pp. 457–67.

Pereira, Michela, *L'Oro dei filosofi: Saggio sulle idee di un alchimista del Trecento* (Spoleto: Centro Italiano di Studi sull'Alto Medioevo, 1992).

Peron, Gianfelice, "Meraviglioso e verosimile nel romanzo francese medievale: Da Benoît de Sainte-Maure a Jean Renart," in Lanza and Longo, *Il meraviglioso*, pp. 293–323.

Pesenti, Tiziana, "Dondi dall'Orologio, Giovanni," *Dizionario biografico*, vol. 43, pp. 96–104.

Pinotti, Patrizia, "Aristotele, Platone e la meraviglia del filosofo," in Lanza and Longo, *Il meraviglioso*, pp. 29–55.

Piponnier, Françoise, *Costume et vie sociale: La Cour d'Anjou, XIVe–XVe siècle* (Paris: Mouton, 1970).

Platen, Magnus von (ed.), *Queen Christina of Sweden: Documents and Studies* (Stockholm: P.A. Norstedt & Söner, 1966).

Poirion, Daniel, *Le Merveilleux dans la littérature française du Moyen Age* (Paris: Presses Universitaires de France, 1982).

Pomata, Gianna, "Uomini mestruati. Somiglianza e differenza fra i sessi in Europa in età moderna," *Quaderni Storici*, n.s. 79/XXVII (1992), pp. 51–103.

———, "'Observatio' ovvero 'Historia': note su empirismo e storia in età moderna," *Quaderni Storici,* n.s. 91 (1996), pp. 173–98.

———, "A New Way of Saving the Phenomenon: From 'Recipe' to 'Historia' in Early Modern Science," forthcoming.

Pomian, Krzysztof, *Collectors and Curiosities: Paris and Venice, 1500–1800* [1987], trans. Elizabeth Wiles-Portier (Cambridge: Polity Press, 1990).

———, "Collezionismo," in *Enciclopedia dell'arte medievale*, vol. 5, pp. 156–60.

———, "Curiosity and Modern Science," Sylvia Lemmie England Memorial Lecture, Victoria and Albert Museum, January 1993.

Popkin, Richard, *The History of Scepticism from Erasmus to Descartes*, rev. ed. (Assen: Van Gorcum, 1964).

Poppi, Antonino (ed.), *Scienza e filosofia all'Università di Padova nel Quattrocento* (Padua: Edizioni Lint, 1983).

Porter, Roy (ed.), *The Medical History of Waters and Spas* (London: Wellcome Institute for the History of Medicine, 1990).

Porter, Roy, and Mikulás Teich (eds.), *The Scientific Revolution in National Context* (Cambridge: Cambridge University Press, 1992).

Porter, V.G., "The West Looks at the East in the Late Middle Ages: The *Livre des merveilles du monde*" (Baltimore: Ph.D. diss., Johns Hopkins University, 1977).

Price, Derek J. de Solla, "Automata and the Origins of Mechanism and Mechanistic Philosophy," *Technology and Culture* 5 (1964), pp. 1–23.

Quiggin, E.C. (ed.), *Essays and Studies Presented to William Ridgeway* (1913; repr. Freeport, NY: Books for Libraries Press, 1966).

Rabb, Theodore, *The Struggle for Stability in Early Modern Europe* (Oxford/New York: Oxford University Press, 1975).

Raby, F.J.E., "*Nuda Natura* and Twelfth-Century Cosmology," *Speculum* 43 (1968), pp. 72–77.

Ragusa, Ida, "The Egg Reopened," *The Art Bulletin* 53 (1971), pp. 435–43.

Raine, James, *Saint Cuthbert, with an Account of the State in Which His Remains Were Found upon the Opening of His Tomb in Durham Cathedral, in the Year MDCCCXXVII* (Durham: George Andrews, 1828).

Ranum, Orest, "Islands and the Self in a Ludovician Fête," in Rubin, *Sun King*, pp. 17–34.

Raven, Charles Earle, *John Ray: Naturalist* [1942], 2nd ed. (Cambridge: Cambridge University Press, 1986).

Rawson, Elizabeth, *Intellectual Life in the Late Roman Republic* (Baltimore: Johns Hopkins University Press, 1985).

Reeds, Karen Meier, *Botany in Medieval and Renaissance Universities* (New York: Garland, 1991).

———, "Renaissance Humanism and Botany," *Annals of Science* 33 (1976), pp. 519–42.

Reliquien. Verehrung und Verklärung: Skizzen und Notizen zur Thematik und Katalog zur Ausstellung der Kölner Sammlung Louis-Peters im Schnütgen-Museum (Cologne: Anton Legner, 1989).

Revel, Jacques, "Forms of Expertise: Intellectuals and 'Popular' Culture in France (1650–1800)," in Kaplan, *Understanding*, pp. 255–73.

Richardson, Linda Deer, "The Generation of Disease: Occult Causes and Diseases of the Total Substance," in Wear et al., *Medical Renaissance*, pp. 175–94.

Riddle, John M., and James A. Mulholland, "Albert on Stones and Minerals," in Weisheipl, *Albertus Magnus*, Ch. 8.

Roberts, Lawrence D. (ed.), *Approaches to Nature in the Middle Ages* (Binghamton NY: Center for Medieval and Early Renaissance Studies, 1982).

Roche, Daniel, *Le Siècle des lumières en province: Académies et académiciens provinciaux 1680–1789*, 2 vols. (Paris/The Hague: Mouton, 1978).

Roe, Shirley A., *Matter, Life, and Generation: Eighteenth-Century Embryology and the Haller-Wolff Debate* (Cambridge: Cambridge University Press, 1981).

Roger, Jacques, *Les Sciences de la vie dans la pensée française du XVIIIe siècle*, 2nd ed. (Paris: Armand Colin, 1971).

Romm, James S., *The Edges of the Earth in Ancient Thought: Geography, Exploration, and Fiction* (Princeton: Princeton University Press, 1992).

Ross, D.J.A., *Illustrated Medieval Alexander-Books in Germany and the Netherlands: A Study in Comparative Iconography* (Cambridge: The Modern Humanities Research Association, 1971).

Ross, Sydney, "'Scientist': The Story of a Word," *Annals of Science* 18 (1962), pp. 65–86.

Rossi, Paolo (ed.), *Cultura popolare e cultura dotta nel Seicento* (Milan: F. Angeli, 1983).

———, *The Dark Abyss of Time: The History of the Earth and the History of Nations from*

Hooke to Vico [1979], trans. Lydia G. Cochrane (Chicago/London: University of Chicago Press, 1984).

———, *Francesco Bacone: Dalla magia alla scienza* [1957] (Torino: Einaudi, 1974).

Rousseau, G.S., and Roy Porter (eds.), *The Ferment of Knowledge: Studies in the Historiography of Eighteenth-Century Science* (Cambridge: Cambridge University Press, 1980).

Rousset, Paul, "Le Sens du merveilleux à l'époque féodale," *Le Moyen Age*, ser. 4, 11 = 62 (1956), pp. 25–37.

Roy, Bruno, "En marge du monde connu: Les Races de monstres," in Allard, *Aspects*, pp. 70–81.

Rubin, David Lee (ed.), *Sun King: The Ascendancy of French Culture during the Reign of Louis XIV* (Washington, D.C.: The Folger Shakespeare Library, 1992).

Ruby, Jane E., "The Origins of Scientific 'Law,'" *Journal of the History of Ideas* 47 (1986), pp. 341–59.

Ruderman, David B., *Kabbalah, Magic, and Science: The Cultural Universe of a Sixteenth-Century Jewish Physician* (Cambridge, MA: Harvard University Press, 1988).

Rudwick, Martin J.S., *The Meaning of Fossils: Episodes in the History of Palaeontology* [1972], 2nd ed. (Chicago/London: University of Chicago Press, 1985).

Ryan, Michael T., "Assimilating New Worlds in the Sixteenth and Seventeenth Centuries," *Comparative Studies in Society and History* 23 (1981), pp. 519–38.

Ryan, W.F., and Charles B. Schmitt (eds.), *Pseudo-Aristotle, The Secret of Secrets: Sources and Influence* (London: Warburg Institute, 1982).

Schaffer, Simon, "Glass Works: Newton's Prisms and the Uses of Experiment," in Gooding et al., *Experiment*, pp. 67–104.

Scheicher, Elisabeth, *Die Kunstkammer. Kunsthistorisches Museum, Sammlungen Schloss Ambras* (Innsbruck: Kunsthistorisches Museum, 1977).

———, *Die Kunst- und Wunderkammern der Habsburger* (Vienna/Munich/Zurich: Molden Edition, 1979).

Schenda, Rudolf, "Die deutschen Prodigiensammlungen des 16. und 17. Jahrhunderts," *Archiv für Geschichte des Buchwesens* 4 (1963), pp. 638–710.

———, *Die französischen Prodigiensammlungen in der zweiten Hälfte des 16. Jahrhunderts*, Münchener Romantistische Arbeiten, Helft 16 (Munich: Hueber, 1961).

———, "Das Monstrum von Ravenna: Eine Studie zur Prodigienliteratur," *Zeitschrift für Volkskunde* 56 (1960), pp. 209–25.

Schepelern, H.D., "The Museum Wormianum Reconstructed. A Note on the Illustration of 1655," *Journal of the History of Collections* 1 (1990), pp. 81–85.

———, "Natural Philosophers and Princely Collectors: Worm, Paludanus, and the Gottorp and Copenhagen Collections," in Impey and MacGregor, *Origins*, pp. 121–27.

Schlosser, Julius von, *Die Kunst- und Wunderkammern der Spätrenaissance: Ein Beitrag zur Geschichte des Sammelwesens* (Leipzig: Klinkhardt & Biermann, 1908).

Schmitt, Charles B., "John Case on Art and Nature," *Annals of Science* 33 (1976), pp. 543–59.

Schmitt, Charles B., Quentin Skinner, Eckhart Kessler, and Jill Kraye (eds.), *The Cambridge History of Renaissance Philosophy* (Cambridge: Cambridge University Press, 1988).

Schnapper, Antoine, *Le Géant, la licorne et la tulipe: Collections et collectionneurs dans la France du XVIIe siècle* (Paris: Flammarion, 1988).

Schneider, Mark A., *Culture and Enchantment* (Chicago/London: University of Chicago Press, 1993).

Schramm, Percy Ernest, and Florentine Mütherich, *Denkmale der deutschen Könige und Kaiser: Ein Beitrag zur Herrschergeschichte von Karl dem Grossen bis Friedrich II., 768–1250* (Munich: Prestal Verlag, 1962).

Schreiner, Klaus, "'Discrimen veri ac falsi.' Ansätze und Formen der Kritik in der Heiligen- und Reliquienverehrung des Mittelalters," *Archiv für Kulturgeschichte* 48 (1966), pp. 1–53.

———, "Zum Wahrheitsverständnis im Heiligen- und Reliquienwesen des Mittelalters," *Saeculum: Jahrbuch für Universalgeschichte* 17 (1966), pp. 131–69.

Schuh, Barbara, *"Von vilen und mancherley seltzamen Wunderzeichen": Die Analyse von Mirakel-büchern und Wallfahrtsquellen*, Halbgraue Reihe, ed. Max-Planck-Institut für Geschichte Göttingen (St. Katharinen: Scripta Mercaturae, 1989).

Schulz, Eva, "Notes on the History of Collecting and of Museums in the Light of Selected Literature of the Sixteenth to Eighteenth Century," *Journal of the History of Collections* 2 (1990), pp. 205–18.

Schupbach, William, "Some Cabinets of Curiosities in European Academic Institutions," in Impey and MacGregor, *Origins*, pp. 169–78.

Scienze, credenze occulte, livelli di cultura: Convegno Internazionale di Studi (Firenze 26–30 giugno 1980) (Florence: Leo S. Olschki, 1982).

Seguin, J.-P., *L'Information en France avant le périodique* (Paris: G.-P. Maisonneuve et Larose, 1964).

———, *L'Information en France de Louis XII à Henri II* (Geneva: Travaux d'Humanisme et Renaissance, 1961).

Séjourné, P., "Superstition," *Dictionnaire de théologie catholique*, vol. 14, Part II, pp. 2763–824.

Serene, Eileen, "Demonstrative Science," in Kretzmann et al., *Cambridge History*, pp. 496–517.

Seymour, M.C., "Some Medieval French Readers of *De proprietatibus rerum*," *Scriptorium* 28 (1974), pp. 100–103.

Seymour, M.C., et al., *Bartholomaeus Anglicus and his Encyclopedia* (Aldershot: Variorum, 1992).

Shapin, Steven, "Of Gods and Kings: Natural Philosophy and Politics in the Leibniz-Clarke Disputes," *Isis* 72 (1981), pp. 187–215.

———, "Robert Boyle and Mathematics: Reality, Representation, and Scientific Practice," *Science in Context* 2 (1988), pp. 373–404.

———, *A Social History of Truth: Civility and Science in Seventeenth-Century England* (Chicago/London: University of Chicago Press, 1994).

Shapin, Steven, and Simon Schaffer, *Leviathan and the Air-Pump: Hobbes, Boyle and The Experimental Life* (Princeton: Princeton University Press, 1985).

Shapiro, Barbara J., *Probability and Certainty in Seventeenth-Century England* (Princeton: Princeton University Press, 1983).

Shearman, John, *Mannerism* (Harmondsworth: Penguin, 1967).

Sheehan, James J., and Morton Sosna (eds.), *The Boundaries of Humanity: Humans, Animals, Machines* (Berkeley: University of California Press, 1991).

Shepard, Odell, *The Lore of the Unicorn* (New York: Avenel Books, 1982).

Sherman, Claire Richter, *Imaging Aristotle: Verbal and Visual Representation in Fourteenth-Century France* (Berkeley: University of California Press, 1995).

Sherwood, Merriam, "Magic and Mechanics in Medieval Fiction," *Studies in Philology* 44 (1947), pp. 567–92.

Shiokawa, Tetsuya, *Pascal et les miracles* (Paris: Librairie A.G. Nizet, 1977).

Simili, Alessandro, *Gerolamo Cardano, lettore e medico a Bologna* (Bologna: Azzoguidi, 1969).

Simmel, Georg, *The Philosophy of Money* [1907], trans. Tom Bottomore and David Frisby (London: Routledge and Kegan Paul, 1978).

Singleton, Charles S. (ed.), *Art, Science, and History in the Renaissance* (Baltimore: Johns Hopkins University Press, 1968).

Siraisi, Nancy G., "Establishing the Subject: Vesalius and Human Diversity in *De humani corporis fabrica*, Books 1 and 2," *Journal of the Warburg and Courtauld Institutes* 57 (1994), pp. 60–88.

———, "The *Expositio Problematum Aristotelis* of Peter of Abano," *Isis* 61 (1970), pp. 321–39.

———, "Girolamo Cardano and the Art of Medical Narrative," *Journal of the History of Ideas* 52 (1991), pp. 581–602.

———, "The Medical Learning of Albertus Magnus," in Weisheipl, *Albertus Magnus*, pp. 379–414.

———, *Medieval and Early Renaissance Medicine: An Introduction to Knowledge and Practice* (Chicago: Chicago University Press, 1990).

———, "'Remarkable' Diseases, 'Remarkable' Cures, and Personal Experience in Renaissance Medical Texts," in idem, *Medicine and Italian Universities, 1250–1600* (Leiden: E.J. Brill, 2001).

———, *Taddeo Alderotti and His Pupils: Two Generations of Italian Medical Learning* (Princeton: Princeton University Press, 1981).

Siraisi, Nancy G., and Luke Demaitre (eds.), *Science, Medicine, and the University, 1200–1550: Essays in Honor of Pearl Kibre* = *Manuscripta* 20/2–3 (1976).

Smith, A.J. (ed.), *John Donne: The Critical Heritage* (London: Routledge, 1975).

Smits, E.R., "Vincent of Beauvais: A Note on the Background of the *Speculum*," in Aerts et al., *Vincent of Beauvais*, pp. 1–9.

Söhring, Otto, "Werke bildender Kunst in altfranzösischen Epen," *Romanische Forschungen* 12 (1900), pp. 491–640.

Soldi Rondinini, Gigliola, "Aspects de la vie des cours de France et de Bourgogne par les dépêches des ambassadeurs milanais (seconde moitié du XVe siècle)," in *Adelige Sachkultur*, pp. 194–214.

Solomon, Howard M., *Public Welfare, Science, and Propaganda in Seventeenth-Century France: The Innovations of Théophraste Renaudot* (Princeton: Princeton University Press, 1972).

Soman, Albert, "Les Procès de sorcellerie au Parlement de Paris (1565–1640)," *Annales. Economies, Sociétés, Civilisations* 32 (1977), pp. 790–814.

South, Malcolm (ed.), *Mythical and Fabulous Creatures* (Westport, CT: Greenwood Press, 1986).

Spallanzani, Marco, and Giovanna Gaeta Bertelà (eds.), *Libro d'inventario dei beni di Lorenzo il Magnifico* (Florence: Amici del Bargello, 1992).

Spargo, John Webster, *Virgil the Necromancer: Studies in Virgilian Legends* (Cambridge: Harvard University Press, 1934).

Spicker, Stuart F. (ed.), *Organism, Medicine and Metaphysics* (Dordrecht: D. Reidel, 1978).

Spiegel, Gabrielle M., "History as Enlightenment: Suger and the *Mos Anagogicus*," in Gerson, *Abbot Suger*, pp. 151–58.

Stafford, Barbara Maria, *Artful Science: Enlightenment Entertainment and the Eclipse of Visual Education* (Cambridge, MA: MIT Press, 1994).

Stagl, Justin, *The History of Curiosity: The Theory of Travel 1550–1800* (Chur: Harwood, 1995)

Stallybrass, Peter, and Allon White, *The Politics and Poetics of Transgression* (Ithaca: Cornell University Press, 1986).

Stannard, Jerry, "Albertus Magnus and Medieval Herbalism," in Weisheipl, *Albertus Magnus*, Ch. 14.

————, "Natural History," in Lindberg, *Science*, pp. 429–60.

Steinle, Friederich, "The Amalgamation of a Concept — Laws of Nature in the New Sciences," in Weinert, *Laws of Nature*, pp. 316–68.

Stock, Brian, *The Implications of Literacy: Written Language and Models of Interpretation in the Eleventh and Twelfth Centuries* (Princeton: Princeton University Press, 1983).

————, "Science, Technology, and Economic Progress in the Early Middle Ages," in Lindberg, *Science*, pp. 1–51.

Straus (Jr.), William L., and Owsei Temkin, "Vesalius and the Problem of Variability," *Bulletin of the History of Medicine* 14 (1943), pp. 609–33.

Stroup, Alice, *A Company of Scientists: Botany, Patronage, and Community at the Seventeenth-Century Parisian Academy of Sciences* (Berkeley/Los Angeles/Oxford: University of California Press, 1990), pp. 65–88.

Sturlese, Loris, "Saints et magiciens: Albert le Grand en face d'Hermès Trismégiste," *Archives de Philosophie* 43 (1980), pp. 615–34.

Sullivan, Penny, "Medieval Automata: The 'Chambre de Beautés' in Benoît's *Roman de Troie*," *Romance Studies* 6 (1985), pp. 1–20.

Summers, David, *The Judgment of Sense: Renaissance Naturalism and the Rise of Aesthetics* (Cambridge: Cambridge University Press, 1987).

Sumption, Jonathan, *Pilgrimage: An Image of Medieval Religion* (Totowa, NJ: Rowman and Littlefield, 1975).

Sutto, Claude, "L'Image du monde à la fin du Moyen Age," in Allard, *Aspects*, pp. 58–69.

Sylla, Edith, and Michael McVaugh (eds.), *Text and Context in Ancient and Medieval Science* (Leiden: Brill, 1997).

Tanner, Marie, *The Last Descendant of Aeneas: The Hapsburgs and the Mythic Image of the Emperor* (New Haven: Yale University Press, 1993).

Taruffi, Cesare, *Storia della teratologia*, 8 vols. (Bologna: Regia Tipografia, 1881–94).

Tayler, Edward William, *Nature and Art in Renaissance Literature* (New York/London: Columbia University Press, 1964).

Thomas, Keith, *Religion and the Decline of Magic* (New York: Scribner's, 1971).

Thorndike, Lynn, *A History of Magic and Experimental Science*, 8 vols. (New York: Columbia University Press, 1923–58).

———, "The Properties of Things of Nature Adapted to Sermons," *Medievalia et Humanistica* 12 (1958), pp. 78–83.

Tongiorgi Tomasi, Lucia, "Il giardino dei semplici dello studio pisano: Collezionismo, scienza e immagine tra Cinque e Seicento," in *Livorno e Pisa*, pp. 514–26.

———, *Immagine e natura: L'immagine naturalistica nei codici e libri a stampa delle Biblioteche Estense e Universitaria, secoli XV–XVII* (Modena: Panini, 1984).

Torrens, Hugh, "Early Collecting in the Field of Geology," in Impey and MacGregor, *Origins*, pp. 204–13.

Tort, Patrick, *L'Ordre et les monstres: Le Débat sur l'origine des déviations anatomiques au XVIIIe siècle* (Paris: Le Sycomore, 1980).

Le Trésor de Saint-Denis: Musée du Louvre, Paris, 12 mars–17 juin 1991 (Paris: Bibliothèque Nationale, Réunion des Musées Nationaux, 1991).

Trexler, Richard, "Ritual Behavior in Renaissance Florence: The Setting," *Medievalia et Humanistica* 4 (1973), pp. 125–44.

Turner, A.J., "Further Documents Concerning Jacques Vaucanson and the Collections of the Hôtel de Mortagne," *Journal of the History of Collections* 2 (1990), pp. 41–45.

Urbain, Charles, and E. Levesque, *L'Eglise et le théâtre Bossuet. Maximes et réflections sur la comédie, précédées d'une introduction historique et accompagnées de documents contemporains et de notes critiques* (Paris: B. Grasset, 1930).

Van Buren, Anne Hagopian, "Reality and Literary Romance in the Park of Hesdin," in Macdougall, *Medieval Gardens*, pp. 117–34.

Van Dülmen, Richard (ed.), *Armut, Liebe, Ehre. Studien zur historischen Kulturforschung* (Frankfurt a. M.: Fischer Taschenbuch Verlag, 1988).

Van Nieuwenhuysen, A., *Les Finances du Duc de Bourgogne Philippe le Hardi (1384–1404): Economie et politique* (Brussels: Editions de l'Université de Bruxelles, 1984).

Vaughan, Richard, *Charles the Bold: The Last Valois Duke of Burgundy* (London: Longmans, 1973).

———, *Philip the Good: The Apogee of Burgundy* (London: Longmans, 1970).

————, *Valois Burgundy* (London: Allen Lane, 1975).

Vet, J.J.V.M. de, "Miracles in the *Journal Littéraire*: A Case-Study on a Delicate Subject for Journalists of the Enlightenment," *Nederlands Archief voor Kerkgeschiednis* 67 (1987), pp. 186–98.

Veyne, Paul, *Did the Greeks Believe in Their Myths? An Essay on the Constitutive Imagination*, trans. Paula Wissing (Chicago: Chicago University Press, 1988).

Vickers, Brian (ed.), *Occult and Scientific Mentalities in the Renaissance* (Cambridge: Cambridge University Press, 1984).

Viguerie, Jean de, "Le Miracle dans la France du XVIIe siècle," *XVIIe Siècle* 35 (1983), pp. 313–31.

Vogel, Klaus A., "'America': Begriff, geographische Konzeption und frühe Entdeckungsgeschichte in der Perspektive der deutschen Humanisten," in Kohut, *Von der Weltkarte zum Kuriositätenkabinett*, pp. 11–43.

Wagner, Robert-Léon, *"Sorcier" et "magicien": Contribution à l'histoire du vocabulaire de la magie* (Paris: E. Droz, 1939).

Walde, A., and J.B. Hoffmann, *Lateinisches etymologisches Wörterbuch*, 3 vols., 3rd ed. (Heidelberg: Carl Winter Universitätsverlag, 1938–56).

Walker, Daniel Pickering, *Spiritual and Demonic Magic from Ficino to Campanella* (London: The Warburg Institute, 1958).

Wallace, William A., "Albertus Magnus on Suppositional Necessity in the Natural Science," in Weisheipl, *Albertus Magnus*, pp. 103–28.

————, *Causality and Scientific Explanation*, 2 vols. (Ann Arbor: University of Michigan, 1972).

———— (ed.), *Reinterpreting Galileo* (Washington, DC: Catholic University Press, 1986).

Ward, Benedicta, *Miracles and the Medieval Mind: Theory, Record and Event, 1000–1215* (Philadelphia: University of Pennsylvania Press, 1982).

Warnke, Martin, *The Court Artist: On the Ancestry of the Modern Artist* [1985], trans. David McLintock (Cambridge: Cambridge University Press, 1993).

Wartburg, Walther von, *Französisches etymologisches Wörterbuch*, 24 vols. (Basel: Helbing & Lichtenhahn, Zbinden Druck und Verlag AG/Tübingen: Paul Siebeck, 1940–83).

Wear, Andrew, "Explorations in Renaissance Writings on the Practice of Medicine," in Wear et al., *Medical Renaissance*, pp. 118–45.

Wear, Andrew, R.K. French, and I.M. Lonie (eds.), *The Medical Renaissance of the Sixteenth Century* (Cambridge: Cambridge University Press, 1985).

Weber, Bruno, *Wunderzeichen und Winkeldrucker 1543–1586* (Zurich: Urs Graf, 1972).

Weber, Max, *From Max Weber: Essays in Sociology* [1946], ed. H.H. Gerth and C. Wright Mills (New York: Oxford University Press, 1958).

Weil, Mark S., "Love, Monsters, Movement, and Machines: The Marvelous in Theaters, Festivals, and Gardens," in Kenseth, *Marvelous*, pp. 158–78.

Weinberg, Bernard, *A History of Literary Criticism in the Italian Renaissance*, 2 vols. (Chicago: University of Chicago Press, 1961).

Weinert, Friedel (ed.), *Laws of Nature: Essays on the Philosophical, Scientific and Historical Dimensions* (Berlin/New York: De Gruyter, 1995).

Weisheipl, James A. (ed.), *Albertus Magnus and the Sciences: Commemorative Essays 1980* (Toronto: Pontifical Institute of Mediaeval Studies, 1980).

———, "The Classification of the Sciences in Medieval Thought," *Mediaeval Studies*, 27 (1965), pp. 54–90.

———, "Curriculum of the Faculty of Arts at Oxford in the Early Fourteenth Century," *Mediaeval Studies* 26 (1984), pp. 143–85.

———, "The Nature, Scope, and Classification of the Sciences," in Lindberg, *Science*, pp. 461–82.

Westfall, Richard S., *Never at Rest: A Biography of Isaac Newton* (Cambridge: Cambridge University Press, 1980).

———, *Science and Religion in Seventeenth-Century England* (New Haven: Yale University Press, 1958).

Westrum, Ron, "Science and Social Intelligence about Anomalies: The Case of Meteorites," *Social Studies of Science* 8 (1978), pp. 461–93.

Whitaker, Katie, "The Culture of Curiosity," in Jardine et al., *Cultures of Natural History*, pp. 75–90.

Wiedemann, E., and F. Hauser, *Über die Uhren im Bereich der islamischen Kultur* (Halle: E. Karras, 1915).

Wilkins, Kay S., "Some Aspects of the Irrational in Eighteenth-Century France," *Studies on Voltaire and the Eighteenth Century* 140 (1975), pp. 107–201.

———, "The Treatment of the Supernatural in the *Encyclopédie*," *Studies on Voltaire and the Eighteenth Century* 90 (1972), pp. 1757–71.

Willard, Charity Cannon, "The Concept of True Nobility at the Burgundian Court," *Studies in the Renaissance* 14 (1967), pp. 33–48.

Wilson, Catherine, "*De Ipsa Natura*: Sources of Leibniz's Doctrine of Force, Activity and Natural Law," *Studia Leibnitiana* 19 (1987), pp. 148–72.

Wilson, Dudley, *Signs and Portents: Monstrous Births from the Middle Ages to the Enlightenment* (London: Routledge, 1993).

Wittkower, Rudolf, *Allegory and the Migration of Symbols* (New York: Thames and Hudson, 1977).

———, "Marco Polo and the Pictorial Tradition of the Marvels of the East," in Wittkower, *Allegory*, pp. 75–92.

———, "Marvels of the East: A Study in the History of Monsters," *Journal of the Warburg and Courtauld Institutes* 5 (1942), pp. 159–97.

Wuttke, Dieter, "Sebastian Brant und Maximilian I: Eine Studie zu Brants Donnerstein-Flugblatt des Jahres 1492," in Herding and Stupperich, *Humanisten*, pp. 141–76.

———, "Sebastian Brants Verhältnis zu Wunderdeutung und Astrologie," in Besch, *Studien*, pp. 272–86.

———, "Wunderdeutung und Politik: Zu den Auslegungen der sogenannten Wormser

Zwillinge des Jahres 1495," in Elm et al., *Landesgeschichte*, pp. 217–44.

Xella, P. (ed.), *Magia: Studi di storia delle religioni in memoria di Raffaela Garosi* (Rome: Bulzoni, 1976).

Yates, Frances A., *Giordano Bruno and the Hermetic Tradition* (Chicago: University of Chicago Press, 1964).

———, "The Hermetic Tradition in Renaissance Science," in Singleton, *Art, Science, and History*, pp. 255–74.

Zacher, Christian K., *Curiosity and Pilgrimage: The Literature of Discovery in Fourteenth-Century England* (Baltimore: Johns Hopkins University Press, 1976).

Zambelli, Paola, "L'immaginazione e il suo potere: Da al-Kindi e Avicenna al Medioevo latino e al Rinascimento," in Albert Zimmermann and Ingrid Craemer-Ruegenberg (eds.), *Orientalische Kultur und Europäisches Mittelalter* (Berlin: Walter de Gruyter, 1985), pp. 188–206.

———, "Scholastic and Humanist Views of Hermeticism and Witchcraft," in Merkel and Debus, *Hermeticism*, pp. 125–53.

Zanier, Giancarlo, *La medicina astrologica e sua teoria: Marsilio Ficino e i suoi critici contemporanei* (Rome: Edizioni dell'Ateneo e Bizarri, 1977).

———, "Miracoli e magia in una *quaestio* di Giacomo da Forlí," *Giornale critico della filosofia italiana*, ser. 4, 7 (1976), pp. 132–42.

———, *Ricerche sulla diffusione e fortuna del "De incantationibus" di Pomponazzi* (Florence: La Nuova Italia, 1975).

———, "Ricerche sull'occultismo a Padova nel secolo XV," in Poppi, *Scienza*, pp. 345–72.

Zapperi, Roberto, "Ein Haarmensch auf einem Gemälde von Agostino Carracci," in Hagner, *Der falsche Körper*, pp. 45–55.

Index

Designed by Bruce Mau with Barr Gilmore
Typeset by Archetype
Printed by Kromar Printing
on Sebago acid-free paper
Bound by York Bookbinding